T0292268

ADVANCED METHODS AND DEEP LEARNING IN COMPUTER VISION

Computer Vision and Pattern Recognition

ADVANCED METHODS AND DEEP LEARNING IN COMPUTER VISION

Edited by

E.R. DAVIES

MATTHEW A. TURK

ACADEMIC PRESS
An imprint of Elsevier

ELSEVIER

Academic Press is an imprint of Elsevier
125 London Wall, London EC2Y 5AS, United Kingdom
525 B Street, Suite 1650, San Diego, CA 92101, United States
50 Hampshire Street, 5th Floor, Cambridge, MA 02139, United States
The Boulevard, Langford Lane, Kidlington, Oxford OX5 1GB, United Kingdom

Copyright © 2022 Elsevier Inc. All rights reserved.

No part of this publication may be reproduced or transmitted in any form or by any means, electronic or mechanical, including photocopying, recording, or any information storage and retrieval system, without permission in writing from the publisher. Details on how to seek permission, further information about the Publisher's permissions policies and our arrangements with organizations such as the Copyright Clearance Center and the Copyright Licensing Agency, can be found at our website: www.elsevier.com/permissions.

This book and the individual contributions contained in it are protected under copyright by the Publisher (other than as may be noted herein).

Notices

Knowledge and best practice in this field are constantly changing. As new research and experience broaden our understanding, changes in research methods, professional practices, or medical treatment may become necessary.

Practitioners and researchers must always rely on their own experience and knowledge in evaluating and using any information, methods, compounds, or experiments described herein. In using such information or methods they should be mindful of their own safety and the safety of others, including parties for whom they have a professional responsibility.

To the fullest extent of the law, neither the Publisher nor the authors, contributors, or editors, assume any liability for any injury and/or damage to persons or property as a matter of products liability, negligence or otherwise, or from any use or operation of any methods, products, instructions, or ideas contained in the material herein.

Library of Congress Cataloging-in-Publication Data
A catalog record for this book is available from the Library of Congress

British Library Cataloguing-in-Publication Data
A catalogue record for this book is available from the British Library

ISBN: 978-0-12-822109-9

For information on all Academic Press publications
visit our website at https://www.elsevier.com/books-and-journals

Publisher: Mara Conner
Acquisitions Editor: Tim Pitts
Editorial Project Manager: Isabella C. Silva
Production Project Manager: Sojan P. Pazhayattil
Designer: Greg Harris

Typeset by VTeX

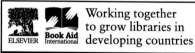

This book is dedicated to my family.

To cherished memories of my parents, Arthur and Mary Davies.

To my wife, Joan, for love, patience, support and inspiration.

To my children, Elizabeth, Sarah and Marion, and grandchildren, Jasper, Jerome, Eva, Tara and Pia, for bringing endless joy into my life!

Roy Davies

This book is dedicated to the students, colleagues, friends, and family who have motivated, guided, and supported me in ways too numerous to mention.

To my wife, Kelly, and my children, Hannah and Matt – special thanks and appreciation for your love and inspiration.

Matthew Turk

Contents

List of contributors

Sathyanarayanan Aakur Computer Science, Oklahoma State University, Stillwater, OK, United States

Yogesh Balaji Department of Computer Science and UMACS, University of Maryland, College Park, MD, United States

Han Cai Massachusetts Institute of Technology, Cambridge, MA, United States

Zhaowei Cai Amazon Web Services, Pasadena, CA, United States

Andrea Cavallaro Centre for Intelligent Sensing, Queen Mary University of London, London, United Kingdom

Rama Chellappa Departments of Electrical and Computer Engineering and Biomedical Engineering, Johns Hopkins University, Baltimore, MD, United States

Dongdong Chen Microsoft Cloud & AI, Redmond, WA, United States

E.R. Davies Royal Holloway, University of London, Egham, Surrey, United Kingdom

Michael Felsberg Computer Vision Laboratory, Department of Electrical Engineering, Linköping University, Linköping, Sweden
School of Engineering, University of KwaZulu-Natal, Durban, South Africa

Cornelia Fermüller University of Maryland, Institute for Advanced Computer Studies, Iribe Center for Computer Science and Engineering, College Park, MD, United States

Efstratios Gavves Informatics Institute, University of Amsterdam, Amsterdam, Netherlands

Deepak Gupta Informatics Institute, University of Amsterdam, Amsterdam, Netherlands

Song Han Massachusetts Institute of Technology, Cambridge, MA, United States

Gang Hua Wormpex AI Research, Bellevue, WA, United States

Ali Krayani DITEN, University of Genoa, Genoa, Italy

Ji Lin Massachusetts Institute of Technology, Cambridge, MA, United States

Lucio Marcenaro DITEN, University of Genoa, Genoa, Italy

Michael Maynord University of Maryland, Computer Science Department, Iribe Center for Computer Science and Engineering, College Park, MD, United States

Umberto Michieli Department of Information Engineering, University of Padova, Padova, Italy

Ramy Mounir Computer Science and Engineering, University of South Florida, Tampa, FL, United States

Hien Nguyen Department of Electrical and Computer Engineering, University of Houston, Houston, TX, United States

Changjae Oh Centre for Intelligent Sensing, Queen Mary University of London, London, United Kingdom

Sujoy Paul Google Research, Bangalore, India

Carlo Regazzoni DITEN, University of Genoa, Genoa, Italy

Amit K. Roy-Chowdhury University of California, Riverside, Electrical and Computer Engineering, Riverside, CA, United States

Sudeep Sarkar Computer Science and Engineering, University of South Florida, Tampa, FL, United States

Giulia Slavic DITEN, University of Genoa, Genoa, Italy

Radu Timofte Computer Vision Lab, ETH Zürich, Zürich, Switzerland

Marco Toldo Department of Information Engineering, University of Padova, Padova, Italy

Hassan Ugail Centre for Visual Computing, University of Bradford, Bradford, United Kingdom

Nuno Vasconcelos   University of California San Diego, Department of Electrical and Computer Engineering, San Diego, CA, United States

Alessio Xompero   Centre for Intelligent Sensing, Queen Mary University of London, London, United Kingdom

Pietro Zanuttigh   Department of Information Engineering, University of Padova, Padova, Italy

Kai Zhang   Computer Vision Lab, ETH Zürich, Zürich, Switzerland

About the editors

Roy Davies is Emeritus Professor of Machine Vision at Royal Holloway, University of London. He has worked on many aspects of vision, from feature detection and noise suppression to robust pattern matching and real-time implementations of practical vision tasks. His interests include automated visual inspection, surveillance, vehicle guidance, and crime detection. He has published more than 200 papers and three books – *Machine Vision: Theory, Algorithms, Practicalities* (1990), *Electronics, Noise and Signal Recovery* (1993), and *Image Processing for the Food Industry* (2000); the first of these has been widely used internationally for more than 25 years and in 2017 came out in a much enhanced fifth edition entitled *Computer Vision: Principles, Algorithms, Applications, Learning*. Roy is a Fellow of the IoP and the IET, and a Senior Member of the IEEE. He is on the Editorial Boards of *Pattern Recognition Letters*, *Real-Time Image Processing*, *Imaging Science* and *IET Image Processing*. He holds a DSc at the University of London: he was awarded *BMVA Distinguished Fellow* in 2005 and *Fellow*

of the International Association of Pattern Recognition in 2008.

Matthew Turk is the President of the Toyota Technological Institute at Chicago (TTIC) and an Emeritus Professor at the University of California, Santa Barbara. His research interests span computer vision and human-computer interaction, including topics such as autonomous vehicles, face and gesture recognition, multimodal interaction, computational photography, augmented and virtual reality, and AI ethics. He has served as General or Program Chair of several major conferences, including the IEEE Conference on Computer Vision and Pattern Recognition, the ACM Multimedia Conference, the IEEE Conference on Automatic Face and Gesture Recognition, the ACM International Conference on Multimodal Interaction, and the IEEE Winter Conference on Applications of Computer Vision. He has received several best paper awards, and he is an ACM Fellow, an IEEE Fellow, an IAPR Fellow, and the recipient of the 2011–2012 Fulbright-Nokia Distinguished Chair in Information and Communications Technologies.

Preface

It is now close to a decade since the explosive growth in the development and application of deep neural networks (DNNs) came about, and their subsequent progress has been little short of remarkable. True, this progress has been helped considerably by the deployment of special hardware in the form of powerful GPUs; and their progress followed from the realization that CNNs constituted a crucial architectural base, to which features such as ReLUs, pooling, fully connected layers, unpooling and deconvolution could also be included. In fact, all these techniques helped to breathe life into DNNs and to extend their use dramatically, so the initial near-exponential growth in their use has been maintained without break for the whole subsequent period. Not only has the power of the approach been impressive but its application has widened considerably from the initial emphasis on rapid object location and image segmentation—and even semantic segmentation—to aspects pertaining to video rather than mere image analysis.

It would be idle to assert that the whole of the development of computer vision since 2012 has been due solely to the advent of DNNs. Other important techniques such as reinforcement learning, transfer learning, self-supervision, linguistic description of images, label propagation, and applications such as novelty and anomaly detection, image inpainting and tracking have all played a part and contributed to the widening and maturing of computer vision. Nevertheless, many such techniques and application areas have been stimulated, challenged, and enhanced by the extremely rapid take-up of DNNs.

It is the purpose of this volume to explore the way computer vision has advanced since these dramatic changes were instigated. Indeed, we can validly ask where we are now, and how solid is the deep neural and machine learning base on which computer vision has recently embarked. Has this been a coherent movement or a blind opportunistic rush forward in which workers have ignored important possibilities, and can we see further into the future and be sure that we are advancing in the right direction? Or is this a case where each worker can take his or her own viewpoint and for any given application merely attend to what appears to be necessary, and if so, is anything lost by employing a limited approach of this sort?

In fact, there are other highly pertinent questions to be answered, such as the thorny one of the extent to which a deep network can only be as powerful as the dataset it is trained on; this question will presumably apply to any alternative learning-based approach, whether describable as a DNN or not. Employing reinforcement learning or self-supervision or other approaches will surely not affect this likely limitation. And note that human beings are hardly examples of how extensive training can in any way be avoided; their transfer learning capabilities will be a vital aspect of how efficient the learning process can be made.

It is the aim of this volume not only to present advanced vision methodologies but also to elucidate the principles involved: i.e., it aims to be pedagogic, concentrating as much on helping the reader to understand as on presenting the latest research. With this in mind, Chapter 1 sets the scene for the remainder of this volume. It starts by looking closely at the legacy of earlier vision work, covering in turn feature detection, object detection, 3D vision and the advent of DNNs; finally, tracking is taken as an important application area which builds on the material of the earlier sections and shows clearly how deep networks can play a crucial role. This chapter is necessarily quite long, as it has to get from ground zero to a formidable attainment level in relatively few pages; in addition, it has to set the scene for the important developments and methodologies described by eminent experts in the remaining chapters.

As is made clear in Chapter 1, object detection is one of the most challenging tasks in computer vision. In particular, it has to overcome problems such as scale-variance, occlusion, variable lighting, complex backgrounds and all the factors of variability associated with the natural world. Chapter 2 describes the various methods and approaches that have been used in recent advances. These include region-of-interest pooling, multitask losses, region proposal networks, anchors, cascaded detection and regression, multiscale feature representations, data augmentation techniques, loss functions, and more.

Chapter 3 emphasizes that the recent successes in computer vision have largely centered around the huge corpus of intricately labeled data needed for training models. It examines the methods that can be used to learn recognition models from such data, while requiring limited manual supervision. Apart from reducing the amount of manually labeled data required to learn recognition models, it is necessary to reduce the level of supervision from strong to weak— at the same time permitting relevant queries from an oracle. An overview is given of theoretical frameworks and experimental results that help to achieve this.

Chapter 4 tackles the computational problems of deep neural networks, which make it difficult to deploy them on resource-constrained hardware devices. It discusses model compression techniques and hardware-aware neural architecture search techniques with the aim of making deep learning more efficient and making neural networks smaller and faster. To achieve all this, the chapter shows how to use parameter pruning to remove redundant weights, low-rank factorization to reduce complexity, weight quantization to reduce weight precision and model size, and knowledge distillation to transfer dark knowledge from large models to smaller ones.

Chapter 5 discusses how deep generative models attempt to recover the lower dimensional structure of the target visual models. It shows how to leverage deep generative models to achieve more controllable visual pattern synthesis via conditional image generation. The key to achieving this is "disentanglement" of the visual representation, where attempts are made to separate different controlling factors in the hidden embedding space. Three case studies, in style transfer, vision-language generation, and face synthesis, are presented to illustrate how to achieve this in unsupervised or weakly supervised settings.

Chapter 6 concentrates on a topical real-world problem—that of face recognition. It discusses state-of-the-art deep learning-based methods that can be used even with partial facial images. It shows (a) how the necessary deep learning architectures are put together; (b) how such models can be trained and tested; (c) how fine tuning of pretrained networks can be utilized for identifying efficient recognition cues with full and partial facial data; (d) the degree of success achieved by the recent developments in deep learning;

(e) the current limitations of deep learning-based techniques used in face recognition. The chapter also presents some of the remaining challenges in this area.

Chapter 7 discusses the crucial question of how to transfer learning from one data domain to another. This involves approaches based on differential geometry, sparse representation and deep neural networks. These fall into the two broad classes—discriminative and generative approaches. The former involve training a classifier model while employing additional losses to make the source and target feature distributions similar. The latter utilize a generative model to perform domain adaptation: typically, a cross-domain generative adversarial network is trained for mapping samples from source domain to target, and a classifier model is trained on the transformed target images. Such approaches are validated on cross-domain recognition and semantic segmentation tasks.

Chapter 8 returns to the domain adaptation task, in the context of semantic segmentation, where deep networks are plagued by the need for huge amounts of labeled data for training. The chapter starts by discussing the different levels at which the adaptation can be performed and the strategies for achieving them. It then moves on to discuss the task of continual learning in semantic segmentation. Although the latter is a relatively new research field, interest in it is rapidly growing, and many different scenarios have been introduced. These are described in detail along with the approaches needed to tackle them.

Following on from Chapter 1, Chapter 9 reemphasizes the importance of visual tracking as one of the prime, classical problems in computer vision. The purpose of this chapter is to give an overview of the development of the field, starting from the Lucas-Kanade and matched filter approaches and concluding with deep learning-based approaches as well as the transition to video segmentation. The overview is limited to holistic models for generic tracking in the image plane, and a particular focus is given to discriminative models, the MOSSE (minimum output sum of squared errors) tracker, and DCFs (discriminative correlation filters).

Chapter 10 takes the concept of visual object tracking one stage further and concentrates on long-term tracking. To be successful at this task, object tracking must address significant challenges that relate to model decay—that is, the worsening of the model due to added bias, and target disappearance and reappearance. The success of deep learning has strongly influenced visual object tracking, as offline learning of Siamese trackers helps to eliminate model decay. However, to avoid the possibility of losing track in cases where the appearance of the target changes significantly, Siamese trackers can benefit from built-in invariances and equivariances, allowing for appearance variations without exacerbating model decay.

If computer vision is to be successful in the dynamic world of videos and action, it seems vital that human cognitive concepts will be required, a message that is amply confirmed by the following two chapters. Chapter 11 outlines an action-centric framework which spans multiple time scales and levels of abstraction. The lower level details object characteristics which *afford* themselves to different actions; the mid-level models individual *actions*, and higher levels model *activities*. By emphasizing the use of grasp characteristics, geometry, ontologies, and physics-based constraints, over-training on appearance characteristics is avoided. To integrate signal-based perception with symbolic knowledge, vectorized knowledge is aligned with visual features. The chapter also includes a discussion on action and activity understanding.

Chapter 12 considers the temporal event segmentation problem. Cognitive science research indicates how to design highly effective computer vision algorithms for spatio-temporal segmentation of events in videos without the need for any annotated data. First, an event segmentation theory model permits event boundaries to be computed: then, temporal segmentation using a perceptual prediction framework, temporal segmentation along with event working models based on attention maps, and spatio-temporal localization of events follow. This approach gives state-of-the-art performance in unsupervised temporal segmentation and spatial-temporal action localization with competitive performance on fully supervised baselines that require extensive amounts of annotation.

Anomaly detection techniques constitute a fundamental resource in many applications such as medical image analysis, fraud detection or video surveillance. These techniques also represent an essential step for artificial self-aware systems that can continually learn from new situations. Chapter 13 presents a semi-supervised method for the detection of anomalies for this type of self-aware agent. It leverages the message-passing capability of generalized dynamic Bayesian networks to provide anomalies at different abstraction levels for diverse types of time-series data. Consequently, detected anomalies could be employed to enable the system to evolve by integrating the new acquired knowledge. A case study is proposed for the description of the anomaly detection method, which will use multisensory data from a semi-autonomous vehicle performing different tasks in a closed environment.

Model- and learning-based methods have been the two dominant strategies for solving various image restoration problems in low-level vision. Typically, those two kinds of method have their respective merits and drawbacks; e.g., model-based methods are flexible for handling different image restoration problems but are usually time-consuming with sophisticated priors for the purpose of good performance; meanwhile, learning-based methods show superior effectiveness and efficiency over traditional model-based methods, largely due to the end-to-end training, but generally lack the flexibility to handle different image restoration tasks. Chapter 14 introduces deep plug-and-play methods and deep unfolding methods, which have shown great promise by leveraging both learning-based and model-based methods: the main idea of deep plug-and-play methods is that a learning-based denoiser can implicitly serve as the image prior for model-based image restoration methods, while the main idea of deep unfolding methods is that, by unfolding the model-based methods via variable splitting algorithms, an end-to-end trainable, iterative network can be obtained by replacing the corresponding subproblems with neural modules. Hence, deep plug-and-play methods and deep unfolding methods can inherit the flexibility of model-based methods, while maintaining the advantages of learning-based methods.

Visual adversarial examples are images and videos purposefully perturbed to mislead machine learning models. Chapter 15 presents an overview of methods that craft adversarial perturbations to generate visual adversarial examples for image classification, object detection, motion estimation and video recognition tasks. The key properties of an adversarial attack and the types of perturbation that an attack generates are first defined; then the main design choices for methods that craft adversarial attacks for images and videos are analyzed and the knowledge they use of the target model is examined. Finally, defense mechanisms that increase the robustness of machine learning models to adversarial attacks or to detect manipulated input data are reviewed.

Together, these chapters provide the interested reader—whether student, researcher, or practitioner—with both breadth and depth with respect to advanced computer vision methodology and state-of-the-art approaches.

Finally, we would like to extend our thanks to all the authors for the huge degree of commitment and dedication they have devoted to producing their chapters, thereby contributing in no small way to making this volume a successful venture for advancing the subject in what is after all a rapidly changing era. Lastly, we are especially indebted to Tim Pitts of Elsevier Science for his constant advice and encouragement, not only from the outset but also while we were in the throes of putting together this volume.

Roy Davies
Royal Holloway, University of London,
London, United Kingdom

Matthew Turk
Toyota Technological Institute at Chicago,
Chicago, IL, United States
May 2021

The dramatically changing face of computer vision

E.R. Davies

Royal Holloway, University of London, Egham, Surrey, United Kingdom

CHAPTER POINTS

- Studies of legacy methods in computer vision, including low-level image processing operators, 2-D and 3-D object detection, location and recognition, tracking and segmentation.

- Examination of the development of deep learning methods from artificial neural networks, including the deep learning explosion.

- Studies of the application of deep learning methods to feature detection, object detection, location and recognition, object tracking, texture classification, and semantic segmentation of images.

- The impact of deep learning methods on preexisting computer vision methodology.

1.1 Introduction – computer vision and its origins

During the last three or four decades, computer vision has gradually emerged as a fully-fledged subject with its own methodology and area of application. Indeed, it has so many areas of application that it would be difficult to list them all. Amongst the most prominent are object recognition, surveillance (including people counting and numberplate recognition), robotic control (including automatic vehicle guidance), segmentation and interpretation of medical images, automatic inspection and assembly in factory situations, fingerprint and face recognition, interpretation of hand signals, and many more. To achieve all this, measurements have to be made from a variety of image sources, including visible and infrared channels, 3-D sensors, and a number of vital medical imaging devices such as CT and MRI scanners. And the measurements have to include position, pose, distances between objects,

Advanced Methods and Deep Learning in Computer Vision
https://doi.org/10.1016/B978-0-12-822109-9.00010-2

1

Copyright © 2022 Elsevier Inc. All rights reserved.

movement, shape, texture, color, and many more aspects. With this plethora of activities and of the methods used to achieve them, it will be difficult to encapsulate the overall situation within the scope of a single chapter: hence the selection of material will necessarily be restricted; nevertheless, we will aim to provide a sound base and a didactic approach to the subject matter.

In the 2020s one can hardly introduce computer vision without acknowledging the enormous advances made during the 2010s, and specifically the 'deep learning explosion', which took place around 2012. This dramatically changed the shape of the subject and resulted in advances and applications that are not only impressive but are also in many cases well beyond what people dreamed about even in 2010. As a result, this volume is aimed particularly at these modern advanced developments: it is the role of this chapter to outline the legacy methodology, to explore the new deep learning methods, and to show how the latter have impacted and improved upon the earlier (legacy) approaches.

At this point it will be useful to consider the origins of computer vision, which can be considered to have started life during the 1960s and 1970s, largely as an offshoot of image processing. At that time it became practical to capture whole images and to store and process them conveniently on digital computers. Initially, images tended to be captured in binary or grey-scale form, though later it became possible to capture them in color. Early on, workers dreamed of emulating the human eye by recognizing objects and interpreting scenes, but with the less powerful computers then available, such dreams were restricted. In practice, image processing was used to 'tidy up' images and to locate object features, while image recognition was carried out using statistical pattern recognition techniques such as the nearest neighbor algorithm. Another of the motivations underlying the development of computer vision was AI and yet another was biological vision. Space will prevent further discussion of these aspects here, except to remark that they sowed the seeds for artificial neural networks and deep learning (for details, see Part F below).

Tidying up images is probably better described as preprocessing: this can include a number of functions, noise elimination being amongst the most important. It was soon discovered that the use of smoothing algorithms, in which the mean value of the intensities in a window around each input pixel is calculated and used to form a separate smoothed image, not only results in reduced levels of noise but also affects the signals themselves (this process can also be imagined as reducing the input bandwidth to exclude much of the noise, with the additional effect of eliminating high spatial frequency components of the input signal). However, by applying median rather than mean filtering, this problem was largely overcome, as it worked by eliminating the outliers at each end of the local intensity distribution—the median being the value least influenced by noise.

Typical mean filtering kernels include the following, the second approximating more closely to the ideal Gaussian form:

$$\frac{1}{9}\begin{bmatrix} 1 & 1 & 1 \\ 1 & 1 & 1 \\ 1 & 1 & 1 \end{bmatrix} \qquad \frac{1}{16}\begin{bmatrix} 1 & 2 & 1 \\ 2 & 4 & 2 \\ 1 & 2 & 1 \end{bmatrix} \tag{1.1}$$

Both of these are linear convolution kernels, which by definition are spatially invariant over the image space. A general 3×3 convolution mask is given by

$$
\begin{bmatrix}
c4 & c3 & c2 \\
c5 & c0 & c1 \\
c6 & c7 & c8
\end{bmatrix}
\tag{1.2}
$$

where the local pixels are assigned labels 0–8. Next, we take the intensity values in a local 3×3 image neighborhood as

$$
\begin{array}{|ccc|}
\hline
P4 & P3 & P2 \\
P5 & P0 & P1 \\
P6 & P7 & P8 \\
\hline
\end{array}
\tag{1.3}
$$

If we now use a notation based approximately on C ++, we can write the complete convolution procedure in the form:

$$
\begin{aligned}
&\text{for all pixels in image do } \{ \\
&\quad Q0 = P0 * c0 + P1 * c1 + P2 * c2 + P3 * c3 + P4 * c4 \\
&\qquad + P5 * c5 + P6 * c6 + P7 * c7 + P8 * c8; \\
&\}
\end{aligned}
\tag{1.4}
$$

So far we have concentrated on convolution masks, which are linear combinations of input intensities: these contrast with nonlinear procedures such as thresholding, which cannot be expressed as convolutions. In fact, thresholding is a very widely used technique, and can be written in the form:

$$
\begin{aligned}
&\text{for all pixels in image do } \{ \\
&\quad \text{if } (P0 < \text{thresh})\, A0 = 1; \ \text{else } A0 = 0; \\
&\}
\end{aligned}
\tag{1.5}
$$

This procedure converts a grey scale image in P-space into a binary image in A-space. Here it is used to identify dark objects by expressing them as 1s on a background of 0s.

We end this section by presenting a complete procedure for median filtering within a 3×3 neighborhood:

$$
\begin{aligned}
&\text{for } (i = 0; \ i <= 255; \ i++) \ \text{hist}[i] = 0; \\
&\text{for all pixels in image do } \{ \\
&\quad \text{for } (m = 0; m <= 8; m++) \ \text{hist}[P[m]] ++; \\
&\quad i = 0; \ \text{sum} = 0; \\
&\quad \text{while } (\text{sum} < 5)\{ \\
&\qquad \text{sum} = \text{sum} + \text{hist}[i]; \\
&\qquad i = i + l; \\
&\quad \}
\end{aligned}
$$

$$Q0 = i - 1;$$
$$\text{for } (m = 0; m <= 8; m + +) \text{ hist}[P[m]] = 0;$$

}   (1.6)

The notation P[0] is intended to denote P0, and so on for P[1] to P[8]. Note that the median operation is computation intensive, so time is saved by only reinitializing the particular histogram elements that have actually been used.

An important point about the procedures covered by Eqs. (1.4)–(1.6) is that they take their input from one image space and output it to another image space—a process often described as parallel processing—thereby eliminating problems relating to the order in which the individual pixel computations are carried out.

Finally, the image smoothing algorithms given by Eqs. (1.1)–(1.4) all use 3×3 convolution kernels, though much larger kernels can obviously be used: indeed, they can alternatively be implemented by first converting to the spatial frequency domain and then systematically eliminating high spatial frequencies, albeit with an additional computational burden. On the other hand, nonlinear operations such as median filtering cannot be tackled in this way.

For convenience, the remainder of this chapter has been split into a number of parts, as follows:

Part A – Understanding low-level image processing perators
Part B – 2-D object location and recognition
Part C – 3-D object location and the importance of invariance
Part D – Tracking moving objects
Part E – Texture analysis
Part F – From artificial neural networks to deep learning methods
Part G – Summary.

Overall, the purpose of this chapter is to summarize vital parts of the early—or 'legacy'—work on computer vision, and to remind readers of their significance, so that they can more confidently get to grips with recent advanced developments in the subject. However, the need to make this sort of selection means that many other important topics have had to be excluded.

1.2 Part A – Understanding low-level image processing operators

1.2.1 The basics of edge detection

No imaging operation is more important or more widely used than edge detection. There are important reasons for this, but ultimately, describing object shapes by their boundaries and internal contours reduces the amount of data required to hold an $N \times N$ image from $O(N^2)$ to $O(N)$, thereby making subsequent storage and processing more efficient. Furthermore, there is much evidence that humans can recognize objects highly effectively, or even with increased efficiency, from their boundaries: the quick responses humans can make from 2-D sketches and cartoons support this idea.

In the 1960s and 1970s, a considerable number of edge detection operators were developed, many of them intuitively, which meant that their optimality was in question. A number of the operators applied 8 or 12 template masks to ensure that edges of different orientations could be detected. Oddly, it was some time before it was fully realized that as edges are vectors, just two masks should be sufficient to detect them. However, this did not immediately eliminate the problem of deciding what mask coefficients should be used in edge detectors—even in the case of 3×3 neighborhoods—and we next proceed to explore this further.

In what follows we initially assume that 8 masks are to be used, with angles differing by $45°$. However, 4 of the masks differ from the others only in sign, which makes it unnecessary to apply them separately. At this point, symmetry arguments lead to the following respective masks for $0°$ and $45°$:

$$\begin{bmatrix} -A & 0 & A \\ -B & 0 & B \\ -A & 0 & A \end{bmatrix} \qquad \begin{bmatrix} 0 & C & D \\ -C & 0 & C \\ -D & -C & 0 \end{bmatrix} \tag{1.7}$$

It is clearly of great importance to design masks so that they give consistent responses in different directions. To find how this affects the mask coefficients, we make use of the fact that intensity gradients must follow the rules of vector addition. If the pixel intensity values within a 3×3 neighborhood are

$$\begin{array}{|ccc|} \hline a & b & c \\ d & e & f \\ g & h & i \\ \hline \end{array} \tag{1.8}$$

the above masks will lead to the following estimates of gradient in the $0°$, $90°$ and $45°$ directions:

$$\begin{aligned} g_0 &= A(c + i - a - g) + B(f - d) \\ g_{90} &= A(a + c - g - i) + B(b - h) \\ g_{45} &= C(b + f - d - h) + D(c - g) \end{aligned} \tag{1.9}$$

If vector addition is to be valid, we also have:

$$g_{45} = (g_0 + g_{90})/\sqrt{2} \tag{1.10}$$

Equating coefficients of $a, b, \ldots, i$ leads to the self-consistent pair of conditions:

$$\begin{aligned} C &= B/\sqrt{2} \\ D &= A\sqrt{2} \end{aligned} \tag{1.11}$$

Next, notice the further requirement—that the $0°$ and $45°$ masks should give equal responses at $22.5°$. In fact, a rather tedious algebraic manipulation (Davies, 1986) shows that

$$B/A = \left(13\sqrt{2} - 4\right)/7 = 2.055 \tag{1.12}$$

If we approximate this value as 2 we immediately arrive at the Sobel operator masks

$$S_x = \begin{bmatrix} -1 & 0 & 1 \\ -2 & 0 & 2 \\ -1 & 0 & 1 \end{bmatrix} \qquad S_y = \begin{bmatrix} 1 & 2 & 1 \\ 0 & 0 & 0 \\ -1 & -2 & -1 \end{bmatrix} \tag{1.13}$$

application of which yields maps of the g_x, g_y components of intensity gradient. As edges are vectors, we can compute the local edge magnitude g and direction θ using the standard vector-based formulae:

$$g = \left[g_x^2 + g_y^2 \right]^{1/2}$$
$$\theta = \arctan\left(g_y / g_x \right) \tag{1.14}$$

Notice that whole-image calculations of g and θ will not be convolutions as they involve nonlinear operations.

In summary, in Sections 1.1 and 1.2.1 we have described various categories of image processing operator, including linear, nonlinear and convolution operators. Examples of (linear) convolutions are mean and Gaussian smoothing and edge gradient component estimation. Examples of nonlinear operations are thresholding, edge gradient and edge orientation computations. Above all, it should be noted that the Sobel mask coefficients have been arrived at in a principled (non ad hoc) way. In fact, they were designed to optimize accuracy of edge orientation. Note also that, as we shall see later, orientation accuracy is of paramount importance when edge information is passed to object location schemes such as the Hough transform.

1.2.2 The Canny operator

The aim of the Canny edge detector was to be far more accurate than basic edge detectors such as the Sobel, and it caused quite a stir when it was published in 1986 (Canny, 1986). To achieve such increases in accuracy, a number of processes are applied in turn:

1. The image is smoothed using a 2-D Gaussian to ensure that the intensity field is a mathematically well-behaved function.
2. The image is differentiated using two 1-D derivative functions, such as those of the Sobel, and the gradient magnitude field is computed.
3. Nonmaximum suppression is employed along the local edge normal direction to thin the edges: this takes place in two stages (1) finding the two noncentral red points shown in Fig. 1.1, which involves gradient magnitude interpolation between two *pairs* of pixels; (2) performing quadratic interpolation between the intensity gradients at the three red points to determine the position of the peak edge signal to subpixel precision.
4. 'Hysteresis' thresholding is performed: this involves applying two thresholds t_1 and t_2 ($t_2 > t_1$) to the intensity gradient field; the result is 'nonedge' if $g < t_1$, 'edge' if $g > t_2$, and otherwise is only 'edge' if next to 'edge'. (Note that the 'edge' property can be propagated from pixel to pixel under the above rules.)

As noted in item 3, quadratic interpolation can be used to locate the position of the gradient magnitude peak. A few lines of algebra shows that, for the g-values g_1, g_2, g_3 of the three red points, the displacement of the peak from the central red point is equal to $(g_3 - g_1)\sec\theta / [2(2g_2 - g_1 - g_3)]$: here, $\sec\theta$ is the factor by which θ increases the distance between the outermost red points.

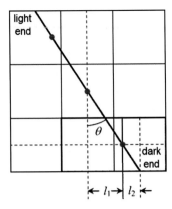

FIGURE 1.1 Using quadratic interpolation to determine the exact position of the gradient magnitude peak.

1.2.3 Line segment detection

In Section 1.2.1 we saw the considerable advantage of edge detectors in requiring only two masks to compute the magnitude and orientation of an edge feature. It is worth considering whether the same vector approach might also be used in other cases. In fact, it is also possible to use a modified vector approach for detecting line segment features. This is seen by considering the following pair of masks:

$$L_1 = A \begin{bmatrix} 0 & -1 & 0 \\ 1 & 0 & 1 \\ 0 & -1 & 0 \end{bmatrix} \qquad L_2 = B \begin{bmatrix} -1 & 0 & 1 \\ 0 & 0 & 0 \\ 1 & 0 & -1 \end{bmatrix} \qquad (1.15)$$

Clearly, two other masks of this form can be constructed, though they differ from the above two only in sign and can be ignored. Thus, this set of masks contains just the number required for a vectorial computation. In fact, if we are looking for dark bars on a light background, the 1 s can usefully denote the bars and the −1 s can represent the light background. (0 s can be taken as 'don't care' coefficients, as they will be ignored in any convolution.) Hence L_1 represents a 0° bar and L_2 a 45° bar. (The term 'bar' is used here to denote a line segment of significant width.) Applying the same method as in Section 1.2.1 and defining the pixel intensity values as in Eq. (1.8), we find

$$l_0 = A(d + f - b - h)$$
$$l_{45} = B(c + g - a - i) \qquad (1.16)$$

However, in this instance there is insufficient information to determine the ratio of A to B, so this must depend on the practicalities of the situation. In fact, given that this computation is being carried out in a 3×3 neighborhood, it will not be surprising if the optimum bar width for detection using the above masks is $\sim$1.0; experimental tests (Davies, 1997) showed that matching the masks to the bar width w (or vice versa) gave optimum orientation accuracy for $w \approx 1.4$, which occurred when $B/A \approx 0.86$. This resulted in a maximum orientation error $\sim$0.4°, which compares favorably with $\sim$0.8° for the Sobel operator.

We now proceed to use formulae similar to those in Section 1.2.1 for pseudo-vectorial computation of the line strength coefficient l and line segment orientation θ:

$$l = \left[l_0^2 + l_{45}^2 \right]^{1/2}$$

$$\theta = \frac{1}{2} \arctan(l_{45}/l_0)$$

(1.17)

Here we have been forced to include a factor of one half in front of the arctan: this is because a line segment exhibits 180° rotation symmetry compared with the usual 360° for ordinary angles.

Note that this is again a case in which optimization is aimed at achieving high orientation accuracy rather than, for example, sensitivity of detection.

It is worth remarking here on two applications of line segment detection. One is the inspection of bulk wheat grains to locate small dark insects which approximate to dark bar-like features: 7×7 masks devised on the above model have been used to achieve this (Davies et al., 2003). Another is the location of artefacts such as telegraph wires in the sky, or wires supporting film actors which can then be removed systematically.

1.2.4 Optimizing detection sensitivity

Optimization of detection sensitivity is a task that is well known in radar applications and has been very effectively applied for this purpose since World War II. Essentially, efficient detection of aircraft by radar systems involves optimization of the signal-to-noise-ratio (SNR). Of course, in radar, detection is a 1-D problem whereas in imaging we need to optimally detect 2-D objects against a background of noise. However, image noise is not necessarily Gaussian white noise, as can normally be assumed in radar, though it is convenient to start with that assumption.

In radar the signals can be regarded as positive peaks (or 'bleeps') against a background of noise which is normally close to zero. Under these conditions there is a well-known theorem that says that the optimum detection of a bleep of given shape is obtained using a 'matched filter' which has the same shape as the idealized input signal. The same applies in imaging, and in that case the *spatial* matched filter has to have the same intensity profile as that of an ideal form of the 2-D object to be detected.

We shall now outline the mathematical basis of this approach. First, we assume a set of pixels at which signals are sampled, giving values S_i. Next, we express the desired filter as an n-element weighting template with coefficients w_i. Finally, we assume that the noise levels at each pixel are independent and are subject to local distributions with standard deviations N_i.

Clearly, the total signal received from the weighting template will be

$$S = \sum_{i=1}^{n} w_i S_i$$

(1.18)

whereas the total noise received from the weighting template will be characterized by its variance:

$$N^2 = \sum_{i=1}^{n} w_i^2 N_i^2 \tag{1.19}$$

Hence the (power) SNR is

$$\rho^2 = S^2 / N^2 = \left(\sum_{i=1}^{n} w_i S_i\right)^2 \bigg/ \sum_{i=1}^{n} w_i^2 N_i^2 \tag{1.20}$$

For optimum SNR, we compute the derivative

$$\begin{aligned}
\partial \rho^2 / \partial w_i &= \left(1/N^4\right) \left[N^2 \left(2SS_i\right) - S^2 \left(2w_i N_i^2\right)\right] \\
&= \left(2S/N^4\right) \left[N^2 S_i - S \left(w_i N_i^2\right)\right]
\end{aligned} \tag{1.21}$$

and then set $\partial \rho^2 / \partial w_i = 0$. This immediately gives:

$$w_i = \frac{S_i}{N_i^2} \times \frac{N^2}{S} \tag{1.22}$$

which can more simply be expressed as:

$$w_i \propto \frac{S_i}{N_i^2} \tag{1.23}$$

though with no loss of generality, we can replace the proportionality sign by an equality.

Note that if N_i is independent of i (i.e., the noise level does not vary over the image), $w_i = S_i$: this proves the theorem mentioned above—that the *spatial* matched filter needs to have the same intensity profile as that of the 2-D object to be detected.

1.2.5 Dealing with variations in the background intensity

Apart from the obvious difference in dimensionality, there is a further important way in which vision differs from radar: for the latter, in the absence of a signal, the system output hovers around, and averages to, zero. However, in vision, the background level will typically vary with the ambient illumination and will also vary over the input image. Basically, the solution to this problem is to employ zero-sum (or zero-mean) masks. Thus, for a mask such as that in Eq. (1.2), we merely subtract the mean value $\bar{c}$ of all the mask components from each component to ensure that the overall mask is zero-mean.

To confirm that using the zero-mean strategy works, imagine applying an unmodified mask to the image neighborhood shown in Eq. (1.3): let us assume we obtain a value K. Now add B to the intensity of each pixel in the neighborhood: this will add $\sum_n Bc_i = B \sum_n c_i = Bn\bar{c}$ to the value K; but if we make $\bar{c} = 0$, we end up with the original mask output K.

Overall, we should note that the zero-mean strategy is only an approximation, as there will be places in an image where the background varies between high and low level, so that zero-mean cancellation cannot occur exactly (i.e., B cannot be regarded as constant over the region of the mask). Nevertheless, assuming that the background variation occurs on a scale significantly larger than that of the mask size, this should work adequately.

It should be remarked that the zero-mean approximation is already widely used—as indeed we have already seen from the edge and line-segment masks in Eqs. (1.7) and (1.15). It must also apply for other detectors we could devise, such as corner and hole detectors.

1.2.6 A theory combining the matched filter and zero-mean constructs

At first sight, the zero-mean construct is so simple that it might appear to integrate easily with the matched filter formalism of Section 1.2.4. However, applying it reduces the number of degrees of freedom of the matched filter by one, so a change is needed to the matched filter formalism to ensure that the latter continues to be an ideal detector. To proceed, we represent the zero-mean and matched filter cases as follows:

$$(w_i)_{\text{z-m}} = S_i - \bar{S}$$
$$(w_i)_{\text{m-f}} = S_i / N_i^2 \qquad (1.24)$$

Next, we combine these into the form

$$w_i = \left(S_i - \tilde{S}\right) / N_i^2 \qquad (1.25)$$

where we have avoided an impasse by trying a hypothetical (i.e., as yet unknown) type of mean for S, which we call $\tilde{S}$. [Of course, if this hypothesis in the end results in a contradiction, a fresh approach will naturally be required.] Applying the zero-mean condition $\sum_i w_i = 0$ now yields the following:

$$\sum_i w_i = \sum_i S_i / N_i^2 - \sum_i \tilde{S} / N_i^2 = 0 \qquad (1.26)$$

$$\therefore \quad \tilde{S} \sum_i \left(1/N_i^2\right) = \sum_i S_i / N_i^2 \qquad (1.27)$$

$$\therefore \quad \tilde{S} = \sum_i \left(S_i / N_i^2\right) / \sum_i \left(1/N_i^2\right) \qquad (1.28)$$

From this, we deduce that $\tilde{S}$ has to be a weighted mean, and in particular the noise-weighted mean $\tilde{S}$. On the other hand, if the noise is uniform, $\tilde{S}$ will revert to the usual unweighted mean $\bar{S}$. Also, if we do not apply the zero-mean condition (which we can achieve by setting $\tilde{S} = 0$), Eq. (1.25) reverts immediately to the standard matched filter condition.

The formula for $\tilde{S}$ may seem to be unduly general, in that N_i should normally be almost independent of i. However, if an ideal profile were to be derived by averaging real object profiles, then away from its center, the noise variance could be more substantial. Indeed, for large objects this would be a distinct limiting factor on such an approach. But for fairly small objects and features, noise variance should not vary excessively and useful matched filter profiles should be obtainable.

On a personal note, the main result proven in this section (cf. Eqs. (1.25) and (1.28)) took me so much time and effort to resolve the various issues that I was never convinced I would solve it. Hence I came to think of it as 'Davies's last theorem'.

1.2.7 Mask design—other considerations

Although the matched filter formalism and the now fully integrated zero-mean condition might seem to be sufficiently general to provide for unambiguous mask design, there are a number of aspects that remain to be considered. For example, how large should the masks be made? And how should they be optimally placed around any notable objects or features? We shall take the following example of a fairly complex object feature to help us answer this. Here region 2 is the object being detected, region 1 is the background, and M is the feature mask region.

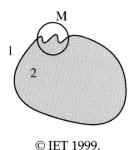

© IET 1999.

On this model we have to calculate optimal values for the mask weighting factors w_1 and w_2 and for the region areas A_1 and A_2. We can write the total signal and noise power from a template mask as:

$$S = w_1 A_1 S_1 + w_2 A_2 S_2$$
$$N^2 = w_1^2 A_1 N_1^2 + w_2^2 A_2 N_2^2$$

(1.29)

Thus, we obtain a power signal-to-noise-ratio (SNR):

$$\rho^2 = \frac{S^2}{N^2} = \frac{(w_1 A_1 S_1 + w_2 A_2 S_2)^2}{w_1^2 A_1 N_1^2 + w_2^2 A_2 N_2^2}$$

(1.30)

It is easy to see that if *both* mask regions are increased in area by the same factor η, ρ^2 will also be increased by this factor. This makes it interesting to optimize the mask by adjusting the *relative* values of A_1, A_2, leaving the total area A unchanged. Let us first eliminate w_2 using the zero-mean condition (which is commonly applied to prevent changes in background intensity level from affecting the result):

$$w_1 A_1 + w_2 A_2 = 0$$

(1.31)

Clearly, the power SNR no longer depends on the mask weights:

$$\rho^2 = \frac{S^2}{N^2} = \frac{(S_1 - S_2)^2}{N_1^2/A_1 + N_2^2/A_2} \tag{1.32}$$

Next, because the total mask area A is predetermined, we have:

$$A_2 = A - A_1 \tag{1.33}$$

Substituting for A_2 quickly leads to a simple optimization condition:

$$A_1/A_2 = N_1/N_2 \tag{1.34}$$

Taking $N_1 = N_2$, we obtain an important result—the equal area rule (Davies, 1999):

$$A_1 = A_2 = A/2 \tag{1.35}$$

Finally, when the equal area rule applies, the zero-mean rule takes the form:

$$w_1 = -w_2 \tag{1.36}$$

Note that many cases, such as those arising when the foreground and background have different textures, can be modeled by taking $N_1 \neq N_2$. In that case the equal area rule does not apply, but we can still use Eq. (1.34).

1.2.8 Corner detection

In Sections 1.2.1 and 1.2.3 we found that only two types of feature have vector (or pseudo-vector) forms—edge and line segments. Hence, whereas these features can be detected using just two component masks, all other features would be expected to require matching to many more templates in order to cope with varying orientations. Corner detectors appear to fall into this category, typical 3×3 corner templates being the following:

$$\begin{bmatrix} -4 & 5 & 5 \\ -4 & 5 & 5 \\ -4 & -4 & -4 \end{bmatrix} \quad \begin{bmatrix} 5 & 5 & 5 \\ -4 & 5 & -4 \\ -4 & -4 & -4 \end{bmatrix} \tag{1.37}$$

(Note that these masks have been adjusted to zero-mean form to eliminate the effects of varying lighting conditions.)

To overcome the evident problems of template matching—not the least amongst which is the need to use limited numbers of digital masks to approximate the underlying analogue intensity variations, which themselves vary markedly from instance to instance—many efforts have been made to obtain a more principled approach. In particular, as edges depend on the first derivatives of the image intensity field, it seemed logical to move to a second-order derivative approach. One of the first such investigations was the Beaudet (1978) approach, which employed the Laplacian and Hessian operators:

$$\begin{aligned} \text{Laplacian} &= I_{xx} + I_{yy} \\ \text{Hessian} &= I_{xx}I_{yy} - I_{xy}^2 \end{aligned} \tag{1.38}$$

These were particularly attractive as they are defined in terms of the determinant and trace of the symmetric matrix of second derivatives, and thus are invariant under rotation.

In fact, the Laplacian operator gives significant responses along lines and edges and hence is not particularly suitable as a corner detector. On the other hand, Beaudet's 'DET' (Hessian) operator does not respond to lines and edges but gives significant signals in the vicinity of corners and should therefore form a useful corner detector—though it responds with one sign on one side of a corner and with the opposite sign on the other side of the corner: on the corner itself it gives a null response. Furthermore, other workers criticized the specific responses of the DET operator and found they needed quite complex analyzes to deduce the presence and exact position of each corner (Dreschler and Nagel, 1981; Nagel, 1983).

However, Kitchen and Rosenfeld (1982) found they were able to overcome these problems by estimating the rate of change of the gradient direction vector along the horizontal edge tangent direction, and relating it to the horizontal curvature κ of the intensity function I. To obtain a realistic indication of the strength of a corner they multiplied κ by the magnitude of the local intensity gradient g:

$$C = \kappa g = \kappa \left(I_x^2 + I_y^2 \right)^{1/2}$$
$$= \frac{I_{xx} I_y^2 - 2 I_{xy} I_x I_y + I_{yy} I_x^2}{I_x^2 + I_y^2} \tag{1.39}$$

Finally, they used the heuristic of nonmaximum suppression along the edge normal direction to localize the corner positions further.

Interestingly, Nagel (1983) and Shah and Jain (1984) came to the view that the Kitchen and Rosenfeld, Dreschler and Nagel, and Zuniga and Haralick (1983) corner detectors were all essentially equivalent. This should not be overly surprising, since in the end the different methods would be expected to reflect the same underlying physical phenomena (Davies, 1988c)—reflecting a second-order derivative formulation interpretable as a horizontal curvature multiplied by an intensity gradient.

1.2.9 The Harris 'interest point' operator

At this point in Harris and Stephens (1988) developed an entirely new operator capable of detecting corner-like features—based not on second-order but on first-order derivatives. As we shall see below, this simplified the mathematics, including the difficulties of applying digital masks to intrinsically analogue functions. In fact, the new operator was able to perform a second-order derivative function by applying first-order operations. It is intriguing how it could acquire the relevant second-order derivative information in this way. To understand this we need to examine its quite simple mathematical definition.

The Harris operator is defined in terms of the local components of intensity gradient I_x, I_y in an image. The definition requires a window region to be defined and averages $\langle . \rangle$ to be taken over this whole window. We start by computing the following matrix:

$$\Delta = \begin{bmatrix} \langle I_x^2 \rangle & \langle I_x I_y \rangle \\ \langle I_x I_y \rangle & \langle I_y^2 \rangle \end{bmatrix} \tag{1.40}$$

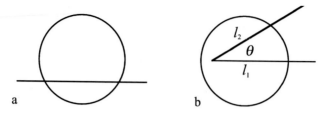

FIGURE 1.2 Geometry for calculating line and corner responses in a circular window. (a) straight edge, (b) general corner. © IET 2005.

We then use the determinant and trace to estimate the corner signal:

$$C = \det \Delta \, / \, \text{trace } \Delta \tag{1.41}$$

(Again, as for the Beaudet operators, the significance of using only the determinant and trace is that the resulting signal will be invariant to corner orientation.)

Before proceeding to analyze the form of C, note that if averaging were not undertaken, $\det \Delta$ would be identically equal to zero: clearly, it is only the smoothing intrinsic in the averaging operation that permits the spread of first-derivative values and thereby allows the result to depend partly on second derivatives.

To understand the operation of the detector in more detail, first consider its response for a single edge (Fig. 1.2a). In fact:

$$\det \Delta = 0 \tag{1.42}$$

because I_x is zero over the whole window region.

Next consider the situation in a corner region (Fig. 1.2b). Here:

$$\Delta = \begin{bmatrix} l_2 g^2 \sin^2 \theta & l_2 g^2 \sin \theta \cos \theta \\ l_2 g^2 \sin \theta \cos \theta & l_2 g^2 \cos^2 \theta + l_1 g^2 \end{bmatrix} \tag{1.43}$$

where l_1, l_2 are the lengths of the two edges bounding the corner, and g is the edge contrast, assumed constant over the whole window. We now find (Davies, 2005):

$$\det \Delta = l_1 l_2 \, g^4 \sin^2 \theta \tag{1.44}$$

and

$$\text{trace } \Delta = (l_1 + l_2)g^2 \tag{1.45}$$

$$\therefore \quad C = \frac{l_1 l_2}{l_1 + l_2} g^2 \sin^2 \theta \tag{1.46}$$

This may be interpreted as the product of (1) a strength factor λ, which depends on the edge lengths within the window, (2) a contrast factor g^2, and (3) a shape factor $\sin^2 \theta$, which depends on the edge 'sharpness' θ. Clearly, C is zero for $\theta = 0$ and $\theta = \pi$, and is a maximum for $\theta = \pi/2$—all these results being intuitively correct and appropriate.

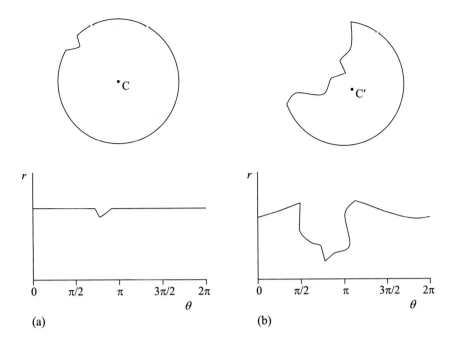

FIGURE 1.3 Problems with the centroidal profile descriptor. (a) shows a circular object with a minor defect on its boundary; its centroidal profile appears beneath it. (b) shows the same object, this time with a gross defect: because the centroid is shifted to C′, the *whole* of the centroidal profile is grossly distorted.

A good many of the properties of the operator can be determined from this formula, including the fact that the peak signal occurs not at the corner itself but at the center of the window used to compute the corner signal—though the shift is reduced as the sharpness of the corner decreases.

1.3 Part B – 2-D object location and recognition

1.3.1 The centroidal profile approach to shape analysis

2-D objects are commonly characterized by their boundary shapes. In this section we examine what can be achieved by tracking around object boundaries and analyzing the resulting shape profiles. Amongst the commonest type of profile used for this purpose is the centroidal profile—in which the object boundary is mapped out using an (r, θ) polar plot, taking the centroid C of the boundary as the origin of coordinates.

In the case of a circle of radius R, the centroidal profile is a straight line a distance R above the θ-axis. Fig. 1.3 clarifies the situation and also shows two examples of broken circular objects. In case (a), the circle is only slightly distorted and thus its centroid C remains virtually unchanged; hence, much of the centroidal plot remains at a distance R above the θ-axis. However, in case (b), even the part of the boundary that is not broken or distorted is far from being a constant distance from the θ-axis: this means that the object is unrecognizable from its profile, though in case (a) there is no difficulty in recognizing it as a slightly damaged circle. In fact, we can trace the relative seriousness of the two cases as being due largely to the fact that in case (b) the centroid has moved so much that even the unmodified part of the shape is not instantly recognizable. Of course, we could attempt to rectify the situation by trying to move the centroid back to its old position, but it would be difficult to do this reliably: in any case, if the original shape turned out not to be a circle, a lot of processing would be wasted before the true nature of the problem was revealed.

Overall, we can conclude that the centroidal profile approach is nonrobust, and is not to be recommended. In fact, this does not mean that it should not be used in practice. For example, on a cheese or biscuit conveyor, any object that is *not* instantly recognizable by its constant R profile should immediately be rejected from the product line; then other objects can be examined to be sure that their R values are acceptable and show an appropriate degree of constancy.

> **Robustness and its importance**
>
> It is not an accident that the idea of robustness has arisen here. It is actually core to much of the discussion on algorithm value and effectiveness that runs right through computer vision. The underlying problem is that of variability of objects or indeed of any entities that appear in computer images. This variability can arise simply from noise, or from varying shapes of even the same types of object, or from variations in size or placement, or from distortions due to poor manufacture, or cracks or breakage, or the fact that objects can be viewed from a variety of positions and directions under various viewing regimes—which tend to be most extreme for full perspective projection. In addition, one object may be partly obscured by another or even only partly situated within a specific image (giving effects that are not dissimilar to the result of breakage).
>
> While noise is well known to affect accuracy of measurement, it might be thought less likely to affect robustness. However, we need to distinguish the 'usual' sort of noise, which we can typify as Gaussian noise, from spike or impulse noise. The latter are commonly described as outlying points or 'outliers' on the noise distribution. (Note that we have already seen that the median filter is significantly better than the mean filter at coping with outliers.) The subject of robust statistics studies the topics of inliers and outliers and how best to cope with various types of noise. It underlies the optimization of accuracy of measurement and reliability of interpretation in the presence of outliers and gross disturbances to object appearance.

Next, it should be remarked that there are other types of boundary plot that can be used instead of the centroidal profile. One is the (s, ψ) plot and another is the derived (s, κ) profile. Here, ψ is the boundary orientation angle, and $\kappa(s)$, which is equal to $d\psi/ds$, is the local curvature function. Importantly, these formulations make no reference to the position of the centroid, and its position need not be calculated or even estimated. In spite of this advantage, all such boundary profile representations suffer from a significant further problem—that if any part of the boundary is occluded, distorted or broken, comparison of the object shape with templates of known shape is rendered quite difficult, because of the different boundary lengths.

In spite of these problems, when it can be employed, the centroidal profile method has certain advantages, in that it contributes ease of measurement of circular radii, ease of identification of squares and other shapes with prominent corners, and straightforward orientation measurement—particularly for shapes with prominent corners.

It now remains to find a method that can replace the centroidal profile method in instances where gross distortions or occlusions can occur. For such a method we need to move on to the following section which introduces the Hough transform approach.

1.3.2 Hough-based schemes for object detection

In Section 1.3.1 we explored how circular objects might be identified from their boundaries using the centroidal profile approach to shape analysis. The approach was found to be non-robust because of its incapability for coping with gross shape distortions and occlusions. In this section we show that the Hough transform provides a simple but neat way of solving this problem. The method used is to take each edge point in the image, move a distance R inwards along the local edge normal, and accumulate a point in a separate image called the *parameter space*: R is taken to be the expected radius of the circles to be located. The result of this will be a preponderance of points (often called 'votes') around the locations of circle centers. Indeed, to obtain accurate estimates of center locations, it is only necessary to find significant peaks in parameter space.

The process is illustrated in Fig. 1.4, making it clear that the method ignores noncircular parts of the boundary and only identifies genuine circle centers: thus the approach focuses on data that correspond to the chosen model and is not confused by irrelevant data that would otherwise lead to nonrobust solutions. Clearly, it relies on edge normal directions being estimated accurately. Fortunately, the Sobel operator is able to estimate edge orientation to within $\sim 1°$ and is straightforward to apply. In fact, Fig. 1.5 shows that the results can be quite impressive.

A disadvantage of the approach as outlined above is that it requires R to be known in advance. The general solution to this problem is to use a 3-D parameter space, with the third dimension representing possible values of R, and then searching for the most significant peaks in this space. However, a simpler solution involves accumulating the results for

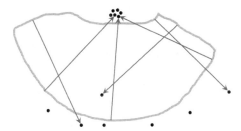

FIGURE 1.4 Robustness of the Hough transform when locating the center of a circular object. The circular part of the boundary gives candidate center points that focus on the true center, whereas the irregular broken boundary gives candidate center points at random positions. In this case the boundary is approximately that of the broken biscuit shown in Fig. 1.5.

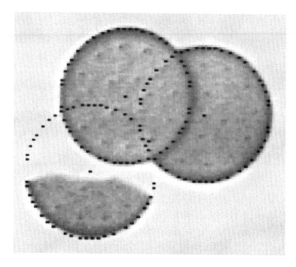

FIGURE 1.5 Location of broken and overlapping biscuits, showing the robustness of the center location technique. Accuracy is indicated by the black dots which are each within 1/2 pixel of the radial distance from the center. © IFS 1984.

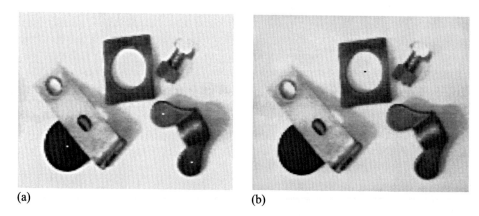

(a)   (b)

FIGURE 1.6 Simultaneous detection of objects with different radii. (a) Detection of a lens cap and a wing nut when radii are assumed to lie in the range 4–17 pixels; (b) hole detection in the same image when radii are assumed to fall in the range −26 to −9 pixels (negative radii are used since holes are taken to be objects of negative contrast): clearly, in *this* image a smaller range of negative radii could have been employed.

a range of likely values of R in the *same* 2-D parameter space—a procedure that results in substantial savings in storage and computation (Davies, 1988a). Fig. 1.6 shows the result of applying this strategy, which works with both positive and negative values of R. On the other hand, note that the information on radial distance has been lost by accumulating all the votes in a single parameter plane. Hence a further iteration of the procedure would be required to identify the radius corresponding to each peak location.

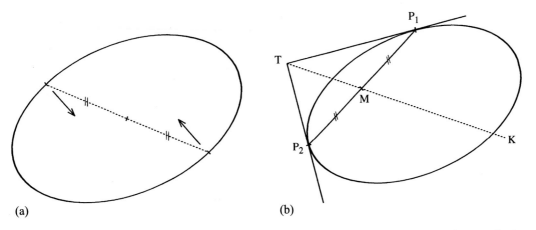

(a)   (b)

FIGURE 1.7 The geometry of two ellipse detection methods. (a) In the diameter-bisection method, a pair of points is located for which the edge orientations are antiparallel. The midpoints of such pairs are accumulated and the resulting peaks are taken to correspond to ellipse centers. (b) In the chord–tangent method, the tangents at P_1 and P_2 meet at T and the midpoint of P_1P_2 is M. The center C of the ellipse lies on the line TM produced.

The Hough transform approach can also be used for ellipse detection: two simple methods for achieving this are presented in Fig. 1.7. Both of these embody an indirect approach in which *pairs* of edge points are employed. Whereas the diameter-bisection method involves considerably less computation than the chord–tangent method, it is more prone to false detections—for example, when two ellipses lie near to each other in an image.

To prove the validity of the chord–tangent method, note that symmetry ensures that the method works for circles: projective properties then ensure that it also works for ellipses, because under orthographic projection, straight lines project into straight lines, midpoints into midpoints, tangents into tangents, and circles into ellipses; in addition, it is always possible to find a viewpoint such that a circle can be projected into a given ellipse.

We now move on to the so-called *generalized Hough transform* (GHT), which employs a more direct procedure for performing ellipse detection than the other two methods outlined above.

To understand how the standard Hough technique is generalized so that it can detect arbitrary shapes, we first need to select a localization point L within a template of the idealized shape. Then, we need to arrange so that, instead of moving from an edge point a *fixed* distance R directly along the local edge normal to arrive at the center, as for circles, we move an appropriate *variable* distance R in a variable direction φ so as to arrive at L: R and φ are now functions of the local edge normal direction θ (Fig. 1.8). Under these circumstances votes will peak at the preselected object localization point L. The functions $R(\theta)$ and $\phi(\theta)$ can be stored analytically in the computer algorithm, or for completely arbitrary shapes they may be stored as lookup tables. In either case the scheme is beautifully simple in principle but an important complication arises because we are going from an isotropic shape (a circle) to an anisotropic shape which may be in a completely arbitrary orientation.

This means adding an extra dimension in parameter space (Ballard, 1981). Each edge point then contributes a set of votes in each orientation plane in parameter space. Finally, the whole

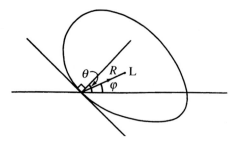

FIGURE 1.8 Computation of the generalized Hough transform.

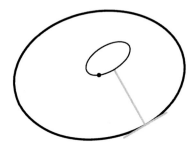

FIGURE 1.9 Use of a PSF shape that takes into account all possible orientations of an ellipse. The PSF is positioned by the grey construction lines so that it passes through the center of the ellipse (see the black dot).

of parameter space is searched for peaks, the highest points indicating both the locations of objects and their orientations. Interestingly, ellipses can be detected by the GHT using a single plane in parameter space, by applying a point spread function (PSF) to each edge point, which takes all possible orientations of the ellipse into account: note that the PSF is applied at some distance from the edge point, so that the center of the PSF can pass through the center of the ellipse (Fig. 1.9). Lack of space prevents details of the computations from being presented here (e.g., see Davies, 2017, Chapter 11).

1.3.3 Application of the Hough transform to line detection

The Hough transform (HT) can also be applied to line detection. Early on, it was found best to avoid the usual slope–intercept equation, $y = mx + c$, because near-vertical lines require near-infinite values of m and c. Instead, the 'normal' (θ, ρ) form for the straight line (Fig. 1.10) was employed:

$$\rho = x \cos\theta + y \sin\theta \tag{1.47}$$

To apply the method using this form, the set of lines passing through each point P_i is represented as a set of sine curves in (θ, ρ) space: e.g., for point $P_1(x_1, y_1)$ the sine curve has equation:

$$\rho = x_1 \cos\theta + y_1 \sin\theta \tag{1.48}$$

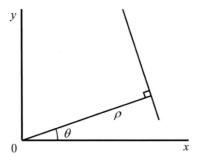

FIGURE 1.10 Normal (θ, ρ) parametrization of a straight line.

After vote accumulation in (θ, ρ) space, peaks indicate the presence of lines in the original image.

A lot of work has been carried out (e.g., see Dudani and Luk, 1978) to limit the inaccuracies involved in line location, which arise from several sources—noise, quantization, the effects of line fragmentation, the effects of slight line curvature, and the difficulty of estimating the exact peak positions in parameter space. In addition, the problem of longitudinal line localization is important. For the last of these processes, Dudani and Luk (1978) developed the method of 'xy–grouping', which involved carrying out connectivity analysis for each line. Segments of a line would then be merged if they were separated by gaps of less than ~5 pixels. Finally, segments shorter than a certain minimum length (also typically ~5 pixels) would be ignored as too insignificant to help with image interpretation.

Overall, we see that all the forms of the HT described above gain considerably by *accumulating evidence* using a voting scheme. This is the source of the method's high degree of robustness. The computation processes used by the HT can be described as inductive rather than deductive as the peaks lead to *hypotheses* about the presence of objects, that need in principle to be confirmed by other evidence, whereas *deduction* would lead to immediate proof of the presence of objects.

1.3.4 Using RANSAC for line detection

RANSAC is an alternative model-based search schema that can often be used instead of the HT. In fact, it is highly effective when used for line detection, which is why the method is introduced here. The strategy can be construed as a voting scheme, but it is used in a different way from that in the HT. It operates by making a sequence of hypotheses about the target objects, and determines the support for each of them by counting how many data points agree with them within reasonable (e.g., $\pm 3\sigma$) limits (see Fig. 1.11). As might be expected, for any potential target object, only the hypotheses with the maximum support are retained at each stage.

We next explain how RANSAC is used for line detection. As in the case of the HT, we start by applying an edge detector and locating all the edge points in the image. As we shall see, RANSAC operates best with a limited number of points, so it is useful to find the edge points that are local maxima of the intensity gradient image. Next, to form a straight line hypothesis,

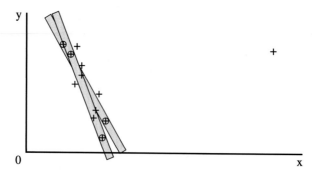

FIGURE 1.11 The RANSAC technique. Here the + signs indicate data points to be fitted, and two instances of pairs of data points (indicated by ⊕ signs) leading to hypothesized lines are also shown. Each hypothesized line has a region of influence of tolerance ±*t* within which the support of maximal numbers of data points is sought. The line with the most support is taken to lead to the best fit.

TABLE 1.1   Basic RANSAC algorithm for finding the line with greatest support. This algorithm only returns one line: in fact it returns the specific line model that has greatest support. Lines with less support are in the end ignored.

```
Mmax=0;
for all pairs of edge points do {
    find equation of line defined by the two points i, j;
    M = 0;
    for all N points in list do
        if (point k is within threshold distance d of line) M ++;
    if (M > Mmax) {
        Mmax = M;
        imax = i;
        jmax = j;
        // this records the hypothesis giving the maximum support so far
    }
}
/* if Mmax > 0, (x[imax], y[imax]) and (x[jmax], y[jmax]) will be the coordinates of
the points defining the line having greatest support */
```

all that is necessary is to take *any* pair of edge points. For each hypothesis we run through the list of N edge points finding how many points M support the hypothesis. Then we take more hypotheses (more pairs of edge points) and at each stage retain only the one giving maximum support M_{max}. This process is shown in Table 1.1.

The algorithm in Table 1.1 corresponds to finding the center of the highest peak in parameter space, as in the case of the HT. To find all the lines in the image, the most obvious strategy is the following: find the first line, then eliminate all the points that gave it support; then find the next line and eliminate all the points that gave it support; and so on until all the points have been eliminated from the list. The process may be written more compactly in the form:

```
repeat {
    find line;
    eliminate support;
}
until no data points remain;
```

As outlined above, RANSAC involves quite considerable computational load—amounting to $O(N^3)$—which should be compared with $O(N)$ for the HT. Hence it is better to reduce N in some way when using RANSAC. This explains why it is useful to concentrate on the local maxima rather than employing the full list of edge points. However, an alternative is to employ repeated random sampling from the full list until sufficient hypotheses have been tested to be confident that all significant lines have been detected. [Note that these ideas reflect the original meaning of the term RANSAC, which stands for RANdom SAmpling Consensus—"consensus" indicating that any hypothesis has to form a consensus with the available support data (Fischler and Bolles, 1981).] Confidence that all significant lines have been detected can be obtained by estimating the risk that a significant line will be missed because no representative pair of points lying on the line was considered.

We are now in a position to consider results obtained by applying RANSAC to a particular case of straight line detection. In the test described, pairs of points were employed as hypotheses, and all edge points were local maxima of the intensity gradient. The case shown in Fig. 1.12 corresponds to detection of a block of wood in the shape of an icosahedron. Note that one line on the right of Fig. 1.12(a) was missed because a lower limit had to be placed on the level of support for each line: this was necessary because below this level of support the number of chance collinearities rose dramatically even for the relatively small number of edge points shown in Fig. 1.12(b), leading to a sharp rise in the number of false positive lines. Overall, this example shows that RANSAC is a highly important contender for location of straight lines in digital images. Not discussed here is the fact that RANSAC is useful for obtaining robust fits to many other types of shape, in 2-D and in 3-D.

Finally, it should be mentioned that RANSAC is less influenced than the HT by aliasing along straight lines. This is because HT peaks tend to be fragmented by aliasing, so the best hypotheses can be difficult to obtain without drastic smoothing of the image. The reason why RANSAC wins in this context is that it does not rely on individual hypotheses being accurate: rather it relies on enough hypotheses easily being generatable, and by the same token, discardable.

1.3.5 A graph-theoretic approach to object location

This section considers a commonly occurring situation involving considerable constraints—objects appearing on a horizontal worktable or conveyor at a known distance from the camera. It is also assumed (a) that objects are flat or can appear in only a restricted number of stances in three dimensions, (b) that objects are viewed from directly overhead, and (c) that perspective distortions are small. In such situations the objects may in principle be

a   b

FIGURE 1.12 Straight line location using the RANSAC technique. (a) shows an original grey-scale image with various straight edges located using the RANSAC technique. (b) shows the edge points fed to RANSAC to obtain (a): these were local maxima of the gradient image. In (a), three edges of the icosahedron have been missed. This is because they are 'roof' edges with low contrast and low intensity gradient. In fact, RANSAC also missed a fourth edge because of a lower limit placed on the level of support (see text).

identified and located from very few point features. Since such features are taken to have no structure of their own, it will be impossible to locate an object uniquely from a single feature, although positive identification and location would be possible using two features if these were distinguishable and if their distance apart were known. For truly indistinguishable point features, an ambiguity remains for all objects not possessing 180° rotation symmetry. Hence at least three point features are in general required to locate and identify objects at known range. Clearly, noise and other artefacts such as occlusions modify this conclusion. In fact, when matching a template of the points in an idealized object with the points present in a real image, we find that:

1. a great many feature points may be present because of multiple instances of the chosen type of object in the image
2. additional points may be present because of noise or clutter from irrelevant objects and structure in the background
3. certain points that should be present are missing because of noise or occlusion, or because of defects in the object being sought.

These problems mean that we should be attempting to match a subset of the points in the idealized template to various subsets of the points in the image. If the point sets are considered to constitute *graphs* with the point features as *nodes*, the task devolves into the mathematical problem of subgraph–subgraph isomorphism, i.e., finding which subgraphs in the image graph are isomorphic to subgraphs of the idealized template graph. [Isomorphic means having the same basic shape and structure.] Clearly, a point feature matching scheme will be most successful if it finds the most likely interpretation by searching for solutions having the greatest internal consistency—i.e., with the greatest number of point matches per object.

Unfortunately, the schema presented above is still too simplistic in many applications as it is insufficiently robust against distortions. In particular, optical (e.g., perspective) distortions

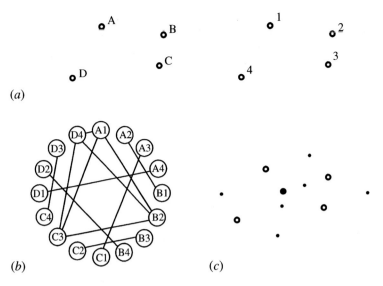

FIGURE 1.13 Matching problem for a general quadrilateral: (a) basic labeling of model (*left*) and image (*right*); (b) match graph; (c) placement of votes in parameter space: small circles indicate hole positions, dots indicate individual votes and the large dot shows the position of the main peak. © AVC 1988.

may arise, or the objects themselves may be distorted, or by resting partly on other objects they may not be quite in the assumed stance: hence distances between features may not be exactly as expected. These factors mean that some tolerance has to be accepted in the distances between pairs of features. Clearly, distortions lay more strain on the point matching technique and make it all the more necessary to seek solutions with the greatest possible internal consistency. Therefore, as many features as possible should be taken into account in locating and identifying objects. The maximal clique approach is intended to achieve this.

As a start, as many features as possible are identified in the original image: typically, they are numbered in order of appearance in a TV raster scan. The numbers then have to be matched against the letters corresponding to the features on the idealized object. A systematic way of achieving this is by constructing a *match graph* (or *association graph*) in which the nodes represent feature assignments, and arcs joining nodes represent pairwise compatibilities between assignments. To find the best match it is then necessary to find regions of the match graph where the cross-linkages are maximized. To achieve this, *cliques* are sought within the match graph. A clique is a *complete subgraph*—i.e., one for which all pairs of nodes are connected by arcs. However, the previous arguments indicate that if one clique is completely included within another clique, it is likely that the larger clique represents a better match—and indeed *maximal cliques* can be taken as leading to the most reliable matches between the observed image and the object model.

Fig. 1.13(a) shows the situation for a general quadrilateral, the match graph being shown in Fig. 1.13(b). In this case there are 16 possible feature assignments, 12 valid compatibilities and 7 maximal cliques. If occlusion of a feature occurs, this will (taken on its own) reduce the number of possible feature assignments and also the number of valid compatibilities: in addition, the number of maximal cliques and the size of the largest maximal clique will be

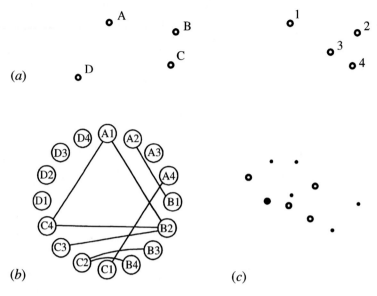

FIGURE 1.14 Matching when one feature is occluded and another is added: (a) basic labeling of model (*left*) and image (*right*); (b) match graph; (c) placement of votes in parameter space (notation as in Fig. 1.13).

reduced. On the other hand, noise or clutter can add erroneous features. If the latter are at arbitrary distances from existing features, then the number of possible feature assignments will be increased but there will not be any more compatibilities in the match graph, so the latter will have only trivial additional complexity. However, if the extra features appear at *allowed* distances from existing features, this will introduce extra compatibilities into the match graph and make it more tedious to analyze. In the case shown in Fig. 1.14, both types of complication—an occlusion and an additional feature—arise: there are now 8 pairwise assignments and 6 maximal cliques, rather fewer overall than in the original case of Fig. 1.13. However, the important factor is that the largest maximal clique still indicates the most likely interpretation of the image, so the technique is inherently highly robust.

Fig. 1.15(a) shows a pair of cream biscuits which are to be located from their "docker" holes—this strategy being advantageous since it has the potential for highly accurate product location prior to detailed inspection. The holes found by a simple template matching routine are indicated in Fig. 1.15(a): the template used is rather small and, as a result, the routine is fairly fast but fails to locate all holes; in addition, it can give false alarms. Hence an "intelligent" algorithm must be used to analyze the hole location data. Analysis of the data in the above example yields two nontrivial maximal cliques, each corresponding correctly to one of the two biscuits in the image.

1.3.6 Using the generalized Hough transform (GHT) to save computation

In these examples, checking which subgraphs are maximal cliques is a simply-stated problem. Unfortunately, the execution time of an optimal maximal clique algorithm is bounded

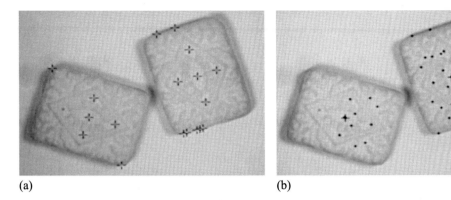

(a)   (b)

FIGURE 1.15 Location of cream sandwich biscuits; (a) two cream sandwich biscuits with crosses indicating the result of applying a simple hole detection routine; (b) the two biscuits reliably located by the GHT from the hole data in (a): the isolated small crosses indicate the positions of single votes. © AVC 1988.

not by a polynomial in M (for a match graph containing maximal cliques of up to M nodes) but by a much faster varying function. Specifically, the task of finding maximal cliques is known to be "NP-complete" and runs in exponential time. Thus, whatever the run-time may be for values of M up to about 6, it will typically be 100 times slower for values of M up to about 10, and 100 times slower again for M greater than ~14.

We shall now see how the GHT can be used as an alternative to the maximal clique approach. To apply the GHT, we first list all the features and then accumulate votes in parameter space at every possible position of a localization point L consistent with each *pair* of features (Fig. 1.16). To proceed it is necessary merely to use the interfeature distance as a lookup parameter in the GHT R-table. For indistinguishable point features this means that there must be two entries for the position of L for each value of the interfeature distance. The procedure is illustrated by the general quadrilateral example of Fig. 1.13: this leads to 7 peaks in parameter space, whose weights are 6, 1, 1, 1, 1, 1, 1 (see Fig. 1.13(c)). A similar situation applies for Fig. 1.14. Close examination of Figs. 1.13 and 1.14 indicates that every peak in parameter space corresponds to a maximal clique in the match graph. Indeed, there is a one-to-one relation between the two, so correct compatibilities all contribute both to a large maximal clique and to a large peak in parameter space. This situation still applies even when occlusions occur or additional features are present (see Fig. 1.14).

Finally, consider again the example of Fig. 1.15(a), this time obtaining a solution using the GHT. Fig. 1.15(b) shows the positions of candidate object centers as found by the GHT. The small isolated crosses indicate the positions of single votes, and those very close to the two large crosses lead to voting peaks of weights 10 and 6 at these respective positions. Hence object location is both accurate and robust, as required (Davies, 1988b).

We next compare the computational requirements of the maximal clique and GHT approaches to object location. For simplicity, imagine an image that contains just one wholly visible example of an object possessing n features and that we are trying to recognize it by seeking all possible pairwise compatibilities.

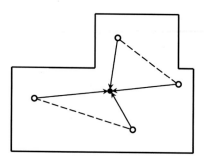

FIGURE 1.16 Method for locating L from pairs of feature positions: each pair of feature points gives two possible voting positions in parameter space, when objects have no symmetries. When symmetries are present, certain pairs of features may give rise to up to 4 voting positions: this is confirmed on careful examination of Fig. 1.15(b).

For an object possessing n features, the match graph contains n^2 nodes (i.e., possible assignments), and there are $^{n^2}C_2 = n^2(n^2 - 1)/2$ possible pairwise compatibilities to be checked in building the graph. The amount of computation at this stage of the analysis is $O(n^4)$. To this must be added the cost of finding the maximal cliques. Since the problem is NP-complete, the load rises at a rate which is close to being exponential in n^2.

Now consider the cost of getting the GHT to find objects via pairwise compatibilities. As has been seen, the total height of all the peaks in parameter space is equal to the number of pairwise compatibilities in the match graph. Hence the computational load is of the same order, $O(n^4)$. Next comes the problem of locating all the peaks in parameter space. For an $N \times N$ image only N^2 points have to be visited in parameter space and the computational load is $O(N^2)$, though keeping a running record of the maximum location during voting can reduce it considerably (Davies, 1988b).

1.3.7 Part-based approaches

Whereas the object location approaches described above tend to rely on objects following quite well-defined geometric models, a totally distinct approach is to use methods such as deformable models to locate and recognize them. The aim of such approaches is to take account of variations in appearance resulting from changes in illumination, viewpoint, and properties such as shape and color. This is particularly important when searching for faces or pedestrians in road scenes, to take two important examples. Methods in this category include rigid templates (Dalal and Triggs, 2005), bag-of-features (Zhang et al., 2007), deformable templates (e.g., Cootes and Taylor, 2001), and part-based models (e.g., Amit and Trouvé, 2007; Leibe et al., 2008). Deformable parts models are trained using collections of parts arranged in deformable configurations. This approach came to the fore in 2010 with the work of Felzenszwalb et al. (2010) when it was shown to lead to efficient, accurate, state-of-the-art results on difficult data sets.

Deformable parts models (DPMs) are based on the idea that objects can be considered as collections of parts. Thus, to detect objects such as faces, it should only be necessary to locate the parts and examine their interrelationships. This can be carried out by identifying parts and their bounding boxes and then making proposals for combining them into larger bounding boxes representing objects. Basically, once object bounding boxes have been found, these regions are protected from further analysis by nonmaximum suppression. In practice, this means giving each potential bounding box a score, keeping the highest score and skipping any that overlap an already existing bounding box by a critical percentage, e.g., 50%. This highly successful approach achieved state-of-the-art results on the PASCAL VOC 2006, 2007, and 2008 benchmarks (Everingham et al., 2006; 2007; 2008) and "established itself as the de-facto standard for generic object detection" (Mathias et al., 2014). Mathias et al. tested the DPM approach very thoroughly and showed that it could achieve top performance for face detection.

Interestingly, the DPM approach permitted the location of overtly 3-D objects—but without their 3-D geometry having to be taken directly into account, this being achieved by sufficiently varied training on the relevant types of object. Another welcome capability is that the approach can also be very effective for locating articulated objects.

The DPM approach is highly significant as it formed the basis for deep learning approaches with even higher performance ratings—as achieved by Bai et al. (2016) and other workers. (See Part F for deep learning methods.)

1.4 Part C – 3-D object location and the importance of invariance

1.4.1 Introduction to 3-D vision

In the earlier parts of this chapter, it has generally been assumed that objects are essentially flat and are viewed in such a way that there are only three degrees of freedom—namely, the two associated with position, and a further one concerned with orientation. While this approach was adequate for carrying out many useful visual tasks, it is inadequate for interpreting most outdoor or indoor scenes or even for helping with quite simple robot assembly and inspection tasks. Indeed, over the past few decades a considerable amount of theory has been developed and backed up by experiment, to find how scenes composed of real 3-D objects can be understood in detail.

In general, this means attempting to interpret scenes in which objects may appear in totally arbitrary positions and orientations—corresponding to six degrees of freedom. Interpreting such scenes, and deducing the translation and orientation parameters of arbitrary sets of objects, takes a substantial amount of computation—partly because of the inherent ambiguity in inferring 3-D information from 2-D images. However, a variety of approaches is available for proceeding with 3-D vision, and subtle combinations of them will often be needed to successfully interpret 3-D scenes.

Before proceeding further, we present the imaging equation for a general point (X, Y, Z) in a scene, under what is known as 'perspective projection'; this yields the image point:

$$(x, y) = (f X/Z, f Y/Z) \tag{1.49}$$

where f is the focal length of the lens being used.

We can now introduce the approach adopted by the human visual system—that of binocular vision. The camera system that is used for this purpose is depicted in Fig. 1.17. With

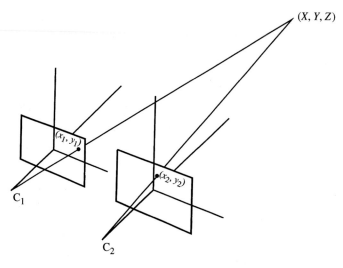

FIGURE 1.17 Stereo imaging using two lenses. The axes of the optical systems are parallel, i.e., there is no 'vergence' between the optical axes.

this geometry, a general point appears in the two images as (x_1, y_1) and (x_2, y_2). In general, the two optical systems need not have parallel optical axes and will exhibit a nonzero 'vergence' angle. However, the zero-vergence case is often employed and for simplicity is the one considered here. Note that the two sets of coordinates corresponding to the general point (X, Y, Z) in the scene will differ because the baseline b between the optical axes causes relative displacement or 'disparity' of the points in the two images.

Next, with a suitable choice of Z-axis on the perpendicular bisector of the baseline b, we obtain two equations:

$$x_1 = (X + b/2)\, f/Z \tag{1.50}$$

$$x_2 = (X - b/2)\, f/Z \tag{1.51}$$

Calculating the disparity $D = x_1 - x_2$ immediately permits the depth Z to be obtained:

$$Z = bf/(x_1 - x_2) \tag{1.52}$$

Whereas this seems to be an ideal way of proceeding with 3-D vision, there is a fundamental problem—that of confirming that both points in a stereo pair actually correspond to the same point in the original scene. Note also that to obtain high accuracy in the determination of depth, a large baseline b is required: unfortunately, as b is increased, the correspondence between the images decreases, so it becomes more difficult to find matching points: this is because the two images become increasingly different and difficult to match.

The standard way of dealing with the stereo correspondence problem mentioned above is the epipolar line approach illustrated in Fig. 1.18. To understand this technique, imagine that we have located a distinctive point in the first image and that we are marking all possible points in the object field which could have given rise to it. This will mark out a line of points

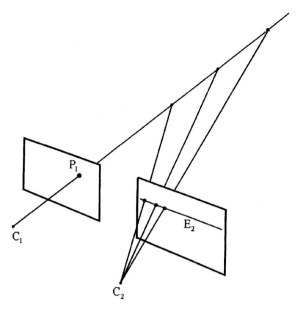

FIGURE 1.18 Geometry of epipolar lines. A point P_1 in one image plane may have arisen from any one of a line of points in the scene, and may appear in the alternate image plane at any point on the so-called epipolar line E_2.

at various depths in the scene and, when viewed in the second image plane, a locus of points can be constructed in that plane. This locus (in the alternate image) is the *epipolar line* corresponding to the original image point. If we now search along the epipolar line for a similarly distinctive point in the second image, the chance of finding the correct match is significantly enhanced. This method not only cuts down the amount of computation required to find corresponding points, but also significantly reduces the incidence of false alarms. In the simple geometry of Fig. 1.17, all epipolar lines are parallel to the x-axis, although this only applies for the case of zero vergence. Note that the correspondence problem is rendered considerably more difficult by the fact that there will be points in the scene that give rise to points in one image but not in the other: this may arise because of occlusion or gross distortion of one of the points. Thus, it is necessary to search for consistent sets of solutions in the form of continuous object surfaces in the scene.

The difficulties caused by the correspondence problem have led to a number of alternative approaches. One of the most prominent has been 'shape from shading'—mapping how the surface orientation varies by analyzing the apparent brightness of the surface. While much has been achieved by this approach, it involves assumptions about the reflectance and specularity of the surfaces and how these vary with the orientation of (a) the surface and (b) the source of illumination. It also requires the application of complex iterative algorithms—a topic that cannot be dealt with in detail here. Similarly, 'photometric stereo', which involves illuminating scenes in turn by separate light sources and analyzing the resulting images, is too complex to be considered in detail here.

Next, 'shape from texture' is an approach that also allows details of surface orientation to be analyzed by examining the relative areas of textural elements. However, this is a specialized method that is not very frequently applied. Another approach is that of 'structured lighting'—typically based on directing patterns of light stripes, or other arrangements of light spots or grids, onto the object field—which has been exceedingly widely used for inspection and assembly on factory lines, though it can hardly be said to be in wide use in other types of application such as surveillance. Again, this has to be regarded as a specialized rather than a general technique for gauging 3-D object shapes.

It should be remarked that all but the last of these approaches lead to the production of surface orientation maps rather than measurement of depth per se, so computation of depth and surface shape has to be deduced from the raw orientation measurements.

Overall, the methods described above employ various means for estimating depth at all places in a scene, and hence are able to map out 3-D surfaces in a fair amount of detail. However, they do not give any clue as to what these surfaces represent. In some situations it may be clear that certain planar surfaces are parts of the background, e.g., the floor and the walls of a room, but in general individual objects will not be inherently identifiable. Indeed, objects tend to merge with each other and with the background, so specific methods are needed to segment the 3-D 'space map' and finally recognize the objects, giving detailed information on their positions and orientations. Clearly, obtaining a depth map of a 3-D object is no closer to identifying it than a boundary map such as a centroidal profile is to recognizing a 2-D object: specific means must be devised to perform the identification. Unfortunately, this task is significantly more difficult in 3-D than it is in 2-D. For example, whereas a Hough transform can in principle be applied in both cases, in 3-D it is hugely more complicated and computation intensive than in 2-D, as the number of free parameters will normally have increased from 3 to 6 for a static shape with no unknown shape parameters—there being 3 degrees of freedom for translation and 3 for rotation. Note also that the computational complexity normally varies not with the number of degrees of freedom but with an exponent of this number.

One further point should be added: over the past few years, sensors have been devised that provide RGB-D (color and depth) outputs: these provide depth information from optical 'time-of-flight'. LIDAR is well known but expensive and works better at long range, whereas matrix-based time-of-flight cameras work better at short range and typically use laser-generated light pulses that are a few nanoseconds apart. These types of advance help to solve the stereo correspondence problem. However, they do not eliminate the problem of interpreting 3-D surfaces, for which the large number of degrees of freedom to be wrestled with by object identification algorithms remains problematic: on the contrary, they serve to highlight this as the most substantial remaining problem.

Given that computational complexity is the primary source of these problems, it is natural to examine each depth map for salient features and to interpret the scenes accordingly: once we have object descriptions based on relatively few salient features rather than bulk surface descriptions, there would appear to be hope of achieving rapid, reliable identifications. We look into this possibility further in the following section.

1.4.2 Pose ambiguities under perspective projection

In this section we define weak and full perspective and aim to understand the *perspective n-point* (P*n*P) *problem*—the problem of finding the pose of objects from *n* features under various forms of perspective.

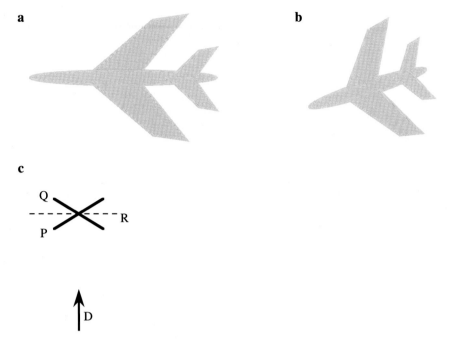

FIGURE 1.19 Perspective inversion for an aeroplane. Here an aeroplane (a) is silhouetted against the sky and appears as in (b). (c) shows the two planes P and Q in which the aeroplane could lie, relative to the direction D of viewing: R is the reflection plane relating the planes P and Q.

Full perspective projection (FPP) is the underlying form of projection between an object and its image, which leads for example to parallel lines no longer appearing parallel, and most shapes appearing distorted—circles even appearing as ellipses. Weak perspective projection (WPP) is the form of perspective projection that occurs for distant objects, for which $\Delta Z \ll Z$. It can be regarded as the same as 'scaled orthographic projection'—orthographic projection being the type of projection that would occur if the object were projected orthogonally by parallel rays onto the image plane; while scaling accounts for the reduced apparent size of an object.

As WPP does not distort object shapes (e.g., the back of a wire cube appears the same size and shape as its front), it is easier to use it to model the imaging process. However, this form of projection is so simple that it can lead to ambiguity when flat objects are viewed. This is demonstrated in Fig. 1.19, which illustrates that a 2-D view of a distant aeroplane can correspond to one of two orientations, as only the cosine of the plane's orientation α is determined from the single view. Interestingly, when α is nonzero, FPP adds an additional distortion to the aeroplane's shape, and the true orientation can then be identified.

Table 1.2 indicates the full extent of this situation, when flat objects are detected via one or more of their features. This table reflects the overall PnP problem mentioned above. In the coplanar case (in which all n features of an object are coplanar), we see from the table that WPP never gives an unambiguous interpretation, whereas FPP does—though only when n is greater than 3. The reason why n has to be greater than 3 for FPP to give an unambiguous

TABLE 1.2 Ambiguities when estimating pose from point features. This table summarizes the numbers of solutions that will be obtained when estimating the pose of a rigid object from point features located in a single image. It is assumed that n point features are detected and identified correctly and in the correct order. The columns WPP and FPP signify weak perspective projection and full perspective projection respectively. The upper half of the table applies when all n points are coplanar; the lower half of the table applies when the n points are noncoplanar. Note that when $n \leq 3$, the results strictly apply only in the coplanar case. However, the top two lines in the lower half of the table are retained for easy comparison.

Arrangement of the points	n	WPP	FPP
	≤ 2	∞	∞
	3	2	4
coplanar	4	2	1
	5	2	1
	≥ 6	2	1
	≤ 2	∞	∞
	3	2	4
noncoplanar	4	1	2
	5	1	2
	≥ 6	1	1

result is that it involves so many parameters that 3 features are insufficient to resolve the situation; however, when 4 or more features are present, the complete situation can be resolved and the ambiguity eliminated. But why can't WPP achieve this too? The reason is that under WPP the location of any additional features (above 3) can be deduced from the first 3, so they can give no additional information: hence WPP cannot lead to the ambiguity being eliminated.

Note that when n is 1 or 2, there is at least one rotational degree of freedom, so there are an infinite number of solutions. At this point we have dealt with all the possibilities in the upper half of the table, where coplanar objects are involved. We next turn to the noncoplanar case, dealt with in the lower half of the table. Instances where $n \leq 3$ have already been dealt with in the upper half of the table, so the noncoplanar case only involves instances where $n > 3$.

Let us now consider what happens when 4 features are viewed under WPP. Taking two of the features with each of the other two in turn, we can generate two planes. When the 3 features on each of these planes are viewed they will generate two solutions with different values of α, and there can only be one consistent solution. This completes our understanding of the WPP entries in Table 1.2, and in particular the noncoplanar cases. The situation is well illustrated in Fig. 1.20(a–c). Note, however, the salutary situation shown in Fig. 1.20(d) showing how an object possessing a special symmetry may be subject to a remanent ambiguity.

To fully understand the situation in which more than 3 noncoplanar points are viewed under FPP, we need to consider the fact that there are 11 camera calibration parameters (see

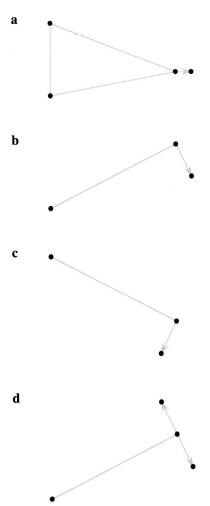

FIGURE 1.20 Determination of pose for 4 points viewed under weak perspective projection. (a) shows an object containing four noncoplanar points, as seen under weak perspective projection. (b) shows a side view of the object. If the first three points (connected by nonarrowed grey lines) were viewed alone, perspective inversion would give rise to a second interpretation (c). However, the fourth point gives additional information about the pose which permits only one overall interpretation. This would not be the case for an object containing an additional symmetry as in (d), since its reflection would be identical to the original view (not shown).

Section 1.4.10) that need to be determined from 12 linear homogeneous equations, which means that at least 6 noncoplanar points (involving 2 × 6 image coordinates) will in general be needed to compute all 11 parameters. Thus, although FPP makes the situation more complex, it also provides more information by which, eventually, to resolve the ambiguity.

Finally, it should be emphasized that the above discussion assumes that the correspondences between object and image features are all known, i.e., that n point features are detected and identified correctly and in the correct order. If this is not so, the number of possible solutions could increase substantially, considering the number of possible permutations of quite small numbers of points. One way of limiting this problem is to note that coplanar points viewed under weak or full perspective projection always appear in the same cyclic order: this is not trivial to check given the possible distortions of an object, though if a convex polygon can be drawn through the points, the cyclic order around its boundary will not change on projection, because planar convexity is an invariant of projection. However, for noncoplanar points, the pattern of the perceived points can reorder itself almost randomly: this means that a considerably greater number of permutations of the points will have to be considered for noncoplanar points than for coplanar points. Another consideration is that the feature points being used for object recognition should not be collinear or in any special pattern and should be describable as being *in general position*: otherwise there is a risk that some ambiguities will not be eliminated as indicated in Table 1.2 (ultimately because noninvertible equations arise when attempting to determine the camera calibration parameters).

1.4.3 Invariants as an aid to 3-D recognition

Invariants are important for object recognition in both 2-D and 3-D. The basic idea of an invariant is to find some parameter or parameters that do not vary between different instances or positions of an object and to use them to facilitate object identification. As we shall see, perspective makes the issue far harder in the general 3-D case.

Let us first consider a flat object being viewed from directly overhead by a camera whose optical axis is normal to the plane on which the object is lying. Consider two point features on the object such as corners or small holes. If we measure the interfeature distance in an image, it will act as an invariant, in that:

1. it has a value independent of the translation and orientation parameters of the object;
2. it will be unchanged for different objects of the same type;
3. it will in general be different from the corresponding parameters of other objects that lie on the object plane.

Thus, measurement of distance provides a certain lookup or indexing quality which will ideally identify the object uniquely, though further analysis will be required to fully locate it and ascertain its orientation. Hence inter-feature distance has all the requirements of a 2-D invariant. Of course, we are here ignoring imprecision in measurement, due to inadequate spatial resolution, noise, lens distortions, and so on; in addition, the effects of partial occlusion or breakage are being ignored. Obviously, there is a limit to what can be achieved with a single invariant measure. In particular, it is not able to cope with object scale variations. Moving the camera closer to the object plane and refocusing totally changes the situation and all values of the distance invariant residing in the object indexing table must be changed and the old values ignored. However, a little thought shows that this last problem can be overcome. All we need to do is to take *ratios* of distances. This requires a minimum of 3 point features to be identified in the image and 2 inter-feature distances measured. If we call these 2 distances d_1 and d_2, then the ratio d_1/d_2 will act as a scale-independent invariant, i.e., we will be able to identify objects using a single indexing operation whatever their 2-D translation, orientation, or apparent size or scale.

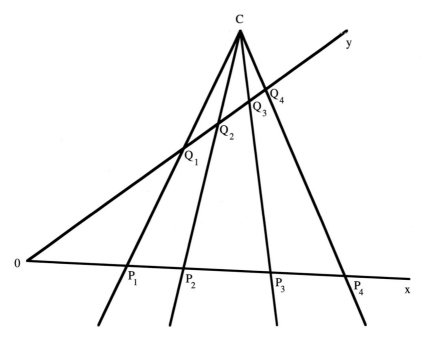

FIGURE 1.21 Perspective transformation of four collinear points. This figure shows four collinear points (P_1, P_2, P_3, P_4) and a transformation of them (Q_1, Q_2, Q_3, Q_4) similar to that produced by an imaging system with optical center C. Such a transformation is called a *perspective transformation*.

Overall, the main motivation for using invariants is to obtain mathematical measures of object feature configurations which are independent of the viewpoint or coordinate system used: indeed, in view of the obvious complexities involved in perspective projection, viewpoint independence is a crucial factor in 3-D object recognition and requires use of perspective invariants.

1.4.4 Cross ratios: the 'ratio of ratios' concept

It would be most useful if we could extend the above ideas to aid the identification of objects when observed under general 3-D transformations. Indeed, an obvious question is whether finding ratios of ratios of distances will lead to invariants providing suitable generalizations. The answer is that ratios of ratios do provide useful further invariants, as we shall now see.

To identify suitable ratios of ratios of distances, we start by examining a set of 4 collinear points on an object. Fig. 1.21 shows such a set of 4 points (P_1, P_2, P_3, P_4) and a transformation of them (Q_1, Q_2, Q_3, Q_4) such as that produced by an imaging system with optical center C (c, d). Choice of a suitable pair of oblique axes permits the coordinates of the two sets of points to be expressed respectively as:

$$(x_1, 0), (x_2, 0), (x_3, 0), (x_4, 0)$$

$$(0, y_1), (0, y_2), (0, y_3), (0, y_4)$$

Taking points P_i, Q_i, $(i = 1, \ldots, 4)$ we can write the ratio $CQ_i : PQ_i$ both as $\frac{c}{-x_i}$ and as $\frac{d-y_i}{y_i}$. Equating these quantities immediately gives:

$$\frac{c}{x_i} + \frac{d}{y_i} = 1 \tag{1.53}$$

After straightforward manipulation of all 4 versions of this relation, and taking suitable differences to eliminate all absolute positions, we eventually obtain the formula:

$$\left(\frac{x_2 - x_4}{x_3 - x_4}\right) \Big/ \left(\frac{x_2 - x_1}{x_3 - x_1}\right) = \left(\frac{y_2 - y_4}{y_3 - y_4}\right) \Big/ \left(\frac{y_2 - y_1}{y_3 - y_1}\right) \tag{1.54}$$

This confirms that a parameter can be constructed that is invariant to perspective transformations. In particular, 4 collinear points viewed from any perspective viewpoint yield the same value of the cross ratio, defined as:

$$C(P_1, P_2, P_3, P_4) = \frac{(x_3 - x_1)(x_2 - x_4)}{(x_2 - x_1)(x_3 - x_4)} \tag{1.55}$$

In what follows, we shall write this particular cross ratio as κ. Note that there are $4! = 24$ possible ways in which 4 collinear points can be ordered on a straight line, and hence there could be 24 cross ratio values for any object. However, they are not all distinct, and in fact there are only 6 different values: it is easily shown that these are κ, $1 - \kappa$, $\kappa/(\kappa - 1)$ and their inverses. Interestingly, numbering the points in reverse (which would correspond to viewing the line from the other side) leaves the cross ratio unchanged. Nevertheless, it is inconvenient that the same invariant has 6 different manifestations, as this implies that 6 different index values have to be looked up before the class of an object can be identified. On the other hand, if points are labeled in order along each line rather than randomly, it is possible to circumvent this situation.

So far we have been able to produce only one projective invariant, and this corresponds to the rather simple case of 4 collinear points. The usefulness of this measure is augmented considerably when it is noted that 4 collinear points, taken in conjunction with another point, define a 'pencil' of concurrent coplanar lines passing through the latter point. Clearly, we can assign a unique cross ratio to this pencil of lines, equal to the cross ratio of the collinear points on any line passing through them. In fact, by considering the angles between the various lines and applying the sine rule 4 times leads to the formula:

$$C(P_1, P_2, P_3, P_4) = \frac{\sin\alpha_{13} \sin\alpha_{24}}{\sin\alpha_{12} \sin\alpha_{34}} \tag{1.56}$$

Thus, the cross ratio depends only on the angles of the pencil of lines.

We can extend this concept to 4 concurrent planes since the concurrent lines can be projected into 4 concurrent planes once a separate concurrency axis has been defined. As there are infinitely many such axes, there are infinitely many ways in which sets of planes can be chosen. Thus, the original simple result on collinear points can be extended to a much more general case.

Finally, note that we started by trying to generalize the case of 4 collinear points, but what we achieved was first to find a dual situation in which points become lines also described by a cross ratio, and then to find an extension in which planes are described by a cross ratio. We now return to the case of 4 collinear points, and see how we can extend it in other ways.

1.4.5 Invariants for noncollinear points

First, imagine that not all the points are collinear: specifically, let us assume that one point is not collinear with the other 3. If this is the case, then there is not enough information to calculate a cross ratio. However, if a further coplanar point is available, we can draw an imaginary line between the noncollinear points to intersect the line through the other 3 points: this will then permit a cross ratio to be computed (Fig. 1.22(a)). Nevertheless, this is some way from a general solution to the characterization of a set of noncollinear points. We might enquire how many point features in general position on a plane will be required to calculate an invariant. In fact, the answer is 5, since the fact that we can form a cross ratio from the angles between 4 lines immediately means that forming a pencil of 4 lines from 5 points defines a cross ratio invariant (Fig. 1.22(b)).

While the value of this cross ratio provides a necessary condition for a match between two sets of 5 general coplanar points, it could be a fortuitous match, as the condition depends only on the relative directions between the various points and the reference point, i.e., any of the nonreference points is only defined to the extent that it lies on a given line. Clearly, two cross ratios formed by taking two reference points will define the directions of all the remaining points uniquely (Fig. 1.22(c)). Interestingly, while at least 5 cross ratios could result from this sort of procedure, it turns out that there are only two functionally independent cross ratios—essentially because the position of any point is defined once its direction relative to two other points is known.

Note that Fig. 1.22 misses out one further interesting case—the situation of two points and two lines. Constructing a line joining the two points and producing it until it meets the two lines, we get 4 points on a single line; thus the configuration is characterized by a single cross ratio. Notice also that the two lines can be extended until they join, and further lines can be constructed from the join to meet the two points: this gives a pencil of lines characterized by a single cross ratio: the latter must have the same value as that computed for the 4 collinear points.

Next, we consider the problem of finding the ground plane in practical situations—e.g., that of egomotion including vehicle guidance. Suppose a set of 4 collinear points can be observed from one frame to the next. If they are on a single plane, then the cross ratio will remain constant, but if one is elevated above the ground plane (as in the case of a bump on the road) then the cross ratio will vary over time. Taking a larger number of points, it should be possible to deduce by a process of elimination which are on the ground plane and which are not: note that all this is possible without any calibration of the camera, this being a major advantage of making use of projective invariants. Note that there is a potential problem regarding irrelevant planes, such as the vertical faces of buildings. The cross ratio test is so resistant to viewpoint and pose that it merely ascertains whether the points being tested are coplanar. But by using a large enough number of independent sets of points, one plane can be discriminated from another.

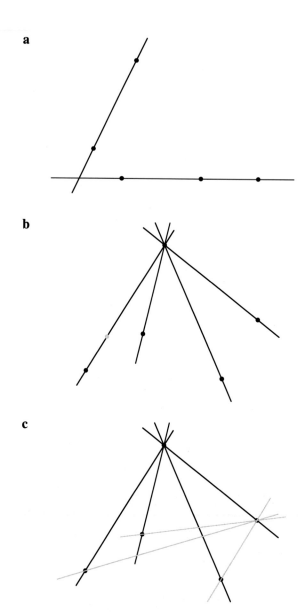

FIGURE 1.22 Calculation of invariants for a set of noncollinear points. (a) shows how the addition of a fifth point to a set of four points, one of which is not collinear with the rest, permits the cross ratio to be calculated. (b) shows how the calculation can be extended to any set of noncollinear points; also shown is an additional (grey) point which a single cross ratio fails to distinguish from other points on the same line. (c) shows how any failure to identify a point uniquely can be overcome by calculating the cross ratio of a second pencil generated from the five original points.

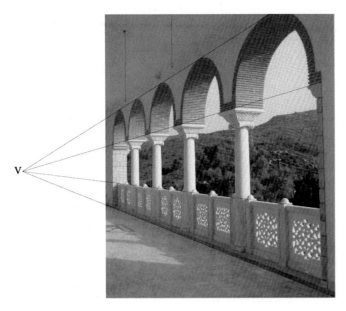

FIGURE 1.23 Position of the vanishing point. In this figure, parallel lines on the arches appear to converge to a vanishing point V outside the image. In general, vanishing points can lie at any distance and may even be situated at infinity.

1.4.6 Vanishing point detection

In this section we consider vanishing points (VPs) and how they can be detected. First, it is useful to understand that the simple cue of a VP gives the human brain a deep understanding of an image and in no small way helps it to globally interpret what is going on in the image: it therefore provides a potential short cut for machines to get started on interpretation. This is valuable in the situation of real 3-D images embodying all the complexities of full perspective projection: hence it is no surprise that much effort has been applied to the detection and use of VPs.

It is usual to carry out VP detection in two stages: first, we locate all the straight lines in the image; next, we find which of the lines pass through common points—the latter being interpreted as VPs. Finding the lines using a Hough transform should be straightforward, though texture edges will sometimes prevent lines from being located accurately and consistently. Basically, locating the VPs requires a second Hough transform in which whole lines are accumulated in parameter space, leading to well defined peaks (the VPs) where multiple lines overlap. In practice, the lines of votes will have to be extended to cover all possible VP locations. This procedure is adequate when the VPs appear within the original image space, but it often happens that they will be outside the original image (Fig. 1.23) and may even be situated at infinity. This means that an image-like parameter space cannot be used successfully, even if it is extended beyond the original image space. Another problem is that for distant VPs, the peaks in parameter space will be spread out over a considerable distance, so detection sensitivity will be poor and accuracy of location will be low.

Fortunately, Magee and Aggarwal (1984) found an improved representation for locating VPs. They constructed a unit sphere G, called a Gaussian sphere, around the center of projection of the camera, and used G instead of an extended image plane as a parameter space. In this representation VPs appear at finite distances even in cases where they would otherwise appear to be at infinity. For this method to work, there has to be a one-to-one correspondence between points in the two representations, and this is clearly valid (note that the back half of the Gaussian sphere is not used). However, the Gaussian sphere representation is not without problems: in particular, many irrelevant votes will be cast from lines that are not parallel in real 3-D space (often, only a small subset of the lines in the image will pass through VPs). To solve this problem, *pairs* of lines are considered in turn, and their crossing points are only accumulated as votes if the lines of each pair are judged likely to originate from parallel lines in 3-D space (e.g., they should have compatible gradients in the image). This procedure drastically limits both the number of votes recorded in parameter space and the number of irrelevant peaks. Nevertheless, the overall cost is still substantial, being proportional to the number of pairs of lines. Thus, if there are N lines, the number of pairs is $^N C_2 = \frac{1}{2}N(N-1)$, so the result is $O(N^2)$.

The above procedure is important as it provides a highly reliable means for performing the search for VPs, and for largely discriminating against isolated lines and image clutter. Note that for a moving robot or other system, the correspondences between the VPs seen in successive images will lead to considerably greater certainty in the interpretation of each image.

1.4.7 More on vanishing points

One advantage of the cross ratio invariant is that it can turn up in many situations and on each occasion provide yet another neat result. An interesting example is when a road or pavement has flagstones whose boundaries are well demarcated and easily measurable. They can then be used to estimate the position of the vanishing point on the ground plane. Imagine viewing the flagstones obliquely from above, with the camera or the eyes aligned horizontally. Then we have the geometry of Fig. 1.24, where the points O, H_1, H_2 lie on the ground plane while O, V_1, V_2, V_3 are in the image plane.

If we take C as a center of projection, the cross ratio formed from the points O, V_1, V_2, V_3 must have the same value as that formed from the points O, H_1, H_2, and infinity in the horizontal direction. Supposing that OH_1 and H_1H_2 have known lengths a and b, equating the cross ratio values gives:

$$\frac{y_1(y_3 - y_2)}{y_2(y_3 - y_1)} = \frac{x_1}{x_2} = \frac{a}{a+b} \tag{1.57}$$

[Note that, in Fig. 1.24, the y values are measured from O rather than from V_3.] This allows us to estimate y_3. Taking $a = b$ (as is likely to be the case with flagstones) we find that:

$$y_3 = \frac{y_1 y_2}{2y_1 - y_2} \tag{1.58}$$

Having found y_3, we have calculated the direction of the vanishing point, whether or not the ground plane on which it lies is actually horizontal, and whether or not the camera axis is

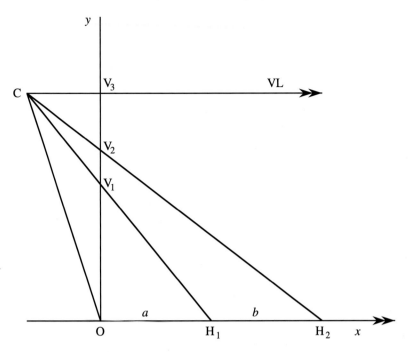

FIGURE 1.24 Geometry for finding the vanishing line from a known pair of spacings. C is the center of projection. VL is the vanishing line direction, which is parallel to the ground plane OH_1H_2. Although the camera plane $OV_1V_2V_3$ is drawn perpendicular to the ground plane, this is not necessary for successful operation of the algorithm (see text).

horizontal. Notice that this proof does not actually assume that points V_1, V_2, V_3 are vertically above the origin, or that line OH_1H_2 is horizontal, just that these points lie along two coplanar straight lines, and that C is in the same plane.

1.4.8 Summary: the value of invariants

Sections 1.4.2–1.4.6 have aimed to give some insight into the important subject of invariants and their application in image recognition. The subject takes off when ratios of ratios of distances are considered, and this idea leads in a natural way to the cross ratio invariant. While its original manifestation lies in its application to recognition of the spacings of points on a line, it generalizes immediately to angular spacings for pencils of lines, and also to angular separations of concurrent planes. A further extension of the idea is the development of invariants which can describe sets of noncollinear points, and two cross ratios suffice to characterize a set of 5 noncollinear points on a plane.

Many other theorems and types of invariant exist, but space prevents more than a mention being made here. As an extension to the point and line examples discussed above, invariants have been produced which cover conics; a conic and two coplanar nontangent lines; a conic and two coplanar points; two coplanar conics. Overall, the value of invariants lies in making computationally efficient checks of whether points or other features might belong to specific objects. In addition, they achieve this without the necessity for camera calibration or knowing the viewpoint of the camera (though there is an implicit assumption that the camera is Euclidean).

1.4.9 Image transformations for camera calibration

When images are obtained from 3-D scenes, the exact position and orientation of the camera sensing device is often unknown and there is a need for it to be related to some global frame of reference. This is especially important if accurate measurements of objects are to be made from their images, e.g., in inspection applications. On the other hand, it may sometimes be possible to dispense with such detailed information—as in the case of a stationary security system for detecting intruders, or a system for counting cars on a motorway. There are also more complicated cases, such as those in which cameras can be rotated or moved on a robot arm, or the objects being examined can move freely in space. In such cases, 'extrinsic' as well as 'intrinsic' camera calibration becomes a central issue (for full explanations of these terms, see Section 1.4.11).

Before we can consider camera calibration, we need to understand in some detail the transformations that can occur between the original world points and the formation of the final image. In particular, we consider rotations and translations of object points relative to a global frame. After a rotation through an angle θ about the Z-axis (Fig. 1.25), the coordinates of a general point (X, Y) change to:

$$X' = X \cos\theta - Y \sin\theta \tag{1.59}$$

$$Y' = X \sin\theta + Y \cos\theta \tag{1.60}$$

We now generalize this result to 3-D and express it as a matrix for a rotation θ about the Z-axis:

$$\mathbf{Z}(\theta) = \begin{bmatrix} \cos\theta & -\sin\theta & 0 \\ \sin\theta & \cos\theta & 0 \\ 0 & 0 & 1 \end{bmatrix} \tag{1.61}$$

Similar matrices apply for rotations of ψ about the X-axis and φ about the Y-axis. Applying sequences of such rotations, we obtain the following general result expressing an arbitrary 3-D rotation $\mathbf{R}$:

$$\begin{bmatrix} X' \\ Y' \\ Z' \end{bmatrix} = \begin{bmatrix} R_{11} & R_{12} & R_{13} \\ R_{21} & R_{22} & R_{23} \\ R_{31} & R_{32} & R_{33} \end{bmatrix} \begin{bmatrix} X \\ Y \\ Z \end{bmatrix} \tag{1.62}$$

Note that the rotation matrix $\mathbf{R}$ is not completely general: it is orthogonal and thus has the property that $\mathbf{R}^{-1} = \mathbf{R}^T$.

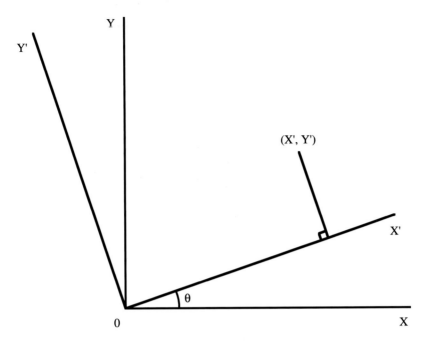

FIGURE 1.25 Effect of a rotation θ about the origin.

In contrast with rotation, translation through a distance (T_1, T_2, T_3) is given by:

$$
\begin{bmatrix} X' \\ Y' \\ Z' \end{bmatrix} = \begin{bmatrix} X \\ Y \\ Z \end{bmatrix} + \begin{bmatrix} T_1 \\ T_2 \\ T_3 \end{bmatrix}
\tag{1.63}
$$

which is not expressible in terms of a multiplicative 3×3 matrix. To combine rotations and translations into a common multiplicative formulation, we have to use *homogeneous coordinates*. To achieve this the matrices must be augmented to 4×4, and the required transformation has to take the form:

$$
\begin{bmatrix} X' \\ Y' \\ Z' \\ 1 \end{bmatrix} = \begin{bmatrix} R_{11} & R_{12} & R_{13} & T_1 \\ R_{21} & R_{22} & R_{23} & T_2 \\ R_{31} & R_{32} & R_{33} & T_3 \\ 0 & 0 & 0 & 1 \end{bmatrix} \begin{bmatrix} X \\ Y \\ Z \\ 1 \end{bmatrix}
\tag{1.64}
$$

This form is sufficiently general to include scaling in object size, and shearing and skewing types of transformation.

In all the cases discussed above it will be observed that the bottom row of the generalized displacement matrix is redundant. In fact, we can put this row to good use in certain other types of transformation. Of particular interest in this context is the case of perspective projection. Following Section 1.4.1, the equations for projection of object points into image points are:

$$x = fX/Z \qquad (1.65)$$
$$y = fY/Z \qquad (1.66)$$
$$z = f \qquad (1.67)$$

To include perspective projection in the above formalism, we need to examine the homogeneous coordinates transformation:

$$
\begin{bmatrix}
1 & 0 & 0 & 0 \\
0 & 1 & 0 & 0 \\
0 & 0 & 1 & 0 \\
0 & 0 & 1/f & 0
\end{bmatrix}
\begin{bmatrix}
X \\ Y \\ Z \\ 1
\end{bmatrix}
=
\begin{bmatrix}
X \\ Y \\ Z \\ Z/f
\end{bmatrix}
\qquad (1.68)
$$

The key to understanding this transformation is to notice that dividing by the fourth coordinate gives the required values of the transformed Cartesian coordinates $(fX/Z, fY/Z, f)$.

Let us now review this result. First, we have found a 4 × 4 matrix transformation which operates on 4-D homogeneous coordinates. These do not correspond directly to real coordinates, but real 3-D coordinates can be calculated from them by dividing the first 3 by the fourth homogeneous coordinate. Thus, there is an arbitrariness in the homogeneous coordinates in that they can all be multiplied by the same constant factor without producing any change in the final interpretation.

The advantage to be gained from use of homogeneous coordinates is the convenience of having a single multiplicative matrix for any transformation, in spite of the fact that perspective transformations are intrinsically nonlinear: thus a quite complex nonlinear transformation can be reduced to a more straightforward linear transformation. This eases computer calculation of object coordinate transformations, and other computations such as those for camera calibration (see below). We may also note that almost every transformation can be inverted by inverting the corresponding homogeneous transformation matrix. The exception is the perspective transformation, for which the fixed value of z leads merely to Z being unknown, and X, Y only being known relative to the value of Z (hence the need for binocular vision or other means of discerning depth in a scene).

1.4.10 Camera calibration

The above discussion has shown how homogeneous coordinate systems are used to help provide a convenient linear 4 × 4 matrix representation for 3-D transformations including rigid body translations and rotations, and nonrigid operations including scaling, skewing and perspective projection. In this last case, it was implicitly assumed that the camera and world coordinate systems are identical, since the image coordinates were expressed in the same frame of reference. However, in general the objects viewed by the camera will have positions which may be known in world coordinates, but which will not *a priori* be known in camera coordinates, since the camera will in general be mounted in a somewhat arbitrary position and will point in a somewhat arbitrary direction. Thus, the camera system will have to be calibrated before the images can be used for practical applications such as robot pick-and-place. A useful approach is to assume a general transformation between the world coordinates and the image seen by the camera under perspective projection, and to locate in the image various calibration points which have been placed in known positions in the scene. If enough such points are available, it should be possible to compute the transformation parameters, and then all image points can be interpreted accurately until recalibration becomes necessary.

In what follows, we will find that there are two types of camera calibration: the first is that of *extrinsic calibration*, in which the position and pose of the camera are determined relative to the world coordinates, via its external parameters; the second is that of *intrinsic calibration*, in which the image (and pixel) locations are determined in relation to the camera's internal parameters. Important factors to be discussed are the numbers of extrinsic and intrinsic parameters and their geometric significance.

To proceed, we need to set up the mathematical formulation in a general way, using the general homogeneous transformation $\mathbf{G}$, which takes the form:

$$
\begin{bmatrix} X_H \\ Y_H \\ Z_H \\ H \end{bmatrix} = \begin{bmatrix} G_{11} & G_{12} & G_{13} & G_{14} \\ G_{21} & G_{22} & G_{23} & G_{24} \\ G_{31} & G_{32} & G_{33} & G_{34} \\ G_{41} & G_{42} & G_{43} & G_{44} \end{bmatrix} \begin{bmatrix} X \\ Y \\ Z \\ 1 \end{bmatrix}
\tag{1.69}
$$

Note that the final Cartesian coordinates appearing in the image are $(x, y, z) = (x, y, f)$, and these are calculated from the first 3 homogeneous coordinates by dividing by the fourth:

$$
x = X_H/H = (G_{11}X + G_{12}Y + G_{13}Z + G_{14})/(G_{41}X + G_{42}Y + G_{43}Z + G_{44}) \tag{1.70}
$$

$$
y = Y_H/H = (G_{21}X + G_{22}Y + G_{23}Z + G_{24})/(G_{41}X + G_{42}Y + G_{43}Z + G_{44}) \tag{1.71}
$$

$$
z = Z_H/H = (G_{31}X + G_{32}Y + G_{33}Z + G_{34})/(G_{41}X + G_{42}Y + G_{43}Z + G_{44}) \tag{1.72}
$$

However, as we know z, there is no point in determining parameters $G_{31}, G_{32}, G_{33}, G_{34}$. Accordingly, we proceed to develop the means for finding the other parameters. In fact, because only the ratios of the homogeneous coordinates are meaningful, only the ratios of the G_{ij} values need be computed, and it is usual to take G_{44} as unity: this leaves only 11 parameters to be determined. Multiplying out the first two equations and rearranging gives:

$$
G_{11}X + G_{12}Y + G_{13}Z + G_{14} - x(G_{41}X + G_{42}Y + G_{43}Z) = x \tag{1.73}
$$

$$
G_{21}X + G_{22}Y + G_{23}Z + G_{24} - y(G_{41}X + G_{42}Y + G_{43}Z) = y \tag{1.74}
$$

Noting that a single world point (X, Y, Z) which is known to correspond to image point (x, y) gives us *two* equations of the above form: it requires a minimum of 6 such points to provide values for all 11 G_{ij} parameters. An important factor is that the world points used for the calculation should lead to independent equations: thus it is important that they should not be coplanar. More precisely, there must be at least 6 points, no 4 of which are coplanar. However, further points are useful in that they lead to overdetermination of the parameters and increase the accuracy with which the latter can be computed. There is no reason why the additional points should not be coplanar with existing points: indeed, a common arrangement is to set up a cube so that 3 of its faces are visible, each face having a pattern of squares with 30–40 easily discerned corner features.

Least squares analysis can be used to perform the computation of the 11 parameters. First, the $2n$ equations (for n points) have to be expressed in matrix form:

$$\mathbf{Ag} = \boldsymbol{\xi} \tag{1.75}$$

where $\mathbf{A}$ is a $2n \times 11$ matrix of coefficients, which multiplies the G-matrix, now in the form:

$$\mathbf{g} = (G_{11}G_{12}G_{13}G_{14}G_{21}G_{22}G_{23}G_{24}G_{41}G_{42}G_{43})^{\mathrm{T}} \tag{1.76}$$

and $\boldsymbol{\xi}$ is a $2n$-element column vector of image coordinates. The pseudo-inverse solution is:

$$\mathbf{g} = \mathbf{A}^{\dagger}\boldsymbol{\xi} \tag{1.77}$$

where

$$\mathbf{A}^{\dagger} = (\mathbf{A}^{\mathrm{T}}\mathbf{A})^{-1}\mathbf{A}^{\mathrm{T}} \tag{1.78}$$

1.4.11 Intrinsic and extrinsic parameters

At this point it is useful to look in more detail at the general transformation leading to camera calibration. When we are calibrating the camera, we are actually trying to bring the camera and world coordinate systems into coincidence. The first step is to move the origin of the world coordinates to the origin of the camera coordinate system. The second step is to rotate the world coordinate system until its axes are coincident with those of the camera coordinate system. The third step is to move the image plane laterally until there is complete agreement between the two coordinate systems (this step is required since it is not known initially which point in the world coordinate system corresponds to the principal point in the image).

There is an important point to be borne in mind during this process. If the camera coordinates are given by $\mathbf{C}$, then the translation $\mathbf{T}$ required in the first step will be $-\mathbf{C}$. Similarly, the rotations that are required will be the inverses of those which correspond to the actual camera orientations. The reason for these reversals is that (for example) rotating an object (here the camera) forwards gives the same effect as rotating the axes backwards. Thus, all operations have to be carried out with the reverse arguments to those indicated above in Section 1.4.1. The complete transformation for camera calibration is hence:

$$\mathbf{G} = \mathbf{PLRT}$$
$$= \begin{bmatrix} 1 & 0 & 0 & 0 \\ 0 & 1 & 0 & 0 \\ 0 & 0 & 1 & 0 \\ 0 & 0 & 1/f & 0 \end{bmatrix} \begin{bmatrix} 1 & 0 & 0 & t_1 \\ 0 & 1 & 0 & t_2 \\ 0 & 0 & 1 & t_3 \\ 0 & 0 & 0 & 1 \end{bmatrix} \begin{bmatrix} R_{11} & R_{12} & R_{13} & 0 \\ R_{21} & R_{22} & R_{23} & 0 \\ R_{31} & R_{32} & R_{33} & 0 \\ 0 & 0 & 0 & 1 \end{bmatrix} \begin{bmatrix} 1 & 0 & 0 & T_1 \\ 0 & 1 & 0 & T_2 \\ 0 & 0 & 1 & T_3 \\ 0 & 0 & 0 & 1 \end{bmatrix} \tag{1.79}$$

where matrix $\mathbf{P}$ takes account of the perspective transformation required to form the image. In fact, it is usual to group together the transformations $\mathbf{P}$ and $\mathbf{L}$ and call them internal camera transformations which include the *intrinsic camera parameters*, while **R** and **T** are taken together as external camera transformations corresponding to *extrinsic camera parameters*. Hence

$$
\mathbf{G}_{\text{internal}} = \mathbf{PL} = \begin{bmatrix} 1 & 0 & 0 & t_1 \\ 0 & 1 & 0 & t_2 \\ 0 & 0 & 1 & t_3 \\ 0 & 0 & 1/f & t_3/f \end{bmatrix} \rightarrow \begin{bmatrix} 1 & 0 & t_1 \\ 0 & 1 & t_2 \\ 0 & 0 & 1/f \end{bmatrix} \tag{1.80}
$$

In the matrix for $\mathbf{G}_{\text{internal}}$ we have assumed that the initial translation matrix **T** moves the camera's center of projection to the correct position, so that the value of t_3 can be made equal to zero, leaving us with a 3×3 matrix.

Although the above treatment gives a good indication of the underlying meaning of **G**, it is not general because we have not so far included scaling and skew parameters in the internal matrix. In fact the generalized form of $\mathbf{G}_{\text{internal}}$ is:

$$
\mathbf{G}_{\text{internal}} = \begin{bmatrix} s_1 & b_1 & t_1 \\ b_2 & s_2 & t_2 \\ 0 & 0 & 1/f \end{bmatrix} \tag{1.81}
$$

Potentially, $\mathbf{G}_{\text{internal}}$ should include transforms for correcting (1) scaling errors, (2) translation errors, (3) sensor skewing errors, (4) sensor shearing errors, (5) unknown sensor orientation within the image plane. Clearly, translation errors are corrected by adjusting t_1 and t_2. All the other adjustments are concerned with the values of the 2×2 submatrix containing parameters s_1, s_2, b_1, b_2.

However, note that application of this matrix performs rotation within the image plane immediately after rotation has been performed in the world coordinates by $\mathbf{G}_{\text{external}}$, and it is virtually impossible to separate the two rotations. This explains why we now have a total of 6 external and 6 internal parameters totaling 12 rather than the expected 11 parameters. As a result it is better to exclude item 5 in the above list of internal transforms and to subsume it into the external parameters. Since the rotational component in $\mathbf{G}_{\text{internal}}$ has been excluded, b_1 and b_2 must now be equal, and the internal parameters will be: s_1, s_2, b, t_1, t_2. Note that the factor $1/f$ provides a scaling which cannot be separated from the other scaling factors during camera calibration, without specific (i.e., separate) measurement of f. Thus, we have a total of 6 parameters from $\mathbf{G}_{\text{external}}$ and 5 parameters from $\mathbf{G}_{\text{internal}}$: this totals 11 and equals the number cited in the previous section.

1.4.12 Multiple view vision

During the 1990s a considerable advance in 3-D vision was made by examining what could be learnt from uncalibrated cameras using multiple views. At first sight, considering the efforts made in earlier sections of this chapter to understand exactly how cameras should be calibrated, this may seem nonsensical. Nevertheless, there are considerable potential advantages in examining multiple views—not least, many thousands of videotapes are available from uncalibrated cameras, including those used for surveillance and those produced in the film industry. In such cases, as much must be made of the available material as possible.

However, the need is deeper than this. Many situations exist in which the camera parameters might vary because of thermal variations, or because the zoom or focus setting has been adjusted: and it is impracticable to keep recalibrating a camera using accurately made test objects. Finally, if multiple cameras are used, each will have to be calibrated separately, and the results compared to minimize the combined error: it is far better to examine the system as a whole, and to calibrate it on the real scenes that are being viewed.

In fact, we have already met some aspects of these aspirations, in the form of invariants that are obtained in sequence by a single camera. For example, if a series of 4 collinear points are viewed and their cross ratio is checked, it will be found to be constant as the camera moves forward, changes orientation or views the points increasingly obliquely—so long as they all remain within the field of view. For this purpose, all that is required to perform the recognition and maintain awareness of the object is an uncalibrated but distortion-free camera.

To understand how image interpretation can be carried out more generally, using multiple views—whether from the same camera moved to a variety of places, or multiple cameras with overlapping views of the world—we need to go back to basics and make a more general attack on concepts such as binocular vision and epipolar constraints. In particular, two important matrices will be called into play—the 'essential' matrix and the 'fundamental' matrix. We start with the essential matrix and then generalize the idea to the fundamental matrix. But first we need to look at the geometry of two cameras with general views of the world.

1.4.13 Generalized epipolar geometry

In Section 1.4.1, we considered the stereo correspondence problem, and had already simplified the task by choosing two cameras whose image planes were not only parallel but in the same plane. This made the geometry of depth perception especially simple, but suppressed possibilities allowed for in the human visual system (HVS), of having a nonzero vergence angle between the two images.

Here we generalize the situation to cover the possibility of disparity coupled with substantial vergence. Fig. 1.26 shows the revised geometry. Note first that observation of a real point P in the scene leads to points P_1 and P_2 in the two images; that P_1 could correspond to any point on the epipolar line E_2 in image 2; and similarly, that point P_2 could correspond to any point on the epipolar line E_1 in image 1. Indeed, the so-called epipolar plane of P is the plane containing P and the projection points C_1 and C_2 of the two cameras: the epipolar lines (see Section 1.4.1) are thus the straight lines in which this plane cuts the two image planes. Furthermore, the line joining C_1 and C_2 cuts the image planes in the so-called epipoles e_1 and e_2: these can be regarded as the images of the alternate camera projection points. Note that all epipolar planes pass through points C_1, C_2 and e_1, e_2: this means that all epipolar lines in the two images pass through the respective epipoles.

1.4.14 The essential matrix

In this section we start with the vectors $\mathbf{P}_1$, $\mathbf{P}_2$, from C_1, C_2 to P, and also the vector $\mathbf{C}$ from C_1 to C_2. Vector subtraction gives:

$$\mathbf{P}_2 = \mathbf{P}_1 - \mathbf{C} \qquad (1.82)$$

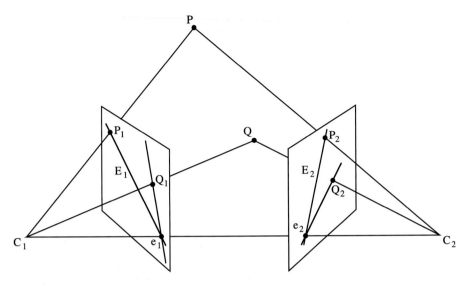

FIGURE 1.26 Generalized imaging of a scene from two viewpoints. In this case there is substantial vergence. All epipolar lines in the left image pass through epipole e_1: of these, only E_1 is shown. Similar comments apply for the right image.

We also know that P_1, P_2 and C are coplanar, the condition of coplanarity being:

$$P_2.C \times P_1 = 0 \tag{1.83}$$

To progress, we need to relate the vectors P_1 and P_2 when these are expressed relative to their own frames of reference. If we take these vectors as having been defined in the C_1 frame of reference, we now reexpress P_2 in its own (C_2) frame of reference, by applying a translation C and a rotation of coordinates expressed as the orthogonal matrix R. This leads to:

$$P_2' = RP_2 = R(P_1 - C) \tag{1.84}$$

so that:

$$P_2 = R^{-1}P_2' = R^{T}P_2' \tag{1.85}$$

Substituting in the coplanarity condition gives:

$$(R^{T}P_2').C \times P_1 = 0 \tag{1.86}$$

At this point it is useful to replace the vector product notation by using a skew-symmetric matrix $C_\times$ to denote $C \times$, where:

$$C_\times = \begin{bmatrix} 0 & -C_z & C_y \\ C_z & 0 & -C_x \\ -C_y & C_x & 0 \end{bmatrix} \tag{1.87}$$

At the same time we observe the correct matrix formulation of all the vectors by transposing appropriately. We now find that:

$$(R^T\mathbf{P}_2')^T C_\times \mathbf{P}_1 = 0 \tag{1.88}$$

$$\therefore \quad \mathbf{P}_2'^T RC_\times \mathbf{P}_1 = 0 \tag{1.89}$$

Finally, we obtain the 'essential matrix' formulation:

$$\mathbf{P}_2'^T E \mathbf{P}_1 = 0 \tag{1.90}$$

where the essential matrix has been found to be:

$$E = RC_\times \tag{1.91}$$

Eq. (1.90) is actually the desired result: it expresses the relation between the observed positions of the same point in the two camera frames of reference. Furthermore, it immediately leads to formulae for the epipolar lines. To see this, first note that in the C_1 camera frame:

$$\mathbf{p}_1 = (f_1/Z_1)\mathbf{P}_1 \tag{1.92}$$

while in the C_2 camera frame (and expressed in terms of that frame of reference):

$$\mathbf{p}_2' = (f_2/Z_2)\mathbf{P}_2' \tag{1.93}$$

Eliminating $\mathbf{P}_1$ and $\mathbf{P}_2'$, and dropping the prime (as within the respective image planes the numbers 1 and 2 are sufficient to specify the coordinates unambiguously), we find:

$$\mathbf{p}_2^T E \mathbf{p}_1 = 0 \tag{1.94}$$

as Z_1, Z_1 and f_1, f_2 can be canceled from this matrix equation.

Now note that writing $\mathbf{p}_2^T E = \mathbf{l}_1^T$ and $\mathbf{l}_2 = E\mathbf{p}_1$ leads to the following relations:

$$\mathbf{p}_1^T \mathbf{l}_1 = 0 \tag{1.95}$$

$$\mathbf{p}_2^T \mathbf{l}_2 = 0 \tag{1.96}$$

This means that $\mathbf{l}_2 = E\mathbf{p}_1$ and $\mathbf{l}_1 = E^T \mathbf{p}_2$ are the epipolar lines corresponding to $\mathbf{p}_1$ and $\mathbf{p}_2$ respectively.

1.4.15 The fundamental matrix

Notice that in the last part of the essential matrix calculation, we implicitly assumed that the cameras are correctly calibrated. Specifically, $\mathbf{p}_1$ and $\mathbf{p}_2$ are corrected (calibrated) image coordinates. However, there is a need to work with uncalibrated images, using the raw pixel measurements—for all the reasons given in Section 1.4.12. Applying the camera intrinsic matrices G_1, G_2 to the calibrated image coordinates (Section 1.4.10), we get the raw image coordinates:

$$q_1 = G_1 p_1 \tag{1.97}$$

$$q_2 = G_2 p_2 \tag{1.98}$$

In fact, we here need to go in the reverse direction, so we use the inverse equations:

$$p_1 = G_1^{-1} q_1 \tag{1.99}$$

$$p_2 = G_2^{-1} q_2 \tag{1.100}$$

Substituting for p_1 and p_2 in Eq. (1.94) we find the desired equation linking the raw pixel coordinates:

$$q_2^{T}(G_2^{-1})^{T} E G_1^{-1} q_1 = 0 \tag{1.101}$$

which can be expressed as:

$$q_2^{T} F q_1 = 0 \tag{1.102}$$

where

$$F = (G_2^{-1})^{T} E G_1^{-1} \tag{1.103}$$

F is defined as the 'fundamental matrix'. Because it contains all the information that would be needed to calibrate the cameras, it contains more free parameters than the essential matrix. However, in other respects the two matrices are intended to convey the same basic information, as is confirmed by the resemblance between the two formulations—Eqs. (1.90) and (1.102).

1.4.16 Properties of the essential and fundamental matrices

Next we consider the composition of the essential and fundamental matrices. In particular, note that $C_\times$ is a factor of E and also, indirectly, of F. In fact, they are homogeneous in $C_\times$, so the scale of C will make no difference to the two matrix formulations (Eqs. (1.90) and (1.102)), only the *direction* of C being important: indeed, the scales of both E and F are immaterial, and as a result only the relative values of their coefficients are of importance. This means that there are at most only 8 independent coefficients in E and F. In fact, in the case of F there are only 7, as $C_\times$ is skew-symmetric, and this ensures that it has rank 2 rather than rank 3—a property that is passed on to F. The same argument applies for E, but the lower complexity of E means that it has only 5 free parameters. In the latter case it is easy to see what they are: they arise from the original 3 translation (C) and 3 rotation (R) parameters, less the one parameter corresponding to scale.

In this context, note that if C arises from a translation of a single camera, the same essential matrix will result whatever the scale of C: only the direction of C actually matters, and the same epipolar lines will result from continued motion in the same direction. In fact, in this case we can interpret the epipoles as foci of expansion or contraction. This underlines the power of this formulation: specifically, it treats motion and displacement a single entity.

Finally, we should try to understand why there are 7 free parameters in the fundamental matrix. The solution is relatively simple. Each epipole requires 2 parameters to specify it. In addition, 3 parameters are needed to map any 3 epipolar lines from one image to the other. But why do just 3 epipolar lines have to be mapped? This is because the family of epipolar lines is a pencil whose orientations are related by cross ratios, so once 3 epipolar lines have been specified, the mapping of any other can be deduced.

1.4.17 Estimating the fundamental matrix

In the previous section we showed that the fundamental matrix has 7 free parameters. This means that it ought to be possible to estimate it by identifying the same 7 features in the two images. However, while this is mathematically possible in principle, and a suitable nonlinear algorithm has been devised by Faugeras et al. (1992) to implement it, it has been shown that the computation can be numerically unstable. Essentially, noise acts as an additional variable boosting the effective number of degrees of freedom in the problem to 8. However, a linear algorithm called the *8-point algorithm* has been devised to overcome the problem. Curiously, this algorithm had been proposed many years earlier by Longuet-Higgins (1981) to estimate the *essential* matrix, but it came into its own when Hartley (1995) showed how to control the errors by first normalizing the values. In addition, by using more than 8 points, increased accuracy can be attained, but then a suitable algorithm must be found that can cope with the now overdetermined parameters. Principal component analysis can be used for this, an appropriate procedure being singular value decomposition (SVD).

1.4.18 Improved methods of triangulation

For some years it was known that there were difficulties in finding accurate numerical solutions of the fundamental matrix, the lack of robustness being due to least squares analysis not coping well when data is corrupted by noise. This problem arises whenever the noise contains outliers. In particular, outliers can arise when noise prevents corresponding lines of sight from meeting in the 3-D scene (i.e., when the two lines of sight are skew). The obvious and widely tested solution to this problem is to choose the mid-point of the common perpendicular to the two lines of sight as the point of intersection. However, this method does not give optimal results, ultimately because the concepts 'common perpendicular' and 'mid-point' are not mathematically valid for FPP. In fact, Kanatani (1996) was able to define a new way of determining the optimal correction—by taking the intersection as the point where the total amount of displacement on the two image planes is a minimum. Though Hartley and Sturm's (1994) idea was similar, it was found that Kanatani's method was several orders of magnitude faster and did not suffer from the epipole singularities arising with the Hartley-Sturm method (Torr and Zisserman, 1997). Subsequently, as late as 2019, interesting improvements to the method are still emerging. In particular, Lee and Civera (2019) proposed a modified mid-point method—'the generalized weighted midpoint' method—in which the two starting points are not assumed to lie on the common perpendicular. They showed that, although their method is not theoretically optimal in the sense of minimizing geometric or algebraic errors, it outperforms existing methods in terms of speed, simplicity and combined 2-D, 3-D and parallax accuracy.

Fathy et al. (2011) clarified the overall situation as follows: the 8-point algorithm is a one-step method which is normally applied after outlier removal to obtain an initial estimate of the fundamental matrix: this is then iteratively refined to produce a more accurate solution.

1.4.19 The achievements and limitations of multiple view vision

The last few sections have discussed the transformations required for camera calibration and have outlined how calibration can be achieved. The camera parameters have been classified as 'internal' and 'external', thereby simplifying the conceptual problem and throwing light on the origins of errors in the system. It has been shown that a minimum of 6 points is required to perform calibration in the general case where 11 transformation parameters are involved. Nevertheless, it is normally important to increase the number of points used for calibration as far as possible, since substantial gains in accuracy can be obtained via the resulting averaging process.

Section 1.4.12 introduced multiple view vision. This important topic was seen to rest on generalized epipolar geometry, and led to the essential and fundamental matrix formulations, which relate the observed positions of any point in two camera frames of reference. The importance of the 8-point algorithm for estimating either of these matrices—and particularly the fundamental matrix, which is relevant when the cameras are uncalibrated—was stressed. In addition, the need for accuracy and robustness in estimating the fundamental matrix is still a research issue, though great strides have been made in recent years (see Section 1.4.18) regarding the outlier removal phase of fundamental matrix estimation.

1.5 Part D – Tracking moving objects

1.5.1 Tracking – the basic concept

In recent years, many algorithms have been devised for interpreting single images and identifying a high proportion of the objects within them: following this success, attention turned to the analysis of image sequences and videos. In fact, if the images in any sequence were simply regarded as sets of separate images or 'frames', this new task could already be regarded as solved; it therefore became just as relevant to interpret image sequences as entities in their own right. Thus, algorithms were needed for identifying and tracking moving objects through any sequence.

We could tackle this task by identifying objects in all the frames, and then computing tracks showing how the objects have moved between frames. This could be implemented by the following algorithm:

```
for all frames in sequence
    find and identify all the objects
link the objects between frames
list all the objects and their tracks.
```

This procedure demands that all objects be detected and recognized, in which case linking them to form tracks would involve linking only objects in the same recognition class (e.g., cars). However, it would be further facilitated if all objects in the same class could be identified individually (e.g., car 1, car 2, etc.), though if some such objects were very similar, it is possible that some confusion would arise when linking them.

This rather exacting procedure could be simplified if objects were characterized by their motion parameters. In fact, tracking information should permit significant savings to be made in computation. Hence we arrive at an alternative strategy:

detect all objects in the first frame
find how these objects have moved in each successive frame
list all the objects and their tracks.

This simplified procedure requires that objects be *detected* (rather than recognized) in the first frame. In addition, there is no need to actually recognize them in subsequent frames, as they should be uniquely identifiable from their relative closeness.

A further saving in effort can in principle be made by avoiding the first stage—that of detecting all objects in the first frame. All we need to do is to study the motions themselves and identify anything that moves as an object. Perhaps the simplest way of approaching this would be to take differences between adjacent frames, in which case any changes should indicate the locations of moving objects. However, this approach tends to locate only limited sections of target outlines: for example, it will ignore the bulk of any object of homogeneous intensity—in accordance with the well-known differencing formula $-\Delta I.\mathbf{v}$. The simplest way out of this difficulty is to model the background; then, by subtracting each frame from the background model, we should be able to locate any moving objects. (Naturally, this will work best if the background remains stationary.)

Though attractive, this idea is not trivial to apply in practice. One of the main problems is that a scene containing a series of moving objects will result in the background model varying continually as moving objects pass over it: a paradigm example is that of a road scene along which vehicles are traveling. In that case, minor changes in the deduced background will have to be eliminated by some sort of averaging process. Note that the deduced background will also have to be updated as time goes on because of changes in ambient illumination and varying weather conditions. So the question is, how to carry out this updating, at the same time compensating for objects that have passed by. Temporal frame averaging is a poor way of achieving this, but some studies (Lo and Velastin, 2001; Cucchiara et al., 2003) showed that temporal *median* filtering can be quite effective, as it is capable of eliminating the effects of outlying intensity values, such as those due to vehicles moving against the stationary background.

One problem with this approach is that finding the temporal median requires the storage of a great many frames—a criticism that does not apply for normal averaging, which can conveniently be carried out by applying a temporally weighted running average. However, Davies (2017) has shown that by implementing the temporal median iteratively, this problem can be solved, and that this solution is also able to update the background model to offset the effects of varying background illumination. Furthermore, he found that an even better approach is to use a 'restrained' median filter, in which extreme intensities and colors in a restricted band relative to the previous iterative approximation are ignored. In a road scene,

FIGURE 1.27 Background subtraction using a temporal median filter. Note the plethora of stationary shadows that are completely ignored during the process of background subtraction. In (a) the 'ghost' of the bus (from its past position) still appears, but in (b) it has started to merge back into the background: the problem is considerably reduced in (c) and (d), which uses a 'restrained' temporal median filter. Overall, foreground object fragmentation and false shapes (including the effects of moving shadows) are the worst problems. The lines of black graphics dots demarcate the relevant road region: almost all of the fluttering vegetation lies outside this region.

this prevents vehicles that are temporarily stationary from excessively distorting the deduced background levels (see Figs. 1.27 and 1.28).

Note that with these types of approach, there is no a priori reason why intensities due to vehicles or other objects should be greater or less than the true background: both possibilities clearly exist. Hence, subtracting the current frame from the background model is liable to break any moving object into several parts, which will subsequently have to be recombined using methods such as morphological processing. Fortunately, this technique can also help to eliminate noise: the effects of fluttering vegetation, such as moving leaves or branches, can also be very successfully suppressed in this way—as may be seen from Figs. 1.27 and 1.28.

Interestingly, whereas the above approach can successfully identify vehicles in road scenes, it regards their shadows as parts of the objects, as it does not involve any high-level reasoning.

a   b

FIGURE 1.28 (a) and (b) show the difficulty of interpreting the immediate results of background subtraction. These two frames show clearly the noise problems that arise during background subtraction: the white pixels indicate where the current frame fails to closely match the background model. Morphological operations (erosions followed by dilations) are used to largely eliminate the noise and to integrate the vehicle shapes as far as possible, as shown by the white graphics outlines in Fig. 1.27. Note that the latter contain not only the vehicle shapes but also the shadows that move with them.

This effect is illustrated in Figs. 1.27 and 1.28. In fact, shadow detection has been widely studied and Horprasert et al. (1999) demonstrated a useful principle for implementing it: this involved noting that shadows have similar chromaticity, but lower brightness, than the background model.

1.5.2 Alternatives to background subtraction

While the background subtraction method described above is intrinsically simple, surprisingly effective, and very fast running, it is also limited in (a) taking the background to be basically unchanging—albeit coping (by temporal averaging and morphological processing) with moving entities in the background, and (b) not using proper models of the foreground objects. A more rigorous approach would be to regard the distributions of intensities and colors for any pixel as the superposition of several distributions corresponding to two or three component sources. Here what is important is that each of the component distributions could be quite narrow and well defined. This means that if each is known from ongoing training, any current intensity **I** can be checked to determine whether it is likely to correspond to background or to a new foreground object.

This makes Gaussian mixture models (GMMs) useful for representing the true background and foreground intensity ranges. In fact, the number of components at any pixel is initially unknown: indeed, a large proportion of pixels will have only a single component, but the number of components required in practice commonly lies in the range 3 to 5. However, determining the GMMs necessitates application of the expectation maximization (EM) algorithm and is computationally burdensome. In fact, while it is usual to use this rigorous approach to *initialize* the background generation process, many workers use simpler, more efficient techniques for updating it, so that the ongoing process can proceed in real time.

Unfortunately, the GMM approach fails when the background has very high frequency variations. Essentially, this is because the algorithm has to cope with rapidly varying distributions, which can change dramatically over very short periods of time, so the statistics become too poorly defined. To tackle this problem, Elgammal et al. (2000) moved away from the parametric approach of the GMM. Their nonparametric method involves taking a kernel smoothing function (typically a Gaussian) and for each pixel, applying it to the N samples of $\mathbf{I}$ for frames appearing during the period Δt prior to the current time t. This approach is able to rapidly adapt to jumps from one intensity value to another, while at the same time obtaining the local variances at each pixel. Thus, its value lies in its capability to forget old intensities and to reflect local variances rather than random intensity jumps. In addition, as it does not employ the EM algorithm, it is able to run highly efficiently in real time and to be capable of sensitive detection of foreground objects coupled with low false alarm rates. To achieve all this, it uses separate Gaussian kernel functions for each color channel, and then uses the chromaticity-based method mentioned earlier for suppressing shadows.

So far, we have seen that background subtraction has the advantages of being straightforward to apply, fast running and highly effective—even to the extent of being able (with suitable algorithmic adjustments) to cope with slowly varying background illumination; to eliminate shadows in the background and the foreground; to suppress 'ghosts' arising from temporarily stationary vehicles; and to suppress the effects of fluttering vegetation. On the downside, it relies on the background being static, which in turn means that it requires the use of fixed cameras. In addition, there is no guarantee that all parts of the foreground objects will have different intensities from the background—a factor that can fragment objects and leads to the need for morphological processing, itself a somewhat ad hoc solution.

A further problem that arises with tracking based on background subtraction is that foreground objects can be partially or completely occluded by other objects: this happens particularly when pedestrians are being viewed in crowded precincts. At best this can lead to fragmented tracks, and at worst to wrongly connected tracks. Note also that the motion of some objects may cease altogether. Clearly, carefully thought-out methods are needed for solving these problems. The traditional way of dealing with broken tracks was to employ predictive filters such as the Kalman filter, but these are of limited use as they are essentially aimed at assessing the probabilities of connecting pairs of tracklets, based on single unimodal Gaussian densities.

Overall, these criticisms and problems require temporal differencing to be backed up by template correlation matching, so that it is known that the same object is still being dealt with. As noted by Lipton et al. (1998), temporal differencing (or related methods such as background subtraction) can be used to detect moving objects, and correlation matching can be used (a) to accurately locate the objects and (b) to help train the correlation template; at each stage the template with the best correlation is used for both (a) and (b). In fact, using the best correlation template is useful even when one object partially occludes another. Furthermore, it is particularly useful when a particular object becomes stationary, as there is then no uncertainty of its identity or location.

Stalder et al. (2009) developed an alternative object tracking strategy which they described as 'tracking by detection'. The aim was to enable the tracker to adapt to any changes in the appearance of an object by updating the object model. However, this leads to the so-called 'template update problem', wherein there is a trade-off between adaptivity and stability (specifically, the tracker could end up with a totally distorted, nonviable version of the object profile—a process called 'drifting'). This difficulty is overcome by reformulating tracking as a *semisupervised* learning problem in which both labeled and unlabeled data can be used during tracking. To make this work, semisupervised boosting is used, each unlabeled sample in the local search region being assigned a pseudo-label y_i and an importance weight λ_i (labeled samples have label y_i and importance 1). After initial detection, leading to a prior H_P, the object classifier H is built, taking account of positive samples from the object and negative ones from the surrounding background: typically, the local maximum of the confidence distribution is taken to indicate the new object position and the class is updated. Thus, we get a sequence of object locations, starting with the first (which resulted from the initial detection), and then proceeding to many further tracked positions. However, in principle the boosting equations can also lead to the prior either vanishing or dominating too much—corresponding respectively to drifting or zero adaptation. But in general drifting is limited as the tracker is not able to get too far away from the prior.

Problems with the above tracking strategy include (a) taking partial occlusions or apparently permissible changes in appearance as unallowed drift; and (b) jumping to similar objects (for example, a face detector might jump from one person's face to another). Clearly, *tracking by detection* is mainly applicable for single target tracking, but when multiple target tracking is required, the three processes, detection, tracking and recognition, must all be taken into account. In particular, recognition has to be taken as distinguishing similar objects in a scene.

Stalder et al.'s (2009) multiple target classifier system proved highly successful for tracking multiple objects. It limited drifting by careful use of supervised updates and the avoidance of feedback loops, thereby preventing the accumulation of small errors which could lead to drift. This was achieved by making the tracker the dominant element in the approach, so that its information flow either went (in a loop) back to itself or (eventually) to the previous recognizer. This system permitted tracking with a moving camera and was also reliable enough to permit long-term tracking over times of ~24 hours.

Another aspect of multiple target tracking is the capability for reidentification of objects that are temporarily hidden from view (e.g., temporarily occluded or outside the field of view of the camera). This involves identification matching and is only permitted if the degree of match achieved is significantly higher than for any other potential matches. Reidentification can fail if the degree of match becomes too different during the relevant temporal gap. Again, Stalder et al.'s system also performed well in these respects.

Kalal et al. (2011) developed a powerful tracking system aimed particularly at eliminating run-time drift errors and problems that occur when tracked objects disappear from view. Their approach was neatly described as 'tracking-learning-detection'. The key aspect of this method is that learning is carried out by a 'P-expert' which estimates missed detections, an 'N-expert' which estimates false alarms, and the means of updating both experts by learning. By keeping tracking and detection separate, they claimed that neither the tracking nor the detection capabilities are compromised, and present strong evidence for the success of their approach—in particular, by scoring an average of 81% accuracy on many datasets, thereby significantly outperforming five earlier approaches, none of which achieved better than 22% accuracy.

Whereas Kalal et al.'s (2011) work only tracked single objects, Wu et al. (2012) showed how to carry out coupling detection and data association for multiple objects to ensure that complete tracks would be determined. This approach employed network flow data association and relied on an earlier paper by Castañón (1990) entitled "Efficient algorithm for finding the k best paths through a trellis". This title shows that the method relies heavily on extensive analysis of graphs: while interesting, space does not permit a full discussion of these techniques to be included here. Suffice it to say that tracking multiple objects is a difficult problem that becomes even more difficult as the number of targets increases and the whole scene fills up with objects and tracks. Thus, in this respect, Wu et al.'s (2012) work can be regarded as more thoroughgoing than that of Stalder et al. (2009). We will study more recent developments on this topic in Part F, Section 1.7.7, after introducing deep learning methods.

1.6 Part E – Texture analysis

1.6.1 Introduction

We have already considered several core aspects of image analysis, including in particular feature detection, object recognition and segmentation. Here we go on to consider texture analysis. To proceed, we start by defining a texture as the characteristic variation in intensity which should allow us to recognize and describe a textured region and to outline its boundaries (Fig. 1.29). Typically, a texture is the intensity pattern that results when light is reflected from a surface having a certain degree of roughness. Clearly, a smooth, uniformly lit surface will exhibit no texture, while a piece of cloth or a sandy beach will have characteristic textures of their own.

Broadly speaking, textures vary in their degrees of randomness and regularity, and in the latter case they may have high or low directionalities. For example, pieces of cloth generally exhibit high degrees of regularity and directionality, while the intensity pattern emanating from the surface of a sandy beach may appear highly random, with negligible directionality. Another factor is the scale of the perceived particle size for the surface, which will be small for sand and much larger for a tray of peas. In fact, the tiny elements composing a textured surface are often called textural elements or *texels*. These considerations lead us to characterize textures in the following ways:

1. The texels will have various sizes and degrees of uniformity.
2. The texels will be orientated in various directions.
3. The texels will be spaced at varying distances in different directions.
4. The contrast will have various magnitudes and variations.
5. Various amounts of background may be visible between texels.
6. The variations composing the texture may each have varying degrees of regularity vis-à-vis randomness.

The significant number of parameters required to characterize a texture are bound to make texture analysis quite complicated: and of course, many of the parameters will have a high

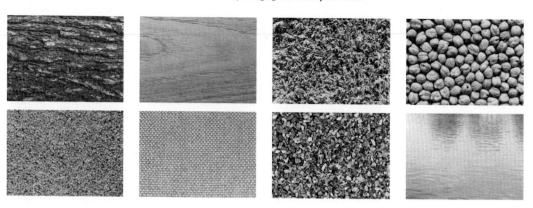

FIGURE 1.29 A variety of textures. These textures demonstrate the wide variety of familiar textures which are easily recognized from their characteristic intensity patterns.

degree of variability, so texture analysis tends to result in statistical descriptions of textures. The following section indicates some of the ways in which this process has been tackled.

1.6.2 Basic approaches to texture analysis

In Section 1.6.1 we defined texture as the characteristic variation in intensity of a region in an image which should allow us to recognize it, describe it and outline its boundaries. In view of the statistical nature of textures, this prompts us to characterize a texture by the variance in intensity values taken over the region of the texture. However, such an approach will not give a rich enough description of the texture for most purposes: it will also be unsuitable in cases where the texels are well defined, or where there is a high degree of periodicity in the texture. On the other hand, for highly periodic textures such as arise with many textiles, it is natural to consider the use of Fourier analysis. Unfortunately, though this approach was long ago tested rigorously, the results were not encouraging.

Autocorrelation is another obvious approach to texture analysis, since it should show up both local intensity variations and also the repeatability of the texture (see Fig. 1.30). An early study was carried out by Kaizer (1955). He examined by how many pixels an image has to be shifted before the autocorrelation function drops to $1/e$ of its initial value, and produced a subjective measure of coarseness on this basis. However, Rosenfeld and Troy (1970a,b) showed that autocorrelation is not a satisfactory measure of coarseness. In addition, autocorrelation is not a good discriminator of isotropy in natural textures. Hence workers were quick to take up the co-occurrence matrix approach introduced by Haralick et al. (1973).

The grey-level co-occurrence matrix approach is based on studies of the statistics of pixel intensity distributions. As hinted above, single pixel statistics do not provide rich enough descriptions of textures for practical applications. Thus, it is natural to consider second order statistics obtained by considering *pairs* of pixels with specific spatial relations to each other. Hence, co-occurrence matrices are used, which express the relative frequencies $P(i, j \mid d, \theta)$ with which two pixels having relative polar coordinates (d, θ) appear with intensities i, j. The co-occurrence matrices provide raw numerical data on the texture, though this data must be

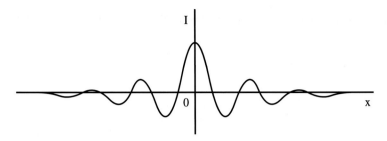

FIGURE 1.30 Use of autocorrelation function for texture analysis. This diagram shows the possible 1-D profile of the autocorrelation function for a piece of material in which the weave is subject to significant spatial variation: notice that the periodicity of the autocorrelation function is damped down over quite a short distance.

condensed to relatively few numbers before it can be used to classify the texture. The early paper by Haralick et al. (1973) gave fourteen such measures, and these were used successfully for classification of many types of material (including, for example, wood, corn, grass and water).

Unfortunately, the amount of data in the co-occurrence matrices is liable to be many times more than in the original image—a situation which is exacerbated in more complex cases by the number of values of d and θ that are required to accurately represent the texture. In addition, the number of grey levels is typically ~256, and the amount of matrix data varies as the square of this number. Finally, co-occurrence matrices merely provide a new representation: they do not themselves solve the recognition problem. As a result of these factors, the 1980s saw a highly significant diversification of methods for the analysis of textures. Of these, Laws' approach (1979; 1980a; 1980b) is important in that it has led to other developments which provide a systematic, adaptive means of tackling texture analysis. This approach is covered in the following section.

1.6.3 Laws' texture energy approach

In 1979 and 1980 Laws presented his novel texture energy approach to texture analysis (Laws, 1979, 1980a,b). This involved the application of simple filters to digital images. The basic filters he used were common Gaussian, edge detector and Laplacian-type filters, and were designed to highlight points of high 'texture energy' in the image. By identifying these high energy points, smoothing the various filtered images, and pooling the information from them, he was able to characterize textures highly efficiently. As remarked earlier, Laws' approach has strongly influenced much subsequent work and it is therefore worth considering it here in some detail.

The Laws' masks are constructed by convolving together just three basic 1×3 masks:

$$L3 = [1 \quad 2 \quad 1] \tag{1.104}$$

$$E3 = [-1 \quad 0 \quad 1] \tag{1.105}$$

$$S3 = [-1 \quad 2 \quad -1] \tag{1.106}$$

TABLE 1.3 The nine 3×3 Laws masks.

$L3^T L3$			$L3^T E3$			$L3^T S3$		
1	2	1	−1	0	1	−1	2	−1
2	4	2	−2	0	2	−2	4	−2
1	2	1	−1	0	1	−1	2	−1
$E3^T L3$			$E3^T E3$			$E3^T S3$		
−1	−2	−1	1	0	−1	1	−2	1
0	0	0	0	0	0	0	0	0
1	2	1	−1	0	1	−1	2	−1
$S3^T L3$			$S3^T E3$			$S3^T S3$		
−1	−2	−1	1	0	−1	1	−2	1
2	4	2	−2	0	2	−2	4	−2
−1	−2	−1	1	0	−1	1	−2	1

The initial letters of these masks indicate *L*ocal averaging, *E*dge detection and *S*pot detection. In fact, these basic masks span the entire 1×3 subspace and form a complete set. Similarly, the 1×5 masks obtained by convolving pairs of these 1×3 masks together form a complete set, only the following five being distinct:

$$L5 = [1 \quad 4 \quad 6 \quad 4 \quad 1] \tag{1.107}$$

$$E5 = [-1 \quad -2 \quad 0 \quad 2 \quad 1] \tag{1.108}$$

$$S5 = [-1 \quad 0 \quad 2 \quad 0 \quad -1] \tag{1.109}$$

$$R5 = [1 \quad -4 \quad 6 \quad -4 \quad 1] \tag{1.110}$$

$$W5 = [-1 \quad 2 \quad 0 \quad -2 \quad 1] \tag{1.111}$$

(Here the initial letters are as before, with the addition of *R*ipple detection and *W*ave detection.) We can also use matrix multiplication to combine the 1×3 and a similar set of 3×1 masks to obtain nine 3×3 masks—for example:

$$\begin{bmatrix} 1 \\ 2 \\ 1 \end{bmatrix} \begin{bmatrix} -1 & 2 & -1 \end{bmatrix} = \begin{bmatrix} -1 & 2 & -1 \\ -2 & 4 & -2 \\ -1 & 2 & -1 \end{bmatrix} \tag{1.112}$$

Again, the resulting set of masks forms a complete set (Table 1.3): note that two of these masks are identical to the Sobel operator masks. The corresponding 5×5 masks are entirely similar but are not considered in detail here as all relevant principles are covered by the 3×3 masks.

All such sets of masks include one whose components do not average to zero. This one is less useful for texture analysis since it will give results dependent more on image intensity than on texture. The remainder is sensitive to edge points, spots, lines and combinations of these.

Having produced images that indicate local edginess, etc., the next stage is to deduce the local magnitudes of these quantities. These magnitudes are then smoothed over a region

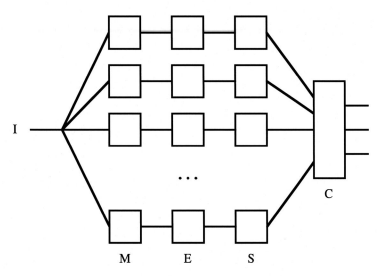

FIGURE 1.31 Basic form for a Laws' texture classifier. Here I is the incoming image, M represents the microfeature calculation, E the energy calculation, S the smoothing, and C the final classification.

rather greater than the basic filter mask size (e.g., Laws used a 15 × 15 smoothing window after applying his 3 × 3 masks): the effect of this is to smooth over the gaps between the texture edges and other micro-features. At this point the image has been transformed into a vector image, each component of which represents energy of a different type. While Laws (1980b) used both squared magnitudes and absolute magnitudes to estimate texture energy, the former corresponding to true energy and giving a better response, the latter are useful in requiring less computation:

$$E(l, m) = \sum_{i=l-p}^{l+p} \sum_{j=m-p}^{m+p} |F(i, j)| \tag{1.113}$$

$F(i, j)$ being the local magnitude of a typical microfeature which is smoothed at a general scan position (l, m) in a $(2p + 1) \times (2p + 1)$ window.

A further stage is required to combine the various energies in a number of different ways, providing several outputs which can be fed into a classifier to decide upon the particular type of texture at each pixel location (Fig. 1.31): if necessary, principal component analysis is used to help select a suitable set of intermediate outputs.

Laws' method resulted in excellent classification accuracy quoted at (for example) 87% compared with 72% for the co-occurrence matrix method, when applied to a composite texture image of grass, raffia, sand, wool, pigskin, leather, water and wood (Laws, 1980b). He also found that the histogram equalization normally applied to images to eliminate first-order differences in texture field grey-scale distributions led to little improvement. In an independent research study, Pietikäinen et al. (1983) confirmed that Laws' texture energy measures are more powerful than measures based on pairs of pixels (notably, co-occurrence matrices).

1.6.4 Ade's eigenfilter approach

In 1983 Ade investigated the theory underlying the Laws' approach, and developed a revised rationale in terms of eigenfilters. He took all possible pairs of pixels within a 3 × 3 window, and characterized the image intensity data by a 9 × 9 covariance matrix. He then determined the eigenvectors required to diagonalize this matrix. These correspond to filter masks similar to the Laws' masks, i.e., use of these 'eigenfilter' masks produces images which are principal component images for the given texture. Furthermore, each eigenvalue gives that part of the variance of the original image that can be extracted by the corresponding filter. Essentially, the variances give an exhaustive description of a given texture in terms of the texture of the images from which the covariance matrix was originally derived. Clearly, the filters that give rise to low variances can be taken to be relatively unimportant for texture recognition.

It will be useful to illustrate the technique for a 3 × 3 window. Here we follow Ade (1983) in numbering the pixels within a 3 × 3 window in scan order:

1	2	3
4	5	6
7	8	9

This leads to a 9 × 9 covariance matrix for describing relationships between pixel intensities within a 3 × 3 window, as stated above. At this point we recall that we are describing a texture, and assuming that its properties are not synchronous with the pixel tessellation, we would expect various coefficients of the covariance matrix $\mathbf{C}$ to be equal. In fact, there are only 12 distinct spatial relationships between pixels if we disregard translations of whole pairs—or 13 if we include the null vector in the set (see Table 1.4). Thus, the covariance matrix, whose components include the 13 parameters a–m, takes the form:

$$\mathbf{C} = \begin{bmatrix} a & b & f & c & d & k & g & m & h \\ b & a & b & e & c & d & l & g & m \\ f & b & a & j & e & c & i & l & g \\ c & e & j & a & b & f & c & d & k \\ d & c & e & b & a & b & e & c & d \\ k & d & c & f & b & a & j & e & c \\ g & l & i & c & e & j & a & b & f \\ m & g & l & d & c & e & b & a & b \\ h & m & g & k & d & c & f & b & a \end{bmatrix} \tag{1.114}$$

TABLE 1.4 Spatial relationships between pixels in a 3 × 3 window.

a	b	c	d	e	f	g	h	i	j	k	l	m
9	6	6	4	4	3	3	1	1	2	2	2	2

Table 1.4 shows the number of occurrences of the spatial relationships between pixels in a 3 × 3 window. Note that a is the diagonal element of the covariance matrix **C**, and that all others appear twice as many times in **C** as indicated in the table.

C is symmetric, and the eigenvalues of a real symmetric covariance matrix are real and positive, and the eigenvectors are mutually orthogonal. In addition, the eigenfilters thus produced reflect the proper structure of the texture being studied and are ideally suited to characterizing it. For example, for a texture with a prominent highly directional pattern, there will be one or more high energy eigenvalues with eigenfilters having strong directionality in the corresponding direction.

1.6.5 Appraisal of the Laws and Ade approaches

At this point, it is useful to compare the Laws and Ade approaches more carefully. In the Laws approach standard filters are used, texture energy images are produced, and *then* principal component analysis may be applied to lead to recognition; whereas in the Ade approach, special filters (the eigenfilters) are applied, incorporating the results of principal component analysis, following which texture energy measures are calculated and a suitable number of these are applied for recognition.

The Ade approach is superior to the extent that it permits low-value energy components to be eliminated early on, thereby saving computation. For example, in Ade's application, the first five of the nine components contain 99.1% of the total texture energy, so the remainder can be ignored; in addition, it would appear that another two of the components containing respectively 1.9% and 0.7% of the energy could also be ignored, with little loss of recognition accuracy. However, in some applications textures could vary continually, and it may well be inadvisable to fine-tune a method to the particular data pertaining at any one time. [For example, these remarks apply (1) to textiles, for which the degree of stretch will vary continuously during manufacture, (2) to raw food products such as beans, whose sizes will vary with the source of supply, and (3) to processed food products such as cakes, for which the crumbliness will vary with cooking temperature and water vapor content.]

In Unser (1986) developed a more general version of the Ade technique. In this approach not only is performance optimized for texture classification but also it is optimized for discrimination between two textures by simultaneous diagonalization of two covariance matrices. The method was developed further by Unser and Eden (1989; 1990): this work makes a careful analysis of the use of nonlinear detectors. As a result, two levels of nonlinearity are employed, one immediately after the linear filters and designed (by employing a specific Gaussian texture model) to feed the smoothing stage with genuine variance or other suitable measures, and the other after the spatial smoothing stage to counteract the effect of the earlier filter, and aiming to provide a feature value that is in the same units as the input signal. In practical terms this means having the capability for providing an r.m.s. texture signal from each of the linear filter channels.

Overall, the originally intuitive Laws approach emerged during the 1980s as a serious alternative to the co-occurrence matrix approach. It is as well to note that alternative methods that are potentially superior have also been devised—see for example the forced-choice method of Vistnes (1989) for finding edges between different textures, which apparently has considerably better accuracy than the Laws approach. Vistnes's (1989) investigation concludes that the Laws approach is limited by (a) the small scale of the masks which can miss larger-scale textural structures, and (b) the fact that the texture energy smoothing operation blurs the texture feature values across the edge. The latter finding (or the even worse situation where a third class of texture appears to be located in the region of the border between two textures) has also been noted by Hsiao and Sawchuk (1989a,b) who applied an improved technique for feature smoothing; they also used probabilistic relaxation for enforcing spatial organization on the resulting data.

1.6.6 More recent developments

Over the 2000s, a trend to scale and rotation invariant texture analysis took place. In particular, the paper by Janney and Geers (2010) described an 'invariant features of local textures' approach, using a strictly circular 1-D array of sampling positions around any given position. The method employs Haar wavelets, and as a result is computationally efficient. It is applied at multiple scales in order to achieve scale invariance; in addition, intensity normalization is used to make the method illumination as well as scale and rotation invariant. Also of note is a book by Mirmehdi et al. (2008) on this rather specialist subject. The latter is an edited volume containing contributions by various researchers and summarizing the position prior to 2010. Later developments on this topic will be deferred to Part F, Sections 1.7.8 and 1.7.9, after introducing deep learning methods.

1.7 Part F – From artificial neural networks to deep learning methods

1.7.1 Introduction: how ANNs metamorphosed into CNNs

The original aim of designing artificial neural networks (ANNs) was to emulate what is known to happen in the human visual system. In fact, vision appeared to happen so straightforwardly in the human brain—whole scenes being analyzed 'at a glance' with no apparent effort—that it seemed worthwhile attempting it in computerized systems. Clearly, an ANN designed to emulate the human visual system should consist of a number of layers, each modifying the data, first locally and then by larger and larger sets of neurons until tasks such as recognition and scene analysis are achieved. However, in the early days, use of ANNs tended to be restricted to very few layers: indeed, a working maximum depth consisted of one input layer, three hidden layers and one output layer—though it was later found that many basic tasks could be tackled using a 3-layer network with a single hidden layer.

One of the reasons for the restriction to few layers was the credit assignment problem, which meant that it became trickier to train many layers 'through' others; at the same time, more layers meant more neurons to be trained and more computation being required to complete the task. ANNs therefore tended to be called on to carry out only the classical recognition process, and to be fed by standard feature detectors applied in previous nonlearning layers. Thus, the standard paradigm was that of an image preprocessor followed by a trained classifier. As very good feature detectors could be designed by hand, this did not cause any obvious problems. However, as time went on, there was pressure to do full-scale analysis of real scenes, which could contain images of many types of object in many positions and poses.

Thus, there was a growing need to move to much more complex multilayer recognition systems for which the early types of ANN were inadequate. It also became desirable to train the preprocessing system itself, so that it closely matched the requirements of the following object analysis system; clearly, it was becoming necessary to produce integrated multilayer neural networks.

In fact, by in the late 1990s ANNs were suffering because other methods such as support-vector machines (SVMs) were challenging their position. In addition, there was no scientifically based rationale for determining how many neurons or layers would be needed, or indeed how ANNs were working internally. As a result, workers who might be in a position to use them did not know how reliable they would be, or have the confidence to employ them in real applications, so they started falling out of favor.

An important reason for this was that their architecture and training gave poor spatial invariance across images. In particular, the neurons were individually trained: each neuron saw different training data from the other neurons in its layer; in addition, the weights needed to be initialized randomly. These factors prevented the same decision from being made about any object wherever it appears in an image. However, a variant on the standard ANN architecture was being developed by a few groups of workers, and during the late 2000s this came to the forefront, and led to the 'deep' learning type of network (a *deep network* is one with more than three nonlinear hidden layers—rather beyond the scope of regular ANNs).

The new type of architecture was the Convolutional Neural Network (CNN). In several ways this was a less demanding architecture, as (a) each CNN neuron does not have to be connected to *all* the outputs from the previous layer of neurons; (b) neurons have the same weight parameters across the whole of any layer. Nevertheless, CNNs still use supervised learning and they still train the network by backpropagation.

Importantly, constraining neurons to have the same weight parameters across any whole layer vastly reduced the total number of parameters in the whole network and made it far easier to train; in addition, increased numbers of layers could be employed. Giving neurons local connectivity improved the situation further. Note that if the neurons and weights are identical across a whole layer, the resulting mathematical operation is by definition a convolution—hence the term *Convolutional* Neural Network.

Another feature of CNNs is that they use ReLU rather than sigmoidal output functions. 'ReLU' means Rectified Linear Unit, and is defined by $\max(0, x)$, where x is the output value of the immediately preceding convolution layer. This is valuable in being less computation intensive than the old sigmoid function, and at the same time it distorts the larger signals less. In fact, the ReLU function avoids the saturation problems that ANNs are subject to (a neuron giving an output close to the limits (± 1) of the tanh function tends to get stuck at the same value because there is no gradient to guide the backpropagation algorithm away from it).

CNNs also incorporate pooling—i.e., taking all the outputs from a locality and deriving a single output from them: usually, this takes the form of a sum or maximum operation on all the inputs, maximum pooling being more common than the sum (or averaging) operation. Pooling is generally carried out in 2×2 or 3×3 windows, the former being more common. These options aimed to modify the data minimally, so as to remove much of the redundancy in a particular layer of the network, while at the same time retaining the most useful data.

Note that several convolutional layers can be placed immediately after one another, making them equivalent to a single larger convolution—a factor that can be useful for allowing larger features to be implemented in the same CNN. Overall, CNNs provide a reasonable alternative to ANNs. In addition, they seem better adapted to the idea of moving steadily from local to global operations on images, and looking for larger and larger features or objects in the process.

Although proceeding through the network takes us from local to more global operations, it is also common for the first few layers of a CNN to look for specific low-level features: hence these will typically have sizes matching those for the particular type of image. Further on in the network it is common for pooling operations to be applied, thereby reducing the sizes of subsequent layers. After several stages of convolution and pooling, the network will have narrowed down considerably, so it is possible to make the final few layers fully-connected— i.e., in any layer each neuron is connected to *all* the outputs of the previous layer. At that stage there are likely to be relatively few outputs, and those that remain will be dictated by whatever parameters need to be supplied by the network: these may include classifications and associated parameters such as absolute or relative positions.

1.7.2 Parameters for defining CNN architectures

When analyzing CNN architectures, there are a number of points that deserve attention. In particular, several quantities and terms need to be defined—*width W, height H, depth N, stride S, 'zero-padding' width P* and *receptive field R*. In fact, the width and height are merely the dimensions of the input image, or else the dimensions of a specific layer of the neural network. The depth N of the network or of a specific block in it is the number of layers it contains.

The width W and height H of a layer are the numbers of neurons it has in each dimension. The stride S is the distance between adjacent neurons in the output field measured in units corresponding to the distance between adjacent neurons in the input field: stride S can be defined along the width and height dimensions but is usually the same for each. If $S = 1$, adjoining layers have the same dimensions (but see below how the size of the receptive field R can modify this). Note that increasing the stride S can be useful, as this saves memory and computation. In principle, it achieves a similar effect to pooling. However, pooling involves some averaging, while increasing the stride merely decreases the number of samples taken.

R_i is the width of the receptive field for each neuron in level i, i.e., the number of inputs for all neurons in that level. Zero-padding is the addition of P 'virtual' neurons providing static inputs at each end of the width dimension: these are given fixed weights of zero, the idea being to ensure that all neurons in the same layer have equal numbers of inputs, thereby facilitating programming. However, it also ensures that successive convolutions don't lead to smaller and smaller active widths; in particular, when $S = 1$, it permits us to make the widths of adjacent layers exactly equal (i.e., $W_{i+1} = W_i$). A simple formula connects several of these quantities:

$$W_{i+1} = (W_i + 2P_i - R_i)/S_i + 1 \qquad (1.115)$$

where the suffices pertain to the layer i inputs and the outputs feeding layer $i + 1$. It is worth underlining the null situation $W_{i+1} = W_i$ that applies when $S_i = 1$, $R_i = 1$ and $P_i = 0$. Overall, the purpose of padding is to allow for end-effects at the extremes of each layer, by ensuring that the number of zeros is adjusted to accommodate the desired stride and receptive field values.

Finally, an important point must be made about the definition of the depth of a number of layers of a CNN. The earlier discussion has implied that a number of adjacent layers of a CNN are normally accessed one after another in sequence—as would indeed be the case if larger and larger convolutions were implemented one after another in an effort to detect larger and larger features or even objects. However, there is another possibility—that the various layers are fed in parallel from a given starting point in the network, for example the input image. This possibility arises typically when an image is to be searched for a variety of different features, such as lines, edges or corners, and the results fed in parallel to a more holistic detector. This strategy was adopted in the LeNet architecture, which LeCun et al. (1998) developed to identify handwritten numerals and zip codes.

1.7.3 Krizhevsky et al.'s AlexNet architecture

AlexNet was designed specifically to target the ImageNet Challenge, (ImageNet Large-Scale Visual Recognition Object Challenge (ILSVRC), 2012) which took place in 2012. The AlexNet designers (Krizhevsky et al., 2012) forced the by then quite old schema based on CNNs into shape as a winning approach. To achieve this they had to radically improve the CNN architecture, and this necessarily gave rise to a very large software machine; they then had to speed it up dramatically with the aid of GPUs—by no means a small task as it meant reoptimizing the software to match the hardware; finally, they had to find how to feed the software system with a very large training set—again no mean task, as an unprecedentedly large number of parameters had to be trained rigorously, and several innovations were required in order to achieve this.

The CNN architecture had 10 hidden layers (counting C, F and S layers)—only 4 more than LeNet. However, these numbers are misleading as the *depths* of the various layers in AlexNet sum to 11,176 compared with 258 for LeNet. Similarly, AlexNet contains ~650,000 neurons compared with 6,508 for LeNet, while the number of trainable parameters is some 60 million compared with 60,000 for LeNet. And when we look at the size of the input image, we find that AlexNet takes a color image of size 224 × 224, whereas LeNet could only manage a bi-level 32 × 32 input image. So overall, AlexNet is larger than LeNet by a factor between 100 and 1000, depending on which factors should be regarded as the most relevant. However, the real change wrought by AlexNet was the possibility of working with huge numbers of layers and managing the credit assignment problem in spite of this—while still using the backpropagation algorithm for training. At the time, this was unprecedented, but it was made possible partly by the already reduced number of parameters required by CNNs, because all the neurons in any given layer of neurons have identical parameters; it was also made possible by employing exceptionally large training sets. However, one prominent feature of the AlexNet architecture is the horizontal split right across the network, above which a single GPU is used for implementation—and similarly below. In principle, this should make the operation of the architecture unduly complex, but in practice it turned out to be practicable. But because of this complexity it will be far easier for us to concentrate on the ZFNet architecture of Zeiler and Fergus (2014), as this is essentially a tidied up, slightly improved version of AlexNet implemented on a single GPU (Fig. 1.32): in particular, it had eight rather than seven hidden

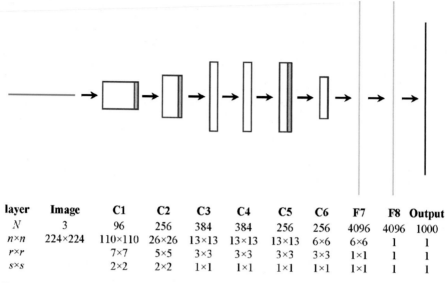

layer	Image	C1	C2	C3	C4	C5	C6	F7	F8	Output
N	3	96	256	384	384	256	256	4096	4096	1000
$n \times n$	224×224	110×110	26×26	13×13	13×13	13×13	6×6	6×6	1	1
$r \times r$		7×7	5×5	3×3	3×3	3×3	3×3	1×1	1	1
$s \times s$		2×2	2×2	1×1	1×1	1×1	1×1	1×1	1	1

FIGURE 1.32 Schematic of the ZFNet architecture. This schematic is very similar to that for AlexNet. Notice that AlexNet contains 7 hidden layers whereas ZFNet contains 8 hidden layers (these figures count S layers as parts of the corresponding C layers). Also, note that ZFNet is implemented using only a single GPU and its architecture is unsplit.

layers [these figures count S layers as parts of the corresponding C layers]. Note also that ZFNet incorporated a more gradual initial narrowing down of the layer dimensions ($n \times n$) than AlexNet (layer C1 being 110 × 110 rather than 55 × 55). On the other hand, both architectures used 'overlapping pooling'—in this case a combination of 3 × 3 pooling and 2 × 2 stride mapping. Notice that the dimensions ($n \times n$) of the layers start at 224 × 224 and gradually fall to 1 × 1. Interestingly, almost all the trainable parameters are in layers F7 and F8 (F6 and F7 for AlexNet), only 1000 connections being left for the final softmax (nonneural) classifier.

In Fig. 1.32, S1, S2 and S3 are shown in blue to the right of C1, C2 and C5 respectively; $n \times n$ indicates the dimensions in the case of a 2-D image format; $r \times r$ is the size of a 2-D neuron input field (a single number indicates abstract 1-D data); $s \times s$ is the 2-D stride. N is the depth within an individual layer: its approximate size is indicated by the vertical scale in the figure.

Very shortly before AlexNet was completed, a new technique called 'dropout' was introduced by Hinton et al. (2012): see also Hinton (2002). The purpose of this was to limit the incidence of overtraining. This was achieved by randomly setting a proportion (typically as high as 50%) of the weights to zero for each training pattern; this rather surprising technique appeared to work well: it did so by preventing hidden layers from relying too much on the specific data fed to them. Krizhevsky et al. (2012) included this feature in AlexNet. To apply it, the output of each neuron is randomly set to zero with probability 0.5. This is done before the forward pass of the input data, and the affected neurons do not contribute to the ensuing backpropagation. On the next forward pass, a different set of neuron outputs is set to zero with probability 0.5, and again the affected neurons do not contribute to backpropagation;

and similarly for later passes. During testing, an alternate procedure occurs, with all neuron outputs being multiplied by 0.5. In fact, multiplying *all* neuron outputs by 0.5 is an approximation to taking the geometric mean of all the local neuron output probability distributions and relies on the geometric mean being not too far from the arithmetic mean. Dropout was incorporated into the first two layers of AlexNet, and significantly reduced the amount of overfitting (overtraining because of too little training data).

AlexNet was trained using the 1.2 million images available from the ImageNet ILSVRC challenge, this number being a subset of the full 15 million in the ImageNet database. In fact, ILSVRC-2010 was the only subset for which test labels were available, there being about 1000 images in each of 1000 categories. However, it was found that these images were far too few to specify a CNN of the complexity required to perform accurate classification of this immense task. Therefore, means were required for expanding the dataset sufficiently to train AlexNet and achieve classification error rates in the 10% to 20% bracket.

Two main means of augmenting the dataset were considered and implemented. One was to apply realistic translations and reflections to the images in order to generate more images of the same type. The transformations even extended to extracting five 224×224 patches and their horizontal reflections from the initial 256×256 ImageNet images, giving a total of ten patches per image. Another was to alter the intensities and colors of the input images. To make this exercise more rigorous, it was carried out by first using principal component analysis (PCA) to identify color principal components for the ImageNet dataset, and then to generate random magnitudes by which to multiply the eigenvalues, thereby producing viable variants of the original images. Together, these two approaches were able to validly generalize, and multiply the size of, the original dataset by a factor of $\sim$2000—the principle being to generate natural changes in position, intensity and color.

At this stage it ought to be emphasized that the aim of the challenge was to find the best vision machine (giving the lowest classification error rates) that is able to recognize an example of a flea, a dog, a car or other common object in any position in an image and in any reasonable pose. Furthermore, the machine should prioritize its classifications to give at least the top five most probable interpretations. Then each machine can be rated not only on the accuracy of its top classification but also on whether the object classified appears within the machine's top-5 classifications. AlexNet was able to achieve a winning top-5 error-rating of 15.3%, compared with 26.2% for the runner up. Another first was the dramatic drop to below 20% error rate for such an exercise, which spelt a new lease of life for neural networks, and brought them sharply into the limelight.

Note that all this was achieved not merely by designing a winning architecture and generating the right dataset to train it adequately, but also it was necessary to bring the training time down to realizable levels. In this respect GPU implementation proved to be crucial. Even with a pair of GPUs, the training time took approximately a week, working 24 hours a day, to manage the task. Without GPUs it would have taken some 50 times longer—most probably, about a year—so the machine would have had to be submitted to a later challenge! [A commonly accepted figure is that a GPU has a speed advantage $\sim$50 relative to a typical host CPU.]

Finally, it should be noted that GPUs provide a very good implementation because of their intrinsic parallelism and thus their capability for handling large data sets in fewer cycles. Note that each layer of a CNN is completely homogeneous and is therefore well adapted for parallel processing. Note also that GPUs are well adapted to working in parallel as they are able to read from and write to each other's memories directly, avoiding the need to move data through the host CPU memory.

1.7.4 Simonyan and Zisserman's VGGNet architecture

In the continued absence of knowledge about the form of an ideal architecture, Simonyan and Zisserman (2015) set out to determine the effect of further increases in depth. To achieve this, they significantly reduced the number of parameters in the basic network by limiting the maximum neuron input field to 3×3. In fact, they restricted the convolution input field and stride to 3×3 and 1×1 respectively, and set both the input field and the stride of each sub-sampling layer to 2×2. In addition, they arranged for the systematic and rapid convergence of the successive layers from 224×224 down to 7×7 in 5 stages, followed by a transition to 1×1 in a single fully-connected stage; this was followed two further fully-connected layers and then a final softmax output layer (Fig. 1.33). All the hidden layers included a ReLU nonlinearity stage (not shown in the figure). Apart from the N 'channels', the 5 convolutional layers C1–C5 respectively contained 2, 2, 3, 3, 3 identical sublayers (not marked in Fig. 1.33). Finally, it should be remarked that, for reasons of experimentation, Simonyan and Zisserman devised 6 variations on the VGGNet architecture, with 11 to 19 weighted hidden layers: here we only cover configuration D (with 16 weighted hidden layers), for which the numbers of identical sublayers in layers C1–C5 are as listed above, and layers F6, F7 and F8 each contain 1 weighted sub-layer. The number of weighted layers obviously strongly influences the number of parameters.

As mentioned above, Simonyan and Zisserman saved on the basic number of parameters by restricting the convolution input field to 3×3. This meant that larger convolutions had to be produced by applying several 3×3 convolutions in sequence. Clearly, a 5×5 input field would necessitate applying two 3×3 convolutions, and a 7×7 field would require three 3×3 convolutions. In the latter case, this would reduce the total number of parameters from $7^2 = 49$ to 3 times $3^2 = 27$. In fact, not only did this way of implementing a 7×7 convolution reduce the number of parameters, but also it forced an additional regularization on the convolution, as a ReLU nonlinearity was interposed between each of the 3×3 component convolutions. It is also relevant that both the input and output of each 3-layer 3×3 convolution stack can have N channels, in which case it will contain a total of $27N^2$ parameters, and it is that figure which should be compared with $49N^2$ parameters.

In spite of its increased depth, VGGNet contains only about 2.4 times as many parameters as AlexNet; also, it is much simpler and doesn't split the architecture to fit it to a 2-GPU system. On the contrary, it immediately obtains a speedup of 3.75 times over a single GPU when using an off-the-shelf 4-GPU system.

Details of the training methodology are similar to those for AlexNet: see the original paper by Simonyan and Zisserman (2015). However, these authors include one interesting innovation: that is to use 'scale jittering' while training—i.e., to augment the training set using objects over a wide range of scales. In fact, random scaling was applied over an image scale factor of 2.

The outcome was that VGGNet achieved top-5 test error results of 7.0% using a single net, compared with 7.9% for GoogLeNet (Szegedy et al., 2014). In fact, GoogLeNet achieved

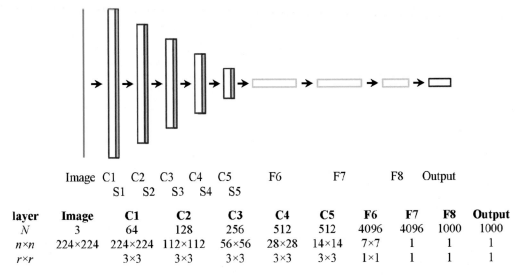

layer	Image	C1	C2	C3	C4	C5	F6	F7	F8	Output
N	3	64	128	256	512	512	4096	4096	1000	1000
$n \times n$	224×224	224×224	112×112	56×56	28×28	14×14	7×7	1	1	1
$r \times r$		3×3	3×3	3×3	3×3	3×3	1×1	1	1	1

FIGURE 1.33 Architecture of VGGNet. This architecture shows a more recent optimization of the standard type of CNN network. Unlike the schematic in Fig. 1.32, this one is arranged to show the relative sizes of the convolution layers, which range from image size down to 1 × 1. Note that the convolution layers all have unit stride, and that their input fields are limited to a maximum size of 3 × 3: the subsampling layers all have 2 × 2 input fields and 2 × 2 strides.

a figure of 6.7%, but only by employing 7 nets. Thus, VGGNet achieved second place in the ILSVRC-2014 challenge. However, after submission the authors managed to decrease the error rate to 6.8% using an ensemble of 2 models—substantially the same performance as for GoogLeNet, but with significantly fewer nets. Interestingly, all this was achieved even though the VGGNet architecture did not depart from the classical LeNet architecture of LeCun et al. (1989), the main improvement being the significantly increased depth of the network.

In spite of being placed second in the ILSVRC-2014 challenge, VGGNet has proved more versatile and adaptable to different datasets, and is a preferred choice in the vision community for extracting features from images. This seems to be because VGGNet actually provides more robust features even though it turned out to have slightly weaker classification performance on a specific dataset. As we shall see in the next section, VGGNet was the network chosen by Noh et al. (2015) for their work on deconvolution networks.

1.7.5 Noh et al.'s DeconvNet architecture

Inspired by the work of Zeiler and Fergus, Noh et al. (2015) produced a 'learning deconvolution network' (DeconvNet) that learnt *from training* how to deconvolve the sets of convolution coefficients in each layer of a CNN. Before examining their network in detail, it is important to understand the motivation they had for designing it. Their purpose was to produce a 'semantic segmentation' network. In fact, image segmentation is aimed at determining the boundaries between the various objects appearing in an image: *semantic* segmentation

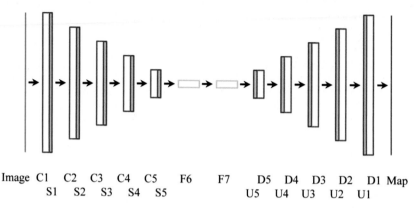

Image C1   C2   C3   C4   C5   F6   F7   D5   D4   D3   D2   D1 Map
        S1   S2   S3   S4   S5   U5   U4   U3   U2   U1

FIGURE 1.34 Schematic of Noh et al.'s Learning deconvolution network. This network contains two networks back to back. On the left is a standard CNN network, and on the right is the corresponding DNN 'deconvolution' network that seems to operate in reverse. The CNN network (on the left) has no output (e.g., softmax) classifier, as the ultimate purpose is not to classify objects but to present a map of where they are on a pixel-by-pixel basis throughout the area of the image. Deconv layers D5 down to D1 are intended to progressively unpick layers C5–C1. Similarly, unpooling layers U5 down to U1 are intended to progressively unpick pooling layers S5–S1. To achieve this, the position parameters from the maxpooling layers have to be fed to corresponding locations in corresponding unpooling layers (i.e., locations from Si should be fed to Ui).

goes further and also classifies all the objects—thereby giving a relevant meaning to each region in the image.

The DeconvNet architecture is shown in Fig. 1.34: note that its initial CNN section is borrowed from layers C1–F7 of VGGNet, though it excludes layer F8 and the Output softmax layer. It will be useful to provide a rationale for this. First, an upflowing CNN is needed for identifying objects in the input image. Second, if objects are to be located in particular parts of an image, another CNN is needed to point to the positions, and this necessarily has to follow the identification process. Undertaking both tasks in one enormous unconstrained CNN would be prohibitive of memory and training, so the two networks have to be linked closely together. The means of connecting them together are to provide feedforward paths from the pooling units to later unpooling units. So the very means by which the CNN output was generalized to *eliminate* the effects of sample variations has led to the second CNN being augmented to *yield* the required location maps. Crucially, we also see that the overall single upflowing data-path makes it obvious why the ReLU units must now all point in the same direction. (Remember that they have now all been turned to face forwards again.) It is also clear that with such a huge network, training will have to be carried out carefully, and it would appear obvious that the originally upflowing (object detection) section should initially be trained on its own.

Overall, the system works by mirroring the input CNN by including a deconvolution network (DNN) after it. The operation may be summarized as follows: unpooling layer Ui is nonlinear and redirects (unpools) the Ci maximum signals; then deconvolution layer Di operates linearly on the data, and therefore has to sum the overlapping inputs, weighted as necessary. However, rather than constructing suitable combination rules to define what happens with the overlapping output windows of each Di layer – and doing this in some very approximate way (such as 'transposed' versions of the convolution filters) – the deconvolution layers are trained as normal parts of the overall network. While this is a rigorous approach, it substantially increases the burden involved in training the network.

It is also useful to have a mental model of the whole process occurring in the DNN. First, each unpooling layer recovers the information from the corresponding pooling layer and reconstitutes the dimensions the dataspace had before pooling. However, it only populates it sparsely, with the local maximum values in appropriate positions. The purpose of the following deconvolutional layer is to reconstruct a dense map in its dataspace. So, whereas the CNN reduces the size of the activations, the following DNN enlarges the activations and makes them dense again. Nevertheless, the situation is not completely unraveled, as only the maximum values are reinserted. As Noh et al. (2015) say in their paper, "unpooling captures *example-specific* structures by tracing the original locations with strong activations back to image space", whereas "learned filters in the deconvolutional layers tend to capture *class-specific* shapes". What this means is that the deconvolutional layers rebuild the example shapes to correspond more accurately to what would be expected for objects of the specific classes.

In spite of this assurance, the network has to be trained appropriately. However, the surmise given above about training in two stages has been improved by Noh et al. as follows: to beat the problem of the space of semantic segmentation being extremely large, the network is first trained on easy examples, and then trained on more challenging examples: this amounts to a sort of bootstrapping approach. More precisely, the first of the training processes involves limiting the variations in object size and location by centering and cropping them in their bounding boxes; the second stage involves ensuring that the more challenging objects are adequately overlapped with ground-truth segmentations: to achieve this, the widely used intersection over union (IoU) measure is used, and is taken to be acceptable only if it is at least 0.5. In fact, the first stage uses a 'tight' bounding box, and this is extended by a factor 1.2 and further expanded into a square in order to include sufficient local context around each object. In this first stage, the box is rated according to the object located at its center, other pixels being labeled as background. However, in the second stage this simplification is not applied and all relevant class labels are used for annotation.

Next, we look at another closely related method—that of Badrinarayanan et al. (2015), which uses much less memory and has several other advantages.

1.7.6 Badrinarayanan et al.'s SegNet architecture

The SegNet architecture strongly resembles that of DeconvNet (Fig. 1.34) and is also aimed at semantic segmentation. However, its authors demonstrated the need for returning to a significantly simpler architecture in order to make it more easily trainable (Badrinarayanan et al., 2015). Basically, it was identical to DeconvNet (Fig. 1.34) but with F6 and F7 excluded. In addition, it was clear to the authors that use of max-pooling and subsampling reduce feature map resolution and thereby reduce location accuracy in the final segmented images. Nevertheless, they start by eliminating the fully connected layers of VGGNet, retain the encoding–decoding (CNN–DNN) structure of DeconvNet, and also retain max-pooling and unpooling. In fact, it is the move away from using fully-connected layers that helps SegNet the most, as this drastically reduces the number of parameters to be learnt, and thereby also drastically reduces the training requirements of the method. Accordingly, the whole network can be considered as a single rather than a dual network and trained efficiently 'end-to-end'. Furthermore, the authors identified a far more efficient way to store object location information: they do so by storing *only* the max-pooling indices, viz. the locations of the maximum feature values in each pooling window in each encoder feature map. As a result only 2 bits of information are needed for each 2×2 pooling window (cf. Fig. 1.33). This means that even for the initial CNN (encoder) layers, it is not necessary to store the feature maps themselves: what has to be stored is the object location information. By this means the encoder storage requirement is reduced from 134M (corresponding to layers C1–F7 of VGGNet) down to 14.7M. The total storage for SegNet is rated at twice this, as the same amount of information has to be saved in the decoder layers. However, the same applies to other deconvnets, so in all cases the total amount of data has to be doubled relative to the contents of the initial CNN encoder.

The smaller size of SegNet makes end-to-end training possible, and thereby far more suitable for real-time applications. The authors acknowledge that larger networks can work better—though at the cost of far more complex training procedures, increased memory and considerably increased inference time. Furthermore, it is difficult to assess their true performance. Basically, the decoders have to be trained via very large and cumbersome encoders, and the latter are general-purpose rather than being targeted at specific applications. In the majority of instances, such networks have been based on a VGGNet front end, typically containing all 13 sublayers of C1–C5, together with a variable (very small) number of fully-connected layers.

These considerations make it no surprise that Badrinarayanan et al. successfully applied SegNet to the CamVid dataset (Brostow et al., 2009) by training it end-to-end for optimum adaptation. They found it outperformed seven conventional (nonneural) methods, including local label descriptors and superparsing (Yang et al., 2012; Tighe and Lazebnik, 2013), obtaining scores averaging to 80.1%, in comparison with 51.2% and 62.0% respectively; the 11 categories to be recognized were building, tree, sky, vehicle, sign, road, pedestrian, fence, pole, pavement, and cyclist, and the accuracies attained for these ranged from 52.9% (cyclist) to 94.7% (pavement). Their success with this task may be judged from the results of their online demo [http://mi.eng.cam.ac.uk/projects/segnet/ (website accessed 7 October 2016).] which was used to generate the pictures in Fig. 1.35: in fact, this demo placed pixels in *twelve* categories, including road markings in addition to the eleven listed above.

They also did a careful comparison of SegNet with other semantic segmentation networks, including FCN (so-called 'fully convolutional networks') and DeconvNet. FCN and DeconvNet have the same encoder size (134M); note that FCN reduces the decoder size down to 0.5M, though DeconvNet continues with a 134M decoder. Class averages for the three methods are 59.1%, 62.2% and 69.6%. Even though SegNet is numerically the worst of the three, its accuracy is still competitive and it has the distinct advantage of being more adaptable by virtue of being trained end-to-end. In fact, it is also easily the fastest running, being ~2.2 times faster than FCN and ~3.3 times faster than DeconvNet, albeit on different sized images.

Overall, the authors state that architectures "which store the encoder network feature maps in full perform best but consume more memory during inference time", which also means that they run more slowly. On the other hand, SegNet is more efficient since it only stores the max-pooling indices; in addition, it has competitive accuracy, and its capability for end-to-end training on currently relevant data make it significantly more adaptable.

| sky | building | pole | road marking | road | pavement | tree | sign | fence | vehicle | pedestrian | bike |

FIGURE 1.35 Two road scenes taken from a front passenger seat. In each case the image on the left is the original, and the image on the right is the segmentation produced by SegNet. The key indicates the 12 possible meanings assigned by SegNet. While location accuracy is not perfect, the assigned meanings are generally reasonable, given the limited number of allowed interpretations and the variety of objects within the fields of view. These pictures were processed using the online demo at http://mi.eng.cam.ac.uk/projects/segnet/ (Badrinarayanan et al., 2015).

1.7.7 Application of deep learning to object tracking

We now return to the subject of tracking moving objects. This is an area in which great gains have been made by the application of deep learning: in fact, deep learning has led to radical improvements over the conventional approaches discussed earlier. We start by taking the paper by Held et al. (2016) as an interesting example.

Held et al. (2016) sought to produce a single-target tracker which would operate in real time, and devised a neural-based way of achieving this at speeds of up to 100 fps. This speed was achieved only for the test version of the network, and reflected very considerable amounts of off-line training. The high test speed is the result of employing a relatively simple architecture in which pairs of frames are fed to a trained neural network, which immediately (i.e., in a single feedforward pass) gives an output image in which the target bounding box is marked.

When testing, the tracker is initialized with a ground-truth bounding box containing the target object, the bounding box being updated after the analysis of each successive pair of frames. However, before detailed examination, each frame is cropped at a size and location sufficiently large to capture the target object for any motion it may reasonably make (typically, this means cropping at double the size of the bounding box): this procedure also permits useful contextual information to be included. The trained neural network then searches the two ($t - 1$ and t) cropped images to find the best match for the position of the moving object. Repetition of the process thus permits the target object to be tracked throughout the video sequence.

For this to be successful, the trained network must contain huge amounts of information about different possible displacements of each pair of images. This is entirely possible using a pair of networks each containing N convolution layers (often called a 'Siamese ConvNet'), which are fed to a set of M fully-connected layers (Held et al. used $N = 5$ and $M = 3$); the final output contained the necessary output bounding box information. To obtain all the required information, the neural network was trained using all permissible shifts and crops of the incoming frames. In fact, the network was trained not only on videos but also on pairs of images, the reason being to teach the network to track a more diverse set of objects, thereby helping to prevent overfitting to the objects in the video data.

As a result of careful training, the tracker was found to be invariant to background motion, out of plane rotations, deformations, lighting changes and minor occlusions. Moreover, the additional training on labeled images helped to provide it with a generic relationship between an object's appearance and motion, enabling it to be able to track objects not appearing in the training set—as well as the capability to perform this task at the unprecedented speed of 100 fps.

The cost of this capability came at the expense of large amounts of training time due to excessive amounts of data augmentation. This was one of the problems that drove Bertinetto et al. (2016) to devise a tracking architecture using a fully convolutional (FC) Siamese network which they called SiamFC. They started with a system having two parallel inputs, one providing an exemplar image and the other providing a search image. The idea is to search the input image for matches with the exemplar image using a suitable similarity operator. In fact, they used correlation for this purpose and a correlation layer was used to generate a score map, which was presented at the output of the network. It was found useful to arrange for the output score map to have reduced dimensionality, from which hits could be related back to the input image by multiplying by the network stride. The latter was taken to be 8, a figure which arises as there are three successive applications of a 2×2 stride, in layers C1, S1 and S2 (see Fig. 1.36): to fully understand the overall effect, note that each 2×2 stride divides the intrinsic image size by 2×2.

A Siamese network is one containing two parallel flows, each having the same intrinsic structure. This is natural for the above architecture as the two flows carry exactly the same type of data, albeit the *sizes* of the images will obviously be different. A fully convolutional network is used for each of the flows as the intention is to achieve translational invariance, so that the signal from any object will not vary with its position in the image. 'Fully convolutional' means that no variation in the convolution parameters is permitted over the whole of

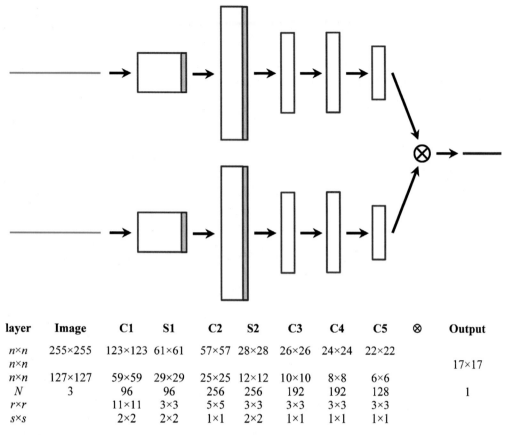

layer	Image	C1	S1	C2	S2	C3	C4	C5	⊗	Output
$n{\times}n$	255×255	123×123	61×61	57×57	28×28	26×26	24×24	22×22		
$n{\times}n$										17×17
$n{\times}n$	127×127	59×59	29×29	25×25	12×12	10×10	8×8	6×6		
N	3	96	96	256	256	192	192	128		1
$r{\times}r$		11×11	3×3	5×5	3×3	3×3	3×3	3×3		
$s{\times}s$		2×2	2×2	1×1	2×2	1×1	1×1	1×1		

FIGURE 1.36 Fully convolutional Siamese tracker. This figure shows the architecture of the FC Siamese tracker devised by Bertinetto et al. (2016). Top: the FC branch containing the *search* flow; middle: the branch containing the *exemplar* flow; bottom: details of the two branches, including no. of channels N, image size $n \times n$, neuron input field $r \times r$, and stride $s \times s$ (N, r and s are identical for the two flows). Note that, as in the case of AlexNet, this architecture uses overlapping pooling in subsampling layers S1 and S2, for both of which $r \times r = 3 \times 3$ and $s \times s = 2 \times 2$.

the image space: note that this means that 'padding' of the outer reaches of the network (typically with zeros) is not allowed, and in practice this means that—unlike the case of AlexNet—the output spaces of successive layers will be curtailed wherever the neuron input fields are larger than 1×1 (see Fig. 1.36). It should be added that the point of including the two fully convolutional networks is to obtain sufficient information on the target and exemplar objects, while ignoring noise, shape variations and irrelevant artefacts such as background detail.

Correlation is carried out using a standard sliding-window computation. The overall system is shown in Fig. 1.36, the two fully convolutional flow networks being adapted from those in Krizhevsky et al.'s (2012) AlexNet architecture: in particular, the fully connected layers have been discarded as they are not fully convolutional and therefore do not permit position invariance. (The purpose of AlexNet was to classify any input image according to the class of the target object appearing within it, object localization information being totally eliminated.)

In Bertinetto et al.'s work, the exemplar image is taken as the initial appearance of the target object and is not updated; neither is a memory of past appearances maintained; nor are predictions of the positions of object bounding boxes computed. Indeed, the main aim was to achieve a simple, reliable tracker for single target objects. However, Bertinetto et al. found that upsampling the score map by a factor 16 using bicubic interpolation (i.e., from 17×17 to 272×272) resulted in more accurate localization. Also, they searched for the object over five scales, from ~ 0.95 to 1.05, in order to cope with scale variations.

Next, we consider the size of the output image space. The correlation layer is a convolution between images of size 22×22 and 6×6, leading to a potential maximum output image size of 27×27 down to a viable image size of 17×17 (these figures arise because $22 + 6 - 1 = 27$ and $22 - 6 + 1 = 17$). Note that the maximum output image size (27×27) would not capture objects that are just outside the search space, whereas the minimum (17×17) would take full account of objects that are completely within the search space; between these limits there would be some chance of detecting partially visible objects. However, SiamFC was aimed at being as successful as possible at detecting fully visible objects.

Finally, an interesting point concerns the calculation of the effects of a 2×2 stride when the previous image has an odd number size (as happens for all three cases in SiamFC). In that case the stride takes into account the first and last pixels in each row and column. Given this, the successive image sizes quoted in Fig. 1.36 all correspond exactly and logically to those given by Bertinetto et al. (2016).

Feichtenhofer et al. (2017) set out to devise a deep learning architecture that learns detection and tracking together in tandem. Their architecture makes use of an object detector and a tracker: in fact, the idea is to carry out detection and tracking simultaneously using a ConvNet, optimized by a combined detection and tracking based loss. Convolutional cross-correlation between the feature responses of adjacent frames is carried out to estimate the local displacement at different feature scales. (Here, correlation is restricted to a small neighborhood, with maximum displacement $d = 8$, to avoid large output dimensionality: this reflects the limitation already noted for the Held et al. (2016) target tracker.) Correlation maps are computed for all positions on a feature map, and region of interest (RoI) pooling (Dai et al., 2016) is applied to these feature maps for track regression. The above approach allows multiple objects to be tracked simultaneously, and the architecture can be trained end-to-end taking input frames from raw videos. Overall, the approach led to significantly improved performance, at the 80% level, relative to previous state-of-the-art methods when run on the ImageNet (2015) VID validation set.

Note that the RoI pooling approach takes fully convolutional design as far as possible, in order to maintain translational invariance, and only the last convolutional layer is modified to deviate from this: this strategy permits the computation of position-sensitive score maps, from which multiple object tracklets can be extracted. However, as the details of the overall architecture depend closely on those of R-FCN (Dai et al., 2016), ResNet-101 (He et al., 2016), Fast R-CNN (Girshick, 2015), and Faster R-CNN (Ren et al., 2015), available space prevents a full description from being presented here.

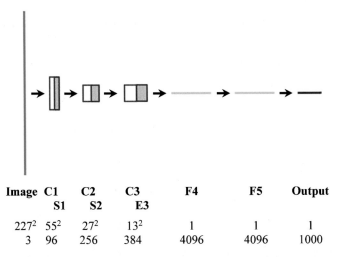

Image	C1 S1	C2 S2	C3 E3	F4	F5	Output
227^2	55^2	27^2	13^2	1	1	1
3	96	256	384	4096	4096	1000

FIGURE 1.37 Andrearczyk and Whelan's T-CNN architecture. This is designed to capture the texture of the whole input image, and to output a vector showing the most probable texture. The numbers under the labels indicate the sizes of the feature spaces being handled by convolution layers C1–C3 and the numbers of feature maps in the corresponding layers. Both of the pooling layers (S1, S2) reduce the image size by a factor 2 × 2. The energy layer E3 is a special form of pooling layer which averages the energies over the whole of each feature map, producing 384 outputs—one for each feature map of C3.

1.7.8 Application of deep learning to texture classification

As we have seen in the immediately preceding sections, deep learning has become a major factor in the design of vision algorithms. This applies no less in the case of texture analysis, which was introduced in Sections 1.6.1–1.6.6. Fig. 1.31 shows the architecture of the Laws texture classifier discussed in Section 1.6.3. This sort of classifier is often described as employing a set of 'filter banks' to extract the input information. However, the previous section shows that the CNN approaches can also be described in this way, so it is not surprising that CNNs (and ANNs) have also been applied to texture analysis. Early on, it was common to train such networks using sets of input images each consisting of a single texture (typically from the Brodatz database). However, this approach is restricted to treating the whole input image as a single region and classifying it accordingly: segmentation of an image into regions with different textures is beyond the capabilities of a simple architecture trained in this way.

Andrearczyk and Whelan (2016) described a basic CNN architecture (T-CNN-3) for classifying textures, which is illustrated in Fig. 1.37. It has some similarity to ZFNet though it contains fewer convolution layers: this does not prevent it from coping with textures because texture features can mostly be described using quite small windows. Note also that the final convolution layer is followed by a pooling layer which averages the texture energy over the whole feature map. Nevertheless, as a considerable number N of feature maps are computed by different parallel layers of the final convolution layer, the result is a vector of N energy values. These are fed to a final texture classifier yielding a single class for any single input texture image.

In fact, Andrearczyk and Whelan (2016) went further than this and devised a hybrid type of texture analyzer (TS-CNN-3) employing both the above approach and one permitting object shapes to be analyzed, the latter acting in a similar way to AlexNet's object recognition system for nontextured images. The overall architecture is shown in Fig. 1.38. An important aspect of this architecture is the concatenation layer M combining the outputs of the texture and shape parts of the system. But it is also important to notice that relevant texture information is mainly derived from relatively small low-level features, whereas shape information is a more global property which requires the input of higher level features. This is why texture information is derived from the output of convolution layer C3, whereas shape information is obtained from the pooled output of C5. (Interestingly, because E3 averages the texture energy over the *whole* of each C3 feature layer, no spatial or shape information remains in the texture channel after E3.) Finally, the output layers F6, F7 and O are used to combine the shape and texture outputs: clearly, there is enough information available in these layers to relate the textures to particular locations in the input image and to present the outputs in terms of the probabilities that the textures used in training actually appear in the input image. However, what this network does not do is to produce an output image map showing the most probable segmentation of the image into various texture regions. It would be good to achieve this, and in principle it ought to be possible by incorporating the encoder–decoder architecture of SegNet (see Section 1.7.6). The main obstacle here is the lack of suitably large texture datasets (comparable with ImageNet): in fact, the underlying problem is that architectures such as SegNet are so deep that they require a much greater amount of training using much larger numbers of training patterns.

The lack of a large texture dataset is a serious factor standing in the way of further progress in texture analysis, whatever training methods and architectures are being devised. In fact, it will be clear that if such a dataset were available, AlexNet (and other nets such as VGGNet) could be trained to perform texture analysis over whole images without any changes being made to its architecture. As remarked by Liu et al. (2019) "The recent success of deep learning in image classification and object recognition is inseparable from the availability of large-scale annotated image datasets such as ImageNet. However, deep learning based texture analysis has not kept pace with the rapid progress witnessed in other fields, partially due to the unavailability of a large-scale texture database."

These statements are serious, because (a) CNN-based architectures are already amongst the most widely used means of performing texture analysis—a trend that seems hardly likely to decrease in the foreseeable future; and (b) the same training restrictions (and need for large texture datasets) must also apply to whatever methods are to be used for analysing textured images. It should also be remembered that the statistical nature of textures implies that, in general, training-based procedures will be preferable to hand-crafted algorithms.

Finally, we consider the performance of Andrearczyk and Whelan's two architectures. When trained on ImageNet, which is an *object* dataset, they found that T-CNN-3 performed less well than AlexNet (51.2% compared with 57.1%): this is not surprising as AlexNet is a much larger network containing 60.9 million trainable parameters, compared to the 23.4 million of T-CNN-3. On the other hand, when training with texture-oriented datasets such as ImageNet-T and KTH-TIPS-2b (Russakovsky et al., 2015; Hayman et al., 2004), T-CNN-3 showed significant improvement over AlexNet (respective accuracies of 71.1% compared with 66.3%, and 48.7% compared with 47.6%). These results correspond to training from

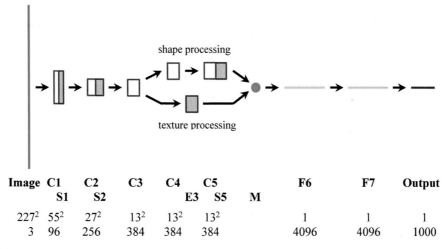

Image	C1 S1	C2 S2	C3	C4 E3	C5 S5	M	F6	F7	Output
227^2	55^2	27^2	13^2	13^2	13^2		1	1	1
3	96	256	384	384	384		4096	4096	1000

FIGURE 1.38 Andrearczyk and Whelan's TS-CNN architecture. This is designed to capture texture *and* shape information for the whole input image, and to output a vector showing the most probable set of textures. The first row of numbers under the labels indicates the sizes of the feature spaces being handled by the convolution layers C1–C5; the second row of numbers indicates the numbers of feature maps in the corresponding layers. All three pooling layers (S1, S2, S5) reduce the image size by a factor 2 × 2, so the image fed to the merger point M has size 6 × 6. At M, the texture and the shape outputs are concatenated, thereby producing a total of $384 \times (1 + 6^2)$ outputs—*all* of which are joined to *all* the inputs of F6, forming a fully connected network. As in Fig. 1.37, the energy layer E3 is a pooling layer which averages the energies over the whole of each feature map, producing 384 outputs, one for each feature map of C3.

scratch on a single database. However, it is also possible to pretrain on one database and fine-tune on another. When this was attempted—specifically, pretraining on ImageNet and fine-tuning on KTH-TIPS-2b—T-CNN-3 performed more accurately than AlexNet (73.2% compared with 71.5%). A further improvement was obtained by using the hybrid TS-CNN-3 architecture, an accuracy of 74.0% then being achieved. Some of this improvement was clearly due to the larger number of trainable parameters of TS-CNN-3 (62.5 million). Nevertheless, when tests were made with architectures with comparable numbers of parameters—e.g., a combination of AlexNet and T-CNN-3, and a combination of VGG-M (Chatfield et al., 2014) with FV-CNN (Cimpoi et al., 2015)—TS-CNN-3 remained superior. In fact, the hallmark of the new TS-CNN-3 texture + shape architecture is that it is highly efficient in its use of training parameters, and by separating texture analysis from shape analysis it achieves more accurate results when processing textured images.

It has been useful to describe the work of Andrearczyk and Whelan (2016) in some detail, as this reveals a good many insights into the application of deep networks to texture analysis—including a number of the subtleties that are involved. Whereas the approach aims at an architecture with relatively modest numbers of training parameters, the less restricted approach of Cimpoi et al. (2015; 2016) appears to be more powerful: it systematically uses global pooling of the CNNs *before* the level of the fully connected layers and thereby captures valuable sets of texture descriptors. (More precisely, it pools local features densely, removing global spatial information, in order to extract dense convolutional features.) Indeed, Cimpoi et al.'s, 2016 paper was able to develop a vocabulary of forty-seven texture attributes that describe a wide range of texture patterns. In fact, the work of Cimpoi et al. (2015; 2016) has been influential in its near-perfect classification accuracy. However, Liu et al. (2019) state that performance is 'saturated' (meaning that it has leveled off, in the vicinity of 100%) "because the datasets are not large enough to allow fine-tuning to obtain improved results"—reflecting the points made earlier about available texture datasets. It should be remarked that Cimpoi et al.'s best results were obtained using the VGG-VD (Very Deep) architecture of Simonyan and Zisserman (2015) containing 19 CNN layers, thereby partly explaining why saturation became a possibility.

One other point raised by Liu at al. (2019) in their review of the development of texture representations is that there is a growing tension between the need for huge image datasets and the corresponding human need for compact, efficient representations. The latter need is increasingly being seen on mobile and embedded platforms with restricted resources—as previously observed by Andrearczyk and Whelan (2016) and Szegedy et al. (2014).

1.7.9 Texture analysis in the world of deep learning

Part E started by exploring the meaning of texture—essentially by asking "What is a texture and how is a texture formed?" Typically, a texture starts with a surface that exhibits local roughness or structure, which is then projected to form a textured image. Such an image exhibits both regularity and randomness, though directionality and orientation can also be important parameters. However, the essential feature of randomness means that textures have to be characterized by statistical techniques and recognized using statistical classification procedures. Techniques that have been used for this purpose include autocorrelation, co-occurrence matrices, texture energy measures, fractal-based measures, Markov random fields, and so on. These aim both to analyze and to model the textures. Indeed, workers in this area have had to spend much time striving to achieve ever-improved models of the textures they are working with in order to better describe, recognize and segment them. However, over less than a decade, DNNs suddenly came into their own, and as we have seen, this led to a further period of rapid development of techniques for texture analysis—taking us even further away from the intuitive approaches of the 1970s.

1.8 Part G – Summary

Parts A–E of this chapter outlined three highly relevant legacy topics on feature detection and object detection, in both 2-D and 3-D. They presented the underlying theory that had become familiar and widely used throughout the subject before the deep learning explosion of 2012. Part F went on to impart an understanding of what this explosion entailed and, in particular, not only how all object features and the objects themselves could be learnt from many, many training examples, but also how semantic segmentation could be achieved with end-to-end training of complete convolutional networks. The fact that such networks can *learn* all features and objects and seemingly don't need legacy methods to back them up is salutary. In fact, semantic segmentation of static scenes is not the only relevant story, but as Section 1.7.7

has demonstrated, a not dissimilar process has been repeated with impressive effect for the tracking of multiple moving objects. Indeed, it was pointed out (Feichtenhofer et al., 2017) that much of what had been achieved in complex ways by hard application of principles has now been achieved in less theoretically intensive ways by deep learning methods; and these have the power to progress performance to even greater heights when even more efforts are applied. In retrospect, it also seems that workers found themselves up against a metaphorical brick wall regarding how to specify sufficiently accurately the nature of the 'soft' data in the *real* (nonidealized) images they wanted to process: and the lack of practicable modeling structures made it problematic to find suitable ways forward; hence it became necessary to at least trial the alternative of deep learning.

At this point it will be useful to summarize quite what it was that enabled the deep learning explosion. In fact, it was a combination of a set of disparate factors. In particular, we can point to: (1) use of CNNs with much lower connectivity than was the case for the old ANN architectures; (2) deployment of ReLUs in place of sigmoidal output functions; (3) application of the 'dropout' procedure to limit the incidence of overtraining; (4) the use of vastly increased amounts of image data for training; (5) extraction of even more image patches from images to further enhance the amount of data available for training; (6) application of GPUs for performing the training with enormously increased speed (it being commonly accepted that a GPU has a speed advantage ~50 relative to a typical host CPU). It must also be remembered that all these changes were brought to bear more or less simultaneously in 2012.

We now cut the discussion short, as the following chapters provide many further insights into what is actually a quite complex and fast-moving situation.

Acknowledgments

The following text and figures have been reproduced with permission from the IET: the in-text figure and associated text in Section 1.2.7—from Electronics Letters (Davies, 1999); Fig. 1.2 and associated text—from Proc. *Visual Information Engineering* Conf. (Davies, 2005); extracts of text—from Proc. *Image Processing and its Applications* Conf. (Davies, 1997). Fig. 1.5 and associated text have been reproduced with permission from IFS Publications Ltd (Davies, 1984). I also wish to acknowledge that Figs. 1.13 and 1.15 and associated text were first published in *Proceedings of the 4th Alvey Vision Conference* (Davies, 1988b).

References

Ade, F., 1983. Characterization of texture by 'eigenfilters'. Signal Processing 5 (5), 451–457.

Amit, Y., Trouvé, A., 2007. POP: patchwork of parts models for object recognition. International Journal of Computer Vision 75 (2), 267–282.

Andrearczyk, V., Whelan, P., 2016. Using filter banks in convolutional neural networks for texture classification. Pattern Recognition Letters 84, 63–69.

Badrinarayanan, V., Kendall, A., Cipolla, R., 2015. SegNet: a deep convolutional encoder-decoder architecture for image segmentation. arXiv:1511.00561v2 [cs.CV].

Bai, Y., Ma, W., Li, Y., Cao, L., Guo, W., Yang, L., 2016. Multi-scale fully convolutional network for fast face detection. In: Proc. British Machine Vision Association Conference. York, 19–22 September. http://www.bmva.org/bmvc/2016/papers/paper051/paper051.pdf.

Ballard, D.H., 1981. Generalizing the Hough transform to detect arbitrary shapes. Pattern Recognition 13, 111–122.

Beaudet, P.R., 1978. Rotationally invariant image operators. In: Proc. 4th Int. Conf. on Pattern Recognition. Kyoto, pp. 579–583.

Bertinetto, L., Valmadre, J., Henriques, J.F., Vedaldi, A., Torr, P.H.S., 2016. Fully-convolutional Siamese networks for object tracking. In: Proc. ECCV Workshops, pp. 850–865.

Brostow, G., Fauqueur, J., Cipolla, R., 2009. Semantic object classes in video: a high-definition ground truth database. Pattern Recognition Letters 30 (2), 88–97.

Canny, J., 1986. A computational approach to edge detection. IEEE Transactions on Pattern Analysis and Machine Intelligence 8, 679–698.

Castañón, D.A., 1990. Efficient algorithms for finding the k best paths through a trellis. IEEE Transactions on Aerospace and Electronic Systems 26 (2), 405–410.

Chatfield, K., Simonyan, K., Vedaldi, A., Zisserman, A., 2014. Return of the devil in the details: delving deep into convolutional nets. In: Proc. BMVC, pp. 1–12.

Cimpoi, M., Maji, S., Vedaldi, A., 2015. Deep filter banks for texture recognition and segmentation. In: IEEE Conf. on Computer Vision and Pattern Recognition, pp. 3828–3836.

Cimpoi, M., Maji, S., Kokkinos, I., Vedaldi, A., 2016. Deep filter banks for texture recognition, description and segmentation. International Journal of Computer Vision 118, 65–94.

Cootes, T.F., Taylor, C.J., 2001. Statistical models of appearance for medical image analysis and computer vision. In: Sonka, M., Hanson, K.M. (Eds.), Proc. SPIE, Int. Soc. Opt. Eng. USA, vol. 4322, pp. 236–248.

Cucchiara, R., Grana, C., Piccardi, M., Prati, A., 2003. Detecting moving objects, ghosts and shadows in video streams. IEEE Transactions on Pattern Analysis and Machine Intelligence 25 (10), 1337–1342.

Dai, J., Li, Y., He, K., Sun, J., 2016. R-FCN: object detection via region-based fully convolutional networks. In: Proc. NIPS.

Dalal, N., Triggs, B., 2005. Histograms of oriented gradients for human detection. In: Proc. Conf. on Computer Vision Pattern Recognition. San Diego, California, USA, pp. 886–893.

Davies, E.R., 1984. Design of cost-effective systems for the inspection of certain food products during manufacture. In: Pugh, A. (Ed.), Proc. 4th Int. Conf. on Robot Vision and Sensory Controls. London, 9–11 October. IFS (Publications) Ltd, Bedford and North-Holland, Amsterdam, pp. 437–446.

Davies, E.R., 1986. Constraints on the design of template masks for edge detection. Pattern Recognition Letters 4 (2), 111–120.

Davies, E.R., 1988a. A modified Hough scheme for general circle location. Pattern Recognition Letters 7 (1), 37–43.

Davies, E.R., 1988b. An alternative to graph matching for locating objects from their salient features. In: Proc. 4th Alvey Vision Conf. Manchester, 31 Aug.–2 Sept., pp. 281–286.

Davies, E.R., 1988c. Median-based methods of corner detection. In: Kittler, J. (Ed.), Proceedings of the Fourth BPRA International Conference on Pattern Recognition. Cambridge, 28–30 March. In: Lecture Notes in Computer Science, vol. 301. Springer-Verlag, Heidelberg, pp. 360–369.

Davies, E.R., 1997. Designing efficient line segment detectors with high orientation accuracy. In: Proc. 6th IEE Int. Conf. on Image Processing and Its Applications. Dublin, 14–17 July. In: IEE Conf. Publication, vol. 443, pp. 636–640.

Davies, E.R., 1999. Designing optimal image feature detection masks: equal area rule. Electronics Letters 35 (6), 463–465.

Davies, E.R., 2005. Using an edge-based model of the Plessey operator to determine localisation properties. In: Proc. IET Int. Conf. on Visual Information Engineering. University of Glasgow, Glasgow, 4–6 April, pp. 385–391.

Davies, E.R., 2017. Computer Vision: Principles, Algorithms, Applications, Learning, 5th edition. Academic Press, Oxford, UK.

Davies, E.R., Bateman, M., Mason, D.R., Chambers, J., Ridgway, C., 2003. Design of efficient line segment detectors for cereal grain inspection. Pattern Recognition Letters 24 (1–3), 421–436.

Dreschler, L., Nagel, H.-H., 1981. Volumetric model and 3D-trajectory of a moving car derived from monocular TV-frame sequences of a street scene. In: Proc. Int. Joint Conf. on Artif. Intell., pp. 692–697.

Dudani, S.A., Luk, A.L., 1978. Locating straight-line edge segments on outdoor scenes. Pattern Recognition 10, 145–157.

Elgammal, A., Harwood, D., Davis, L., 2000. Non-parametric model for background subtraction. In: Proc. European Conf. on Computer Vision. In: LNCS, vol. 1843, pp. 751–767.

Everingham, M., Van Gool, L., Williams, C.K.I., Winn, J., Zisserman, A., 2007. The Pascal visual object classes challenge 2007. (VOC2007) Results. http://www.pascalnetwork.org/challenges/VOC/voc2007/.

Everingham, M., Van Gool, L., Williams, C.K.I., Winn, J., Zisserman, A., 2008. The Pascal visual object classes challenge 2008. (VOC2008) Results. http://www.pascalnetwork.org/challenges/VOC/voc2008/.

Everingham, M., Zisserman, A., Williams, C.K.I., Van Gool, L., 2006. The Pascal visual object classes challenge 2006. (VOC2006) Results. http://www.pascalnetwork.org/challenges/VOC/voc2006/.

Fathy, M.E., Hussein, A.S., Tolba, M.F., 2011. Fundamental matrix estimation: a study of error criteria. Pattern Recognition Letters 32 (2), 383–391.

Faugeras, O., Luong, Q.-T., Maybank, S.J., 1992. Camera self-calibration: theory and experiments. In: Sandini, G. (Ed.), Proc. 2nd European Conf. on Computer Vision. In: Lecture Notes in Computer Science, vol. 588. Springer-Verlag, Berlin Heidelberg, pp. 321–334.

Feichtenhofer, C., Pinz, A., Zisserman, A., 2017. Detect to track and track to detect. In: Proc. ICCV, pp. 3038–3046.

Felzenszwalb, P.F., Girshick, R.B., McAllester, D., Ramanan, D., 2010. Object detection with discriminatively trained part based models. IEEE Transactions on Pattern Analysis and Machine Intelligence 32 (9), 1627–1645.

Fischler, M.A., Bolles, R.C., 1981. Random sample consensus: a paradigm for model fitting with applications to image analysis and automated cartography. Communications of the ACM 24 (6), 381–395.

Girshick, R.B., 2015. Fast R-CNN. In: Proc. ICCV. 2015.

Haralick, R.M., Shanmugam, K., Dinstein, I., 1973. Textural features for image classification. IEEE Transactions on Systems, Man and Cybernetics 3 (6), 610–621.

Harris, C., Stephens, M., 1988. A combined corner and edge detector. In: Proc. 4th Alvey Vision Conf., pp. 147–151.

Hartley, R.I., 1995. A linear method for reconstruction from lines and points. In: Proc. Int. Conf. on Computer Vision, pp. 882–887.

Hartley, R.I., Sturm, P., 1994. Triangulation. In: American Image Understanding Workshop, pp. 957–966.

Hayman, E., Caputo, B., Fritz, M., Eklundh, J.-O., 2004. On the significance of real-world conditions for material classification. In: ECCV, pp. 253–266.

He, K., Zhang, X., Ren, S., Sun, J., 2016. Deep residual learning for image recognition. In: Proc. CVPR.

Held, D., Thrun, S., Savarese, S., 2016. Learning to track at 100 fps with deep regression networks. In: Proc. ECCV. Springer, pp. 749–765.

Hinton, G.E., 2002. Training products of experts by minimizing contrastive divergence. Neural Computation 14 (8), 1771–1800.

Hinton, G.E., Srivastava, N., Krizhevsky, A., Sutskever, I., Salakhutdinov, R.R., 2012. Improving neural networks by preventing co-adaptation of feature detectors. arXiv:1207.0580v1 [cs.NE].

Horprasert, T., Harwood, D., Davis, L.S., 1999. A statistical approach for real-time robust background subtraction and shadow detection. In: Proc. IEEE ICCV Frame-Rate Applications Workshop, pp. 1–19.

Hsiao, J.Y., Sawchuk, A.A., 1989a. Supervised textured image segmentation using feature smoothing and probabilistic relaxation techniques. IEEE Transactions on Pattern Analysis and Machine Intelligence 11 (12), 1279–1292.

Hsiao, J.Y., Sawchuk, A.A., 1989b. Unsupervised textured image segmentation using feature smoothing and probabilistic relaxation techniques. Computer Vision, Graphics, and Image Processing 48, 1–21.

ImageNet, 2015. ImageNet large scale visual recognition challenge (ILSVRC2015). http://image-net.org/challenges/LSVRC/2015/.

Janney, P., Geers, G., 2010. Texture classification using invariant features of local textures. IET Image Processing 4 (3), 158–171.

Kaizer, H., 1955. A quantification of textures on aerial photographs. MS thesis. Boston Univ.

Kalal, Z., Mikolajczyk, K., Matas, J., 2011. Tracking-learning-detection. IEEE Transactions on Pattern Analysis and Machine Intelligence 34 (7), 1409–1422.

Kanatani, K., 1996. Statistical Optimization for Geometric Computation: Theory and Practice. Elsevier, Amsterdam, the Netherlands.

Kitchen, L., Rosenfeld, A., 1982. Gray-level corner detection. Pattern Recognition Letters 1, 95–102.

Krizhevsky, A., Sutskever, I., Hinton, G.E., 2012. ImageNet classification with deep convolutional neural networks. In: Proc. 26th Annual Conf. on Neural Information Processing Systems. Lake Tahoe, Nevada, pp. 3–8.

Laws, K.I., 1979. Texture energy measures. In: Proc. Image Understanding Workshop, Nov., pp. 47–51.

Laws, K.I., 1980a. Rapid texture identification. In: Proc. SPIE Conf. on Image Processing for Missile Guidance, 238. San Diego, Calif. 28 July – 1 Aug., pp. 376–380.

Laws, K.I., 1980b. Textured Image Segmentation. PhD thesis. Univ. of Southern California, Los Angeles.

LeCun, Y., Boser, B., Denker, J.S., Henderson, D., Howard, R.E., Hubbard, W., Jackel, L.D., 1989. Backpropagation applied to handwritten zip code recognition. Neural Computation 1 (4), 541–551.

LeCun, Y., Bottou, L., Bengio, Y., Haffner, P., 1998. Gradient-based learning applied to document recognition. Proceedings of the IEEE 86, 2278–2324.

Lee, S.H., Civera, J., 2019. Triangulation: why optimize? vol. 23. arXiv:1907.11917v2 [cs.CV].

Leibe, B., Leonardis, A., Schiele, B., 2008. Robust object detection with interleaved categorization and segmentation. International Journal of Computer Vision 77 (1), 259–289.

Lipton, A.J., Fujiyoshi, H., Patil, R.S., 1998. Moving target classification and tracking from real-time video. In: Proc. 4th IEEE Workshop on Applications of Computer Vision, pp. 8–14.

Liu, L., Chen, J., Fieguth, P., Zhao, G., Chellappa, R., Pietikäinen, M., 2019. From BoW to CNN: two decades of texture representation for texture classification. International Journal of Computer Vision 127, 74–109.

Lo, B.P.L., Velastin, S.A., 2001. Automatic congestion detection system for underground platforms. In: Proc. Int. Symposium on Intelligent Multimedia, Video and Speech Processing, pp. 158–161.

Longuet-Higgins, H.C., 1981. A computer algorithm for reconstructing a scene from two projections. Nature 293, 133–135.

Magee, M.J., Aggarwal, J.K., 1984. Determining vanishing points from perspective images. Computer Vision, Graphics, and Image Processing 26 (2), 256–267.

Mathias, M., Benenson, R., Pedersoli, M., Van Gool, L., 2014. Face detection without bells and whistles. In: Proc. 13th European Conf. on Computer Vision. Zurich, Switzerland, 8–11 September.

Mirmehdi, M., Xie, X., Suri, J. (Eds.), 2008. Handbook of Texture Analysis. Imperial College Press, London.

Nagel, H.-H., 1983. Displacement vectors derived from second-order intensity variations in image sequences. Computer Vision, Graphics, and Image Processing 21, 85–117.

Noh, H., Hong, S., Han, B., 2015. Learning deconvolution network for semantic segmentation. In: Proc. IEEE Int. Conf. on Computer Vision. Santiago, Chile, 13–16 December, pp. 1520–1528.

Pietikäinen, M., Rosenfeld, A., Davis, L.S., 1983. Experiments with texture classification using averages of local pattern matches. IEEE Transactions on Systems, Man and Cybernetics 13 (3), 421–426.

Ren, S., He, K., Girshick, R., Sun, J., 2015. Faster R-CNN: towards real-time object detection with region proposal networks. arXiv:1506.01497 [cs.CV].

Rosenfeld, A., Troy, E.B., 1970a. Visual Texture Analysis. Computer Science Center, Univ. of Maryland. Techn. Report TR-116.

Rosenfeld, A., Troy, E.B., 1970b. Visual texture analysis. In: Conf. Record for Symposium on Feature Extraction and Selection in Pattern Recognition, IEEE Publication 70C-51C, Argonne, Ill., pp. 115–124.

Russakovsky, O., Deng, J., Su, H., Krause, J., Satheesh, S., Ma, S., Huang, Z., Karpathy, A., Khosla, A., Bernstein, M., Berg, C., Fei-Fei, L., 2015. ImageNet large scale visual recognition challenge. International Journal of Computer Vision 115 (3), 211–252.

Shah, M.A., Jain, R., 1984. Detecting time-varying corners. Computer Vision, Graphics, and Image Processing 28, 345–355.

Simonyan, K., Zisserman, A., 2015. Very deep convolutional networks for large-scale image recognition. arXiv:1409.1556v6.

Stalder, S., Grabner, H., van Gool, L., 2009. Beyond semi-supervised tracking: tracking should be as simple as detection, but not simpler than recognition. In: IEEE 12Th Int. Conf. on Workshop on On-Line Learning for Computer Vision.

Szegedy, C., Liu, W., Jia, Y., Sermanet, P., Reed, S., Anguelov, D., Erhan, D., Vanhoucke, V., Rabinovich, A., 2014. Going deeper with convolutions. arXiv:1409.4842v1 [cs.CV].

Tighe, J., Lazebnik, S., 2013. Finding things: image parsing with regions and per-exemplar detectors. In: Proc. IEEE Conf. on Computer Vision and Pattern Recognition. Portland, Oregon, 23–28 June, pp. 3001–3008.

Torr, P., Zisserman, A., 1997. Performance characterization of fundamental matrix estimation under image degradation. Machine Vision and Applications 9 (5), 321–333.

Unser, M., 1986. Local linear transforms for texture measurements. Signal Processing 11, 61–79.

Unser, M., Eden, M., 1989. Multiresolution feature extraction and selection for texture segmentation. IEEE Transactions on Pattern Analysis and Machine Intelligence 11 (7), 717–728.

Unser, M., Eden, M., 1990. Nonlinear operators for improving texture segmentation based on features extracted by spatial filtering. IEEE Transactions on Systems, Man and Cybernetics 20 (4), 804–815.

Vistnes, R., 1989. Texture models and image measures for texture discrimination. International Journal of Computer Vision 3, 313–336.

Wu, Z., Thangali, A., Sclaroff, S., Betke, M., 2012. Coupling detection and data association for multiple object tracking. In: Proc. IEEE Conf. CVPR, pp. 1948–1955.

Yang, Y., Li, Z., Zhang, L., Murphy, C., Ver Hoeve, J., Jiang, H., 2012. Local label descriptor for example based semantic image labelling. In: Proc. 12th European Conf. on Computer Vision. Florence, Italy, 7–13 October, pp. 361–375.

Zeiler, M., Fergus, R., 2014. Visualizing and understanding convolutional neural networks. In: Proc. 13th European Conf. on Computer Vision. Zurich, Switzerland, 8–11 September.

Zhang, L., Wu, B., Nevatia, R., 2007. Pedestrian detection in infrared images based on local shape features. In: Proc. 3rd Joint IEEE Int. Workshop on Object Tracking and Classification in and Beyond the Visible Spectrum.

Zuniga, O.A., Haralick, R.M., 1983. Corner detection using the facet model. In: Proc. IEEE Conf. on Computer Vision and Pattern Recognition, pp. 30–37.

Biographies

Roy Davies is Emeritus Professor of Machine Vision at Royal Holloway, University of London. He has worked on many aspects of vision, from feature detection and noise suppression to robust pattern matching and real-time implementations of practical vision tasks. His interests include automated visual inspection, surveillance, vehicle guidance and crime detection. He has published more than 200 papers and three books – *Machine Vision: Theory, Algorithms, Practicalities* (1990), *Electronics, Noise and Signal Recovery* (1993), and *Image Processing for the Food Industry* (2000); the first of these has been widely used internationally for more than 25 years and is now out in a much enhanced fifth edition. Roy is a Fellow of the IoP and the IET, and a Senior Member of the IEEE. He is on the Editorial Boards of *Pattern Recognition Letters*, *Real-Time Image Processing*, *Imaging Science* and *IET Image Processing*. He holds a DSc at the University of London: he was awarded *BMVA Distinguished Fellow* in 2005 and *Fellow of the International Association of Pattern Recognition* in 2008.

Advanced methods for robust object detection

Zhaowei Cai[a] *and Nuno Vasconcelos*[b]

[a]Amazon Web Services, Pasadena, CA, United States [b]University of California San Diego, Department of Electrical and Computer Engineering, San Diego, CA, United States

CHAPTER POINTS

- This chapter introduces the problem of object detection in computer vision.
- This chapter reviews some advanced object detectors based on deep neural networks and some techniques that have become important in the detection literature in recent years.

2.1 Introduction

Object detection is one of the most fundamental and challenging problems of computer vision. It generalizes the more commonly studied problem of object classification. Given an image, object recognition aims to answer the question of "what". What are the objects in the image? For example, the image of Fig. 2.1 includes a person and a boat. Beyond "what", object detection also aims to answer the question of "where". What regions of the image contain the object? This is illustrated in Fig. 2.1(b), where bounding boxes are used to demarcate the region of each object.

Object detection has many practical applications. For example, autonomous vehicles rely on object detection to localize objects, understand the surrounding environment, and help make safe decisions. In medical imaging, object detectors can help locate lesions in medical scans, alleviating the burden of radiologists and other medical specialists. However, object

Advanced Methods and Deep Learning in Computer Vision
https://doi.org/10.1016/B978-0-12-822109-9.00011-4

Copyright © 2022 Elsevier Inc. All rights reserved.

(a) Object Classification   (b) Object Detection

FIGURE 2.1 The difference between object classification and object detection.

detection is also an upstream task in computer vision, supporting many other downstream tasks, like visual question answering, captioning, visual navigation, robot grasping, pose estimation, etc. For example, object detection can not only help a robot accurately grasp objects in the physical world, but also enable it to understand the semantics of that object, how it relates to the other objects in the scene, and how it can play a part in solving a task, either alone or in a team of robots. Hence, the advancement of object detection will benefit many other domains of computer vision and enable more effective computer vision systems in general.

An object detector faces many challenges. For example, it is required to accurately detect objects of multiple categories, scales, aspect ratios, etc, sometimes under severe lighting conditions, occlusion, and background distractors. This makes it challenging to develop detectors robust enough to work well under most conditions, a necessary condition to mimic the human visual system.

Due to the importance of the problem, there is a long history of research in object detection (Sung and Poggio, 1998; Rowley et al., 1996; Papageorgiou et al., 1998). Early work focused on the detection of specific objects, namely faces and humans, of importance for many applications. The Viola-Jones (VJ) detector (Viola and Jones, 2001, 2004) was a milestone among these works. It was the first real-time object detector for unconstrained environments and much faster than all other competitive detectors at the time. The idea was to formulate the detector as a cascade of classifiers, which reject object hypotheses stage by stage, using very simple Haar wavelet features. By adding more features in later stages, the cascade can form a powerful detector. However, because most hypotheses can be rejected with simple features (early stages) the average computation is small. Although wavelet features are fast, they are not very accurate. Later works proposed the histogram of oriented gradients (HOG) (Dalal and Triggs, 2005), as an important improvement of the scale-invariant feature transform (SIFT) feature (Lowe, 1999, 2004). HOG has shown very impressive performance, originally on human detection, and was later widely adopted in various object detection problems. A breakthrough on the detection of general object was the introduction of the deformable part-based model (DPM) (Felzenszwalb et al., 2010). This built on HOG features, representing each object as a combination of a root model and deformable parts, where the configurations of part filters were latent variables, learned automatically. DPM was the winner of the Pascal VOC object detection challenge (Everingham et al., 2010), which addresses the detection of 20 object categories such as desk, bus, human, or bike, in 2007, 2008 and 2009. This made it the default framework for object detection research before the introduction of deep learning.

In recent years, it has been shown that learned feature representations extracted by deep convolutional neural networks (CNN) are vastly superior to even the best handcrafted features, like SIFT, HOG, and Haar wavelets. Although the vanilla CNN feature representations have strong classification performance, their application to object detection requires nontrivial extensions. Unlike classification, object detection requires the solution of two tasks. First, the detector must solve the *recognition* problem, distinguishing foreground objects from background and assigning them the proper object class labels. Second, the detector must solve the *localization* problem, assigning accurate bounding boxes to different objects. In this chapter, we review CNN-based object detection frameworks proposed in the past few years. These can be split into two major groups: two-stage object detectors, such as the R-CNN (Girshick et al., 2014), SPP-Net (He et al., 2014), Fast R-CNN (Girshick, 2015), Faster R-CNN (Ren et al., 2017), MS-CNN (Cai et al., 2016), FPN (Lin et al., 2017a), and Cascade R-CNN (Cai and Vasconcelos, 2021), and single-stage object detectors, including YOLO (Redmon et al., 2016), SSD (Liu et al., 2016) and RetinaNet (Lin et al., 2017b).

2.2 Preliminaries

Most modern object detectors implement a combination of classification and bounding box regression. Classification attempts to predict the class of the object in a image region, bounding box regression attempts to determine the region, by predicting the tightest box that contains the object. Consider a ground truth object of bounding box $\mathbf{g}$ associated with class label y, and a detection hypothesis $\mathbf{x}$ of bounding box $\mathbf{b}$. Since $\mathbf{b}$ usually includes an object and some amount of background, it can be difficult to determine if the detection is correct. This is usually addressed by the intersection over union (IoU) metric

$$IoU(\mathbf{b}, \mathbf{g}) = \frac{\mathbf{b} \cap \mathbf{g}}{\mathbf{b} \cup \mathbf{g}}. \tag{2.1}$$

If the IoU is above a threshold u, $\mathbf{x}$ is considered an example of the class of the object of bounding box $\mathbf{g}$ and denoted a "positive" example. Thus, the class label of a hypothesis $\mathbf{x}$ is a function of u,

$$y_u = \begin{cases} y, & IoU(\mathbf{b}, \mathbf{g}) \geq u \\ 0, & \text{otherwise.} \end{cases} \tag{2.2}$$

If the IoU does not exceed the threshold for any object, $\mathbf{x}$ is assigned to the background and denoted a "negative" example.

Although there is no need to define positive/negative examples for the bounding box regression task, an IoU threshold u is also required to select the set of samples

$$\mathcal{G} = \left\{ (\mathbf{g}_i, \mathbf{b}_i) | IoU(\mathbf{b}_i, \mathbf{g}_i) \geq u \right\} \tag{2.3}$$

used to train the regressor. While the IoU thresholds used for the two tasks do not have to be identical, this is usual in practice. Hence, the IoU threshold u defines the *quality* of a detector.

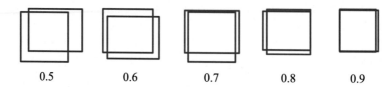

0.5   0.6   0.7   0.8   0.9

FIGURE 2.2 Examples of increasing qualities. The numbers are the IoU of (2.1) between two bounding boxes, indicating how well they are overlapped with each other.

Large thresholds encourage detected bounding boxes to be tightly aligned with their ground truth counterparts. Small thresholds reward detectors that produce loose bounding boxes, of small overlap with the ground truth. Some examples of hypotheses of increasing quality are shown in Fig. 2.2.

2.3 R-CNN

The R-CNN (Girshick et al., 2014) (Regions with CNN features) was a pioneering effort on the use of deep neural networks for general object detection. It was the first work to outperform DPM-style methods by leveraging powerful CNN feature representations. It also demonstrated that CNN features pretrained for classification on ImageNet (Russakovsky et al., 2015) could be successfully finetuned to other downstream tasks, e.g., detection, segmentation, etc.

2.3.1 System design

The R-CNN consists of three modules, shown in Fig. 2.3. Since CNN computations are expensive, the first step is to generate category-independent region proposals, using Selective Search (van de Sande et al., 2011). These proposals define the set of candidate detections available to the detector, reducing the number of detection hypotheses from the order of millions to thousands. The second step is feature extraction from each proposal region, using a CNN trained for recognition, e.g., AlexNet (Krizhevsky et al., 2012) or VGG-Net (Simonyan and Zisserman, 2014). The detected proposals, of arbitrary scales and sizes, are first cropped and warped to the size of the network input. The resized images are then forwarded through the CNN and the output of the penultimate network layer is used as feature representation for each proposal. Finally, the third module, implemented with class specific linear SVMs, produces class predictions for the proposals. For better localization, additional bounding-box regressors are applied to refine the bounding boxes of detected objects.

2.3.2 Training

The CNN used for feature extraction is pretrained on the ImageNet (Russakovsky et al., 2015) classification task, and finetuned to the warped proposal regions used in the detection task. Finetuning is a $K + 1$-way classification problem, with K object categories and one back-

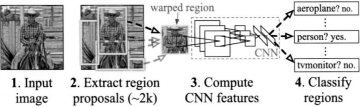

R-CNN: *Regions with CNN features*

1. Input image   **2.** Extract region proposals (~2k)   **3.** Compute CNN features   **4.** Classify regions

FIGURE 2.3 System overview of R-CNN.

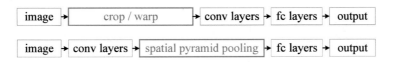

FIGURE 2.4 The pipeline comparison of R-CNN (top) and SPP-Net (bottom).

ground class (e.g., $K = 20$ for the VOC dataset (Everingham et al., 2010) and $K = 80$ for the COCO dataset (Lin et al., 2014)). Since CNNs are data hungry, it is not sufficient to use the ground truth only as positive examples. The solution is to use all proposals with IoU above 0.5 with the nearest ground-truth box as positives, and the remainder as negatives. During training, it is important to keep a balanced ratio between positives and negatives. In practice, positives and negatives are sampled uniformly from the sample pool with a ratio of 1:3 in each training batch. Finetuning minimizes the cross-entropy loss

$$L_{cls}\big(h(\mathbf{x}), y\big) = -\log h_y(\mathbf{x}) \tag{2.4}$$

where $\mathbf{x}$ is the proposal to classify, y the class label, and $h(\mathbf{x})$ the classifier. After finetuning, the category-specific linear SVM classifiers and bounding box regressors are trained on the proposal features generated by the CNN. This multistage training procedure of feature extraction, CNN fine-tuning with cross-entropy loss, SVM training, and fitting bounding-box regressors, is slow, tedious and inelegant.

2.4 SPP-Net

While the R-CNN significantly boosted general object detection performance, it is a complex detector, since the expensive CNN computation is repeated for the thousands of proposals derived from each single image. As a result, it can take more than 30 seconds to run the R-CNN detector on each image. Since the proposals extracted from an image share most pixels, most of these computations are redundant. This redundancy was reduced by the SPP-Net (He et al., 2014), which shared computation among proposals.

Different from the R-CNN pipeline, which crops proposals before CNN computation, as shown in Fig. 2.4 (top), the SPP-Net forwards the whole image through the convolutional

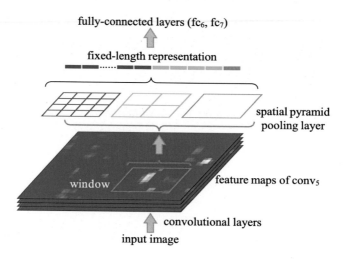

FIGURE 2.5 The pipeline of SPP-Net for object detection.

layers of the network. Spatial pyramid pooling (Lazebnik et al., 2006) is then used to extract the fixed-length features of the cropped feature maps associated with each proposal. These fixed-length features are finally input to a set of fully connected layers for final prediction, as shown in Fig. 2.4 (bottom). This simple change enables the sharing of the expensive CNN computations among proposals, and the whole image is processed only once. The important operation is the spatial pyramid pooling (SPM), operation of Fig. 2.5, which maps the instance-wise features of arbitrary scale and size into a fixed-length vector. This was the first showing that features from of a convolutional feature map can be pooled over a spatial region to produce an instance-wise feature representation with good properties for instance recognition. This inspired latter works, such as the Fast R-CNN.

2.5 Fast R-CNN

The SPP-Net inherits the tedious multistage training procedure of R-CNN. This is significantly simplified in the Fast R-CNN. In this approach, the extraction of features, fine-tuning of the network to a new task, instance-wise classification and bounding-box regression are all integrated into a unified framework, enabling easy to use deep learning based object detection.

2.5.1 Architecture

The pipeline of the Fast R-CNN is shown in Fig. 2.6. Similar to the SPP-Net, the Fast R-CNN forwards the whole image through the convolutional layers of a CNN to generate feature maps. Next, a region of interest (RoI) pooling layer is used to extract a vector of fixed-length features for each object proposal. Finally, two fully connected (FC) layers are used to

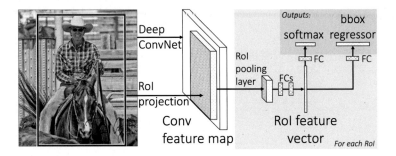

FIGURE 2.6 Pipeline of Fast R-CNN.

make the final predictions: classification probabilities for $K + 1$ classes and regression of four bounding box coordinates. Unlike the R-CNN and the SPP-Net, the Fast R-CNN is trained end-to-end with a multitask loss, avoiding the tedious multistage training procedure.

2.5.2 RoI pooling

The RoI pooling operation is a simpler version of the spatial pyramid pooling of Fig. 2.5. Instead of pooling a RoI with height and width (h, w) to different spatial resolutions (1×1, 2×2 and 4×4) and concatenating them together, as in SPP, RoI pooling uses a single $H \times W$ resolution, e.g., 7×7. Given a RoI window of size $h \times w$, RoI pooling divides it into $H \times W$ subwindows, each of size $h/H \times w/W$. Max-pooling is then used inside each subwindow to extract the largest feature value. This process is applied independently to each feature map channel. Although RoI pooling is simpler than SPM, it can still effectively extract a powerful feature representation for proposals of arbitrary size and scale from the precomputed convolutional feature maps. This is the critical requirement for object detection.

2.5.3 Multitask loss

The Fast R-CNN is trained with two learning tasks: classification and bounding box regression. These are jointly optimized during training with recourse to a multitask loss function

$$L = L_{cls}\big(h(\mathbf{x}), y\big) + \lambda[y \geq 1]L_{loc}\big(f(\mathbf{x}, \mathbf{b}), \mathbf{g}\big), \tag{2.5}$$

where λ controls the balance between the two requirements. $[y \geq 1]$ equals to 1 when $y \geq 1$ and 0 otherwise, meaning that there is no bounding box regression loss for background class.

Classification

The classifier is a function $h(\mathbf{x})$ that assigns an image patch $\mathbf{x}$ to one of $K + 1$ classes, where class 0 contains background and the remaining classes the objects to detect. $h(\mathbf{x})$ is a $K + 1$-dimensional estimate of the posterior distribution over classes, i.e., $h_k(\mathbf{x}) = p(y = k|\mathbf{x})$, where y is the class label. L_{cls} is the cross-entropy loss of (2.4).

Bounding box regression

A bounding box $\mathbf{b} = (b_x, b_y, b_w, b_h)$ contains the four coordinates of an image patch $\mathbf{x}$. Bounding box regression aims to regress a candidate bounding box $\mathbf{b}$ into a target bounding box $\mathbf{g}$, using a regressor $f(\mathbf{x}, \mathbf{b})$. The regression loss function is

$$L_{loc}(\mathbf{a}, \mathbf{b}) = \sum_{i \in \{x,y,w,h\}} smooth_{L_1}(a_i - b_i), \tag{2.6}$$

where

$$smooth_{L_1}(x) = \begin{cases} 0.5x^2, & |x| < 1, \\ |x| - 0.5, & \text{otherwise,} \end{cases} \tag{2.7}$$

is the smooth L_1 loss function. It is a combination of L_1 and L_2 losses, which behaves as L_1 loss when $|x| < 1$ and L_2 loss otherwise. It fixes the nonsmooth behavior of L_1 loss, i.e., the gradient is -1 when x is negative and $+1$ otherwise. The smooth L_1 loss could enable more stable learning behavior.

To encourage invariance to scale and location, $smooth_{L_1}$ operates on the distance vector $\Delta = (\delta_x, \delta_y, \delta_w, \delta_h)$ defined by

$$\begin{aligned} \delta_x &= (g_x - b_x)/b_w, & \delta_y &= (g_y - b_y)/b_h \\ \delta_w &= \log(g_w/b_w), & \delta_h &= \log(g_h/b_h). \end{aligned} \tag{2.8}$$

Since bounding box regression usually performs minor adjustments on $\mathbf{b}$, the numerical values of (2.8) can be very small. This usually makes the regression loss much smaller than the classification loss. To improve the effectiveness of multitask learning, Δ is normalized by its mean and variance, e.g., δ_x is replaced by

$$\delta'_x = \frac{\delta_x - \mu_x}{\sigma_x}. \tag{2.9}$$

2.5.4 Finetuning strategy

Sampling

Both the R-CNN and SPP-Net sample RoIs. This could lead to very inefficient training, since RoIs are extracted from different images and each image requires a full CNN forward computation. To avoid the problem, the Fast R-CNN first samples N images, from which it then samples R RoIs per image. By choosing $N \ll R$ it is possible to only require CNN computations for a small number (N) of images. However, this raises the concern that the sampled RoIs are correlated, which could slow down training convergence. In practice, however, this strategy has been found to work well (Girshick, 2015; Ren et al., 2017). The RoIs are sampled from each image so as to produce 25% positive and 75% negative training examples.

Back-propagation through RoI pooling

Another important improvement of the Fast R-CNN over the SPP-net was the backpropagation of the gradient through the RoI pooling layer. In the absence of this, the convolutional

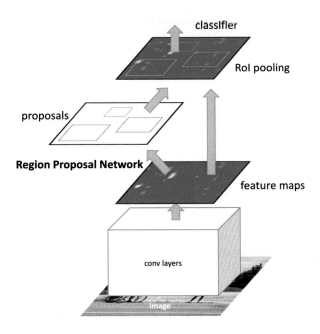

FIGURE 2.7 The architecture of Faster R-CNN.

layers below the RoI pooling layer will not be finetuned to the detection task, as is the case for the SPP-net. Since, in RoI pooling, each output feature is the max-pooling of the corresponding subwindow on the feature map, the back-propagation computations reduce to those of the max-pooling operation. Namely, the output gradient is only back-propagated to the position of largest max feature value in the subwindow. This strategy is applied to every RoI feature of every region proposal.

2.6 Faster R-CNN

The SPP-Net and Fast R-CNN significantly improved the running speed of the R-CNN, from about 30 to about 2 seconds per image. They made the generic proposal detection stage, which relied on low-level features, like pixels and edge, and ran on a CPU, the speed bottleneck. For example, the selective search proposal detector requires about 2 seconds per image. The Faster R-CNN addressed this problem, by introducing a region proposal network (RPN) that runs GPUs and shares feature computations with the Fast R-CNN network.

2.6.1 Architecture

As shown in Fig. 2.7, the Faster R-CNN consists of two modules: a region proposal network (RPN) that proposes regions and is fully convolutional, and the Fast R-CNN detector that classifies these proposals. Unlike the R-CNN and Fast R-CNN frameworks, the entire system

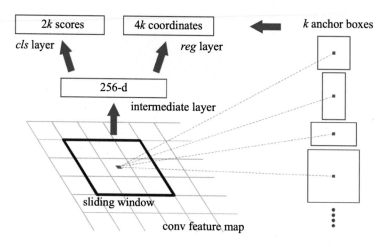

FIGURE 2.8 The illustration of Region Proposal Network.

is a single, unified and end-to-end network for object detection. Since the RPN shares most of its computations with the Fast R-CNN network, the RPN adds little additional computational cost. This allows the Faster R-CNN to eliminate the proposal generation time and runs in real-time on a modern GPU.

2.6.2 Region proposal networks

Region proposals are detected by sliding a small network over the convolutional feature map, as shown in Fig. 2.8. This small network is implemented with a 256-dim 3×3 convolutional layer, a ReLU layer and two fully-connected sibling output layers. Similar to the final output layers of Fast R-CNN, the first output layer is for binary (foreground v.s. background) classification, and the second one for bounding box regression. This produces an "objectness" score and 4 coordinates for a given anchor. According to the objectness score, the top 300 proposals are generated by RPN, and will be used in the later Fast R-CNN stage.

Anchors

Each sliding-window location should, in principle, generate a single prediction, since each location on the feature map corresponds to a single location on the input image. However, the RPN simultaneously predicts k region proposals per each sliding-window location, to account for different object sizes and aspect ratios. This is possible with the concept of *anchors*. At a single location, a proposal prediction is associated with an anchor, which is centered at the center of the sliding window and has its own scale and aspect ratio. A common practice is to use $k = 9$ anchors, of 3 different scales and 3 different aspect ratios, per single sliding window location. Each anchor produces a 6-dim proposal, with four dimensions encoding coordinates for bounding box regression and the remaining two foreground and background class probabilities.

Training

The loss function of the RPN is the same as (2.5). Anchor bounding boxes are required to regress to the associated ground truth boxes. To balance the learning, anchors are sampled during training so that there is a ratio of 1:1 between positive and negative anchors. Note that the ratio here is different the ratio of 1:3 in the training of Fast R-CNN in Section 2.5, because the task of RPN is to detect proposals as many as possible. With a higher ratio of positives, the model will be encouraged to detect more positives. With the top 300 proposals generated by RPN, the training of the Fast R-CNN detector remains as above. The convolutional feature computations are shared between the RPN and the Fast R-CNN, and the whole network can be trained end-to-end by standard backpropagation and stochastic gradient descent (SGD).

2.7 Cascade R-CNN

The detection problem is difficult, partly due to the fact that there are many "close" false positives, corresponding to "close but not correct" bounding boxes. An effective detector must find all true positives in an image, while suppressing these close false positives. The quality of a detection hypothesis is defined by its IoU with the ground truth, and the quality of a detector as the IoU threshold used to train it.

High quality detection

The challenge is that, no matter the choice of IoU threshold u, the detection setting is highly adversarial. When u is high, positives contain little background but it is difficult to assemble large positive training sets. When u is low, richer and more diverse positive training sets are possible, but the trained detector has little incentive to reject close false positives. In general, it is very difficult to guarantee that a single classifier performs uniformly well over all IoU levels. Furthermore, since at inference time the majority of the hypotheses produced by proposal detectors (such as the RPN or selective search) have low quality, the object detector at the top of the network must be discriminant for lower quality hypotheses. A standard compromise between these conflicting requirements is to settle on $u = 0.5$, which is used in almost *all* modern object detectors. This, however, is a relatively low threshold, making it difficult to train detectors that can effectively reject close false positives.

In general, the detector will only achieve high quality if presented with high quality proposals. This, however, cannot be guaranteed by simply increasing the threshold u during training. On the contrary, forcing a high value of u usually degrades detection performance. This problem, i.e., that training a detector with higher threshold leads to poorer performance, is referred as the *paradox of high-quality detection*. It has two causes. First, object proposal mechanisms tend to produce hypotheses distributions heavily imbalanced towards low quality. As a result, the use of larger IoU thresholds during training exponentially reduces the number of positive training examples. This is particularly problematic for neural networks, which are very example intensive, making the "high u" training strategy very prone to overfitting. Second, there is a mismatch between the quality of the detector and that of the hypotheses available at inference time. Since high quality detectors are only optimal for high quality hy-

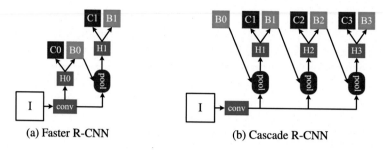

(a) Faster R-CNN   (b) Cascade R-CNN

FIGURE 2.9 Cascade R-CNN is the multistage extension of Faster R-CNN. "I" is input image, "conv" backbone convolutions, "pool" region-wise feature extraction, "H" network head, "B" bounding box, and "C" classification. "B0" is proposals in all architectures.

potheses, detection performance can degrade substantially for hypotheses of lower quality. The Cascade R-CNN addresses this problem, so as to enable object detectors of high quality.

2.7.1 Architecture

The Cascade R-CNN (Cai and Vasconcelos, 2021) is a multistage extension of the Faster R-CNN, as shown in Fig. 2.9. Rather than a single detector, it uses a cascade of detector stages, which are sequentially more selective against close false positives. The IoU thresholds are typically 0.5, 0.6 and 0.7 for the different detection heads. The cascade of R-CNN stages is trained sequentially, using the output of one stage to train the next. This leverages the observation that the output IoU of a bounding box regressor is almost always better than its input IoU. As a result, the output of a detector trained with a certain IoU threshold is a good hypothesis distribution to train the detector of the next higher IoU threshold. By adjusting bounding boxes, each stage aims to find a good set of close false positives for training the next stage. The main outcome of this resampling is that the quality of the detection hypotheses increases gradually from one stage to the next. As a result, the sequence of detectors addresses the two problems underlying the paradox of high-quality detection. First, because the resampling operation guarantees the availability of a large number of examples for the training of all detectors in the sequence, it is possible to train detectors of high IoU without overfitting. Second, the use of the same cascade procedure at inference time produces a set of hypotheses of progressively higher quality, well matched to the increasing quality of the detector stages. This enables higher detection accuracies.

2.7.2 Cascaded bounding box regression

Since it is difficult for a single regressor to perform uniformly well over all quality levels, the regression task is decomposed into a sequence of simpler steps in the Cascade R-CNN. This consists of a cascade of specialized regressors

$$f(\mathbf{x}, \mathbf{b}) = f_T \circ f_{T-1} \circ \cdots \circ f_1(\mathbf{x}, \mathbf{b}), \tag{2.10}$$

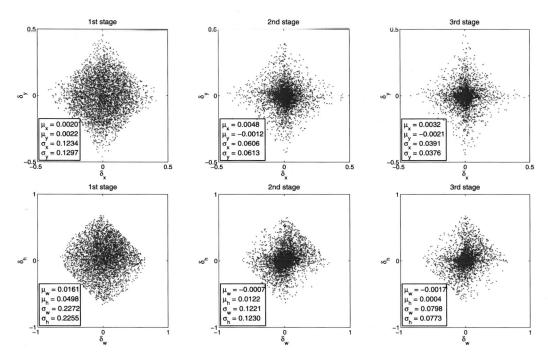

FIGURE 2.10 Distribution of the distance vector Δ of (2.8) (without normalization) at different cascade stages. Top: plot of (δ_x, δ_y). Bottom: plot of (δ_w, δ_h). Red dots are outliers for the increasing IoU thresholds of later stages, and the statistics shown are obtained after outlier removal.

where T is the total number of cascade stages. The key point is that each regressor f_t is optimized for the bounding box distribution $\{b^t\}$ generated by the previous regressor, rather than the initial distribution $\{b^1\}$. In this way, the hypotheses are improved *progressively*. The efficacy of cascade regression is illustrated in Fig. 2.10, which presents the distribution of the regression distance vector $\Delta = (\delta_x, \delta_y, \delta_w, \delta_h)$ of (2.8) at different cascade stages. Note that most hypotheses become closer to the ground truth as they progress through the cascade.

2.7.3 Cascaded detection

It is difficult to train a high quality detector directly. The Cascade R-CNN addresses the problem by leveraging cascade regression as a resampling mechanism. Starting from examples $\{(x_i, b_i)\}$, cascade regression is used to successively resample an example distribution $\{(x'_i, b'_i)\}$ of higher IoU. This enables the sets of positive examples of the successive stages to keep a roughly *constant* size, as the detector quality u is increased.

At each stage t, the R-CNN head includes a classifier h_t and a regressor f_t optimized for the corresponding IoU threshold u^t, where $u^t > u^{t-1}$. These are learned with loss

$$L(x^t, g) = L_{cls}(h_t(x^t), y^t) + \lambda[y^t \geq 1]L_{loc}(f_t(x^t, b^t), g), \tag{2.11}$$

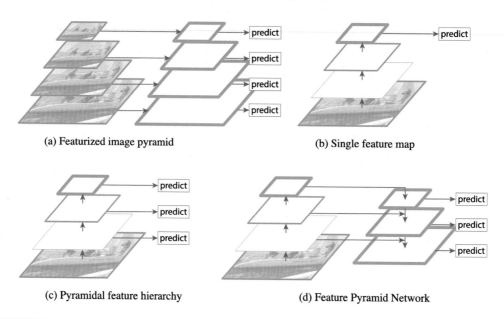

(a) Featurized image pyramid   (b) Single feature map

(c) Pyramidal feature hierarchy   (d) Feature Pyramid Network

FIGURE 2.11 (a) Image pyramid: features are computed on each image scale independently. (b) Single feature map: object detection operates on only single scale feature map in CNN. (c) Feature pyramid: a pyramid of features for multiscale detection but with a single scale of image input. (d) Feature Pyramid Network (FPN): FPN adds top-down connections to feature pyramid of (c), enabling more scale-invariant semantic feature representation at different scales.

where $\mathbf{b}^t = f_{t-1}(\mathbf{x}^{t-1}, \mathbf{b}^{t-1})$, $\mathbf{g}$ is the ground truth object for $\mathbf{x}^t$, $\lambda = 1$ the trade-off coefficient, y^t the label of $\mathbf{x}^t$ under the u^t criterion, according to (2.2), and $[\cdot]$ the indicator function. Note that the use of $[\cdot]$ implies that the IoU threshold u of bounding box regression is identical to that used for classification. This cascaded learning has two important consequences for detector training. First, the potential for overfitting at large IoU thresholds u is reduced, since positive examples become plentiful at all stages. Second, detectors of deeper stages are optimal for higher IoU thresholds. This simultaneous improvement of hypotheses and detector quality enables the Cascade R-CNN to beat the paradox of high quality detection. At inference, the same cascade is applied. The quality of the hypotheses is improved sequentially, and higher quality detectors are only required to operate on higher quality hypotheses, for which they are optimal.

2.8 Multiscale feature representation

Recognizing objects at various scales is a fundamental challenge in computer vision. A classical solution in the literature is to rely on image pyramids, such as those shown in Fig. 2.11(a), where the original image is resized to different scales from which features are extracted (Viola and Jones, 2004; Felzenszwalb et al., 2010; Dollár et al., 2014). By applying a

FIGURE 2.12 In natural images, objects can appear at very different scales, as illustrated by the yellow bounding boxes. A single receptive field of a fixed-size filter (shown in the shaded area) cannot match this variability.

fixed scale detector to all the pyramid feature representations it is then possible to detect objects at different scales, without loss of accuracy. Small (large) objects are detected in the large (small) resolution channels of the feature pyramid. Nevertheless, building a pyramid of CNN features is computationally expensive, making this solution impractical for most real applications. The design of effective CNN feature representations of various scales is an important research direction for object detection.

Despite the great success of deep learning based object detectors (Girshick et al., 2014; Girshick, 2015; He et al., 2014; Ren et al., 2017), there has been limited progress towards the detection of objects at multiple scales. As discussed above, R-CNN, SPP-Net and Fast R-CNN sample object proposals at multiple scales, using a preliminary attention stage, e.g., Selective Search (van de Sande et al., 2011), and then warp these proposals to a fixed size supported by the CNN. This pushes the scale invariance problem to the attention stage, which is not trained jointly with the CNN. The Faster R-CNN (Ren et al., 2017) addresses the issue of joint training, using the RPN to generate proposals of multiple scales. However, as shown in (Fig. 2.11(b)), this is done by sliding a fixed set of filters over a single set of convolutional feature maps. This creates an inconsistency between objects of variable size and filters of fixed receptive field. As shown in Fig. 2.12, a fixed receptive field cannot cover the multiple scales at which objects appear in natural scenes. As a result, detection performance is compromised, in particular for small objects like that in the center of Fig. 2.12, which are quite difficult to detect. Several networks have been proposed to extend the two-stage detector architecture by introducing multiscale extensions of the RPN.

2.8.1 MS-CNN

The MS-CNN was proposed to address the problem of multiscale object detection. It pursues an alternative strategy to the expensive computation of image pyramids, leveraging the fact that deep neural networks already compute a feature hierarchy layer by layer. Given that higher-level layers are subsampled, this hierarchy even has a multiscale pyramidal structure. Hence, the inconsistency between the sizes of objects and receptive fields

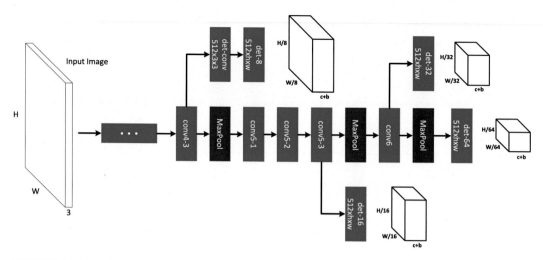

FIGURE 2.13 The feature pyramid architecture of the MS-CNN. The cuboids are the output tensors of the network (except the input image). $h \times w$ is the filter size, c the number of classes, and b the number of bounding box coordinates.

can be addressed by simply adding output layers at multiple stages of the network, as shown in Fig. 2.11(c). In this way, the MS-CNN implements several detectors that specialize in different scale ranges. While detectors based on lower network layers, such as "conv-3," have smaller receptive fields and are better matched to detect small objects, those based on higher layers, such as "conv-5," are best suited for the detection of large objects. The complementary detectors at the outputs of different layers are combined into a strong multiscale detector.

2.8.1.1 Architecture

The detailed architecture of the MS-CNN proposal network is shown in Fig. 2.13. The network detects objects through several detection branches, which are merged into a final set of proposals. It has a standard CNN trunk, depicted in the center of the figure, and a set of output branches, which emanate from different layers of the trunk. These branches consist of a single detection layer. Note that a buffer convolutional layer is introduced on the branch that emanates after layer "conv4-3". Since this branch is close to the lower layers of the trunk network, it affects their gradients more than the other detection branches. This can lead to some instability during learning. The buffer convolution prevents the gradients of the detection branch from being back-propagated directly to the trunk layers.

During training, the parameters $\mathbf{W}$ of the multiscale proposal network are learned from a set of training samples $S = \{(X_i, Y_i)\}_{i=1}^{N}$, where X_i is a training image patch, and $Y_i = (y_i, b_i)$ the combination of its class label $y_i \in \{0, 1, 2, \cdots, K\}$ and bounding box coordinates $b_i = (b_i^x, b_i^y, b_i^w, b_i^h)$. This is achieved with a multitask loss

$$\mathcal{L}(\mathbf{W}) = \sum_{m=1}^{M} \sum_{i \in S^m} \alpha_m l^m (X_i, Y_i | \mathbf{W}), \tag{2.12}$$

where M is the number of detection branches, l^m the multitask loss combining classification and bounding box regression of (2.5), α_m the weight of loss l^m, and $S = \{S^1, S^2, \cdots, S^M\}$, where S^m contains the examples of scale m. Note that only a subset S^m of the training samples, selected by scale, contributes to the loss of detection layer m.

2.8.2 FPN

Although the feature pyramid of Fig. 2.11(c) improves the representation of different object scales, it introduces a large semantic gap between these representations. While high-resolution maps contain information about features of low-level semantics, such as edges, corners, etc., low-resolution maps convey semantically rich information, such as object category. Hence, high-resolution (low-resolution) maps are mostly informative of object location (identity). By asking all feature representations to contribute to the localization and classification tasks, the architecture of Fig. 2.11(c) can have suboptimal detection performance.

To reduce these semantic gaps, the feature pyramid network (FPN) adds a top-down connection from the high-level (semantically richer) feature maps to the low-level (semantically poorest) feature maps, as shown in Fig. 2.11(d). This ensures that the feature pyramid has strong semantics at all pyramid levels. Like Fig. 2.11(c), the FPN is an in-network pyramid and thus efficient, but the top-down connections increase its representation power.

2.8.2.1 Architecture

The major difference between the FPN and standard bottom-up classification networks, like the ResNet (He et al., 2016), is the addition of a top-down pathway, to enable semantically rich feature pyramids. This is implemented with a simple set of building blocks, shown in Fig. 2.14.

Bottom-up pathway

A standard feed-forward network is naturally a bottom-up pyramid, due to the use of downsampling operations like pooling, convolution with stride of 2, etc. In general, the resolution of feature maps is reduced by 2 times at every network stage, where a stage is defined as a sequence of layers with the same resolution. The FPN builds on the ResNet, extracting the bottom-up pyramid from the activations of the last layer of each ResNet stage. Specifically, the outputs of layers conv2, conv3, conv4, and conv5 of the ResNet, denoted as $\{C_2, C_3, C_4, C_5\}$, are used to create a pyramid of stride $\{4, 8, 16, 32\}$ pixels with respect to the input image.

Top-down pathway and lateral connections

The goal of the FPN is to enrich the semantics of the low-level feature maps. A simple way to do this is to add the high-level feature maps, of strong semantics, to the low-level ones. Since higher level feature maps have lower resolution, they are first upsampled spatially by a factor of two, using nearest neighbor sampling. Before the summation, the lower layer feature maps are fed to a 1×1 lateral convolutional layer, in order to ensure that both feature maps have the same channel dimensions. The feature maps are then summed element-wise and a 3×3 convolution is applied to generate the final feature map, so as to avoid potential aliasing due to the upsampling operation. There are no nonlinearities in these

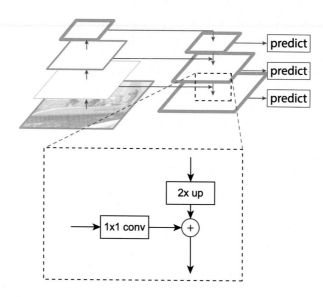

FIGURE 2.14 FPN building block.

extra layers. The procedure is iterated from the top to the bottom of the pyramid, i.e., layers $\{C_2, C_3, C_4, C_5\}$, to produce the final FPN pyramid of layers $\{P_2, P_3, P_4, P_5\}$, each containing 256 channel dimensions. Each layer P_i corresponds to layer C_i of the same resolution. To ensure that all FPN pyramid levels possess the same semantics, the classification and bounding box regression layers are shared across different levels, to produce the final predictions for all scales.

Since two-stage object detectors, such as those discussed above, require RoI pooling to extract the instance-wise features processed by the second stage, they are not fully convolutional, which complicates their hardware implementation. Although accurate, these detectors can only achieve speeds of 10–20 frames per second (fps) on modern GPUs. Higher detection speeds usually require more hardware friendly architectures, typically fully convolutional and containing a single-stage. A number of such architectures have been proposed in the literature, including YOLO (Redmon et al., 2016), SSD (Liu et al., 2016), and RetinaNet (Lin et al., 2017b). Single-stage detectors usually trade-off accuracy for speed.

2.9 YOLO

You only look once (YOLO) (Redmon et al., 2016) was one of the first and is still the most popular single-stage object detector. It gained popularity mainly due to its high speed, more than 50 fps on a modern GPU. However, its accuracy is significantly lower than the state of the art for two-stage detectors. The first version of YOLO did not use anchors, as in Faster R-CNN, which were only introduced in later versions. Its computations are illustrated in Fig. 2.15. The input image is divided into $S \times S$ cells, and B predictions are

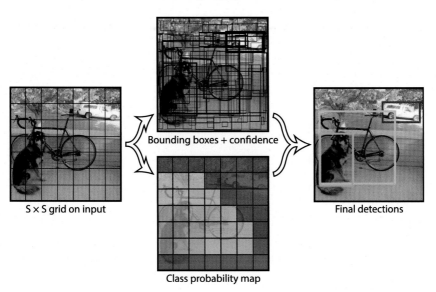

FIGURE 2.15 Detection system of YOLO.

made per cell **x**. A cell is considered responsible for an object if and only if the center of the ground truth bounding box of the object is located inside it. Each prediction consists of four bounding box coordinates x, y, w and h and an objectness score $p(o = 1|\mathbf{x})$. The latter reflects the confidence that the predicted box includes an object and is ideally equal to the IoU between the predicted box and the object ground truth box. If no object exists in the cell, the objectness score should be 0. The confidence prediction for class k is then defined as

$$p(y = k|\mathbf{x}) = p(y = k|o = 1, \mathbf{x})p(o = 1|\mathbf{x}), \qquad (2.13)$$

where $p(y = k|o = 1, \mathbf{x})$ is the class conditional probability of class k appearing in cell **x** given that the cell contains an object. However, a single set of C class conditional probabilities is shared by the B predictions of the cell, i.e., all predictions in a cell have the same class conditional probabilities. A typical implementation of YOLO, for the detection of the 20 object classes of the Pascal VOC dataset (Everingham et al., 2010), uses $S = 7$, $B = 2$ and $C = 20$, for a total of $7 \times 7 \times 30$ predictions.

Backbone design

One of the reasons behind YOLO's efficiency is its backbone network, called DarkNet. This is inspired by the GoogLeNet (Szegedy et al., 2015) architecture, containing 24 convolutional layers followed by 2 fully connected layers, implemented with a combination of 1×1 channel reduction and 3×3 convolutional layers, whose implementation is optimized in modern GPUs.

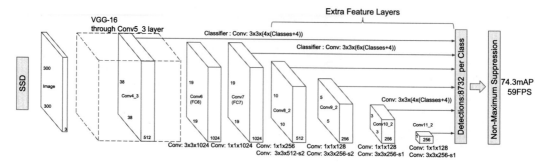

FIGURE 2.16 Detection system of SSD.

2.10 SSD

SSD (Liu et al., 2016) is another popular single-stage object detector. It is as fast as YOLO but has much higher accuracy, especially for small objects. The major differences are the use of the multiscale feature pyramid of Fig. 2.11(c) and RPN anchors.

2.10.1 Architecture

The backbone network is a standard image classification network (without the final classification layer), e.g., the VGG-Net (Simonyan and Zisserman, 2014). Some auxiliary layers are added to this backbone network to produce detection predictions. The overall architecture is shown in Fig. 2.16.

Multiscale detection

Similarly to the MS-CNN, SSD uses the hierarchical feature representations of Fig. 2.11(c) for multiscale detection. As shown in Fig. 2.16, detections are generated from the feature maps of layers *Conv4_3*, *Conv6*, *Conv7*, *Conv8_2*, *Conv9_2*, *Conv10_2* and *Conv11_2*, which have different resolutions. For the reasons discussed above, this enables the detection of more objects and higher accuracy for the detection of small objects. Similar to the RPN (Ren et al., 2017), the SSD predictor applied to each convolution feature map is composed of an additional 3×3 convolution layer, whose outputs are class scores and bounding box offsets relative to anchor positions.

Anchor boxes

Similar to the RPN, there are k anchors at a given location, and c class scores and 4 coordinate offsets are predicted per anchor. Hence, a feature map of $h \times w$ resolution produces $(c + 4) \times k \times h \times w$ outputs. This is equivalent to applying RPN on multiple feature maps and helps understand why single-stage detectors are in general less accurate than two-stage detectors: they are similar to the proposal generation stage of the latter. When comparing the SSD to the RPN implemented by the MS-CNN, the main difference is that the RPN is class-agnostic, i.e., $c = 2$, while the SSD is class specific, making $c + 1$ class predictions, e.g., $c = 21$ for the VOC dataset (Everingham et al., 2010).

2.10.2 Training

SSD uses a multitask loss function similar to (2.5), combining classification and bounding box regression terms. For more accurate detection, it also uses hard negative mining and strong data augmentation during training.

Hard negative mining

A difficulty of object detection is that most negative samples, e.g., sky, are easy to classify. If too many simple negatives are included in the training, the detector will underperform for hard negatives (negatives that are visually similar to positives). This problem is more serious for single-stage detectors, which lack the effective resampling implemented by the second stage. Hard negative mining is a sampling mechanism, widely used in object detection prior to the emergence of deep learning (Viola and Jones, 2004; Felzenszwalb et al., 2010; Dollár et al., 2014), designed to combat this problem. To create the pool of negatives, SSD sorts negative samples by higher to lower confidence scores (higher meaning that the example is harder to classify), and selects the top ones needed to achieve a ratio of 1:3 between positives and negatives, similar to the sampling of Fast R-CNN. This makes the detector more discriminative of hard negative samples.

Data augmentation

Insufficiency of training data is another problem for deep learning based object detection, which can be addressed by data augmentation. In SSD training, each image is randomly augmented by either 1) keeping the original image; 2) cropping a patch of minimum IoU with the object in $\{0.1, 0.3, 0.5, 0.7, 0.9\}$; or 3) cropping a random patch. The size (aspect ratio) of the random patch is chosen randomly to be in $[0.1, 1]$ of the original image size (in $\{\frac{1}{2}, 2\}$). After cropping, the patch is resized to the fixed square size (e.g., 512×512) with randomly horizontal flipping, to be forwarded to the network. Beyond spatial augmentation, some augmentations are also applied in the color space.

2.11 RetinaNet

While hard negative sampling addresses the imbalance between easy, e.g., sky, and hard negatives, it is only mildly effective for deep learning detectors. The RetinaNet (Lin et al., 2017b) instead proposes a generalization of the cross-entropy loss, denoted as the focal loss, that downweighs easy negatives and emphasizes hard ones.

2.11.1 Focal loss

The cross entropy (CE) loss for binary classification is

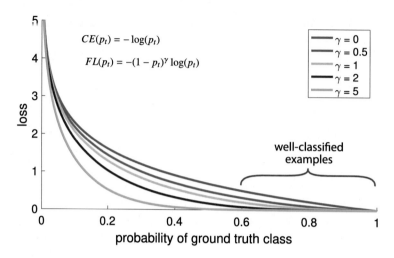

FIGURE 2.17 *Focal Loss* visualizations of different γ. Setting $\gamma > 0$ downweighs the losses of well-classified examples ($p_t > 0.5$), putting more focus on hard misclassified examples.

$$CE(p, y) = \begin{cases} -\log(p) & \text{if } y = 1 \\ -\log(1 - p) & \text{otherwise,} \end{cases} \qquad (2.14)$$

where $y \in \{\pm 1\}$ is the ground-truth label and $p = p(y = 1|x) \in [0, 1]$ the probability for class $y = 1$. Define p_t as

$$p_t = \begin{cases} p & \text{if } y = 1 \\ 1 - p & \text{otherwise.} \end{cases} \qquad (2.15)$$

This can be expressed as $CE(p, y) = CE(p_t) = -\log(p_t)$ and is plotted in Fig. 2.17 (top blue curve, with $\gamma = 0$). It can be observed that a sample easily classified (e.g., with $p_t \gg 0.5$) still has nontrivial loss magnitude. When the majority of the training samples are easy, they dominate the overall loss, overwhelming the contributions of hard samples.

The *Focal Loss* generalizes the CE loss by inserting a modulating factor $(1 - p_t)^\gamma$ of tunable parameter $\gamma \geq 0$,

$$FL(p_t) = -(1 - p_t)^\gamma \log(p_t). \qquad (2.16)$$

This factor changes the loss as shown in Fig. 2.17. While for small p_t (hard examples) the loss does not change much, it decays to zero much faster for large p_t (easy examples), especially for large values of γ. Hence, easy samples contribute less to the overall loss. For example, when $\gamma = 2$, an easy example with $p_t = 0.9$ will have a 100-times lower contribution under the focal than under the cross-entropy loss. The downweighing rate is controlled by the hyperparameter γ, with larger γ inducing heavier downweighing, as shown in Fig. 2.17. In practice, it has been found that $\gamma = 2$ tends to work best.

In general, the focal loss is balanced by another hyperparameter α,

$$FL(p_t) = -\alpha_t (1 - p_t)^\gamma \log(p_t). \qquad (2.17)$$

TABLE 2.1   Performances of the advanced object detectors on COCO.

	backbone	speed (fps)	epoch	AP	AP_{50}	AP_{75}	AP_S	AP_M	AP_L
YOLOv3	DarkNet-53	48.1	273	33.4	56.3	35.2	19.5	36.4	43.6
SSD512	VGG16	30.7	24	29.4	49.3	31.0	11.7	34.1	44.9
RetinaNet	ResNet-50	24.4	36	38.7	58.0	41.5	23.3	42.3	50.3
Faster R-CNN	ResNet-50	9.6	36	38.4	58.7	41.3	20.7	42.7	53.1
FPN	ResNet-50	26.3	36	40.2	61.0	43.8	24.2	43.5	52.0
Cascade R-CNN	ResNet-50	21.8	36	43.6	61.6	47.4	26.2	47.1	56.9

When this loss is used, the problem of overwhelming easy negatives can be circumvented during training.

2.12 Detection performances

We finish by briefly comparing the performances of some of the object detectors discussed above. The R-CNN, SPP-Net, and Fast R-CNN are not included in this comparison since they are now obsolete and rarely used in practice. Table 2.1 presents a summary of the performance of the remaining methods on the COCO dataset (Lin et al., 2014), in terms of speed, training epochs, and Average Precision (AP). In COCO, AP is averaged over 10 IoU thresholds of 0.50:0.05:0.95, and AP_{50} (at threshold 0.5), AP_{75} (at threshold 0.75) AP_S (for small objects), AP_M (for medium objects) and AP_L (for large objects) provide more performance details. Note that the more comprehensive metric of AP, averaged over IoU thresholds of 0.50:0.05:0.95, rewards detectors with better localization than the traditional metric of AP_{50} at IoU threshold of 0.5. It is clear that single-stage detectors (YOLO and SSD) are faster but much less accurate than their two-stage counterparts (Faster R-CNN, FPN and Cascade R-CNN). Among the two-stage detectors, speeds are comparable but the Cascade R-CNN achieves the highest accuracies. Comparisons of this type can be found in object detection papers and allow practitioners to choose the detector with the trade-off between complexity and accuracy most suited for a given application.

2.13 Conclusion

In this chapter, we have reviewed recent advances in deep learning based object detection. Broadly speaking, existing methods can be divided into two categories, single-stage and two-stage. Single-stage methods are faster but less accurate. Two stage-methods combine a proposal network, which is similar to a single-stage detector but class-insensitive, and a second stage that classifies objects into different classes and refines their bounding boxes. Many contributions have been made through the years to improve on the performance of the pioneering R-CNN (Girshick et al., 2014), both in terms of speed and accuracy. These include concepts that have become important in the detection literature, such as RoI pooling, multitask losses, the RPN, anchors, cascaded detection and regression, multiscale feature representations, data augmentation techniques, loss functions, etc. While, as shown in Table 2.1, these contributions have enabled substantial performance improvements, object detectors are still far from perfect. A large literature also exists in topics not covered in this review, such as instance segmentation, domain adaptation, or low complexity deep learning architectures, among several others. Finally, object detection is frequently used as a preliminary stage of many other vision tasks, including pose estimation, image and video captioning, or visual question answering, among others. In summary, while current object detectors are orders of magnitude more effective than those of just a decade ago, much research remains to be done in this problem of fundamental importance for computer vision.

References

Cai, Z., Fan, Q., Feris, R.S., Vasconcelos, N., 2016. A unified multi-scale deep convolutional neural network for fast object detection. In: ECCV, pp. 354–370.

Cai, Z., Vasconcelos, N., 2021. Cascade R-CNN: high quality object detection and instance segmentation. IEEE Transactions on Pattern Analysis and Machine Intelligence 43 (5), 1483–1498.

Dalal, N., Triggs, B., 2005. Histograms of oriented gradients for human detection. In: CVPR, pp. 886–893.

Dollár, P., Appel, R., Belongie, S.J., Perona, P., 2014. Fast feature pyramids for object detection. IEEE Transactions on Pattern Analysis and Machine Intelligence 36 (8), 1532–1545.

Everingham, M., Gool, L.J.V., Williams, C.K.I., Winn, J.M., Zisserman, A., 2010. The Pascal visual object classes (VOC) challenge. International Journal of Computer Vision 88 (2), 303–338.

Felzenszwalb, P.F., Girshick, R.B., McAllester, D.A., Ramanan, D., 2010. Object detection with discriminatively trained part-based models. IEEE Transactions on Pattern Analysis and Machine Intelligence 32 (9), 1627–1645.

Girshick, R.B., 2015. Fast R-CNN. In: ICCV, pp. 1440–1448.

Girshick, R.B., Donahue, J., Darrell, T., Malik, J., 2014. Rich feature hierarchies for accurate object detection and semantic segmentation. In: CVPR, pp. 580–587.

He, K., Zhang, X., Ren, S., Sun, J., 2014. Spatial pyramid pooling in deep convolutional networks for visual recognition. In: ECCV, pp. 346–361.

He, K., Zhang, X., Ren, S., Sun, J., 2016. Deep residual learning for image recognition. In: CVPR, pp. 770–778.

Krizhevsky, A., Sutskever, I., Hinton, G.E., 2012. Imagenet classification with deep convolutional neural networks. In: NIPS, pp. 1106–1114.

Lazebnik, S., Schmid, C., Ponce, J., 2006. Beyond bags of features: spatial pyramid matching for recognizing natural scene categories. In: CVPR. IEEE Computer Society, pp. 2169–2178.

Lin, T., Dollár, P., Girshick, R.B., He, K., Hariharan, B., Belongie, S.J., 2017a. Feature pyramid networks for object detection. In: CVPR. IEEE Computer Society, pp. 936–944.

Lin, T., Goyal, P., Girshick, R.B., He, K., Dollár, P., 2017b. Focal loss for dense object detection. In: ICCV. IEEE Computer Society, pp. 2999–3007.

Lin, T., Maire, M., Belongie, S.J., Hays, J., Perona, P., Ramanan, D., Dollár, P., Zitnick, C.L., 2014. Microsoft COCO: common objects in context. In: ECCV, pp. 740–755.

Liu, W., Anguelov, D., Erhan, D., Szegedy, C., Reed, S.E., Fu, C., Berg, A.C., 2016. SSD: Single Shot Multibox Detector. ECCV, vol. 9905. Springer, pp. 21–37.

Lowe, D.G., 1999. Object recognition from local scale-invariant features. In: ICCV. IEEE Computer Society, pp. 1150–1157.

Lowe, D.G., 2004. Distinctive image features from scale-invariant keypoints. International Journal of Computer Vision 60 (2), 91–110.

Papageorgiou, C., Oren, M., Poggio, T.A., 1998. A general framework for object detection. In: ICCV. IEEE Computer Society, pp. 555–562.

Redmon, J., Divvala, S.K., Girshick, R.B., Farhadi, A., 2016. You only look once: unified, real-time object detection. In: CVPR, pp. 779–788.

Ren, S., He, K., Girshick, R.B., Sun, J., 2017. Faster R-CNN: towards real-time object detection with region proposal networks. IEEE Transactions on Pattern Analysis and Machine Intelligence 39 (6), 1137–1149.

Rowley, H.A., Baluja, S., Kanade, T., 1996. Neural network-based face detection. In: CVPR. IEEE Computer Society, pp. 203–208.

Russakovsky, O., Deng, J., Su, H., Krause, J., Satheesh, S., Ma, S., Huang, Z., Karpathy, A., Khosla, A., Bernstein, M.S., Berg, A.C., Li, F., 2015. Imagenet large scale visual recognition challenge. International Journal of Computer Vision 115 (3), 211–252.

Simonyan, K., Zisserman, A., 2014. Very deep convolutional networks for large-scale image recognition. CoRR. arXiv: 1409.1556 [abs].

Sung, K.K., Poggio, T.A., 1998. Example-based learning for view-based human face detection. IEEE Transactions on Pattern Analysis and Machine Intelligence 20 (1), 39–51.

Szegedy, C., Liu, W., Jia, Y., Sermanet, P., Reed, S., Anguelov, D., Erhan, D., Vanhoucke, V., Rabinovich, A., 2015. Going deeper with convolutions. In: CVPR, pp. 1–9.

van de Sande, K.E.A., Uijlings, J.R.R., Gevers, T., Smeulders, A.W.M., 2011. Segmentation as selective search for object recognition. In: ICCV, pp. 1879–1886.

Viola, P.A., Jones, M.J., 2001. Rapid object detection using a boosted cascade of simple features. In: CVPR. IEEE Computer Society, pp. 511–518.

Viola, P.A., Jones, M.J., 2004. Robust real-time face detection. International Journal of Computer Vision 57 (2), 137–154.

Biographies

Zhaowei Cai is an Applied Scientist at Amazon Web Services. He received the B.S. degree in Automation from Dalian Maritime University in 2011, and the M.S. and Ph.D. degrees from the University of California, San Diego, in 2019. From 2011 to 2013, he worked as research assistant at Institute of Automation, Chinese Academy of Sciences. His current research interests are in computer vision and machine learning, including object detection and recognition.

Nuno Vasconcelos received the licenciatura in electrical engineering and computer science from the Universidade do Porto, Portugal, and the MS. and Ph.D. degrees from the Massachusetts Institute of Technology. He is a Professor in the Electrical and Computer Engineering Department at the University of California, San Diego, where he heads the Statistical Visual Computing Laboratory. He has received a NSF CAREER award, a Hellman Fellowship, several best paper awards and is a Fellow of the IEEE.

Learning with limited supervision
Static and dynamic tasks

Sujoy Paul[a,c] and Amit K. Roy-Chowdhury[b]

[a]Google Research, Bangalore, India [b]University of California, Riverside, Electrical and Computer Engineering, Riverside, CA, United States

CHAPTER POINTS

- Reducing supervision for computer vision models is important for scalability and adaptability.

- We review different methods that have been proposed for learning with limited supervision, and provide results that justify the methods.

- We specifically focus on active learning for recognition, weakly supervised learning for event localization, domain adaptation for semantic segmentation, and reinforcement learning for subgoal discovery in training robots for dynamical tasks.

3.1 Introduction

The recent successes in computer vision have been mostly around using a huge corpus of intricately labeled data for training recognition models. But, in real-world cases, acquiring such large datasets will require a lot of manual annotation, which may be strenuous, out of budget, or even prone to errors. However, a lot of real data that are generated daily can be acquired at low to no annotation cost. Such data can be unlabeled or contain tag/meta-data information, termed as weak annotation. Our goal is to develop methods that can learn recognition models from such data involving limited manual supervision. In this chapter, we will look into two dimensions of learning with limited supervision – first, reducing the *number* of manually labeled data required to learn recognition models, and second, reducing

---

[c] This work was done when the author was at UC Riverside.

Copyright © 2022 Elsevier Inc. All rights reserved.

the *level* of supervision from strong to weak which can be mined from the web, easily queried from an oracle, or imposed as rule-based labels derived from domain knowledge.

In the first dimension of learning with limited supervision, we show that context information, often present in natural data, can be used to reduce the number of annotations required. In the second dimension – reducing the level of supervision – we use weak labels instead of dense strong labels, for learning dense prediction tasks. We discuss frameworks to learn using weak labels for action detection in videos and domain adaptation of semantic segmentation models on images. All of these tasks discussed are static in nature. Continuing in the direction of learning from weak labels, we explore sequential decision-making problems, where the next input depends on the current output. We look into the problem of learning robotics tasks with a small set of expert human demonstrations via decomposing the complex task into subgoals. Detailed explanation of these approaches follows.

3.2 Context-aware active learning

In recent years, due to advances in technology, a huge amount of visual and text data are generated daily, which are mostly unlabeled for the purpose of learning machine learning models. Also, machine learning algorithms are becoming more commonplace in human life. A large proportion of these algorithms are based on supervised learning which requires a large quantity of data to be labeled. Moreover, these models need to be updated over time as new data becomes available in order to dynamically adapt to the different semantic concepts which may drift with time. Manually labeling this continuous flow of data is not only a tedious task for humans but also prone to incorrect labeling. In such a scenario, it is maybe advantageous to label only the informative data points, and not label the data points carrying redundant information. This intuition is supported in works (Lapedriza et al., 2013) which show that not all data points carry the same amount of information, and choosing the most informative ones may even lead to better performance than labeling all the data points in the unlabeled dataset. Active Learning, which has been studied in the literature for the last few decades has shown immense potential in choosing the informative data points and reducing the manual labeling effort.

3.2.1 Active learning

Active Learning (Settles, 2012) has been proposed as a solution to the problem of reducing the amount of manual labeling, without compromising recognition performance. A pictorial overview of active learning is given in Fig. 3.1. Given a large unlabeled dataset, active learning methods first choose a small random subset of data and query the human expert to obtain their labels. These data points are used to learn a prediction model for the task at hand. This is the initial model. The next job is to choose only a small subset of data points from the unlabeled set, to get the maximum amount of information possible. Informativeness utility scores are defined on these unlabeled samples based on the uncertainty of the current model, data density, etc. These scores are used to choose the samples to label. Generally, active learning methods are iterative with human-in-the-loop, i.e., computing the informativeness scores, querying the human to obtain labels, followed by updating the model with the new labeled

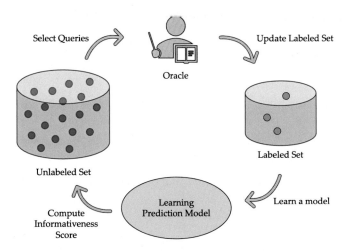

FIGURE 3.1 This figure presents a pictorial overview of the iterative process in Active Learning. It starts with choosing a few samples from an unlabeled set to query the oracle for manual labeling. The labeled set is then used to update the prediction model, which is further used to compute the informativeness measure of the remaining unlabeled samples and in turn select a small subset to annotate.

data points, and again computing the informativeness scores of the unlabeled data points, as shown in Fig. 3.1. This loop continues either until the annotation budget is exhausted, or forever, in case of continuous learning where concepts may drift over time and a constant update of the model is required.

Notations   Before looking into the details of each of these steps, let us formalize the notations to be used hereafter. Consider a classification task with c categories. We learn a model which given a data point feature x predicts a probability mass function (pmf) $p_\theta(y|x)$ over the c categories, which is parameterized by θ, which we need to learn. θ can be a single vector for linear models or a group of matrices for deep neural networks. To learn that, we have a labeled set of tuples $\mathcal{L} = \{(x_i, y_i)_{i=1}^l\}$ and an unlabeled set $\mathcal{U} = \{(x_j)_{j=1}^u\}$.

Informativeness measures

Most active learning approaches formulate a utility score for each unlabeled sample, based on which they are chosen for manual labeling. The entropy of the predictions is one of the most common informativeness measures used in the literature (Settles, 2012; Li and Guo, 2014; Paul et al., 2016). Given a data point x, the entropy can be represented as follows:

$$H(x) = \sum_{i=1}^{c} -p_\theta(y = i|x) \log p_\theta(y = i|x) \tag{3.1}$$

A higher value of entropy signifies that the classifier is uncertain about the prediction and thus should be chosen for manual labeling.

Expected change in model parameter gradients (Settles, 2012) is another utility score which measures the amount of change in gradients possible in the model when a sample x is included in the training set. It is calculated as

$$G(x) = \sum_{i=1}^{c} p_\theta(y=i|x) ||\nabla_\theta l(\mathcal{L} \cup (x,i))||_2 \qquad (3.2)$$

where $l(.,.)$ is the loss function used to learn the classification model. A higher amount of expected gradient change would signify a high amount of information carried in the data point. Expected model output change (Käding et al., 2016) and expected error rate (Cuong et al., 2013; Li and Guo, 2013) are also similar measures used in the literature.

Data density (Li and Guo, 2013), which considers the density of the data points in the feature space, is another important measure for active learning. It can be defined as follows:

$$D(x) = \frac{1}{|nei(x)|} \sum_{x_i \in nei(x)} 1 - dist(x, x_i) \qquad (3.3)$$

where $nei(x)$ are the neighboring data points of x. A higher value of $D(x)$ would signify that the data points around x are very close to each other and thus obtaining the label of x would be beneficial to get an understanding of the surrounding data points as well. Note that this measure is generally used in conjunction with other measures discussed above. Given these measures, active learning methods choose the data points for manual annotation as discussed next.

Selecting informative samples

Selecting informative samples in active learning depends on the application and annotation budget at hand. There are two possible ways of choosing informative samples – *serial* where one data point is chosen at a time or *batch-mode*, where multiple data points are chosen at a time. After choosing the samples, the model is updated and the new scores are obtained on the unlabeled data points. Thus, in scenarios where computation is restricted, batch-mode may turn out to be better, but the serial method may result in better performance.

Most active learning methods consider that the unlabeled dataset is fixed, which may not be the case in general. Data can be streaming, i.e., appear in small batches. In that case, we have to select either a certain prespecified portion for manual labeling which is generally designed based on the annotation budget per batch. A more difficult but useful scenario is when the total budget for labeling is specified and the algorithm needs to decide on its own to choose the suitable subset to label for each batch of streaming data. It is interesting to note that most of the methods discussed above define informativeness measures independently for the samples without considering the interrelationships that may occur between data points. Moreover, these methods consider active learning of one task at a time, and they cannot sample for multiple recognition tasks at once. These are addressed below.

3.2.2 Context in active learning

In this section we discuss how relationships between data points can be useful in further reducing the supervision required to learn recognition tasks. We also discuss how we can devise active learning for multiple recognition tasks simultaneously.

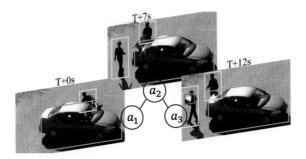

FIGURE 3.2 A sequence of a video stream from Oh et al. (2011) shows three new unlabeled activities – person getting out of a car (a_1) at T + 0 s, person opening a car trunk (a_2) at T + 7 s, and person carrying an object (a_3) at T + 12 s. These activities are spatio-temporally correlated, and this can provide contextual information for a holistic understanding.

Context in learning

Relationships between data points often occur in the real-world. Such relationships are often termed as *context* in the literature. For example, consider the relationship between scenes and objects. It is unlikely to find a 'cow' in a 'bedroom', but, the probability of finding 'bed' and 'lamp' in the same scene may be high. Thus gaining information about a scene can help in the enhanced prediction of objects and vice versa. Similarly, events/activities in videos may be spatiotemporally correlated as shown in Fig. 3.2. Note in the figure that without knowledge about activity a_2 and a_3, it is hard to predict whether activity a_1 is a person getting out of the car or getting in the car. The relationships with the other activities help to understand that particular activity. Relationships also occur between documents via hyperlinks or citations. Several works have shown that in many applications such as activity recognition (Yao and Fei-Fei, 2010; Wang et al., 2013), object recognition (Galleguillos et al., 2008; Choi et al., 2010), text classification (Sen and Getoor, 2003; Settles and Craven, 2008), etc, the relationships between data points can be exploited to get better recognition performance. In many of these works, Probabilistic Graphical Models (Koller and Friedman, 2009), Structural Support Vector Machines (SVM) (Cristianini and Ricci, 2008), etc are used for a holistic understanding. Even in the era of deep learning, Conditional Random Fields (Koller and Friedman, 2009) are used for better scene understanding.

Context for query selection: motivation

Keeping in mind the global understanding that is offered by utilizing context in data, an interesting direction is to look at whether context can also be useful for the purpose of reducing the number of labeled samples. The motivation behind this is that in the presence of context, as the predictions are correlated, can we gather information about a large number of unlabeled data points while labeling only a few of them. As discussed previously, unlike methods in the literature that do not consider the contextual information in the informativeness measures, context-aware active learning takes into account the context to reduce the number of annotations.

Although there have been some works that consider relationships between data points in active learning (Bilgic and Getoor, 2009; Mac Aodha et al., 2014; Hasan and Roy-Chowdhury,

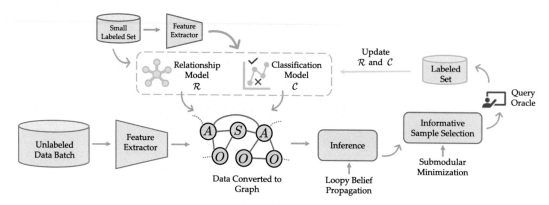

FIGURE 3.3 This figure presents the flow of the proposed framework. 1. A small set of labeled data is used to obtain the initial relationship ($\mathcal{R}$) and the classification model ($\mathcal{C}$). 2. As a new unlabeled batch of data becomes available sequentially over time, we first extract features from the raw data. Then the current $\mathcal{C}$ and $\mathcal{R}$ models are used to construct a graph from the data to represent the relationships between them. Then inference on the graph is used to obtain the node and edge probabilities, which are used to choose the informative samples for manual labeling. The newly labeled instances are then used to update the models $\mathcal{C}$ and $\mathcal{R}$.

2015; Hu et al., 2013), they do not consider the flow of beliefs between data points to have a joint understanding of their predictions, which may be helpful for choosing the most informative ones. Moreover, most of them are problem-specific algorithms and deal with the active learning of a single recognition task. A general approach for active learning that considers the interrelationships between data points, and which can be used across a variety of application domains, is necessary. Joint learning of tasks such as scene-object (Yao et al., 2012; Wang et al., 2016a) or activity-object (Jain et al., 2015; Koppula et al., 2013) classification may be required to be learned actively, to reduce the manual labeling effort. In such scenarios, it is challenging to choose the informative samples for manual labeling as they may belong to different recognition tasks. We next present a framework for such context-aware active learning, including jointly for multiple tasks.

Context for query selection: overview

Given an unlabeled set, we describe a framework (Paul et al., 2017) that chooses a small informative subset of data points for manual annotation while exploiting the context information, i.e., the structural relationships between them. The flow of the framework is pictorially presented in Fig. 3.3. The framework starts with a small set of labeled data and uses it to build the classification ($\mathcal{C}$) and relationship ($\mathcal{R}$) models. $\mathcal{R}$ represents the underlying relationship between the data points via categorical cooccurrence probabilities. Note that the classification models may contain multiple classifiers for multiple recognition tasks. After learning the initial models, given a new batch of unlabeled samples, the goal is to select a subset of informative samples for manual labeling which can be used to update the current classification and relationship models.

As new batches of data become available, the samples in the batches are separated into different sets based on the recognition task to which they belong, followed by feature extraction. Using the current classifiers, a probability mass function over the possible categories is obtained for each unlabeled sample. It is used along with $\mathcal{R}$ to construct a graph whose nodes represent the samples. A message-passing algorithm is used to infer on the graph to obtain the beliefs of each node and the edges of the graphs. An information-theoretic objective function is derived, which utilizes the beliefs to select the informative nodes for manual labeling. The submodular nature of this optimization function allows us to achieve this in a computationally efficient manner. The newly labeled nodes are used to update the models $\mathcal{C}$ and $\mathcal{R}$. It may be noted that the number of samples selected per batch is *nonuniform*, dependent on the information content of each batch.

3.2.3 Framework for context-aware active learning

In this section, we present a framework to utilize contextual information, specifically, cooccurrence based relationships between data points, in an active learning framework to reduce the number of manual annotations to learn recognition models.

Data representation

Consider that the unlabeled data points have some underlying structure, i.e., relationships among them. We take a probabilistic graphical model approach to build a graph whose nodes represent the unlabeled samples and edges represent the relationships between them and help as pathways for the flow of information between the nodes. The nodes are represented using node potentials, which are probability mass function (pmf) predictions from the current prediction models (single or multiple for joint tasks). The edges are represented using edge potentials which are represented as a matrix with cooccurrence frequencies between categories, which is represented as relationship model $\mathcal{R}$. The computation of the cooccurrence frequency is dependent on the application at hand and will be discussed subsequently. Note that this framework can be applied to any application containing relationships which can be modeled as edge potentials.

We construct a graph $G = (V, E)$ with the instances in $\mathcal{U}$. Each node in $V = \{v_1, \ldots, v_N\}$ represents each data point. The edges $E = \{(i, j) | v_i \text{ and } v_j \text{ are linked}\}$ represent the relationships between the data points. The node and edge potentials are assigned using the current classification model $\mathcal{C}$ and relationship model $\mathcal{R}$. A message-passing algorithm can be used to infer the node and edge beliefs which are the marginal node probabilities and the pair-wise joint distribution of the edges respectively. Loopy Belief Propagation (LBP) (Ugm, 2007) can be used for this purpose.

Selection of informative samples

Using the node and edge probabilities, the goal is to choose a small subset $V^{l*} \subset V$ for manual labeling, which will improve the current models $\mathcal{C}$ and $\mathcal{R}$. We wish to select a subset of the nodes such that the joint entropy of all the nodes $H(V)$ is minimized. The joint entropy of all the nodes in the graph can be approximated as follows:

$$H(V) \approx \sum_{v_i \in V} H(v_i) - \sum_{(i,j) \in E} I(v_j; v_i) \tag{3.4}$$

Note that the joint entropy of the graph can only be approximated for a cyclic graph in general, but the above expression is an exact representation for acyclic graphs. Consider we have two disjoint vertex subgraphs with vertex V^l and V^{nl} and edges divided accordingly into E^l and E^{nl}. Then using Eq. (3.4), the entropy of the graph can be expressed as follows:

$$H(V) \approx H\big(V^l\big) + H\big(V^{nl}\big) - \sum_{\substack{(i,j)\in E \\ v_i \in V^l, v_j \in V^{nl}}} I(v_j; v_i) \qquad (3.5)$$

If we choose V^l for manual annotation, it can be shown that the first and last terms of the above equation become zero, which will be the reduction in entropy. Thus we need to choose the optimal subset V^{l*} such that the entropy is minimized by the maximum extent. Thus, the optimization problem can be formulated as follows:

$$V^{l*} = \underset{\substack{V^l \\ s.t.|V^l|=K}}{\arg\max} \left[H\big(V^l\big) - \sum_{\substack{(i,j)\in E \\ v_i \in V^l, v_j \in V^{nl}}} I(v_j; v_i) \right] \qquad (3.6)$$

The above optimization problem is NP-Hard and difficult to solve. Heuristics techniques such as Branch and Bound can be used to efficiently search for a solution. This method of optimization is necessary when there is a strict requirement for budget constraints per batch of data, and we addressed this problem in one of our works (Hasan et al., 2018). However, each batch of data may contain a nonuniform amount of information, and choosing the same number of budget-constrained samples (i.e., K) from each batch may not be a good idea in general. Instead, the number of samples could be determined based on the information content of each batch. With this motivation, we can modify the above optimization problem to be unconstrained with a cardinality regularizer as follows:

$$V^{l*} = \underset{V^l}{\arg\min} \left[\sum_{\substack{(i,j)\in E \\ v_i \in V^l, v_j \in V^{nl}}} I(v_j; v_i) - H\big(V^l\big) + \lambda|V^l| \right] \qquad (3.7)$$

where λ is a positive trade-off parameter. The objective function in Eq. (3.7) can be proved to be submodular which makes the optimization problem simpler compared to Eq. (3.6). Submodular Function Minimization (SFM) often arises in fields of machine learning, game theory, information theory, etc. Detailed description may be found here (McCormick, 2005). There exist some algorithms which can be used to solve SFM in polynomial time. The popular Fujishige-Wolfe Min Norm Point algorithm (Fujishige et al., 2006) can be used to solve the optimization problem.

Model update

After the chosen samples are labeled by a human annotator, we perform inference on the graph, conditioned on the acquired labels to update the beliefs of the nodes and then we apply the concept of weak teacher (Zhang and Chaudhuri, 2015), which does not involve the human. We choose those nodes having the confidence in classification greater than ϵ, with

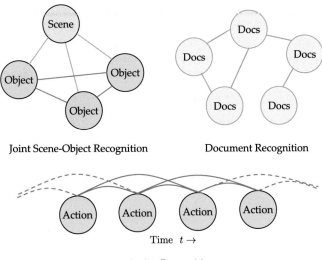

Joint Scene-Object Recognition   Document Recognition

Action Recognition

FIGURE 3.4 The semantic context graphs for three tasks discussed in Section 3.2.4 – joint scene-object recognition where we utilize the context present between scene and objects in an image, document recognition where the context relationship information is shared via citations/weblinks, and activity recognition where action sequences share spatio-temporal relationships.

the corresponding label, to be in the labeled set $\mathcal{L}$. ϵ should be high enough to avoid incorrect labeling. The classification model $\mathcal{C}$ is updated by retraining the classifier using $\mathcal{L}$. Model $\mathcal{R}$ is comprised of only the cooccurrence matrix ψ and it is incremented using the new labeled instances.

3.2.4 Applications

In this section, we discuss a few applications where context-aware active learning can be utilized to reduce the manual labeling effort. We primarily discuss three different applications – joint scene-object classification, activity recognition, and document classification. As shown in Fig. 3.4, the applications have data that share relationships among them. We use linear classifiers such as Support Vector Machine (SVM) (Chang and Lin, 2011) as a baseline classifier in all the applications discussed next. In active learning, methods are generally compared against the full-set performance, i.e., the performance obtained when all the data points (except the test set) are labeled and used for training. We also compare with other popular active learning methods in literature, i.e., Batch Rank (Chakraborty et al., 2015), BvSB (Li et al., 2012), Entropy (Settles, 2012; Holub et al., 2008), Density-Based Sampling (DENS) (Settles, 2012), Expected Gradient Length (GRL) (Settles and Craven, 2008) and Random Sampling.

Scene-object classification

Scene and objects tend to cooccur in images. Although scene and object classifiers are separate, their joint understanding can be beneficial (Yao et al., 2012), which can be exploited in an active learning framework to reduce manual labeling. This is a special feature of our framework, as it can actively learn with multiple recognition tasks at once. The SUN dataset (Choi et al., 2010; Xiao et al., 2010) is good for experimenting with this framework as they have annotations for both the entire scene as well as the objects in the scene. We extract features from pretrained networks VGG-net (Zhou et al., 2014) and Alex-net (Krizhevsky et al., 2012) for scenes and objects respectively. We use selective search used in the RCNN pipeline to obtain object proposals. As shown in Fig. 3.4, in joint scene-object recognition, we represent each image as a graph with a single scene node and multiple object nodes corresponding to the different object proposals in the image. The graph is considered to be fully connected with two different types of edges – scene-object and object-object. The edge potential for the scene-object links is computed as the cooccurrence frequency of a scene category with an object category; and for the object-object links, the edge potential is the cooccurrence frequency between object categories in an image.

In the case of scene classification, the context-aware active learning method requires only 35% manual annotation to reach almost the full-set performance. The other methods in the literature require about 60% manual annotation to obtain similar performance. For object recognition, the context-aware active learning method requires 45% manual annotation whereas the methods in the literature need 65% annotation to reach almost the full-set performance. Our work (Bappy et al., 2016) on model adaptation of scene-object recognition showcases results specific to this application.

Document classification

Documents are generally interlinked by citations and hyperlinks, which may be exploited using our active learning approach to reduce manual labeling effort. We use the CORA dataset (Sen et al., 2008) for our experiments on document classification. It consists of 2708 scientific publications divided into seven categories. There are a total of 5429 links (citations) between the publications. The publications are represented using a dictionary of 1433 unique words and the feature vectors $\in \{0, 1\}^{1433}$ indicate the absence or presence of these words. As shown in Fig. 3.4, in document classification, we represent all the documents as the nodes of a graph which are linked if one document cites another document. We consider the edge potential to be the cooccurrence frequency that a publication of category i cited a publication of category j.

For document classification, the context-aware active learning method requires only 33% manual annotation to reach almost the full-set performance (Paul et al., 2017). The other methods in the literature require about 50% manual annotation to obtain similar performance. This shows that using the context information helps in reducing the amount of manual annotation by utilizing information which is readily available with the data points.

Activity classification

Activities are generally spatiotemporally related which can be exploited to reduce the number of instances chosen for manual labeling. We use the VIRAT dataset (Oh et al., 2011) on human activity for our experiments on activity classification. The dataset consists of 11 videos segmented into 329 activity sequences. We extracted features using the pretrained model of 3D convolutional networks (Tran et al., 2015). We extract the features for 16 frames at a time with a temporal stride of 8 and then apply max pooling along the temporal dimension to ob-

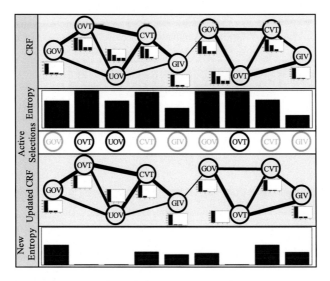

FIGURE 3.5 An example run of our proposed active learning framework on a part of an activity sequence from VIRAT dataset (Oh et al., 2011). Circles are activity nodes along with their class probability distribution. Edges have different thicknesses based on the pairwise mutual information. The node labels are – getting out of the vehicle (GOV), opening vehicle trunk (OVT), unloading from the vehicle (UOV), closing vehicle trunk (CVT), and getting into the vehicle (GIV). Inference on the graph (top) gives us marginal probability distribution of the nodes and edges. We use these distributions to compute entropy and mutual information. Relative mutual information is shown by the thickness of the edges, whereas entropy of the nodes is plotted below the top CRF. Eq. (3.14) exploits entropy and mutual information criteria in order to select the most informative nodes (2-OVT, 3-UOV, and 7-OVT). We condition upon these nodes (filled) and perform inference again, which provides us more accurate recognition and a system with lower entropy (bottom plot).

tain a single vector for each activity. We consider that there exists a link between two activities if they occurred within a certain spatio-temporal distance. We consider the edge potential to be the spatio-temporal cooccurrence between the two activities.

In the case of activity classification, the context-aware active learning method requires only about 18% manual annotation to reach almost the full-set performance. The other methods in the literature require about 40% manual annotation to obtain similar performance. Note that although we utilize the relationships between the action categories for this experiment, we can also utilize the context information imparted by the objects in the activities, which are especially important in human-object interaction related activities. Fig. 3.5 provides an example visualizing the working of the context-aware active learning method. As can be seen, gaining knowledge about some of the nodes helps to reduce the entropy of the other nodes, thus reducing manual annotation cost.

3.3 Weakly supervised event localization

Temporal activity localization is a core problem in computer vision, where given a long video, the algorithm needs to temporally localize the portions of the videos corresponding

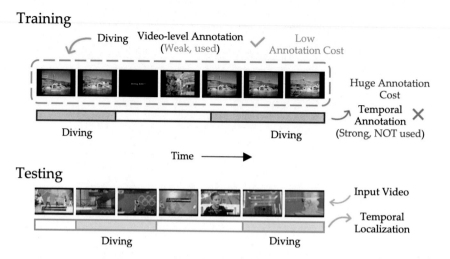

FIGURE 3.6 This figure presents the train-test protocol of weakly supervised action localization. The training set consists of videos with their video-level activity tags and NOT the temporal annotation. Whereas, while testing, the network not only estimates the labels of the activities in the video but also temporally locates their occurrence.

to different event categories of interest (Aggarwal and Ryoo, 2011). Its recent success (Xu et al., 2017; Zhao et al., 2017b) has evolved around a *fully* supervised setting, which considers the availability of frame-wise activity labels. However, acquiring such precise frame-wise information requires enormous manual labor. This may not scale efficiently with a growing set of cameras and activity categories. On the other hand, it is much easier for a person to provide a few categorical labels which encapsulate the content of a video. Moreover, videos available on the web are often accompanied by tags that provide semantic discrimination. Such video-level labels are generally termed as *weak* labels, which may be utilized to learn models with the ability to classify and localize activities in videos, as presented in Fig. 3.6.

In computer vision, researchers have utilized weak labels to learn models for several tasks including semantic segmentation (Hartmann et al., 2012; Khoreva et al., 2017; Yan et al., 2017), visual tracking (Zhong et al., 2014), reconstruction (Tulyakov et al., 2017; Kanazawa et al., 2016), video summarization (Panda et al., 2017), learning robotic manipulations (Singh et al., 2017), video captioning (Shen et al., 2017), object boundaries (Khoreva et al., 2016), place recognition (Arandjelovic et al., 2016), and so on. The weakly supervised localization problem is analogous to weak object detection in images, where object category labels are provided at the image level. There have been several works in this domain mostly utilizing the techniques of Multiple Instance Learning (MIL) (Zhou, 2004) due to their close relation in terms of the structure of information available for training. The positive and negative bags required for MIL are generated by state-of-the-art region proposal techniques (Li et al., 2016; Jie et al., 2017). In spite of its similarities, temporal localization using weak labels is a much more challenging task compared to weakly-supervised object detection. The key reason is the additional variation in content as well as the length along the temporal axis in videos. Few works in the literature (Bojanowski et al., 2015; Huang et al., 2016) have considered the availability of temporal order of activities, apart from the video-level labels during training.

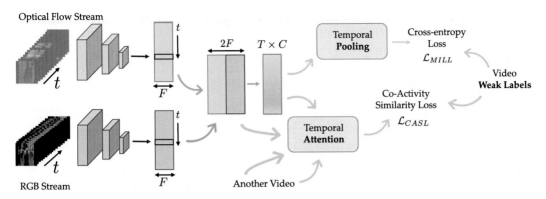

FIGURE 3.7 This figure presents the proposed framework for weakly-supervised activity localization and classification. Given a video, we extract features from two streams – RGB and Optical Flow. After concatenating the feature vectors from the two streams, we learn a few layers specific to the task of weak localization and finally project to the category space to obtain a $T \times C$ matrix where T and C are the number of time steps and categories respectively. We utilize two loss functions to learn the network parameters – Cross-entropy loss on the temporally pooled predictions, and CoActivity Loss obtained using a pair of videos containing at least one category in common.

We next formally present the weakly supervised temporal localization task, followed by a solution to it.

Problem statement   Consider that we have a training set of n videos $\mathcal{X} = \{x_i\}_{i=1}^{n}$ with variable temporal duration denoted by $L = \{l_i\}_{i=1}^{n}$ (after feature extraction) and activity label set $\mathcal{A} = \{a_i\}_{i=1}^{n}$, where $a_i = \{a_i^j\}_{j=1}^{m_i}$ are the $m_i (\geq 1)$ labels for the ith video. We also define the set of activity categories as $\mathcal{S} = \bigcup_{i=1}^{n} a_i = \{\alpha_i\}_{i=1}^{C}$. During test time, given a video x, we need to predict a set $x_{det} = \{(s_j, e_j, c_j, p_j)\}_{j=1}^{n(x)}$, where $n(x)$ is the number of detections for x. s_j, e_j are the start time and end time of the jth detection, c_j represents its predicted activity category with confidence p_j. Fig. 3.7 presents an overview of a framework for weakly supervised temporal event localization, the details of which are progressively discussed next.

3.3.1 Network architecture

We focus particularly on two-stream networks, as they encapsulate the information from both the appearance features in the RGB stream and motion features in the Optical Flow stream. We utilize two networks – UntrimmedNets (Wang et al., 2017) pretrained on Imagenet and I3D (Carreira and Zisserman, 2017) for feature extraction. Please note that the rest of our framework is agnostic to the features used. The number of frames to send as input to these networks depends on its architecture. The RGB stream of UntrimmedNets takes 1 frame as input, whereas its Flow stream takes five frames for every feature vector. In the case of the I3D network, both the RGB and the Flow stream take in 16 frames for every feature vector.

Natural videos may have large variations in length, from a few seconds to more than an hour. In the weakly-supervised setting, we have information about the labels for the video as a whole, thus requiring it to process the entire video at once. This may be problematic for very long videos due to GPU memory constraints. As a solution to this problem, we send the entire video as input, if its length is less than the predefined length T determined by the GPU bandwidth. However, if the length of the video is greater than T, we randomly extract from it a clip of length T with contiguous frames and assign all the labels of the entire video to the extracted video clip. It may be noted that although this may introduce some errors in the labels, this way of sampling does have the advantages of data augmentation and performs well in practice.

After extracting the video features from the two-stream networks, we obtain a matrix of dimension $X_i \in \mathbb{R}^{T \times 2F}$, where T is the number of time steps of the video and F are the feature dimension of RGB and Flow streams, which are concatenated to obtain the features of dimension $2F$. We then pass these features through a fully-connected layer with ReLU nonlinearity and dropout, followed by a classification layer to finally obtain a matrix of categorical predictions $\mathcal{A} \in \mathbb{R}^{T \times C}$, where C is the number of categories.

3.3.2 k-max multiple instance learning

The weakly-supervised activity localization and classification problem as described above can be directly mapped to the problem of Multiple Instance Learning (MIL) (Zhou, 2004). In MIL, individual samples are grouped into two bags, namely positive and negative bags. A positive bag contains at least one positive instance and a negative bag contains no positive instance. Using these bags as training data, we need to learn a model, which will be able to distinguish each instance to be positive or negative, besides classifying a bag. In our case, we consider the entire video as a bag of instances, where each instance is represented by a feature vector at a certain time instant. In order to compute the loss for each bag, i.e., video in our case, we need to represent each video using a single confidence score per category.

For a given video, we compute the activation score corresponding to a particular category as the average of k-max activation over the temporal dimension for that category. As the number of videos in a bag varies widely, we set k proportional to the number of elements in the bag. Thereafter, a softmax nonlinearity is applied to obtain the probability mass function over all the categories, which allows one to compute a vector of predictions p_i over the categories. We need to compare this pmf with the ground truth distribution of labels for each video in order to compute the MIL loss (MILL). As each video can have multiple activities occurring in it, we represent the label vector for a video with ones at the temporal positions if that activity occurs in the video, else zero. We then normalize this ground truth vector in order to convert it to a legitimate pmf. The MILL is then the cross-entropy between the predicted pmf p_i and ground-truth, which can then be represented as follows,

$$\mathcal{L}_{MILL} = \frac{1}{n} \sum_{i=1}^{n} \sum_{j=1}^{C} -y_i^j \log\left(p_i^j\right) \tag{3.8}$$

where $y_i = [y_i^1, \dots, y_i^C]^T$ is the normalized ground truth vector representing the weak labels.

3.3.3 Coactivity similarity

The CoActivity Similarity Loss (CASL) enforces constraints to learn better network parameters for activity localization. The Weakly supervised Temporal Activity Localization and

Classification (W-TALC) problem motivates us to identify the correlations between videos of similar categories. Before discussing in more detail, let us define category-specific sets for the jth category as, $\mathcal{S}_j = \{x_i \mid \exists\, a_i^k \in a_i, \text{s.t. } a_i^k = \alpha_j\}$, i.e., the set $\mathcal{S}_j$ contains all the videos of the training set, which has activity α_j as one of its labels. Ideally, we may want the following properties in the learned feature representations X_i discussed in Section 3.3.1.

- A video pair belonging to the set $\mathcal{S}_j$ (for any $j \in \{1, \ldots, C\}$) should have similar feature representations in the portions of the video where the activity α_j occurs.
- For the same video pair, the feature representation of the portion where α_j occurs in one video should be different from that of the other video where α_j does not occur.

These properties are not directly enforced in the MILL. Thus, we introduce CoActivity Similarity Loss to embed the desired properties in the learned feature representations. As we do not have frame-wise labels, we use the category-wise activations $\mathcal{A}$ to identify the required activity portions. The loss function is designed in a way that helps to learn simultaneously the feature representation as well as the label space projection. We first normalize the per-video category-wise activations scores along the temporal axis using softmax nonlinearity to obtain $\hat{\mathcal{A}}$ at time t and category j as $\hat{\mathcal{A}}_i[j, t] = \exp(\mathcal{A}_i[t, j]) / \sum_{t'=1}^{l_i} \exp(\mathcal{A}_i[t', j])$ where t indicates the time instants and $j \in \{1, \ldots, C\}$. We refer to these as *attention*, as they attend to the portions of the video where the activity of a certain category occurs. A high value of attention for a particular category indicates its high occurrence-probability of that category. In order to formulate the loss function, let us first define the category-wise feature vectors of regions with high and low attention as follows:

$$^H f_i^j = X_i \hat{\mathcal{A}}_i[:, j]$$
$$^L f_i^j = \frac{1}{l_i - 1} X_i (1 - \hat{\mathcal{A}}_i[:, j]) \tag{3.9}$$

where $^H f_i^j, {}^L f_i^j \in \mathbb{R}^{2048}$ represents the high and low attention region aggregated feature representations respectively of video i for category j. In order to enforce the two properties discussed above, we use the ranking hinge loss. Given a pair of videos $x_m, x_n \in \mathcal{S}_j$, the loss function may be represented as follows:

$$\mathcal{L}_j^{mn} = \frac{1}{2}\Big\{ \max\big(0, d[^H f_m^j, {}^H f_n^j] - d[^H f_m^j, {}^L f_n^j] + \delta\big)$$
$$+ \max\big(0, d[^H f_m^j, {}^H f_n^j] - d[^L f_m^j, {}^H f_n^j] + \delta\big)\Big\} \tag{3.10}$$

where $d[]$ is the cosine distance and δ is the margin parameter and we set it to 0.5 in our experiments. The two terms in the loss function are equivalent in meaning, and they represent that the high attention region features in both the videos should be more similar than the high attention region feature in one video and the low attention region feature in the other video. The total loss for the entire training set is computed for every pair of videos having at least one category in common. The two loss functions in Eq. (3.8) and (3.10) can be optimized jointly to learn the network parameters.

Localization   After learning the weights of the network, we use them to localize the events in a video during test time. Given a video, we obtain the category-wise confidence scores $\mathcal{A}$. For every category, we obtain a threshold, which is the middle point between the maximum and minimum activations for that category, and use that to threshold the activations to obtain the localizations.

3.3.4 Applications

In this section, we understand the efficacy of the proposed framework for activity localization and classification from weakly labeled videos. We first discuss the datasets, followed by the implementation details, quantitative, and some qualitative results.

Datasets

We perform experimental analysis on two datasets namely ActivityNet v1.2 (Heilbron et al., 2015) and Thumos14 (Idrees et al., 2017). These two datasets contain untrimmed videos with frame-wise labels of activities occurring in the video. However, as our algorithm is weakly-supervised, we use only the activity tags associated with the videos. The ActivityNet v1.2 dataset has 4819 videos for training, 2383 videos for validation, which we use for testing. The number of categories involved is 100, with an average of 1.5 temporal activity segments per video. The Thumos14 dataset has 200 validation videos, which we use for training, and 212 test videos divided into 20 categories. Among these videos, 200 validation videos and 213 test videos have temporal annotations belonging to 20 categories. Although this is a smaller dataset than ActivityNet1.2, the temporal labels are very precise and have an average of 15.5 temporal activity segments per video. This dataset has several videos where multiple activities occur, thus making it even more challenging. The length of the videos also varies widely from a few seconds to more than an hour. The lower number of videos makes it challenging to efficiently learn the weakly-supervised network.

Activity localization

To compare the performance of activity localization, Mean Average Precision (mAP) at different Intersection over Union (IoU) thresholds between the predicted and the ground truth localization is used. For Thumos14, we discuss the average mAP for IoU thresholds $\in \{0.1, 0.2, 0.3, 0.4, 0.5\}$. The results are presented in Table 3.1. The table is divided into three rows: (a) methods using strong supervison, i.e., using temporal annotations for every action that appears in the videos, (b) methods using weak supervision in literature, and finally (c) results of the method proposed in Paul et al. (2018). So, even with weak labels, which are much easier to acquire, our algorithm in Paul et al. (2018) is able to perform close to methods with strong labels, which involves a huge amount of manual labeling cost. It is important to note that although the Kinetics pretrained I3D features (I3DF) have some knowledge about activities, using only MILL as in Wang et al. (2017) along with I3DF performs much worse than combining it with CASL, viz. 33.1 vs 39.8. In the case of ActivityNet v1.2, our method performs on average 18, whereas methods utilizing strong supervision such as SSN perform 24.8.

TABLE 3.1 Detection performance comparisons on Thumos14. UNTF & I3DF are abbreviations for UntrimmedNet features (ImageNet pretrained features) and I3D features respectively.

Supervision	Methods	Avg. IoU 0.1:0.1:0.5
Strong	R-C3D (Xu et al., 2017)	43.1
	SSN (Zhao et al., 2017b)	47.4
Weak	UntrimmedNets (Wang et al., 2017)	29.0
	STPN (UNTF) (Nguyen et al., 2018)	30.9
	STPN (I3DF) (Nguyen et al., 2018)	34.9
Weak (Paul et al., 2018)	MILL+CASL+UNTF	33.8
	MILL+CASL+I3DF	39.8

TABLE 3.2 Classification performance comparisons on Thumos14. UNTF & I3DF are abbreviations for UntrimmedNet features (ImageNet pretrained features) and I3D features respectively.

Supervision	Methods	Thumos14	ActivityNet-1.2
Strong	TSN (Wang et al., 2016b)	72.0	86.3
Weak	UntrimmedNets (Wang et al., 2017)	82.2	91.3
	MILL+CASL (Paul et al., 2018)	85.6	93.2

Activity classification

We now present the performance of our framework for activity classification. We use mean average precision (mAP) to compute the classification performance from the predicted videos-level scores p after applying softmax. The results are presented in Table 3.2. It shows that the proposed method (Paul et al., 2018) performs significantly better than other state-of-the-art approaches video classification methods. This can be partially attributed to the features used, but most importantly to the way learning is considered in weakly supervised localization techniques, which ignore the background regions while classifying the videos with their content activity.

Qualitative results

We present a few interesting example localizations with ground truths in Fig. 3.8. The figure has two examples from Thumos14 and two from the ActivityNet1.2 dataset. To test how the proposed framework performs on videos outside the aforementioned datasets, we tested the learned networks on randomly collected videos from YouTube. We present two such example detections in Fig. 3.8, using the model trained on Thumos14.

The first example in Fig. 3.8 is quite challenging as the localization should precisely be the portions of the video, where Golf Swing occurs, which has very similar features in the RGB domain to portions of the video where the player prepares for the swing. In spite of this, our model is able to localize the relevant portions of Golf Swing, potentially based on the flow features. In the second example from Thumos14, the detections of Cricket Shot and Cricket Bowl appear to be correlated in time. This is because Cricket Shot and Bowl are two activities that generally cooccur in videos. To have fine-grained localization for such activities, videos

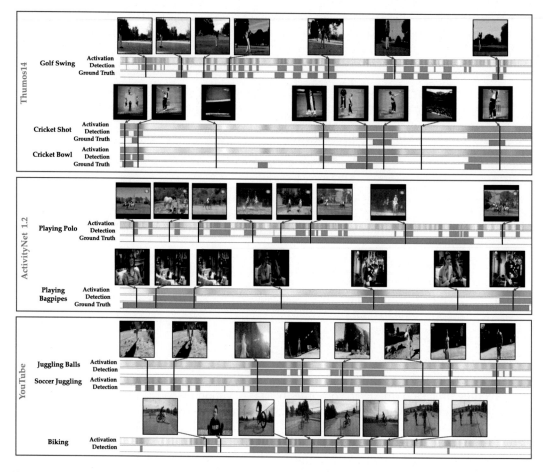

FIGURE 3.8 This figure presents some detection results for qualitative analysis on Thumos14, ActivityNet1.2, and a couple of random videos from YouTube.

that have only one of these activities are required. However, in the Thumos14 dataset, very few training examples contain only one of these two activities, which explains the behavior noted in the figure.

In the third example, which is from ActivityNet1.2, although 'Playing Polo' occurs in the first portion of the video, it is absent in the ground truth. However, our model is able to localize those activity segments as well. The same discussion is also applicable to the fourth example, where 'Bagpiping' occurs in the frames in a sparse manner, and our model's response is aligned with its occurrence, but the ground truth annotations are for almost the entire video. These two examples are motivations behind weakly-supervised localization because obtaining precise unanimous ground truths from multiple labelers is difficult, costly, and sometimes even infeasible.

The fifth example is on a randomly selected video from YouTube. It has a person, who is juggling balls in an outdoor environment. But, most of the examples in Thumos14 of the same category are indoors, with the person taking up a significant portion of the frames spatially. Despite such differences in data, our model is able to localize some portions of the activity. However, the model also predicts some portions of the video to be 'Soccer Juggling', which may be because its training samples in Thumos14 contains a combination of feet, hand, and head, and a subset of such movements are present in 'Juggling Balls'. Moreover, it is interesting to note that the first two frames show some maneuver of a ball with feet and it is detected as 'Soccer Juggling' as well.

3.4 Domain adaptation of semantic segmentation using weak labels

Semantic segmentation is a task where given an input image, we need to learn a model, to predict the category of every pixel in the image. In current state-of-the-art methods (Chen et al., 2016; Zhao et al., 2017a), the model is generally a convolutional neural networks, which learns using pixel-level annotations. However, the segmentation model learned on one dataset, say the source, may not generalize well to images from a different distribution, the target, due to the domain gap between them. Thus, the model needs to be adapted to the images originating from the target distribution. But as annotations may be expensive on the target side, we want to adapt the model from source to target with minimal or even no annotation cost.

Unsupervised domain adaptation (UDA) methods for semantic segmentation have been developed to tackle the issue of domain gap, requiring no annotation cost on the target images. Methods in the literature aim to adapt a model learned on the source domain with pixel-wise ground truth annotations, e.g., from a simulator that requires the least annotation effort, to the target domain that does not have any form of annotations. These UDA methods in the literature for semantic segmentation are developed mainly using two mechanisms: pseudo-label self-training and distribution alignment between the source and target domains. For the first mechanism, pixel-wise pseudo-labels are generated via strategies such as confidence scores (Li et al., 2019; Hung et al., 2018) or self-paced learning (Zou et al., 2018), but such pseudo-labels are specific to the target domain and do not consider the alignment between domains. For the second mechanism, numerous spaces could be considered to operate the alignment procedure, such as pixel (Hoffman et al., 2018; Murez et al., 2018), feature (Hoffman et al., 2016; Zhang et al., 2017), output (Tsai et al., 2018; Chen et al., 2018), and patch (Tsai et al., 2019) spaces. However, the alignment performed by these methods is agnostic to the category, which may be problematic as the domain gap may vary across categories.

The issue of lacking annotations in the target domain can be alleviated, by introducing the concept of utilizing *weak labels* on the target dataset for adaptation. Such weak labels can be used for category-wise alignment between the source and target domain, and also to enforce constraints on the categories present in an image. There can be multiple forms of weak label – image-level labels, point labels, which we explore in this text, as well as other forms of weak label such as pixel density of certain categories, object counts, etc, which is quite easy to acquire from an annotator. It is important to note that our weak labels could be estimated from the model prediction in the UDA setting, or provided by the human oracle in

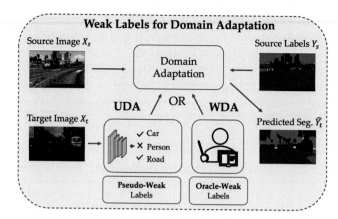

FIGURE 3.9 This figure presents how we can use image-level weak annotations for domain adaptation in two different ways – either estimated, i.e., pseudo-weak labels (Unsupervised Domain Adaptation, UDA) or acquired from a human oracle (Weakly-supervised Domain Adaptation (WDA)).

the weakly-supervised domain adaptation (WDA) paradigm as shown in Fig. 3.9. The target annotation cost in UDA is zero, and in the WDA case is very low, as we will see subsequently.

The literature in domain adaptation for semantic segmentation models has been only around unsupervised methods (UDA), and it can be summarized under three categories – pixel-level adaptation (Hoffman et al., 2018; Murez et al., 2018; Wu et al., 2018), which aims at aligning the input image space, pseudo-label training (Zou et al., 2018; Sadat Saleh et al., 2018; Lian et al., 2019), which aims at labeling the unlabeled target data using the source model and using it for adaptation, and feature or output space adaptation (Tsai et al., 2018; Chen et al., 2018; Tsai et al., 2019; Du et al., 2019), which aims at aligning the output space between the source and target domains. The performance of these methods is quite low compared to methods with strong supervision. In the text below we present our method (Paul et al., 2020) which can be used for both unsupervised and weakly-supervised domain adaptation, and bridging the performance gap with minimal or no annotation cost. We start with the formal problem definition.

Problem definition   We have two domains – source and target. The goal is to adapt a segmentation model learned on the source domain to the target domain. In the source domain, we have images and pixel-wise labels denoted as $\mathcal{I}_s = \{X_s^i, Y_s^i\}_{i=1}^{N_s}$. Our target dataset contains images and only image-level labels as $\mathcal{I}_t = \{X_t^i, y_t^i\}_{i=1}^{N_t}$. $X_s, X_t \in \mathbb{R}^{H \times W \times 3}$, $Y_s \in \mathbb{B}^{H \times W \times C}$ are pixel-wise one-hot vectors, $y_t \in \mathbb{B}^C$ is a multihot vector representing the categories present in the image and C is the number of categories, for both the source and target datasets. Note that in a one-hot vector, only one of the element of the vector is 1, and remaining are zero, while in multihot vectors, multiple elements of the vector can be 1. Such image-level labels y_t are termed as weak labels, as they are a much weaker form of label compared to pixel-wise labels. We can either estimate them, in which case we call them pseudo-weak labels (Unsupervised Domain Adaptation, UDA), or acquire them from a human oracle and call them oracle-weak labels (Weakly-supervised Domain Adaptation, WDA). We will further discuss

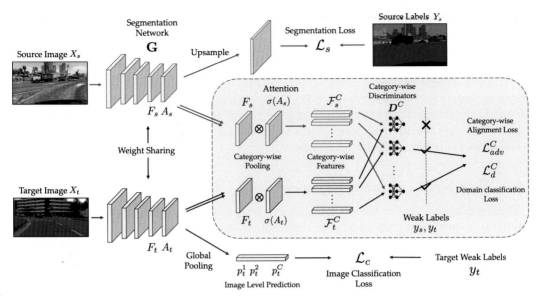

FIGURE 3.10 The proposed architecture consists of the segmentation network G and the weak label module. We compute the pixel-wise segmentation loss $\mathcal{L}_s$ for the source images and image classification loss $\mathcal{L}_c$ using the weak labels y_t for the target images. Note that the weak labels can be estimated as pseudo-weak labels or provided by a human oracle. We then use the output prediction A, convert it to an attention map $\sigma(A)$ and pool category-wise features $\mathcal{F}^C$. Next, these features are aligned between source and target domains using the category-wise alignment loss $\mathcal{L}^C_{adv}$ guided by the category-wise discriminators D^C learned via the domain classification loss $\mathcal{L}^C_d$.

the details of acquiring weak labels in Section 3.4.4. Given such data, the problem is to adapt a segmentation model G learned on the source dataset $\mathcal{I}_s$ to the target dataset $\mathcal{I}_t$.

3.4.1 Weak labels for category classification

We use the weak labels y_t and learn to predict the categories present/absent in the target images. We first feed the target images X_t through G to obtain the predictions $A_t \in \mathbb{R}^{H' \times W' \times C}$ and then apply a global pooling layer to obtain a single vector of predictions for each category:

$$p_t^c = \sigma_s \left[\frac{1}{k} \log \frac{1}{H'W'} \sum_{h',w'} \exp k A_t^{(h',w',c)} \right], \tag{3.11}$$

where σ_s is the sigmoid function such that p_t represents the probability that a particular category appears in an image. Note that Eq. (3.11) is a smooth approximation of the `max` function. The higher the value of k, the better it approximates to `max`. We set $k = 1$ as we do not want the network to focus only on the maximum value of the prediction, which may be noisy, but also on other predictions that may have high values. Using p_t and the weak labels y_t, we can compute the category-wise binary cross-entropy loss:

$$\mathcal{L}_c(X_t; G) = \sum_{c=1}^{C} -y_t^c \log(p_t^c) - (1 - y_t^c) \log(1 - p_t^c). \tag{3.12}$$

This is shown at the bottom stream of Fig. 3.10. This loss function $\mathcal{L}_c$ helps to identify the categories which are absent/present in a particular image and enforces the segmentation network G to pay attention to those objects/stuff that are partially identified when the source model is used directly on the target images.

3.4.2 Weak labels for feature alignment

The classification loss using weak labels introduced in (3.12) regularizes the network focusing on certain categories. However, distribution alignment across the source and target domains is not considered yet. As discussed in the previous section, methods in the literature either align feature space (Hoffman et al., 2016) or output space (Tsai et al., 2018) across domains. However, such alignment is agnostic to the category, so it may align features of categories that are not present in certain images. Moreover, features belonging to different categories may have different domain gaps. Thus, performing category-wise alignment could be beneficial but has not been widely studied in UDA for semantic segmentation. To alleviate these issues, we use image-level weak labels to perform category-wise alignment in the feature space. Specifically, we obtain the category-wise features for each image via an attention map, i.e., segmentation prediction, guided by our classification module using weak labels, and then align these features between the source and target domains. We next discuss the category-wise feature pooling mechanism followed by the adversarial alignment technique.

Category-wise feature pooling. Given the last layer features $F \in \mathbb{R}^{H' \times W' \times 2048}$ and the segmentation prediction $A \in \mathbb{R}^{H' \times W' \times C}$, we obtain the category-wise features for the cth category as a 2048-dimensional vector by using the prediction as an attention over the features as follows:

$$\mathcal{F}^c = \sum_{h', w'} \sigma(A)^{(h', w', c)} F^{(h', w')}, \tag{3.13}$$

where $\sigma(A)$ is a tensor of dimension $H' \times W' \times C$, with each channel along the category dimension representing the category-wise attention obtained by the softmax operation σ over the spatial dimensions. As a result, $\sigma(A)^{(h', w', c)}$ is a scalar and $F^{(h', w')}$ is a 2048-dimensional vector, while $\mathcal{F}^c$ is the summed feature of $F^{(h', w')}$ weighted by $\sigma(A)^{(h', w', c)}$ over the spatial map $H' \times W'$. Note that we drop the subscripts s, t for source and target, as we employ the same operation to obtain the category-wise features for both domains. We next present the mechanism to align these features across domains. Note that we will use $\mathcal{F}^c$ to denote the pooled feature for the cth category and $\mathcal{F}^C$ to denote the set of pooled features for all the categories. Category-wise feature pooling is shown in the middle of Fig. 3.10.

Category-wise feature alignment. To learn the segmentation network G such that the source and target category-wise features are aligned, we use an adversarial loss while using category-specific discriminators $D^C = \{D^c\}_{c=1}^{C}$. The reason for using category-specific discriminators is to ensure that the feature distribution for each category could be aligned independently, which avoids the noisy distribution modeling from a mixture of categories. In practice, we train C distinct category-specific discriminators, to distinguish between category-wise features drawn from the source and target images. The loss function to train the discriminators $\mathbf{D}^C$ is as follows:

$$\mathcal{L}_d^C\left(\mathcal{F}_s^C, \mathcal{F}_t^C; \mathbf{D}^C\right) = \sum_{c=1}^{C} -y_s^c \log \mathbf{D}^c\left(\mathcal{F}_s^c\right) - y_t^c \log\left(1 - \mathbf{D}^c\left(\mathcal{F}_t^c\right)\right). \tag{3.14}$$

Note that, while training the discriminators, we only compute the loss for those categories which are present in the particular image via the weak labels $y_s, y_t \in \mathbb{B}^C$ that indicate whether a category occurs in an image or not. Then, the adversarial loss for the target images to train the segmentation network $\mathbf{G}$ can be expressed as follows:

$$\mathcal{L}_{adv}^C\left(\mathcal{F}_t^C; \mathbf{G}, \mathbf{D}^C\right) = \sum_{c=1}^{C} -y_t^c \log \mathbf{D}^c\left(\mathcal{F}_t^c\right). \tag{3.15}$$

Similarly, we use the target weak labels y_t to align only those categories present in the target image. By minimizing $\mathcal{L}_{adv}^C$, the segmentation network tries to fool the discriminator by maximizing the probability of the target category-wise feature being considered as drawn from the source distribution. These loss functions in (3.14) and (3.15) are obtained in the right of the middle box in Fig. 3.10.

3.4.3 Network optimization

Discriminator training. We learn a set of C distinct discriminators for each category c. We use the source and target images to train the discriminators, which learn to distinguish between the category-wise features drawn from either the source or the target domain. The optimization problem to train the discriminator can be expressed as: $\min_{\mathbf{D}^C} \mathcal{L}_d^C(\mathcal{F}_s^C, \mathcal{F}_t^C)$. Note that each discriminator is trained only with features pooled specific to that particular category. Therefore, given an image, we update only those discriminators corresponding to those categories which are present in the image.

Segmentation network training. We train the segmentation network with the pixel-wise cross-entropy loss $\mathcal{L}_s$ on the source images, image classification loss $\mathcal{L}_c$ and adversarial loss $\mathcal{L}_{adv}^C$ on the target images. We combine these loss functions to learn $\mathbf{G}$ as follows:

$$\min_{\mathbf{G}} \mathcal{L}_s(X_s) + \lambda_c \mathcal{L}_c(X_t) + \lambda_d \mathcal{L}_{adv}^C\left(\mathcal{F}_t^C\right). \tag{3.16}$$

We follow the standard GAN training procedure (Goodfellow et al., 2014) to alternatively update $\mathbf{G}$ and $\mathbf{D}^C$. Note that, computing $\mathcal{L}_{adv}^C$ involves the category-wise discriminators $\mathbf{D}^C$. Therefore, we fix $\mathbf{D}^C$ and backpropagate gradients only for the segmentation network $\mathbf{G}$.

3.4.4 Acquiring weak labels

In the above sections, we have proposed a mechanism to utilize image-level weak labels of the target images and adapt the segmentation model between source and target domains. In this section, we explain two methods to obtain such image-level weak labels.

Pseudo-weak labels (UDA). One way of obtaining weak labels is to directly estimate them using the data we have, i.e., source images/labels and target images, which is the unsupervised domain adaptation (UDA) setting. In this work, we utilize the baseline model (Tsai et al., 2018) to adapt a model learned from the source to the target domain, and then obtain the weak labels of the target images as follows:

$$y_t^c = \begin{cases} 1, & \text{if } p_t^c > T, \\ 0, & \text{otherwise} \end{cases} \tag{3.17}$$

where p_t^c is the probability for category c as computed in (3.11) and T is a threshold, which we set to 0.2 in all the experiments unless specified otherwise. In practice, we compute the weak labels online during training and avoid any additional inference step. Specifically, we forward a target image, obtain the weak labels using (3.17), and then compute the loss functions in (3.16). As the weak labels obtained in this manner do not require human supervision, adaptation using such labels is unsupervised.

Oracle-weak labels (WDA). In this form, we obtain the weak labels by querying a human oracle to provide a list of the categories that occur in the target image. As we use supervision from an oracle on the target images, we refer to this as weakly-supervised domain adaptation (WDA). It is worth mentioning that the WDA setting could be practically useful, as collecting such human annotated weak labels is much easier than pixel-wise annotations. Also, there has not been any prior research involving this setting for domain adaptation.

To show that our method can use different forms of oracle-weak labels, we further introduce the point supervision as in Bearman et al. (2016), which only increases effort by a small amount compared to the image-level supervision. In this scenario, we randomly obtain one pixel coordinate of each category that belongs in the image, i.e., the set of tuples $\{(h^c, w^c, c) | \forall y_t^c = 1\}$. For an image, we compute the loss as follows: $\mathcal{L}_{point} = -\sum_{\forall y_t^c=1} y_t^c \log(O_t^{(h^c, w^c, c)})$, where $O_t \in \mathbb{R}^{H \times W \times C}$ is the output prediction of target after pixel-wise softmax.

3.4.5 Applications

In this section, we perform an evaluation of our domain adaptation framework for semantic segmentation where the source is a dataset made of simulated street scene images (GTA5 (Richter et al., 2016)) and the target is made of real-world images (Cityscapes (Cordts et al., 2016)). We use the Intersection-over-Union (IoU) ratio as the metric. For the segmentation network G, we use the DeepLab-v2 framework (Chen et al., 2016) with the ResNet-101 (He et al., 2016) architecture. We extract features F_s, F_t before the Atrous Spatial Pyramid Pooling (ASPP) layer. For the category-wise discriminators $D^C = \{D^c\}_{c=1}^C$, we use C separate networks, where each consists of three fully-connected layers, having number of nodes

{2048, 2048, 1} with ReLU activation. Fig. 3.11 presents the results obtained from the literature and our methods for the variety of annotations we use.

State-of-the-art methods

When we look at the state-of-the-art methods in the literature, all of them are unsupervised, i.e., requiring no annotation cost on the target side. In such a setting, the performance levels obtained is quite low. For example, in the current setting mentioned above, i.e., with GTA5 as the source and Cityscapes as the target, the performance obtained by state-of-the-art methods is 45.4 (Chang et al., 2019), 46.5 (Tsai et al., 2019) and 47.2 (Li et al., 2019). Note that this percentage is higher than the performance obtained by the model trained on the source and applied directly on the target, which is only 36.6. However, it is much lower than the performance obtained when all the target images are labeled with pixel-level annotations, which is 64.4, but requiring about 90 minutes of annotation time per image.

Unsupervised domain adaptation (UDA)

In comparison to the methods in the literature, when we use a basic method (Tsai et al., 2018) to align just the output space, we obtain performance of 41.4. Then, when we use the classification loss as in Eq. (3.12), we obtain a performance of only 46.7. On top of this, when we include the category-wise alignment loss as in Eq. (3.15), we obtain a performance of 48.2. Note that in this method, we use the pseudo-weak labels, which are estimated by the network itself as in Eq. (3.17).

Weakly-supervised domain adaptation (WDA)

We use two different types of annotation in the WDA and they are discussed as follows. Note that other forms of weak supervision can be count of the objects, very coarse estimate of area coverage of the categories, scribbles, and so on.

Image-level supervision   When we use the image-level annotations from the user and learn the segmentation network using only the classification loss in Eq. (3.12), we obtain a performance of 52. On top of this, when we include the category-wise domain alignment loss, we obtain a performance of 53. Note that there exists no work in the literature which uses oracle-weak labels from humans to perform WDA. From the results, it is interesting to note that the major boost in performance using WDA compared to UDA occurs for categories such as truck, bus, train, and motorbike. One reason is that those categories are most underrepresented in both the source and the target datasets. Thus, they are not predicted in most of the target images, but using the oracle-weak labels helps to identify them better.

Point supervision   We introduce another interesting setting of point supervision as in Bearman et al. (2016), which adds only a slight increase of annotation time compared to that for image-level supervision. We follow (Bearman et al., 2016) and randomly sample one pixel per category in each target image as the supervision. Note that, all the details and the modules are the same during training in this setting. In this setting with one point labeled per category per image, we obtain a performance of 56.4. When we increase the number of annotations to three or five point labeled per category per image, we obtain performance of 58.4 and 59.4 respectively. Note that the annotation cost required to obtain these annotations is quite low as shown in Fig. 3.11, but yet they are able to obtain comparable performance as with pixel-wise annotations which incur a huge amount of annotation cost.

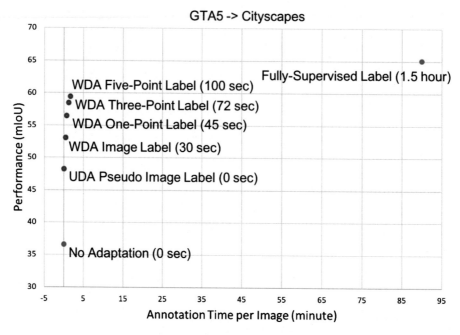

FIGURE 3.11 Performance comparison on GTA5→Cityscapes with different levels of supervision on target images: no target labels ("No Adapt." and "UDA"), weak image labels (30 seconds), one point labels (45 seconds), and fully-supervised setting with all pixels labeled ("All Labeled") that takes 1.5 hours per image according to Cordts et al. (2016).

3.4.6 Output space visualization

We present some visualizations of the segmentation prediction probability for each category in Fig. 3.12. Before using any weak labels (third row), the probabilities may be low, even though there is a category present in that image. However, based on these initial predictions, our model can estimate the categories and then enforce their presence/absence explicitly in the proposed classification loss and alignment loss. The fourth row in Fig. 3.12 shows that such pseudo-weak labels help the network discover object/stuff regions towards better segmentation. For example, the fourth and fifth column shows that, although the original prediction probabilities are quite low, results using pseudo-weak labels are estimated correctly. Moreover, the last row shows that the predictions can be further improved when we have oracle-weak labels.

3.5 Weakly-supervised reinforcement learning for dynamical tasks

Until now in this chapter, we primarily looked into static problems in the light of limited supervision. However, in real-world scenarios, we may have to build a system that interacts with the environment to achieve a task and thus to be dynamic in nature as its current

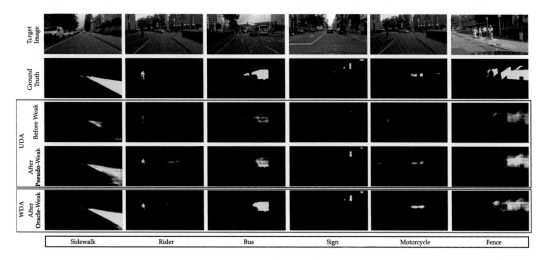

FIGURE 3.12 Visualizations of category-wise segmentation prediction probability before and after using the pseudo-weak labels on GTA5→Cityscapes. Before adaptation, the network only highlights the areas partially with low probability, while using the pseudo-weak labels helps the adapted model obtain much better segments, and is closer to the model using oracle-weak labels.

action determines the data point it is going to receive next. Similar to static tasks such as classification, localization, and segmentation, learning dynamic tasks are also quite challenging in nature. Reinforcement Learning (RL) using Deep Neural Networks (DNNs) has shown tremendous success in several such dynamical tasks such as playing games (Mnih et al., 2015; Silver et al., 2016), solving complex robotics tasks (Levine et al., 2016; Duan et al., 2016), etc. However, with sparse rewards, these algorithms often require a huge number of interactions with the environment, which is costly in real-world applications such as self-driving cars (Bojarski et al., 2016), and manipulations using real robots (Levine et al., 2016). Manually designed dense reward functions could mitigate such issues; however, in general, it is difficult to design detailed reward functions for complex real-world tasks.

Imitation Learning (IL) using demos generated by an expert can potentially be used to learn the policies faster (Argall et al., 2009). But, the performance of IL algorithms (Ross et al., 2011) is not only dependent on the performance of the expert providing the demos, but also on the state-space distribution represented by the demos, especially in the case of high dimensional states. In order to avoid such dependencies on the expert, some methods in the literature (Sun et al., 2017; Cheng et al., 2018) take the path of combining RL and IL. However, these methods assume access to the expert value function, which may become impractical in real-world scenarios.

Keeping the shortcomings of the methods in the literature in mind, we present a strategy (Paul et al., 2019) that starts with IL and then switches to RL. In the IL step, our framework performs supervised pretraining which aims at learning a policy that best describes the expert demonstrations. However, due to the limited availability of expert demos, the policy trained with IL will have errors, which can then be alleviated using RL. However, note that the reward function in RL is still sparse, making it difficult to learn. With this in mind, we

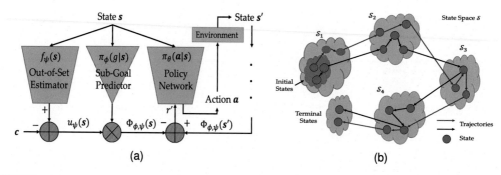

(a)   (b)

FIGURE 3.13 (a) This shows an overview of our proposed framework to train the policy network along with subgoal based reward function with out-of-set augmentation. (b) An example state-partition with two independent demos in black and red. Note that the terminal state is shown as a separate state partition because we assume it to be indicated by the environment and not learned.

present a method that uses the human demos to divide the entire task into smaller subgoals and use that as a reward function in the RL step.

Given a set of demos, humans can quickly identify waypoints, which need to be completed in order to achieve the goal. We tend to break down the entire complex task into subgoals and try to achieve them in the best order possible. Prior knowledge of humans helps to achieve tasks much faster (Andreas et al., 2017; Dubey et al., 2018) than using only the demos for learning. The human psychology of divide-and-conquer has been crucial in several applications and it serves as a motivation behind our algorithm which learns to partition the state-space into subgoals using expert demos. The learned subgoals provide a discrete reward signal, unlike value-based continuous reward (Ng et al., 1999; Sun et al., 2018), which can be erroneous, especially with a limited number of demos in long time horizon tasks. As the expert demos set may not contain all the states where the agent may visit during exploration in the RL step, we augment the subgoal predictor via one-class classification to deal with such underrepresented states. We perform experiments on three goal-oriented tasks on MuJoCo (Todorov, 2014) with sparse terminal-only reward, which state-of-the-art RL, IL, or their combinations are not able to solve.

Problem definition Consider a standard RL setting where an agent interacts with an environment which can be modeled by a Markov Decision Process (MDP) $\mathcal{M} = (\mathcal{S}, \mathcal{A}, \mathcal{P}, r, \gamma, \mathcal{P}_0)$, where $\mathcal{S}$ is the set of states, $\mathcal{A}$ is the set of actions, r is a scalar reward function, $\gamma \in [0, 1]$ is the discount factor and $\mathcal{P}_0$ is the initial state distribution. Our goal is to learn a policy $\pi_\theta(a|s)$, with $a \in \mathcal{A}$, which optimizes the expected discounted reward $\mathbb{E}_\tau[\sum_{t=0}^{\infty} \gamma^t r(s_t, a_t)]$, where $\tau = (\dots, s_t, a_t, r_t, \dots)$ and $s_0 \sim \mathcal{P}_0$, $a_t \sim \pi_\theta(a|s_t)$ and $s_{t+1} \sim \mathcal{P}(s_{t+1}|s_t, a_t)$.

With sparse rewards, optimizing the expected discounted reward using RL may be difficult. In such cases, it may be beneficial to use a set of state-action demonstrations $\mathcal{D} = \{\{(s_{ti}, a_{ti}^*)\}_{t=1}^{n_i}\}_{i=1}^{n_d}$ generated by an expert to guide the learning process. n_d is the number of demos in the dataset and n_i is the length of the ith demo. We propose a methodology to efficiently use $\mathcal{D}$ by discovering subgoals from these demos and use them to develop an extrinsic reward function.

Subgoal definition

Consider that the state-space S is partitioned into sets of states as $- \{S_1, S_2, \ldots, S_{n_g}\}$, s.t., $S = \cup_{i=1}^{n_g} S_i$ and $\cap_{i=1}^{n_g} S_i = \emptyset$, n_g being the number of subgoals specified by the user. For each (s, a, s'), we say that the particular action takes the agent from one subgoal to another iff $s \in S_i$, $s' \in S_j$ for some $i, j \in G = \{1, 2, \ldots, n_g\}$ and $i \neq j$.

Let's assume that there is an ordering in which groups of states appear in the demos as shown in Fig. 3.13(b). However, the states within these groups of states may appear in any random order in the demos. These groups of states are not defined a priori and our algorithm aims at estimating these partitions. Note that such orderings are natural in several real-world applications where a certain subgoal can only be reached after completing one or more previous subgoals. We may consider that states in the demos of $\mathcal{D}$ appear in increasing order of subgoal indices, i.e., achieving subgoal j is harder than achieving subgoal i ($i < j$). This gives us a natural way of defining an extrinsic reward function, which would help towards faster policy search. Also, all the demos in $\mathcal{D}$ should start from the initial state distribution and end at the terminal states.

3.5.1 Learning subgoal prediction

We use $\mathcal{D}$ to partition the state-space into n_g subgoals, with n_g being a hyperparameter. We learn a neural network to approximate $\pi_\phi(g|s)$, which given a state $s \in S$ predicts a probability mass function (pmf) over the possible subgoal partitions $g \in G$. The order in which the subgoals occur in the demos, i.e., $S_1 < S_2 < \cdots < S_{n_g}$, which can be derived from our assumption mentioned above, acts as a supervisory signal. The framework to learn $\pi_\phi(g|s)$ is iterative and alternated between the learning step and the inference/correction step as explained next.

Learning step In this step, we consider that we have a set of tuples (s, g), which we use to learn the function π_ϕ. This can be posed as a multiclass classification problem with n_g categories. We optimize the following cross-entropy loss function,

$$\pi_\phi^* = \arg\min_{\pi_\phi} \frac{1}{N} \sum_{i=1}^{n_d} \sum_{t=1}^{n_i} \sum_{k=1}^{n_g} -\mathbf{1}\{g_{ti} = k\} \log \pi_\phi(g = k|s_{ti}) \tag{3.18}$$

where $\mathbf{1}$ is the indicator function and N is the number of states in the dataset $\mathcal{D}$. To begin with, we do not have any labels g, and thus we consider equipartition of all the subgoals in G along each demo. That is, given a demo of states $\{s_{1i}, s_{2i}, \ldots, s_{n_i i}\}$ for some $i \in \{1, 2, \ldots, n_d\}$, the initial equi-partition subgoals are,

$$g_{ti} = j, \quad \forall \left\lfloor \frac{(j-1)}{n_g} n_i \right\rfloor < t <= \left\lfloor \frac{j}{n_g} n_i \right\rfloor, \ j \in G \tag{3.19}$$

Using this initial labeling scheme, similar states across demos may have different labels, but the network is expected to converge at the Maximum Likelihood Estimate (MLE) of the entire dataset. We also optimize CASL (Paul et al., 2018) presented in Section 3.3.3, for stable learning as the initial labels can be erroneous. In the next iteration of the learning step, we use the inferred subgoal labels, which we obtain as follows.

Inference step   Although the equipartition labels in Eq. (3.19) may have similar states across different demos mapped to dissimilar subgoals, the learned network modeling π_ϕ maps similar states to the same subgoal. But, Eq. (3.18), and thus the predictions of π_ϕ do not account for the natural temporal ordering of the subgoals. Even when using architectures such as Recurrent Neural Networks (RNN), it may be better to impose such temporal order constraints explicitly rather than relying on the network to learn them. We inject such order constraints using Dynamic Time Warping (DTW).

Formally, for the ith demo in $\mathcal{D}$, we obtain the following set: $\{(s_{ti}, \pi_\phi(g|s_{ti})\}_{t=1}^{n_i}$, where π_ϕ is a vector representing the pmf over the subgoals G. However, as the predictions do not consider temporal ordering, the constraint that subgoal j occurs after subgoal i, for $i < j$, is not preserved. To impose such constraints, we use DTW between the two sequences $\{e_1, e_2, \ldots, e_{n_g}\}$, which are the standard basis vectors in the n_g dimensional Euclidean space and $\{\pi_\phi(g|s_{1i}), \pi_\phi(g|s_{2i}), \ldots, \pi_\phi(g|s_{n_i i})\}$. We use the $l1$-norm of the difference between two vectors as the similarity measure in DTW. In this process, we obtain a subgoal assignment for each state in the demos, which become the new labels for training in the learning step.

We then invoke the learning step using the new labels (instead of Eq. (3.19)), followed by the inference step to obtain the next subgoal labels. We continue this process until the number of subgoal labels changed between iterations is less than a certain threshold. This method is presented in Algorithm 1, where the superscript k represents the iteration number in learning-inference alternates.

Reward using subgoals   The ordering of the subgoals, as discussed before, provides a natural way of designing a reward function as follows:

$$r'(s, a, s') = \gamma * \arg\max_{j \in G} \pi_\phi(g = j|s') - \arg\max_{k \in G} \pi_\phi(g = k|s) \tag{3.20}$$

where the agent in state s takes action a and reaches state s'. The augmented reward function would become $r + r'$. Considering that we have a function of the form $\Phi_\phi(s) = \arg\max_{j \in G} \pi_\phi(g = j|s)$, and without loss of generality that $G = \{0, 1, \ldots, n_g - 1\}$, so that for the initial state $\Phi_\phi(s_0) = 0$, it follows from Ng et al. (1999) that every optimal policy in $\mathcal{M}' = (\mathcal{S}, \mathcal{A}, \mathcal{P}, r + r', \gamma, \mathcal{P}_0)$, will also be optimal in $\mathcal{M}$, the original MDP. However, the new reward function may help to learn the task faster.

Algorithm 1 Learning subgoal prediction.

Input: Expert demo set $\mathcal{D}$
Output: subgoal predictor $\pi_\phi(g|s)$
$k \leftarrow 0$
Obtain g^k for each $s \in \mathcal{D}$ using Eq. (3.19)
repeat
    Optimize Eq. (3.18) to obtain π_ϕ^k
    Predict pmf of G for each $s \in \mathcal{D}$ using π_ϕ^k
    Obtain new subgoals g^{k+1} using the pmf in DTW
    done = True, if $|g^k - g^{k+1}| < \epsilon$, else False
    $k \leftarrow k + 1$
until done is True

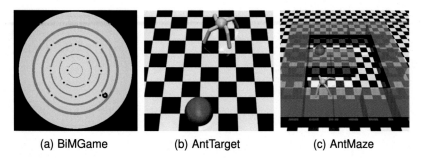

(a) BiMGame   (b) AntTarget   (c) AntMaze

FIGURE 3.14 This figure presents the three environments we use – (a) Ball-in-Maze Game (BiMGame) (b) Ant locomotion in an open environment with an end goal (AntTarget) (c) Ant locomotion in a maze with an end goal (AntMaze).

3.5.2 Supervised pretraining

As discussed previously, an initial way to utilize the demos is by pretraining the policy network π_θ using the demo set $\mathcal{D}$ in a supervised learning framework. We pretrain the network by optimizing the following:

$$\theta^* = \arg\min_{\theta} \sum_{i=1}^{n_d} \sum_{t=1}^{n_i} l\big(\pi_\theta(\boldsymbol{a}|\boldsymbol{s}_{ti}), \boldsymbol{a}_{ti}^*\big) + \lambda||\theta||_F^2 \tag{3.21}$$

where l is the loss function which can be cross-entropy or regression loss depending on discrete or continuous actions. Note that the continuous actions comprise of (μ, σ) which are parameters of a Gaussian distribution. The second part of Eq. (3.21) is the l_2 regularization loss. The policy obtained after optimizing Eq. (3.21) possesses the ability to take actions with low error rates at the states sampled from the distribution induced by the demo set $\mathcal{D}$. However, as shown in Ross et al. (2011), a small error at the beginning would compound quadratically with time as the agent starts visiting states which are not sampled from the distribution of $\mathcal{D}$. Algorithms like DAgger can be used to fine-tune the policy by querying expert actions at states visited after executing the learned policy. This query to the expert is often very costly and even may not be feasible in some applications. More importantly, as DAgger aims to mimic the expert, it can only reach its performance and not improve on it. For this reason, we fine-tune the policy using RL with the extrinsic reward function obtained after identifying the subgoals.

3.5.3 Applications

In this section, we present three challenging dynamical tasks: we use our framework to solve them and compare these with other state-of-the-art methods in the literature.

Tasks   We perform experiments on three challenging environments as shown in Fig. 3.14. The first environment is the Ball-in-Maze Game (BiMGame) introduced in van Baar et al. (2018), where the task is to move a ball from the outermost to the innermost ring using a set of five discrete actions – clock-wise and anticlockwise rotation by 1° along the two principal dimensions of the board and "no-op" where the current orientation of the board is maintained. The

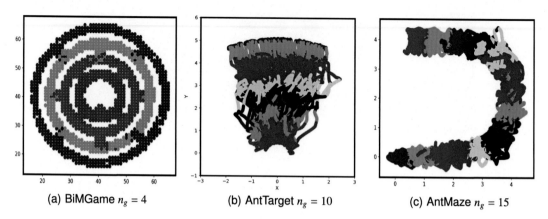

(a) BiMGame $n_g = 4$ (b) AntTarget $n_g = 10$ (c) AntMaze $n_g = 15$

FIGURE 3.15 (a) This figure presents the learned subgoals for the three tasks which are color-coded. Note that for (b) and (c), multiple subgoals are assigned the same color, but they can be distinguished by their spatial locations.

states are images of size 84×84. The second environment is AntTarget which involves the Ant (Schulman et al., 2015). The task is to reach the center of a circle of radius 5 m with the Ant being initialized on a 45° arc of the circle. The state and action are continuous with 41 and 8 dimensions respectively. The third environment, AntMaze, uses the same Ant, but in a U-shaped maze used in Held et al. (2017). The Ant is initialized on one end of the maze with the goal being the other end indicated as red in Fig. 3.14(c). For all tasks, we use sparse terminal-only reward, i.e., $+1$ only after reaching the goal state and 0 otherwise. Standard RL methods such as A3C (Mnih et al., 2016) are not able to solve these tasks with such sparse rewards.

Visualization   We visualize the subgoals discovered by our algorithm and plot it on the x–y plane in Fig. 3.15. As can be seen in BiMGame, with 4 subgoals, our method is able to discover the bottle-neck regions of the board as different subgoals. For AntTarget and AntMaze, the path to the goal is more or less equally divided into subgoals. This shows that our method of subgoal discovery can work for both environments with and without bottle-neck regions.

Comparison with baselines   We primarily compare our method with other RL methods which utilize demo or expert information – AggreVaTeD (Sun et al., 2017) and value-based reward shaping (Ng et al., 1999), equivalent to the $K = \infty$ in THOR (Sun et al., 2018). The comparisons are in Fig. 3.16. As may be observed, none of the baselines show any sign of learning for the tasks, except for ValueReward, which performs comparably with the proposed method for AntTarget only. Our method, on the other hand, is able to learn and solve the tasks consistently over multiple runs. The expert cumulative rewards are also drawn as straight lines in the plots and imitation learning methods like DAgger (Ross et al., 2011) can only reach that mark. Our method is able to surpass the expert for all the tasks. In fact, for AntMaze, even with a rather suboptimal expert (an average cumulative reward of only 0.0002), our algorithm achieves about 0.012 cumulative reward at 100 million steps.

The poor performance of the ValueReward and AggreVaTeD can be attributed to the imperfect value function learned with a limited number of demos. Specifically, with an increase in the demo length, the variations in cumulative reward in the initial set of states are quite high. This introduces a considerable amount of error in the estimated value function in the

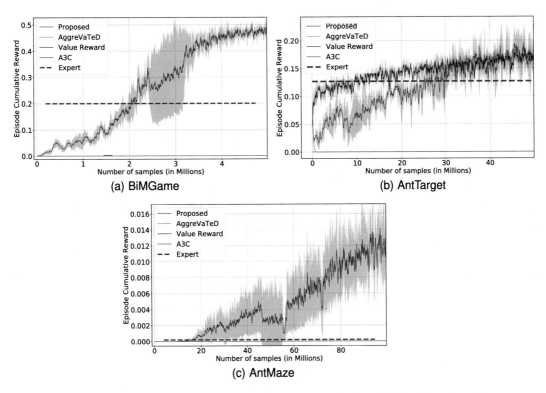

FIGURE 3.16 This figure shows the comparison of our proposed method with the baselines. Some lines may not be visible as they overlap. For tasks (a) and (c) our method clearly outperforms others. For task (b), although value reward initially performs better, our method eventually achieves the same performance. For a fair comparison, we do not use the out-of-set augmentation to generate these plots. Note that as some of the algorithms do not learn at all, thus accumulating zero reward, such lines overlap with the x-axis of the plots and are not visible.

initial states, which in turn traps the agent in some local optima when such value functions are used to guide the learning process.

Discussions   Here are the key points which can be inferred from the results: First, the method to discover subgoals works both for tasks with inherent bottlenecks (e.g. BiMGame) and without any bottlenecks (e.g. AntTarget and AntMaze), but with temporal order between groups of states in the expert demos, which occurs in many applications. Second, discrete rewards using subgoals perform much better than value function based continuous rewards. This may be because the value functions learned from the limited number of demos may be erroneous, whereas segmenting the demos based on temporal ordering may still work well.

3.6 Conclusions

The widely-heralded successes of machine learning have been largely driven by the availability of large volumes of labeled data used to train the learning-based systems. This is simply unrealistic as these tools are deployed in real-life applications. We cannot assume that all scenarios likely to be encountered during the operation of a system have been seen during the training phase, and the expertise to provide labels just may not be available. While cats and dogs can be labeled by anyone, the same cannot be said when trying to obtain training data for different species of birds or vegetation types or medical diagnosis. Thus, learning with limited to no supervision is extremely critical.

This chapter provides an overview of machine learning with limited supervision for applications in computer vision and robotics. Supervision can be of different forms and depend on the application domain. Hence, we review a host of different methods. The first approach is related to active learning, where we show how contextual information available in the data sources can be used to predict many of the labels, which may otherwise need to be provided manually. Next, we consider situations where detailed labels are not available, but textual descriptions accompanying a video can be used as a form of weak supervision. Here, the goal would be to learn how to localize the relevant video segment from these descriptions. Both of these are in a single application domain. In the next section, we considered the problem of semantic segmentation in an image and how learned models in one domain can be adapted to another with minimal or even no additional supervision. Finally, we considered the problem of how to train a robot for certain tasks from limited demonstration examples using a combination of imitation and reinforcement learning, through the identification of subgoals. Overall, we have explored different learning paradigms and demonstrated them in multiple problem settings in computer vision and robotics.

The future of machine learning should focus on how to learn with limited amounts of data. Researchers and practitioners should consider different settings for this purpose. Few examples are provided below. Can we learn from vast quantities of data that is uploaded on social media but are error-prone and incomplete? Can we transfer learned models in easy-to-label domains with coarse classification to domains that require significant expertise to label (e.g., labeling birds and trees are easy, but fine-grained labeling of species of birds and trees requires significant knowledge that few people may have)? Can we adapt learned models to unseen conditions that may be safety critical (e.g., can an autonomous driving system identify a dangerous condition on a road that it may have never encountered)? Can machine learning be used in applications where there are large volumes of data, e.g., medicine, but supervision is coarse and very expensive (maybe impossible) to obtain?

This chapter has touched upon some methods that take a step in the direction of solving these problems. There is a huge amount of work to be done still. Although it may seem that learning methods are able to best human performance in many cases, they are largely in controlled settings. The ability to reason through and identify general principles that can be transferred across application scenarios remains elusive. They provide a very rich tapestry of problems for researchers to work on.

Acknowledgments

The work was partially supported by US National Science Foundation through grant 1724341, US Office of Naval Research through grant N00014-19-1-2264 and Mitsubishi Electric Research Laboratories.

References

Aggarwal, Jake K., Ryoo, Michael S., 2011. Human activity analysis: a review. ACM Computing Surveys (CSUR) 43 (3), 16.

Andreas, Jacob, Klein, Dan, Levine, Sergey, 2017. Modular multitask reinforcement learning with policy sketches. In: ICML, pp. 166–175.

Arandjelovic, Relja, Gronat, Petr, Torii, Akihiko, Pajdla, Tomas, Netvlad, Josef Sivic, 2016. Cnn architecture for weakly supervised place recognition. In: CVPR, pp. 5297–5307.

Argall, Brenna D., Chernova, Sonia, Veloso, Manuela, Browning, Brett, 2009. A survey of robot learning from demonstration. RAS 57 (5), 469–483.

Bappy, Jawadul H., Paul, Sujoy, Roy-Chowdhury, Amit K., 2016. Online adaptation for joint scene and object classification. In: ECCV. Springer, pp. 227–243.

Bearman, Amy, Russakovsky, Olga, Ferrari, Vittorio, Fei-Fei, Li, 2016. What's the point: semantic segmentation with point supervision. In: ECCV.

Bilgic, Mustafa, Getoor, Lise, 2009. Link-based active learning. In: NIPS-Workshop.

Bojanowski, Piotr, Lajugie, Rémi, Grave, Edouard, Bach, Francis, Laptev, Ivan, Ponce, Jean, Schmid, Cordelia, 2015. Weakly-supervised alignment of video with text. In: ICCV. IEEE, pp. 4462–4470.

Bojarski, Mariusz, Del Testa, Davide, Dworakowski, Daniel, Firner, Bernhard, Flepp, Beat, Goyal, Prasoon, Jackel, Lawrence D., Monfort, Mathew, Muller, Urs, Zhang, Jiakai, et al., 2016. End to end learning for self-driving cars. arXiv preprint. arXiv:1604.07316.

Carreira, Joao, Zisserman, Andrew, 2017. Quo vadis, action recognition? A new model and the kinetics dataset. In: CVPR. IEEE, pp. 4724–4733.

Chakraborty, Shayok, Balasubramanian, Vineeth, Sun, Qian, Panchanathan, Sethuraman, Ye, Jieping, 2015. Active batch selection via convex relaxations with guaranteed solution bounds. TPAMI 37 (10), 1945–1958.

Chang, Chih-Chung, Lin, Chih-Jen, 2011. Libsvm: a library for support vector machines. TIST 2 (3), 27.

Chang, Wei-Lun, Wang, Hui-Po, Peng, Wen-Hsiao, Chiu, Wei-Chen, 2019. All about structure: adapting structural information across domains for boosting semantic segmentation. In: CVPR.

Chen, Liang-Chieh, Papandreou, George, Kokkinos, Iasonas, Murphy, Kevin, Deeplab, Alan L. Yuille, 2016. Semantic image segmentation with deep convolutional nets, atrous convolution, and fully connected crfs. CoRR. arXiv:1606.00915 [abs].

Chen, Yuhua, Li, Wen, Van Gool Road, Luc, 2018. Reality oriented adaptation for semantic segmentation of urban scenes. In: CVPR.

Cheng, Ching-An, Yan, Xinyan, Wagener, Nolan, Boots, Byron, 2018. Fast Policy Learning Through Imitation and Reinforcement. UAI.

Choi, Myung Jin, Lim, Joseph J., Torralba, Antonio, Willsky, Alan S., 2010. Exploiting hierarchical context on a large database of object categories. In: CVPR. IEEE, pp. 129–136.

Cordts, Marius, Omran, Mohamed, Ramos, Sebastian, Rehfeld, Timo, Enzweiler, Markus, Benenson, Rodrigo, Franke, Uwe, Roth, Stefan, Schiele, Bernt, 2016. The cityscapes dataset for semantic urban scene understanding. In: CVPR.

Cristianini, N., Ricci, Elisa, 2008. Support vector machines. Encyclopedia of algorithms.

Cuong, Nguyen Viet, Lee, Wee Sun, Ye, Nan, Chai, Kian Ming A., Chieu, Hai Leong, 2013. Active learning for probabilistic hypotheses using the maximum Gibbs error criterion. In: NIPS, pp. 1457–1465.

Du, Liang, Tan, Jingang, Yang, Hongye, Feng, Jianfeng, Xue, Xiangyang, Zheng, Qibao, Ye, Xiaoqing, Zhang, Xiaolin, 2019. Ssf-dan: separated semantic feature based domain adaptation network for semantic segmentation. In: ICCV.

Duan, Yan, Chen, Xi, Houthooft, Rein, Schulman, John, Abbeel, Pieter, 2016. Benchmarking deep reinforcement learning for continuous control. In: ICML, pp. 1329–1338.

Dubey, Rachit, Agrawal, Pulkit, Pathak, Deepak, Griffiths, Thomas L., Efros, Alexei A., 2018. Investigating human priors for playing video games. In: ICML.

Fujishige, Satoru, Hayashi, Takumi, Isotani, Shigueo, 2006. The minimum-norm-point algorithm applied to submodular function minimization and linear programming. In: Citeseer.

Galleguillos, Carolina, Rabinovich, Andrew, Belongie, Serge, 2008. Object categorization using co-occurrence, location and appearance. In: CVPR. IEEE, pp. 1–8.

Goodfellow, Ian J., Pouget-Abadie Mehdi Mirza, Jean, Xu, Bing, Warde-Farley, David, Ozair, Sherjil, Courville, Aaron, Bengio, Yoshua, 2014. Generative adversarial nets. In: NIPS.

Hartmann, Glenn, Grundmann, Matthias, Hoffman, Judy, Tsai, David, Kwatra, Vivek, Madani, Omid, Vijaya-narasimhan, Sudheendra, Essa, Irfan, Rehg, James, Sukthankar, Rahul, 2012. Weakly supervised learning of object segmentations from web-scale video. In: ECCVW. Springer, pp. 198–208.

Hasan, Mahmudul, Paul, Sujoy, Mourikis, Anastasios I., Roy-Chowdhury, Amit K., 2018. Context-aware query selection for active learning in event recognition. T-PAMI.

Hasan, Mahmudul, Roy-Chowdhury, Amit K., 2015. Context aware active learning of activity recognition models. In: ICCV. IEEE, pp. 4543–4551.

He, Kaiming, Zhang, Xiangyu, Ren, Shaoqing, Sun, Jian, 2016. Deep residual learning for image recognition. In: CVPR.

Heilbron, Fabian Caba, Escorcia, Victor, Ghanem, Bernard, Niebles, Juan Carlos, 2015. Activitynet: a large-scale video benchmark for human activity understanding. In: CVPR. IEEE, pp. 961–970.

Held, David, Geng, Xinyang, Florensa, Carlos, Abbeel, Pieter, 2017. Automatic goal generation for reinforcement learning agents. In: ICML.

Hoffman, Judy, Tzeng, Eric, Park, Taesung, Zhu, Jun-Yan, Isola, Phillip, Saenko, Kate, Efros, Alexei A., Cycada, Trevor Darrell, 2018. Cycle-consistent adversarial domain adaptation. In: ICML.

Hoffman, Judy, Wang, Dequan, Yu, Fisher, Darrell, Trevor, 2016. Fcns in the wild: pixel-level adversarial and constraint-based adaptation. CoRR. arXiv:1612.02649 [abs].

Holub, Alex, Perona, Pietro, Burl, Michael C., 2008. Entropy-based active learning for object recognition. In: CVPR-Workshops. IEEE, pp. 1–8.

Hu, Xia, Tang, Jiliang, Gao, Huiji, Liu Actnet, Huan, 2013. Active learning for networked texts in microblogging. In: SDM. SIAM, pp. 306–314.

Huang, De-An, Fei-Fei, Li, Niebles, Juan Carlos, 2016. Connectionist temporal modeling for weakly supervised action labeling. In: ECCV. Springer, pp. 137–153.

Hung, Wei-Chih, Tsai, Yi-Hsuan, Liou, Yan-Ting, Lin, Yen-Yu, Yang, Ming-Hsuan, 2018. Adversarial learning for semi-supervised semantic segmentation. In: BMVC.

Idrees, Haroon, Zamir, Amir R., Jiang, Yu-Gang, Gorban, Alex, Laptev, Ivan, Sukthankar, Rahul, Shah, Mubarak, 2017. The thumos challenge on action recognition for videos "in the wild". CVIU 155, 1–23.

Jain, Ashesh, Zamir, Amir R., Savarese, Silvio, Saxena, Ashutosh, 2015. Structural-rnn: deep learning on spatio-temporal graphs. In: CVPR.

Jie, Zequn, Wei, Yunchao, Jin, Xiaojie, Feng, Jiashi, Liu, Wei, 2017. Deep self-taught learning for weakly supervised object localization. In: CVPR.

Käding, Christoph, Freytag, Alexander, Rodner, Erik, Perino, Andrea, Denzler, Joachim, 2016. Large-scale active learning with approximations of expected model output changes. In: GCPR. Springer, pp. 179–191.

Kanazawa, Angjoo, Jacobs, David W., Chandraker, Manmohan, 2016. Warpnet: weakly supervised matching for single-view reconstruction. In: CVPR, pp. 3253–3261.

Khoreva, Anna, Benenson, Rodrigo, Hosang, Jan, Hein, Matthias, Schiele, Bernt, 2017. Simple does it: weakly supervised instance and semantic segmentation. In: CVPR.

Khoreva, Anna, Benenson, Rodrigo, Omran, Mohamed, Hein, Matthias, Schiele, Bernt, 2016. Weakly supervised object boundaries. In: CVPR, pp. 183–192.

Koller, Daphne, Friedman, Nir, 2009. Probabilistic Graphical Models: Principles and Techniques. MIT Press.

Koppula, Hema Swetha, Gupta, Rudhir, Saxena, Ashutosh, 2013. Learning human activities and object affordances from rgb-d videos. IJRR 32 (8), 951–970.

Krizhevsky, Alex, Sutskever, Ilya, Hinton, Geoffrey E., 2012. Imagenet classification with deep convolutional neural networks. In: NIPS, pp. 1097–1105.

Lapedriza, Agata, Pirsiavash, Hamed, Bylinskii, Zoya, Torralba, Antonio, 2013. Are all training examples equally valuable? arXiv preprint. arXiv:1311.6510.

Levine, Sergey, Finn, Chelsea, Darrell, Trevor, Abbeel, Pieter, 2016. End-to-end training of deep visuomotor policies. JMLR 17 (1), 1334–1373.

Li, Dong, Huang, Jia-Bin, Li, Yali, Wang, Shengjin, Yang, Ming-Hsuan, 2016. Weakly supervised object localization with progressive domain adaptation. In: CVPR, pp. 3512–3520.

Li, Xianglin, Guo, Runqiu, Cheng, Jun, 2012. Incorporating Incremental and Active Learning for Scene Classification. In: ICMLA, vol. 1. IEEE, pp. 256–261.

Li, Xin, Guo, Yuhong, 2013. Adaptive active learning for image classification. In: CVPR, pp. 859–866.

Li, Xin, Guo, Yuhong, 2014. Multi-level adaptive active learning for scene classification In: ECCV. Springer, pp. 234–249.

Li, Yunsheng, Yuan, Lu, Vasconcelos, Nuno, 2019. Bidirectional learning for domain adaptation of semantic segmentation. In: CVPR.

Lian, Qing, Lv, Fengmao, Duan, Lixin, Gong, Boqing, 2019. Constructing self-motivated pyramid curriculums for cross-domain semantic segmentation: a non-adversarial approach. In: ICCV.

Mac Aodha, Oisin, Campbell, Neill, Kautz, Jan, Brostow, Gabriel, 2014. Hierarchical subquery evaluation for active learning on a graph. In: CVPR. IEEE, pp. 564–571.

McCormick, S. Thomas, 2005. Submodular function minimization. Handbooks in Operations Research and Management Science 12, 321–391.

Mnih, Volodymyr, Badia Mehdi Mirza, Adria Puigdomenech, Graves, Alex, Lillicrap, Timothy, Harley, Tim, Silver, David, Kavukcuoglu, Koray, 2016. Asynchronous methods for deep reinforcement learning. In: ICML, pp. 1928–1937.

Mnih, Volodymyr, Kavukcuoglu, Koray, Silver, David, Rusu, Andrei A., Veness, Joel, Bellemare, Marc G., Graves, Alex, Riedmiller, Martin, Fidjeland, Andreas K., Ostrovski, Georg, et al., 2015. Human-level control through deep reinforcement learning. Nature 518 (7540), 529.

Murez, Zak, Kolouri, Soheil, Kriegman, David, Ramamoorthi, Ravi, Kim, Kyungnam, 2018. Image to image translation for domain adaptation. In: CVPR.

Ng, Andrew Y., Harada, Daishi, Russell, Stuart, 1999. Policy invariance under reward transformations: theory and application to reward shaping. In: ICML, vol. 99, pp. 278–287.

Nguyen, Phuc, Liu, Ting, Prasad, Gautam, Han, Bohyung, 2018. Weakly supervised action localization by sparse temporal pooling network. In: CVPR.

Oh, Sangmin, Hoogs, Anthony, Perera, Amitha, Cuntoor, Naresh, Chen, Chia-Chih, Taek Lee, Jong, Mukherjee, Saurajit, Aggarwal, J.K., Lee, Hyungtae, Davis, Larry, et al., 2011. A large-scale benchmark dataset for event recognition in surveillance video. In: CVPR. IEEE, pp. 3153–3160.

Panda, Rameswar, Das, Abir, Wu, Ziyan, Ernst, Jan, Roy-Chowdhury, Amit K., 2017. Weakly supervised summarization of web videos. In: ICCV, pp. 3657–3666.

Paul, Sujoy, Bappy, Jawadul H., Roy-Chowdhury, Amit K., 2016. Efficient selection of informative and diverse training samples with applications in scene classification. In: ICIP. IEEE, pp. 494–498.

Paul, Sujoy, Bappy, Jawadul H., Roy-Chowdhury, Amit K., 2017. Non-uniform subset selection for active learning in structured data. In: CVPR, pp. 6846–6855.

Paul, Sujoy, Roy, Sourya, Roy-Chowdhury, Amit K., 2018. W-talc: weakly-supervised temporal activity localization and classification. In: ECCV, pp. 563–579.

Paul, Sujoy, Tsai, Yi-Hsuan, Schulter, Samuel, Roy-Chowdhury, Amit K., Chandraker, Manmohan, 2020. Domain adaptive semantic segmentation using weak labels. In: ECCV.

Paul, Sujoy, Vanbaar, Jeroen, Roy-Chowdhury, Amit, 2019. Learning from trajectories via subgoal discovery. In: NeurIPS, pp. 8411–8421.

Richter, Stephan R., Vineet, Vibhav, Roth, Stefan, Koltun, Vladlen, 2016. Playing for data: ground truth from computer games. In: ECCV.

Ross, Stéphane, Gordon, Geoffrey, Bagnell, Drew, 2011. A reduction of imitation learning and structured prediction to no-regret online learning. In: AISTATS, pp. 627–635.

Sadat Saleh, Fatemeh, Aliakbarian, Mohammad Sadegh, Salzmann, Mathieu, Petersson, Lars, Alvarez, Jose M., 2018. Effective use of synthetic data for urban scene semantic segmentation. In: ECCV.

Ugm, Mark Schmidt, 2007. A Matlab toolbox for probabilistic undirected graphical models.

Schulman, John, Moritz, Philipp, Levine, Sergey, Jordan, Michael, Abbeel, Pieter, 2015. High-dimensional continuous control using generalized advantage estimation. In: ICLR.

Sen, Prithviraj, Getoor, Lise, 2003. Link-based classification. In: ICML.

Sen, Prithviraj, Namata, Galileo, Bilgic, Mustafa, Getoor, Lise, Galligher, Brian, Eliassi-Rad, Tina, 2008. Collective classification in network data. AI Magazine 29 (3), 93.

Settles, Burr, 2012. Active learning. Synthesis Lectures on Artificial Intelligence and Machine Learning 6 (1), 1–114.

Settles, Burr, Craven, Mark, 2008. An analysis of active learning strategies for sequence labeling tasks. In: EMNLP. Association for Computational Linguistics, pp. 1070–1079.

Shen, Zhiqiang, Li, Jianguo, Su, Zhou, Li, Minjun, Chen, Yurong, Jiang, Yu-Gang, Xue, Xiangyang, 2017. Weakly supervised dense video captioning. In: CVPR, vol. 2, p. 10.

Silver, David, Huang, Aja, Maddison, Chris J., Guez, Arthur, Sifre, Laurent, Van Den Driessche, George, Schrittwieser, Julian, Antonoglou, Ioannis, Panneershelvam, Veda, Lanctot, Marc, et al., 2016. Mastering the game of go with deep neural networks and tree search. Nature 529 (7587), 484.

Singh, Avi, Yang, Larry, Gplac, Sergey Levine, 2017. Generalizing vision-based robotic skills using weakly labeled images. In: ICCV.

Sun, Wen, Bagnell, J. Andrew, Boots, Byron, 2018. Truncated horizon policy search: combining reinforcement learning & imitation learning. In: ICLR.

Sun, Wen, Venkatraman, Arun, Gordon, Geoffrey J., Boots, Byron, Bagnell, J. Andrew, 2017. Deeply aggrevated: differentiable imitation learning for sequential prediction. In: ICML, pp. 3309–3318.

Todorov, Emanuel, 2014. Convex and analytically-invertible dynamics with contacts and constraints: theory and implementation in mujoco. In: ICRA, pp. 6054–6061.

Tran, Du, Bourdev, Lubomir, Fergus, Rob, Torresani, Lorenzo, Paluri, Manohar, 2015. Learning spatiotemporal features with 3d convolutional networks. In: ICCV. IEEE, pp. 4489–4497.

Tsai, Yi-Hsuan, Hung, Wei-Chih, Schulter, Samuel, Sohn, Kihyuk, Yang, Ming-Hsuan, Chandraker, Manmohan, 2018. Learning to adapt structured output space for semantic segmentation. In: CVPR.

Tsai, Yi-Hsuan, Sohn, Kihyuk, Schulter, Samuel, Chandraker, Manmohan, 2019. Domain adaptation for structured output via discriminative patch representations. In: ICCV.

Tulyakov, Stepan, Ivanov, Anton, Fleuret, Francois, 2017. Weakly supervised learning of deep metrics for stereo reconstruction. In: CVPR, pp. 1339–1348.

van Baar, Jeroen, Sullivan, Alan, Cordorel, Radu, Jha, Devesh, Romeres, Diego, Nikovski, Daniel, 2018. Sim-to-real transfer learning using robustified controllers in robotic tasks involving complex dynamics. In: ICRA.

Wang, Botao, Lin, Dahua, Xiong, Hongkai, Zheng, Y.F., 2016a. Joint inference of objects and scenes with efficient learning of text-object-scene relations. TMM 18 (3), 507–520.

Wang, Limin, Xiong, Yuanjun, Wang, Zhe, Qiao, Yu, Lin, Dahua, Tang, Xiaoou, Van Gool, Luc, 2016b. Temporal segment networks: towards good practices for deep action recognition. In: ECCV. Springer, pp. 20–36.

Wang, Limin, Xiong, Yuanjun, Lin, Dahua, Van Gool, Luc, 2017. Untrimmednets for weakly supervised action recognition and detection. In: CVPR.

Wang, Zhenhua, Shi, Qinfeng, Shen, Chunhua, Van Den Hengel, Anton, 2013. Bilinear programming for human activity recognition with unknown mrf graphs. In: CVPR, pp. 1690–1697.

Wu, Zuxuan, Han, Xintong, Lin Mustafa Gkhan Uzunbas, Yen-Liang, Goldstein, Tom, Nam Lim, Ser, Dcan, Larry S. Davis, 2018. Dual channel-wise alignment networks for unsupervised scene adaptation. In: ECCV.

Xiao, Jianxiong, Hays, James, Ehinger, Krista A., Oliva, Aude, Torralba, Antonio, 2010. Sun database: large-scale scene recognition from abbey to zoo. In: CVPR. IEEE, pp. 3485–3492.

Xu, Huijuan, Das, Abir, Saenko, Kate, 2017. R-c3d: region convolutional 3d network for temporal activity detection. In: ICCV, vol. 6, p. 8.

Yan, Yan, Xu, Chenliang, Cai, Dawen, Corso, Jason, 2017. Weakly supervised actor-action segmentation via robust multi-task ranking. CVPR 48, 61.

Yao, Bangpeng, Fei-Fei, Li, 2010. Modeling mutual context of object and human pose in human-object interaction activities. In: CVPR. IEEE, pp. 17–24.

Yao, Jian, Fidler, Sanja, Urtasun, Raquel, 2012. Describing the scene as a whole: joint object detection, scene classification and semantic segmentation. In: CVPR. IEEE, pp. 702–709.

Zhang, Chicheng, Chaudhuri, Kamalika, 2015. Active learning from weak and strong labelers. In: NIPS, pp. 703–711.

Zhang, Yang, David, Philip, Gong, Boqing, 2017. Curriculum domain adaptation for semantic segmentation of urban scenes. In: ICCV.

Zhao, Hengshuang, Shi, Jianping, Qi, Xiaojuan, Wang, Xiaogang, Jia, Jiaya, 2017a. Pyramid scene parsing network. In: CVPR.

Zhao, Yue, Xiong, Yuanjun, Wang, Limin, Wu, Zhirong, Tang, Xiaoou, Lin, Dahua, 2017b. Temporal action detection with structured segment networks. In: ICCV.

Zhong, Bineng, Yao, Hongxun, Chen, Sheng, Ji, Rongrong, Chin, Tat-Jun, Wang, Hanzi, 2014. Visual tracking via weakly supervised learning from multiple imperfect oracles. Pattern Recognition 47 (3), 1395–1410.

Zhou, Bolei, Lapedriza, Agata, Xiao, Jianxiong, Torralba, Antonio, Oliva, Aude, 2014. Learning deep features for scene recognition using places database. In: NIPS, pp. 487–495.

Zhou, Zhi-Hua, 2004. Multi-Instance Learning: A Survey. Tech. Rep. Department of Computer Science & Technology, Nanjing University.

Zou, Yang, Yu, Zhiding, Vijaya Kumar, B.V.K., Wang, Jinsong, 2018. Domain adaptation for semantic segmentation via class-balanced self-training. In: ECCV.

Biographies

Sujoy Paul is currently at Google Research. He received his PhD in Electrical and Computer Engineering from University of California, Riverside and his Bachelor's degree in Electronics and Telecommunication Engineering from Jadavpur University. His broad research interest includes Computer Vision and Machine Learning with focus on semantic segmentation, human action recognition, domain adaptation, weak supervision, active learning, reinforcement learning.

Amit Roy-Chowdhury is Professor and Bourns Family Faculty Fellow of Electrical and Computer Engineering, Director of the Center for Robotics and Intelligent Systems, and Cooperating Faculty in the department of Computer Science and Engineering at the University of California, Riverside (UCR). He leads the Video Computing Group, working on foundational principles of computer vision, image processing, and statistical learning, with applications in cyber-physical, autonomous and intelligent systems. He has published over 200 papers in peer-reviewed journals and conferences. He is a Fellow of the IEEE and IAPR, received the Doctoral Dissertation Advising/Mentoring Award 2019 from UCR, and the ECE Distinguished Alumni Award from University of Maryland, College Park.

4

Efficient methods for deep learning [☆]

Han Cai, Ji Lin, and Song Han

Massachusetts Institute of Technology, Cambridge, MA, United States

CHAPTER POINTS

- We first introduce various model compression approaches, such as pruning, factorization, quantization, and efficient model design.

- To reduce the design cost, we then describe neural architecture search, automated pruning and quantization, which can outperform the manual design with minimal human efforts.

- Finally, we describe the once-for-all technique to efficiently handle many hardware platforms and efficiency constraints without repeating the costly search and retraining phases.

4.1 Model compression

Compressing an existing deep neural network is an effective way to improve the inference efficiency. Compression methods include *parameter pruning* to remove the redundant weights, *low-rank factorization* to reduce the complexity, *weight quantization* to reduce the weight precision and model size, and *knowledge distillation* to transfer the dark knowledge from large models to smaller ones. Finally, we discuss *automated methods* to automatically find a good compression policy without human effort.

4.1.1 Parameter pruning

Deep neural networks are usually overparameterized. Pruning removes the redundant elements in neural networks (Fig. 4.1) to reduce the model size and computation.

[☆] All student authors have contributed equally to this work and are listed in the alphabetical order.

159

Copyright © 2022 Elsevier Inc. All rights reserved.

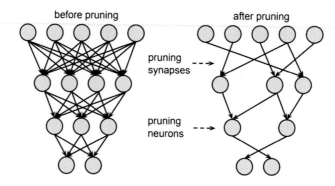

FIGURE 4.1 Synapses and neurons before and after pruning (Han et al., 2015b).

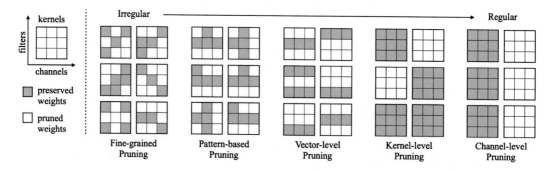

FIGURE 4.2 Different granularities for weight pruning (figure modified from Mao et al. (2017)).

Notations

We analyze the convolutional layers in deep neural networks, which are the most computationally expensive components. In a single convolutional layer, the weights compose a 4-dimensional tensor of shape $n \times c \times k_h \times k_w$, where n is the number of filters (*i.e.*, output channels), c is the number of channels (*i.e.*, input channels), and k_h, k_w is the kernel size (usually symmetric, *i.e.*, $k_h = k_w$). One layer's weights can be viewed as multiple *filters* (3-dimensional tensors of shape $c \times k_h \times k_w$), each corresponding to an output channels; or viewed as multiple *channels* (3-dimensional tensors of shape $n \times k_h \times k_w$), each corresponding to an input channel. Each $k_h \times k_w$ tensor is a *kernel*; there are $n \times c$ kernels in a convolutional layer.

Granularity

Pruning can be performed at different *granularities* (Mao et al., 2017) (Fig. 4.2).

Fine-grained pruning removes individual elements from the weight tensor. An early approach was Optimal Brain Damage (LeCun et al., 1989) and Optimal Brain Surgeon (Hassibi and Stork, 1993), which reduced the number of connections based on the Hessian of the loss function. Han et al. (2015b) proposed a three-step method, train-prune-retrain, to prune the redundant connections in a deep neural network. It reduced the number of parameters of AlexNet by a factor of 9×, and VGG-16 by 13×, with no loss of accuracy. Srinivas and Babu (2015) proposed a data-free pruning method to remove the redundant neurons. In fine-grained pruning, the set of weights to be pruned can be chosen arbitrarily, it can achieve a very high compression ratio on CNN (Han et al., 2015b), RNN (Giles and Omlin, 1994), LSTM (Han et al., 2017), Transformers (Cheong and Daniel, 2019), *etc.*, without hurting accuracy.

Pattern-based pruning is a special kind of fine-grained pruning which has better hardware acceleration with compiler optimization (Ma et al., 2020; Tan et al., 2020b; Niu et al., 2020). Taking 3 × 3 convolutions as an example, pattern-based pruning assigns a fixed set of masks to each of the 3 × 3 kernels. The number of the masks is usually limited (4–6) to ensure hardware efficiency. Each mask template has a fixed number of pruned elements for each kernel (5 pruned out of 9 in Fig. 4.2). The pattern is determined by heuristics (Ma et al., 2020) or clustering from pretrained weights (Niu et al., 2020). Despite the intra-kernel fine-grained pruning pattern, pattern-based pruning can be accelerated with compiler optimization by reordering the computation loops, reducing the control-flow overhead.

Coarse-grained pruning or *structured pruning* removes a regular tensor block for better hardware efficiency. Depending on the block size, entire vectors (Mao et al., 2017), kernels (Mao et al., 2017; Niu et al., 2020), or channels (He et al., 2017; Wen et al., 2016; Li et al., 2016b; Molchanov et al., 2016) are removed (Fig. 4.2). Coarse-grained pruning like channel pruning can bring direct hardware acceleration on GPUs using standard deep learning libraries, but it usually comes at noticeable accuracy drop compared with fine-grained sparsity, as indicated by Li et al. (2016b). Pruning using a smaller granularity usually brings a smaller accuracy drop at the same compression rate (Mao et al., 2017).

Hardware acceleration

Generally speaking, more regular pruning schemes are more hardware friendly, making it easier for inference acceleration on existing hardware such as GPUs; while more irregular pruning schemes better preserve the accuracy at the same compression rate. With specialized hardware accelerators (Han et al., 2016; Chen et al., 2016; Han et al., 2017; Chen et al., 2019a; Zhang et al., 2016a; Yu et al., 2017) and compiler-based optimization techniques (Ma et al., 2020; Niu et al., 2020), it is also possible to gain a considerable acceleration speed for more irregular pruning methods.

Importance criteria

After choosing a pruning granularity, determining which weights should be pruned is also essential to the pruned model's performance. There have been several importance criteria heuristics to estimate the importance of each weights *after* the model is trained; the less important weights are pruned according to the criteria. The most straightforward heuristic is based on the magnitudes, *i.e.*, absolute values, of the weights (Han et al., 2015b,a):

$$\text{Importance} = |w|,$$

where the weights of larger magnitude are considered more important. It also extends to coarse-grained pruning like channel pruning, where the tensor norm is used as the criterion:

$$\text{Importance} = ||\mathbf{W}||_2.$$

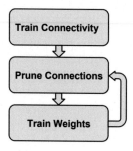

FIGURE 4.3 Three-Step training pipeline with iterative pruning (Han et al., 2015b).

Other criteria include second-order derivatives (*i.e.*, the Hessian of the loss function) (LeCun et al., 1989; Hassibi and Stork, 1993), loss-approximating Taylor expansion (Molchanov et al., 2017), output sensitivity (Engelbrecht, 2001), *etc.*.

Recently, Frankle and Carbin proposed Lottery Ticket Hypothesis (Frankle and Carbin, 2018) to find a sparse subnetwork within the dense, randomly-initialized deep networks *before* training, which can be trained to achieve the same accuracy. Experiments show that the method can find sparse subnetworks with less than 10–20% of the weights while reaching the same level of accuracy on MNIST (LeCun et al., 2010) and CIFAR (Krizhevsky and Hinton, 2009). It was later scaled up to larger-scale setting (*e.g.*, ResNet-50 and Inception-v3 on ImageNet), where the sparse subnetwork can be found at the early phase of training (Frankle et al., 2020) instead of on initialization.

Training methods

Directly removing the weights in deep neural networks will significantly hurt the accuracy at a large compression ratio. Therefore, some training/fine-tuning is needed to recover the performance loss. Fine-tuning can be done after pruning to recover the performance drop (He et al., 2017). It can be extended to iterative pruning (Han et al., 2015b,a) (Fig. 4.3), where multiple iterations of pruning and fine-tuning are performed to further boost the accuracy. To avoid incorrect pruning of weights, dynamic pruning (Guo et al., 2016) incorporates connection splicing into the whole process and provides continual network maintenance. Runtime pruning (Lin et al., 2017) chooses the pruning ratio according to each input sample, assigning a more aggressive pruning strategy for easier samples, which further improves the accuracy-computation trade-off.

Another implementation trains compact DNNs using sparsity constraints. The sparsity constraints are usually implemented using L_0, L_1, or L_2-norm regularization applied to the weights, which are added to the training loss for joint optimization. Han et al. (2015b) applied L_1/L_2 regularization to each individual weights during training. Lebedev and Lempitsky (2016) applied group sparsity constraints on convolutional filters to achieve structured sparsity.

4.1.2 Low-rank factorization

Low-rank factorization uses matrix/tensor decomposition to reduce the complexity of convolutional or fully-connected layers in deep neural networks. The idea of using low-rank filters to accelerate convolution has long been investigated in signal processing area.

The most widely used decomposition is Truncated SVD (Golub and Van Loan, 1996), which is effective for accelerating fully connected layers (Xue et al., 2013; Denton et al., 2014; Girshick, 2015). Given a fully connected layer with weight $W \in \mathbb{R}^{m \times k}$, the SVD is defined as $W = USV^T$, where $U \in \mathbb{R}^{m \times m}$, $S \in \mathbb{R}^{m \times k}$, $V \in \mathbb{R}^{k \times k}$. S is a diagonal matrix with the singular values on the diagonal. If the weight falls in a low-rank structure, it can be approximated by keeping only the t largest entries of S, $t << \min(m, k)$. The computation Wx can be reduced from $O(mk)$ to $O(mt + tk)$ for each sample.

For 4D convolutional weights, Jaderberg et al. (2014) proposed to factorize $k \times k$ kernels into $1 \times k$ and $k \times 1$ kernels; this approach was also adopted in the Inception-V3 design (Szegedy and Vanhoucke, 2016). Zhang et al. (2016b) proposed to factorize a convolution weight of $n \times c \times k \times k$ into $n' \times c \times k \times k$ and $n \times n' \times 1 \times 1$, where $n' << n$. Canonical Polyadic (CP) decomposition can be used to decompose higher dimensional kernels like convolutional weights (Lebedev et al., 2014). This approach computes a low-rank CP-decomposition of the 4D convolution kernel tensor into the sum of a small number of rank-one tensors. At inference time, the original convolution is replaced with a sequence of four convolutional layers with smaller kernels. Kim et al. (2015) used Tucker Decomposition (known as the higher order extension of SVD) to factorize the convolutional kernels, getting higher compression ratio compared to using SVD.

4.1.3 Quantization

Network quantization compresses the network by reducing the bits per weight required to represent the deep network. The quantized network can also have a faster inference speed with hardware support.

Rounding schemes

To quantize a weight of full precision (32-bit float-point value) into a lower precision, rounding is performed to map the float-point value into one of the quantization buckets.

Early work (Han et al., 2015a; Gong et al., 2014; Wu et al., 2016) applied *k-means clustering* to find the shared weights for each layer of a trained network; all the weights that fall into the same cluster will share the same weight. Specifically, when partitioning n original weights $W = \{w_1, w_2, ..., w_n\}$ into k clusters $C = \{c_1, c_2, ..., c_k\}$, $n >> k$, we minimize the within-cluster sum of squares (WCSS):

$$\arg \min_{C} \sum_{i=1}^{k} \sum_{w \in c_i} |w - c_i|^2 \qquad (4.1)$$

k-means clustering based quantization can be combined with pruning and Huffman coding to perform model compression (Han et al., 2015a) (Fig. 4.4). It can compress the model size of VGG-16 by $49 \times$ with no loss of accuracy.

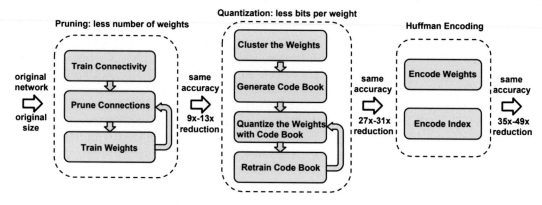

FIGURE 4.4 The three stage compression pipeline: pruning, quantization and Huffman coding (Han et al., 2015a).

Linear/uniform (Vanhoucke et al., 2011; Jacob et al., 2017) quantization directly rounds the float-point value into the nearest quantized values after range truncation; the gradient is propagated using the STE approximation (Bengio et al., 2013). Suppose the clipping range is $[a, b]$, and the number of quantization levels is n, the forward process of quantizing float-point value x into quantized value q is:

$$\text{clamp}(r, a, b) = \min\big(\max(x, a), b\big), \tag{4.2}$$

$$\text{s}(a, b, n) = \frac{b - a}{n - 1}, \tag{4.3}$$

$$q = \text{round}\left(\frac{\text{clamp}(r, a, b) - a}{\text{s}(a, b, n)}\right)\text{s}(a, b, n) + a. \tag{4.4}$$

The back-propagation gradient is computed using:

$$\frac{\partial \mathcal{L}}{\partial q} = \frac{\partial \mathcal{L}}{\partial x}. \tag{4.5}$$

Apart from using the truncation value a, b, some work (Zhou et al., 2016) uses activation functions like tanh to map the range of the weights into $[-1, 1]$, making quantization easier.

Bit precision

We can trade-off the model size and model accuracy by using different bit precisions. A lower bit precision can lead to a smaller model size, but it may come at the cost of accuracy reduction. Full-precision networks use *FP32* for both weights and activations. Half-precision networks use *FP16* to reduce the model size by half. *INT8* quantization for both weights and activations (Jacob et al., 2017) is widely used for integer-arithmetic-only inference, which can be accelerated on GPUs, CPUs, smartphones, *etc.*.

Lower precisions include Ternary Weight Networks (Li et al., 2016a), where the weights are quantized to $\{-1, 0, +1\}$ or $\{-E, 0, +E\}$ (E is the mean of the absolute weight value, but is not trained). Trained Ternary Quantization (Zhu et al., 2016) uses two learnable full-precision scaling coefficients W_l^p, W_l^n for each layer l, and quantizes the weights to $\{-W_l^n, 0, +W_l^P\}$. It can quantize AlexNet on ImageNet with no loss of accuracy.

The extreme case for low-bit quantization is binary weight neural networks (*e.g.*, BinaryConnect (Courbariaux et al., 2015), BinaryNet (Courbariaux and Bengio, 2016), XNOR (Rastegari et al., 2016), *etc.*), where the weights are represented using only 1 bit. The binary weights or activations are usually directly learned during network training using certain forward and backward rules. For example, BinaryConnect (Courbariaux et al., 2015) discusses both deterministic binarization:

$$w_b = \text{sign}(w) = \begin{cases} +1 & \text{if } x \geq 0, \\ -1 & \text{otherwise}, \end{cases} \tag{4.6}$$

and stochastic binarization:

$$w_b = \begin{cases} +1 & \text{with probability } p = \sigma(w), \\ -1 & \text{with probability } 1 - p, \end{cases} \tag{4.7}$$

where σ is the "hard sigmoid" function:

$$\sigma(x) = \max\left(0, \min\left(1, \frac{x+1}{2}\right)\right). \tag{4.8}$$

This is a piece-wise linear approximation to a standard sigmoid function.

Quantization schemes

For quantization of higher precisions (*e.g.*, INT8), it is possible to perform *post-training quantization*, where the weights and activations are quantized after the full-precision model training. The quantization range for activations is determined by computing the distribution on the training set, and Batch Normalization (Ioffe and Szegedy, 2015) layers are folded. Applying post-training INT8 quantization usually leads to minor or no loss of accuracy.[1] Recent work (Banner et al., 2019) studied the post-training quantization of INT4 models.

Quantization-ware training can reduce the quantization accuracy loss by emulating inference-time quantization during training time (Jacob et al., 2017). The forward pass during training is consistent with testing time, which helps on-device deployment. During training, the "fake quantization operator" is injected into the convolutional layers, and the Batch Normalization (Ioffe and Szegedy, 2015) layers are folded.

Both post-training quantization and quantization-aware training require access to the training data to get a good quantization performance, which is not always feasible on some privacy-sensitive applications. *Data-free quantization* aims to reduce the bit precisions with no access to the training data. Nagel et al. (2019) proposed to perform INT8 quantization in a data-free manner equalizing the weight ranges in the network. ZeroQ (Cai et al., 2020b) optimizes for a Distilled Dataset to match the statistics of batch normalization across different layers of the network for data-free quantization.

---

[1] https://www.tensorflow.org/lite/performance/post_training_quantization.

Low-bit training

Apart from inference, *training* with quantized weights, activations, and gradients can reduce the cost of deep learning training. Training with mixed 16-bit and 32-bit floating-point types in a model has been widely supported by deep learning frameworks like TensorFlow, PyTorch, TensorCores, *etc.* With techniques like loss scaling, such mixed-precision training can reduce the memory consumption and improve training speed with no loss of accuracy. DoReFa-Net uses 1-bit weights, 2-bit activations, and 6-bit gradients for faster training and inference, which can obtain comparable accuracy compared to FP32 AlexNet (Krizhevsky et al., 2012) on ImageNet (Deng et al., 2009). The work in Lin et al. (2015) stochastically binarizes the weights to reduce the time on floating-point multiplication in training.

Hardware support for low-precision acceleration

Quantized models can reduce the model size and storage for deployment, but they require hardware support on low-precision arithmetic for inference acceleration. INT8 quantization is supported on mobile ARM CPUs (*e.g.*, Qualcomm Hexagon, ARM Neon), x86 CPUs, NVIDIA GPUs with TensorRT, Xilinx FPGAs with DNNDK, *etc.*. Binary quantized network can also be accelerated with bit operations. Lower bit precisions (*e.g.*, ternary, INT4) are less supported on existing hardware. NVIDIA's Turing architecture supports INT4 inference,[2] which brings an additional 59% speedup compared to INT8. There are efforts designing specialized hardware accelerators for accelerating low-bit quantized models (Zhang et al., 2015; Sharify et al., 2018), which achieve superior energy efficiency compared to full-precision models.

Recently, hardware support for *mixed-precision* quantization reveals a new opportunity for improving accuracy *vs.* cost trade-off. NVIDIA's Turing Tensor Core supports 1-bit, 4-bit, 8-bit and 16-bit arithmetic operations; Imagination launched a flexible neural network IP that supports per-layer bit-width adjustment for both weights and activations. Recent specialized hardware accelerator design also enables mixed-precision support: hardware based on bit-serial multiplier units (Judd et al., 2016; Umuroglu et al., 2018) supports multiplications of 1 to 8 bits in a temporal manner; BitFusion (Sharma et al., 2018) supports multiplications of 2, 4, 8 and 16 bits in a spatial manner.

4.1.4 Knowledge distillation

Knowledge distillation (KD) (Buciluǎ et al., 2006; Hinton et al., 2015) can transfer the "dark knowledge" learned in a large model (denoted as the teacher) to a smaller model (denoted as the student) to improve the performance of the smaller one. The small model is either a compressed model or a shallower/narrower model. (Buciluǎ et al., 2006) achieves the goal by training the student network to match output logits; (Hinton et al., 2015) introduced the idea of temperature in the softmax output and trained the student to mimic the softened distribution of the teacher model's Softmax output. KD shows promising results in various image classification tasks despite the simple implementation.

Apart from the final output logits, the intermediate activations also contain useful information. FitNet (Romero et al., 2014) trains the student to mimic the full feature map of the

---

[2] https://developer.nvidia.com/blog/int4-for-ai-inference/.

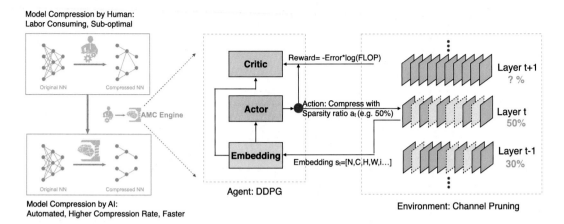

FIGURE 4.5 Overview of AutoML for Model Compression (AMC) engine. Left: AMC replaces human and makes model compression fully automated while performing better than human. Right: Form AMC as a reinforcement learning problem (He et al., 2018).

teacher model through regression. Attention Transfer (AT) (Zagoruyko and Komodakis, 2016) transfers the attention map of the activation from teacher to student, using the summation of the feature map across the channel dimension. Both methods require the intermediate activation to share the same spatial resolution, which limits the choice of the student model architecture.

KD-based methods are also applicable to other applications beyond classification, like object detection (Chen et al., 2017), semantic segmentation (Liu et al., 2019a), language modeling (Sanh et al., 2019), image synthesis (Li et al., 2020), *etc.*.

4.1.5 Automated model compression

Model compression methods can improve the efficiency of the deployed models. However, the performance of model compression is largely affected by the hyperparameters. For example, different layers in deep networks have different capacities and sensitivities (*e.g.*, the first layer in CNN is usually very sensitive to pruning). Therefore, we should apply different pruning ratios for different layers of the network to achieve optimal performance. The design space is so large that human heuristics will usually be suboptimal, and manual model compression is time-consuming. To this end, automated model compression is proposed to find good compression policy without human effort.

Automated pruning

Conventional model pruning techniques rely on hand-crafted features and require domain experts to explore the large design space trading off among model size, speed, and accuracy: this is usually suboptimal and time-consuming. AutoML for Model Compression (AMC) (He et al., 2018) leverages reinforcement learning to efficiently sample the design space and find the optimal pruning policy for a given network (Fig. 4.5). We process a pretrained network (*e.g.*, MobileNet-V1) in a layer-by-layer manner. Our reinforcement learning

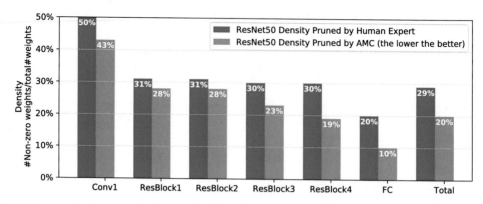

FIGURE 4.6 AMC can prune the model to a lower density compared with human experts without losing accuracy. (Human expert: 3.4× compression on ResNet50. AMC: 5× compression on ResNet50.)

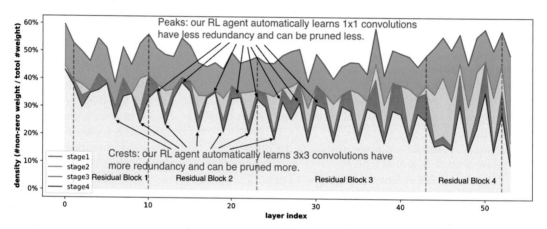

FIGURE 4.7 The pruning policy (sparsity ratio) given by AMC agent for ResNet-50. With 4 stages of iterative pruning, AMC finds very salient sparsity pattern across layers: peaks are 1 × 1 convolution, crests are 3 × 3 convolution. **The reinforcement learning agent automatically learns that 3 × 3 convolution has more redundancy than 1 × 1 convolution and can be pruned more.**

agent (DDPG (Lillicrap et al., 2015)) receives the embedding s_t from a layer t, and outputs a sparsity ratio a_t. After the layer is compressed with a_t, it moves to the next layer L_{t+1}. The accuracy of the pruned model with all layers compressed is evaluated. Finally, as a function of accuracy and MACs (Multiply-Accumulate Operations), reward R is returned to the reinforcement learning agent.

On fine-grained pruning of ResNet-50 (He et al., 2016), AMC can outperform human experts in a fully automated manner: it pushes the expert-tuned compression ratio of ResNet-50 on ImageNet from 3.4 × to 5 × (see Fig. 4.6) without loss of performance. AMC can also find pruning patterns similar to human heuristics. The density of each layer during each stage is displayed in Fig. 4.7. The peaks and crests show that the RL agent automatically learns to prune 3 × 3 convolutional layers with larger sparsity, since they generally have larger redun-

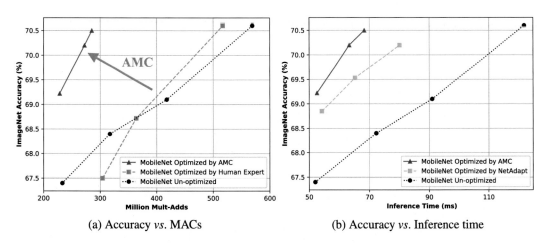

(a) Accuracy *vs.* MACs   (b) Accuracy *vs.* Inference time

FIGURE 4.8 (a) Comparing the accuracy and MACs trade-off among AMC, human expert, and unpruned MobileNet-v1. AMC strictly dominates human expert in the Pareto optimal curve. (b) Comparing the accuracy and latency trade-off among AMC, NetAdapt, and unpruned MobileNet-V1. AMC significantly improves the Pareto curve of MobileNet-V1. Reinforcement-learning based AMC surpasses heuristic-based NetAdapt on the Pareto curve (inference time both measured on Google Pixel 1).

dancy; whereas it prunes more compact 1×1 convolutions with lower sparsity. The density statistics of each block are provided in Fig. 4.6. We can find that the density distribution of AMC is quite different from the human expert's result, shown in Table 3.8 of (Han, 2017), suggesting that AMC can fully explore the design space and allocate sparsity in a better way.

AMC is hardware-aware: it can improve not only the computation (*i.e.*, MACs) but also the actual latency tested on device (Fig. 4.8(b)). We use a highly compact network MobileNet-V1 (Howard et al., 2017) as an example to measure how much we can improve its inference speed. Previous attempts using *hand-crafted* policy to prune MobileNet-V1 led to significant accuracy degradation (Li et al., 2016b): pruning MobileNet-V1 to 75.5% original parameters results in 67.2% top-1 accuracy,[3] which is even worse than the original 0.75 MobileNet-V1 (61.9% parameters with 68.4% top-1 accuracy). However, AMC pruning policy significantly improves the pruning quality on ImageNet, achieving better Pareto curve on accuracy *vs.* computation trade-off (*i.e.*, getting better accuracy at the same computation). As illustrated in Fig. 4.8(a), the human expert's hand-crafted policy achieves slightly worse performance than that of the original MobileNet-V1 under $2\times$ MACs reduction. AMC also outperforms another heuristic-based policy (Yang et al., 2018) trading off accuracy and latency.

A recent work MetaPruning (Liu et al., 2019b) first trains a PruningNet, a kind of meta network, which is able to generate weight parameters for any pruned structure given the target network, and then use it to search for the best pruning policy under different constraints.

Automated quantization

Mixed-precision quantization also requires extensive effort deciding the optimal bit-width for each layer to achieve the best accuracy-performance trade-off. Hardware-Aware Auto-

---

[3] http://machinethink.net/blog/compressing-deep-neural-nets/.

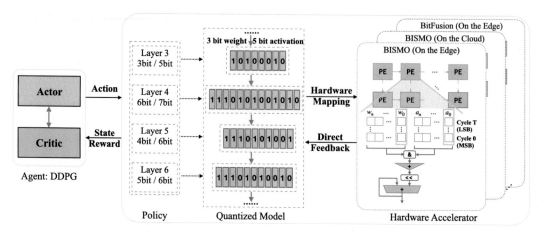

FIGURE 4.9 An overview of HAQ framework. It leverages the reinforcement learning to automatically search over the huge quantization design space with hardware in the loop. The agent proposes an optimal bitwidth allocation policy given the amount of computation resources (*i.e.*, latency, power, and model size). The RL agent integrates the hardware accelerator into the exploration loop so that it can obtain direct feedback from the hardware, instead of relying on indirect proxy signals (Wang et al., 2018).

mated Quantization (HAQ) (Wang et al., 2018) (Fig. 4.9) is proposed to automate the process. HAQ leverages the reinforcement learning to automatically determine the quantization policy. It takes the hardware accelerator's feedback in the design loop, rather than relying on proxy signals such as MACs and model size. Compared with conventional methods, HAQ is fully automated and can specialize the quantization policy for different neural network architectures and hardware architectures.

HAQ uses quite as different quantization policy for edge and cloud accelerators. The quantization policy of MobileNet-V1 on the BISMO accelerator (Umuroglu et al., 2018) (both edge and cloud configuration) is plotted in Fig. 4.10. On the edge accelerator, the RL agent allocates *less* activation bits to the depthwise convolutions, which echoes that the depthwise convolutions are memory bounded and that the activations dominate the memory access. On the cloud accelerator, our agent allocates *more* bits to the depthwise convolutions and allocates *less* bits to the pointwise convolutions, as the cloud device has more memory bandwidth and high parallelism, so the network appears to be computation bounded.

4.2 Efficient neural network architectures

In addition to compressing an existing deep neural network, another widely adopted approach to improve efficiency is to design new neural network architectures. A CNN model typically consists of convolution layers, pooling layers, and fully-connected layers, where most of the computation comes from the convolution layers. For example, in ResNet-50 (He et al., 2016), more than 99% of the multiply-accumulate operations (MACs) are from convo-

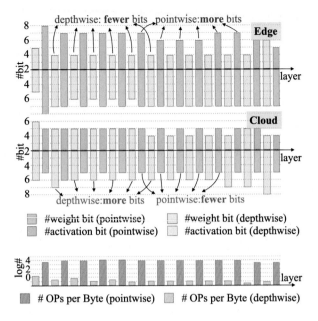

FIGURE 4.10 Quantization policy under latency constraints for MobileNet-V1 (Wang et al., 2018).

lution layers. Therefore, designing efficient convolution layers is the core of building efficient CNN architectures.

This section first describes the standard convolution layer and then describes three efficient variants of the standard convolution layer. Next, we present three representative manually design efficient CNN architectures, including SqueezeNet (Iandola et al., 2016), MobileNets (Howard et al., 2017; Sandler et al., 2018), and ShuffleNets (Ma et al., 2018; Zhang et al., 2017). Finally, we describe automated methods for designing efficient CNN architectures.

4.2.1 Standard convolution layer

A standard convolution layer is parameterized by convolution kernel $\mathbf{K}$ of size $O_c \times I_c \times K \times K$ where O_c is the number of output channels, I_c is the number of input channels, K is the spatial dimension of the kernel (Fig. 4.11a). Here for simplicity, we assume that the convolution kernel's width and height are the same. It is also possible to have asymmetric convolution kernels (Szegedy and Vanhoucke, 2016).

Given input feature map $\mathbf{F_i}$ of size $I_c \times H \times W$, the output feature map $\mathbf{F_o}$ of size $O_c \times H \times W$ is computed as follows[4]:

$$\mathbf{F_o}[n, h, w] = \sum_{m,i,j} \mathbf{K}[n, m, i, j] \times \mathbf{F_i}\big[m, h + i - \lfloor K/2 \rfloor, w + j - \lfloor K/2 \rfloor\big]. \tag{4.9}$$

---

[4] Assuming the stride is 1 and zero-padding is applied to preserve the spatial dimension of the feature map.

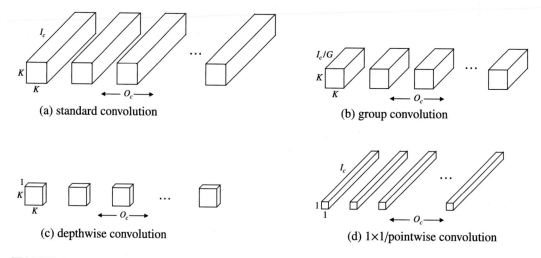

(a) standard convolution   (b) group convolution

(c) depthwise convolution   (d) 1×1/pointwise convolution

FIGURE 4.11 Illustration of the standard convolution and three commonly used efficient variants.

In the following discussions, we use $\mathbf{F_o} = Conv_{K \times K}(\mathbf{F_i}; \mathbf{K})$ to represent a standard convolution layer with kernel size K. According to Eq. (4.9), the computational cost of a standard convolution is

$$\#\text{MACs}(Conv_{K \times K}) = H \times W \times O_c \times I_c \times K \times K, \tag{4.10}$$

while the number of parameters is given as

$$\#\text{Params}(Conv_{K \times K}) = O_c \times I_c \times K \times K. \tag{4.11}$$

4.2.2 Efficient convolution layers

1 × 1 (pointwise) convolution

1×1 convolution (also called pointwise convolution) is a special kind of standard convolution layer, where the kernel size K is 1 (Fig. 4.11d). According to Eq. (4.10) and Eq. (4.11), replacing a $K \times K$ standard convolution layer with a 1×1 convolution layer will reduce the number of MACs (#MACs) and the number of parameters (#Params) by K^2 times. In practice, as the 1×1 convolution itself cannot aggregate spatial information, it is combined with other convolution layers to form CNN architectures. For example, a 1×1 convolution is usually used to reduce/increase the channel dimension of the feature map in CNN.

Group convolution

Different from a 1×1 convolution, which reduces the cost by decreasing the kernel size dimension, group convolution reduces the cost by decreasing the channel dimension. Specifically, the input feature map $\mathbf{F_i}$ is split into G groups along the channel dimension (Fig. 4.11b):

$$\text{split}(\mathbf{F_i}) = \left(\mathbf{F_i}[0:c,:,:], \mathbf{F_i}[c:2c,:,:], \cdots, \mathbf{F_i}[I_c - c : I_c,:,:]\right), \quad \text{where } c = I_c/G.$$

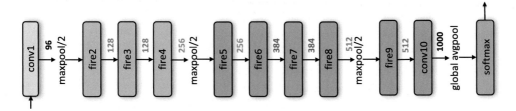

FIGURE 4.12 SqueezeNet Architecture (Iandola et al., 2016).

Then each group is fed to a standard $K \times K$ convolution of size $\frac{O_c}{G} \times \frac{I_c}{G} \times K \times K$. Finally, the outputs are concatenated along the channel dimension. Compared to a standard $K \times K$ convolution, #MACs and #Params are reduced by $G\times$ in a group convolution.

Depthwise convolution

The number of groups G is an adjustable hyperparameter in group convolutions. A larger G leads to lower computational cost and fewer parameters. An extreme case is that G equals the number of input channels I_c. In that case, the group convolution layer is called a depthwise convolution (Fig. 4.11 c). #MACs and #Params of a depthwise convolution are

$$\#MACs(DWConv_{K \times K}) = H \times W \times O_c \times K \times K, \tag{4.12}$$

$$\#Params(DWConv_{K \times K}) = O_c \times K \times K, \tag{4.13}$$

$$\text{where } O_c = I_c.$$

4.2.3 Manually designed efficient CNN models

SqueezeNet (Fig. 4.12)

SqueezeNet (Iandola et al., 2016) targets extremely compact model sizes for mobile applications. It has only 1.2 million parameters but achieved an accuracy similar to AlexNet (Table 4.1). SqueezeNet has 26 convolution layers and no fully-connected layer. The last feature map goes through a global average pooling and forms a 1000-dimension vector to feed the softmax layer. SqueezeNet has eight *Fire* modules. Each fire module contains a squeeze layer with 1×1 convolution and a pair of 1×1 and 3×3 convolutions. The SqueezeNet caffemodel achieved a top-1 accuracy of 57.4% and a top-5 accuracy of 80.5% on ImageNet 2012 (Deng et al., 2009). SqueezeNet is widely used in mobile applications in which model size is a large constraint.

MobileNets (Fig. 4.13)

MobileNetV1 (Howard et al., 2017) is based on a building block called *depthwise separable convolution* (Fig. 4.13a), which consists of a 3×3 depthwise convolution layer and a 1×1 convolution layer. The input image first goes through a 3×3 standard convolution layer with stride 2, then 13 depthwise separable convolution blocks. Finally, the feature map goes through a global average pooling and forms a 1280-dimension vector fed to the final

TABLE 4.1 Summarized results of manually designed CNN architectures on ImageNet.

Network	#Params	#MACs	ImageNet	
			Top-1 Acc	Top-5 Acc
AlexNet (Krizhevsky et al., 2012)	60M	720M	57.2%	80.3%
GoogleNet (Szegedy et al., 2015)	6.8M	1550M	69.8%	89.5%
VGG-16 (Simonyan and Zisserman, 2014)	138M	15300M	71.5%	–
ResNet-50 (He et al., 2016)	25.5M	4100M	76.1%	92.9%
SqueezeNet (Iandola et al., 2016)	1.2M	1700M	57.4%	80.5%
MobileNetV1 (Howard et al., 2017)	4.2M	569M	70.6%	89.5%
MobileNetV2 (Sandler et al., 2018)	3.4M	300M	72.0%	–
MobileNetV2-1.4 (Sandler et al., 2018)	6.9M	585M	74.7%	–
ShuffleNetV1-1.5x (Zhang et al., 2017)	3.4M	292M	71.5%	–
ShuffleNetV2-1.5x (Ma et al., 2018)	3.5M	299M	72.6%	–
ShuffleNetV2-2x (Ma et al., 2018)	7.4M	591M	74.9%	–

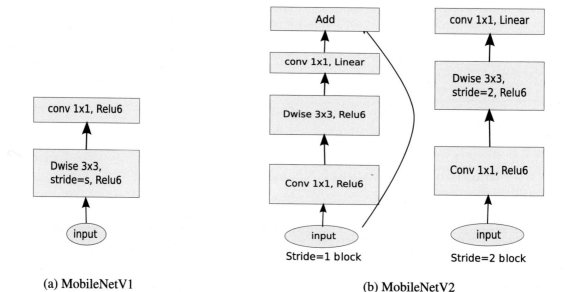

(a) MobileNetV1   (b) MobileNetV2

FIGURE 4.13 (a) Building block of MobileNetV1 (Howard et al., 2017). It consists of a 3×3 depthwise convolution layer and a 1×1 convolution layer. (b) Building blocks of MobileNetV2 (Sandler et al., 2018). Each block consists of a 3×3 depthwise convolution layer and two 1×1 convolution layers. When the stride is 1, the block will have a skip connection.

fully-connected layer with 1000 output units. With 569M MACs and 4.2M parameters, MobileNetV1 achieves 70.6% top-1 accuracy on ImageNet 2012 (Table 4.1).

MobileNetV2 (Sandler et al., 2018), an improved version of MobileNetV1, also uses 3×3 depthwise convolution and 1×1 convolution to compose its building blocks. Unlike MobileNetV1, the building block in MobileNetV2 has three layers, including a 3×3 depthwise

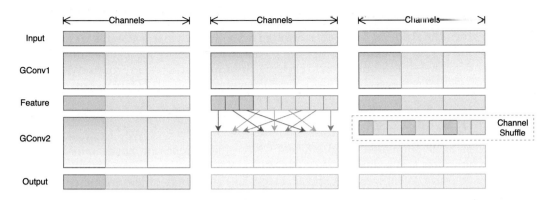

FIGURE 4.14 Illustration of the channel shuffle operation (Zhang et al., 2017).

convolution layer and two 1×1 convolution layers (Fig. 4.13b). The intuition is that the capacity of depthwise convolution is much lower than the standard convolution, and thus needs more channels to improve its capacity. From a cost perspective, the #MACs and #Params of a depthwise convolution only grow linearly (rather than quadratically as for a standard convolution) as the number of channels increases. Thus, even with a large channel number, the cost of a depthwise convolution layer is still moderate. Therefore, in MobileNetV2, the input feature map first goes through a 1×1 convolution to increase the channel dimension by a factor called *expand ratio*. Then the expanded feature map is fed to a 3×3 depthwise convolution, followed by another 1×1 convolution to reduce the channel dimension back to the original value. This structure is called *inverted bottleneck* and the block is called *mobile inverted bottleneck block*. Apart from the mobile inverted bottleneck block, MobileNetV2 has another two improvements over MobileNetV1. First, MobileNetV2 has skip connections[5] for blocks in which the stride is 1. Second, the activation function of the last 1×1 convolution in each block is removed. Combining these improvements, MobileNetV2 achieves 72.0% top-1 accuracy on ImageNet 2012 with only 300M MACs and 3.4M parameters (Table 4.1).

ShuffleNets

Similar to MobileNets, ShuffleNetV1 utilizes 3×3 depthwise convolution rather than standard convolution. Besides, ShuffleNetV1 introduces two new operations, pointwise group convolution and channel shuffle. The pointwise group convolution's motivation is to reduce the computational cost of 1×1 convolution layers. However, it has a side effect: a group cannot see information from other groups. This will significantly hurt accuracy. The channel shuffle operation is thus introduced to address this side effect by exchanging feature maps between different groups. An illustration of the channel shuffle operation is provided in Fig. 4.14. After shuffling, each group will contain information from all groups. On ImageNet 2012, ShuffleNetV1 achieves 71.5% top-1 accuracy with 292M MACs (Table 4.1).

In ShuffleNetV2, the input feature map is divided into two groups at the beginning of each building block. One group goes through the convolution branch that consists of a 3×3 depth-

---

[5] With a skip connection, $output = \mathcal{F}(input) + input$. Without a skip connection, $output = \mathcal{F}(input)$.

wise convolution layer and two 1×1 convolution layers. The other group goes through a skip connection when the stride is 1 and goes through a 3×3 depthwise separable convolution when the stride is 2. In the end, the outputs are concatenated along the channel dimension, followed by a channel shuffle operation to exchange information between groups. With 299M MACs, ShuffleNetV2 achieves 72.6% top-1 accuracy on ImageNet 2012 (Table 4.1).

4.2.4 Neural architecture search

The success of the aforementioned efficient CNN models relies on hand-crafted neural network architectures that require domain experts to explore the large design space, trading off among model size, latency, energy, and accuracy. This is not only time-consuming but also suboptimal. Thus, there is a growing interest in developing automated methods to tackle this challenge.

Neural Architecture Search (NAS) refers to using machine learning techniques to automatically design neural network architectures. In the conventional NAS formulation (Zoph and Le, 2016), designing neural network architectures is modeled as a sequence generation problem, where an auto-regressive RNN controller is introduced to generate neural network architectures. This RNN controller is trained by repeatedly sampling neural network architectures, evaluating the sampled neural network architectures, and updating the controller based on the feedback. To find a good neural network architecture in the vast search space, this process typically has to train and evaluate tens of thousands of neural networks (e.g., 12,800 in (Zoph et al., 2017)) on the target task, leading to prohibitive computational cost (10^4 GPU hours). To address this challenge, many techniques are proposed that try to improve different components of NAS, including search space, search algorithm, and performance evaluation strategy.

Search space

All the NAS methods need a predefined search space that contains basic network elements and how they connect with each other. For example, the typical basic elements of CNN models consist of (1) convolutions (Zoph et al., 2017; Real et al., 2018): standard convolutions (1×1, 3×3, 5×5), asymmetric convolutions (1×3 and 3×1, 1×7 and 7×1), depthwise-separable convolutions (3×3, 5×5), dilated convolutions (3×3); (2) poolings: average pooling (3×3), max pooling (3×3); (3) activation functions (Ramachandran et al., 2017). Then these basic elements are stacked sequentially (Baker et al., 2016) with identity connections (Zoph and Le, 2016). The full network-level search space grows exponentially as the network deepens (Fig. 4.15(a)). When the depth is 20, this search space contains more than 10^{36} different neural network architectures in Zoph et al. (2017).

Instead of directly searching on such an exponentially large space, restricting the search space is a very effective approach for improving the search speed. Specifically, Zoph et al. (2017); Zhong et al. (2017) propose to search for basic building cells (Fig. 4.15(b)) that can be stacked to construct neural networks, rather than the entire neural network architecture. As such, the architecture complexity is independent of the network depth, and the learned cells are transferable across different datasets. This enables NAS to search on a small proxy dataset (e.g., CIFAR-10), and then transfer to another large-scale dataset (e.g., ImageNet) by adapting the number of cells. Within the cell, the complexity is further reduced by supporting

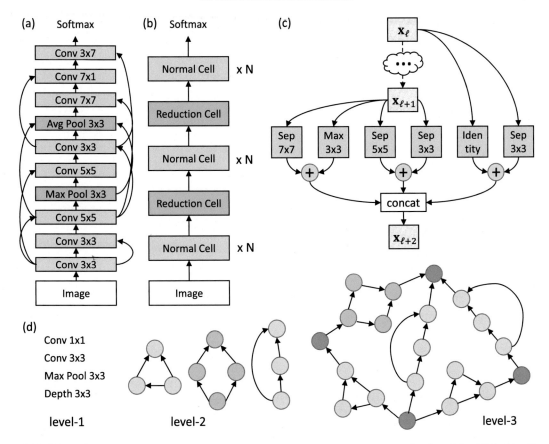

FIGURE 4.15 NAS search space (Deng et al., 2020): (a) network-level search space (Zoph and Le, 2016); (b) cell-level search space (Zoph et al., 2017); (c) an example of learned cell structure (Liu et al., 2017); (d) three-level hierarchical search space (Liu et al., 2018a).

hierarchical topologies (Liu et al., 2018a), or increasing the number of elements (blocks) in a progressive (simple to complex) manner (Liu et al., 2017).

Search algorithm

NAS methods usually have two stages at each search step: (1) the generator produces an architecture, and then (2) the evaluator trains the network and obtains the performance. As getting the performance of a sampled neural network architecture involves training a neural network, which is very expensive, search algorithms that affect the sample efficiency play an important role in improving the search speed of NAS. Most of the search algorithms used in NAS fall into five categories: random search, reinforcement learning (RL), evolutionary algorithms, Bayesian optimization, and gradient-based methods. Among them, RL, evolutionary algorithms, and gradient-based methods provide the most competitive results.

RL-based methods model the architecture generation process as a Markov Decision Process, treat the validation accuracy of the sampled architecture as the reward and update the architecture generation model using RL algorithms, including Q-learning (Baker et al., 2016; Zhong et al., 2017), REINFORCE (Zoph and Le, 2016), PPO (Zoph et al., 2017), etc. Instead of training an architecture generation model, evolutionary methods (Real et al., 2018; Liu et al., 2018a) maintain a population of neural network architectures. This population is updated through mutation and recombination. While both RL-based methods and evolutionary methods optimize neural network architectures in the discrete space, DARTS (Liu et al., 2018b) proposes continuous relaxation of the architecture representation:

$$y = \sum_i \alpha_i o_i(x), \quad \text{where } \alpha_i \geq 0, \sum_i \alpha_i = 1, \tag{4.14}$$

where $\{\alpha_i\}$ denote architecture parameters, $\{o_i\}$ denote candidate operations, x is the input, and y is the output. Such continuous relaxation allows neural network architectures to be optimized in the continuous space using gradient descent, which greatly improves the search efficiency. Apart from the above techniques, the search efficiency of NAS can be improved by exploring the architecture space with network transformation operations, starting from an existing network, and reusing the weights (Cai et al., 2018a,b; Elsken et al., 2018).

Performance evaluation

To guide the search process, NAS methods need to get the performances (typically accuracy on the validation set) of sampled neural architectures. The trivial approach to get these performances is to train sampled neural network architectures on the training data and measure their accuracy on the validation set. However, this would result in excessive computational cost (Zoph and Le, 2016; Zoph et al., 2017; Real et al., 2018). This motivates many techniques that aim at speeding up the performance evaluation step.

Alternatively, the evaluation step can be accelerated using Hypernetwork (Brock et al., 2017), which can directly generate weights of a neural architecture without training it. Though the model's accuracy will degrade significantly using the generated weights, this accuracy can be used as a proxy metric to select neural architectures. As such, only a single Hypernetwork needs to be trained, which greatly saves the search cost. Similarly, One-shot NAS methods (Pham et al., 2018; Liu et al., 2018b; Cai et al., 2019) focus on training a single supernet, from which small subnetworks directly inherit weights without training cost.

Auto-designed vs. human-designed

Fig. 4.16 reports the summarized results of auto-design CNN models and human-design CNN models on ImageNet. NAS not only saves engineer labor costs but also provides better CNN models over human-designed CNNs. Apart from ImageNet classification, auto-design CNN models have outperformed manually designed CNN models on object detection (Zoph et al., 2017; Chen et al., 2019b; Ghiasi et al., 2019; Tan et al., 2020a) and semantic segmentation (Liu et al., 2019; Chen et al., 2018).

4.2.5 Hardware-aware neural architecture search

While NAS has shown promising results, achieving significant MACs reduction without sacrificing accuracy, in real-world applications, we care about the real hardware efficiency (e.g., latency, energy) rather than #MACs. Unfortunately, MAC-efficiency does not

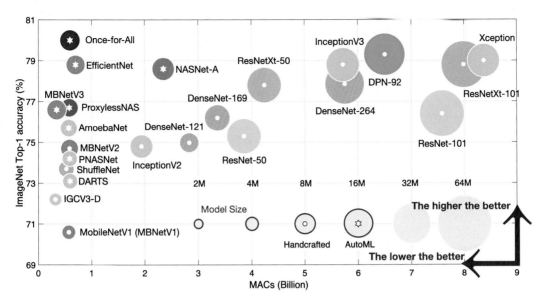

FIGURE 4.16 Summarized results of auto-design CNN models and human-design CNN models on ImageNet (Cai et al., 2020a).

directly translate to real hardware efficiency. Fig. 4.17 shows the comparison between auto-designed CNN models (NASNet-A and AmoebaNet-A) and human-designed CNN models (MobileNetV2-1.4). Although NASNet-A and AmoebaNet-A have fewer MACs than MobileNetV2-1.4, they actually run slower than MobileNetV2-1.4 on hardware. It is because #MACs only reflect the computation complexity of convolution operations. Other factors like data access cost, parallelism, and cost of element-wise operations that significantly affect real hardware efficiency are not taken into consideration.

This problem motivates hardware-aware NAS techniques (Tan et al., 2018; Cai et al., 2019; Wu et al., 2019) that directly incorporate hardware feedback into the architecture search process. An example of the hardware-aware NAS framework is shown in Fig. 4.18. Besides accuracy, each sampled neural network architecture is measured on the target hardware to collect its latency information. A multiobjective reward is defined based on accuracy ACC and latency LAT:

$$\text{reward} = ACC \times \left(\frac{LAT}{T}\right)^{\omega}, \tag{4.15}$$

where T is the target latency and ω is a hyperparameter.

Latency prediction

Measuring the latency on-device is accurate but not ideal for scalable neural architecture search. There are two reasons: (i) *Slow speed*. As suggested in TensorFlow-Lite,[6] we need to

---

[6] https://www.tensorflow.org/lite.

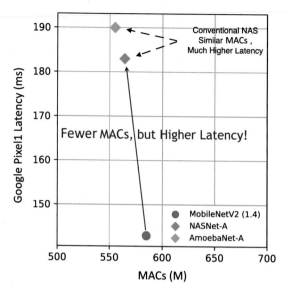

FIGURE 4.17 #MACs does not reflect real hardware efficiency. NASNet-A and AmoebaNet-A (auto-designed CNN models) have fewer MACs than MobileNetV2-1.4 (human-designed CNN model). However, they run slower than MobileNetV2-1.4 on Google Pixel1.

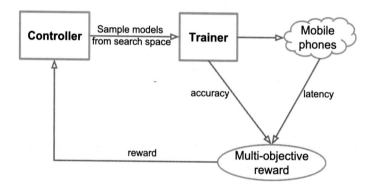

FIGURE 4.18 An example of the hardware-aware NAS framework (Tan et al., 2018).

average hundreds of runs to produce a precise measurement, approximately 20 seconds. This is far more slower than a single forward / backward execution. (ii) *High cost*. A lot of mobile devices and software engineering work are required to build an automatic pipeline to gather the latency from a mobile farm.

Instead of direct measurement, an economical solution is to build a prediction model to estimate the latency (Cai et al., 2019). In practice, this is implemented by sampling neural network architectures from the candidate space and profiling their latency on the target hardware platform. The collected data is then used to build the latency prediction model. For hardware platforms that sequentially execute operations, like mobile and FPGA, a simple la-

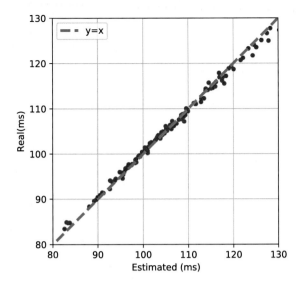

FIGURE 4.19 Predicted latency v.s. real latency on Google Pixel 1 (Cai et al., 2019).

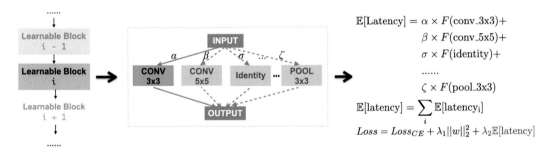

FIGURE 4.20 Making latency differentiable by introducing latency loss (Cai et al., 2019).

tency lookup table that maps each operation to its estimated latency is sufficient to provide very accurate latency predictions (Fig. 4.19). Another benefit of this approach is that it allows modeling the latency of a neural network as a regularization loss (Fig. 4.20), enabling trade-off between accuracy and latency to be optimized in a differentiable manner.

Specialized models for different hardware

Given the high cost of building a new neural network model, it is common to deploy the same model for all hardware platforms. However, it is suboptimal, as different hardware platforms have different properties, such as the number of arithmetic units, memory bandwidth, cache size, etc. Using hardware-aware NAS techniques, it is possible to have a specialized neural network architecture for each target hardware.

Fig. 4.21 demonstrates the detailed architectures of specialized CNN models on GPU and Mobile. We notice that the architecture shows different preferences when targeting different

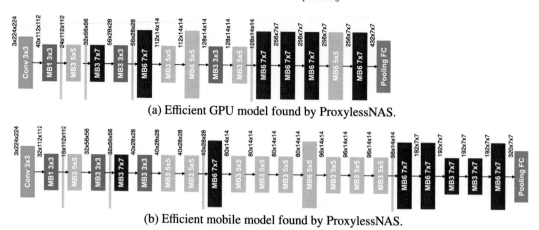

(a) Efficient GPU model found by ProxylessNAS.

(b) Efficient mobile model found by ProxylessNAS.

FIGURE 4.21 Efficient models optimized for different hardware. "MBConv3" and "MBConv6" denote mobile inverted bottleneck block with an expand ratio of 3 and 6 respectively. Insights: GPU prefers shallow and wide model with early pooling; Mobile prefers deep and narrow model with late pooling. Pooling layers prefer large and wide kernel. Early layers prefer small kernel. Late layers prefer large kernel (Cai et al., 2019).

TABLE 4.2   Hardware prefers specialized models (Cai et al., 2019). With a similar accuracy, the specialized model (ProxylessNAS-Mobile) reduces the latency by 1.8 × compared to the nonspecialized CNN model (MobileNetV2-1.4). Besides, models optimized for GPU do not run fast on Mobile, and vice versa.

Network	ImageNet Top-1 (%)	GPU latency	Mobile latency
MobileNetV2-1.4 (Sandler et al., 2018)	74.7	–	143 ms
ProxylessNAS-GPU (Cai et al., 2019)	75.1	5.1 ms	124 ms
ProxylessNAS-Mobile (Cai et al., 2019)	74.6	7.2 ms	78 ms

platforms: (i) The GPU model is shallower and wider, especially in the early stages where the feature map has higher resolution; (ii) The GPU model prefers large MBConv operations (e.g., 7 × 7 MBConv6), while the Mobile model would go for smaller MBConv operations. This is because GPU has much higher parallelism than Mobile so it can take advantage of large MBConv operations. Another interesting observation is that the searched models on all platforms prefer larger MBConv operations in the first block within each stage where the feature map is downsampled. This might be because larger MBConv operations are beneficial for the network to preserve more information when downsampling.

Table 4.2 shows the summarized results of specialized models on GPU and Mobile. An interesting observation is that models optimized for GPU do not run fast on Mobile, and vice versa. Therefore, it is essential to learn specialized neural networks for different hardware architectures to achieve the best efficiency on different hardware.

Handling many hardware platforms and efficiency constraints

Although specialized CNN models are superior to nonspecialized counterparts, designing specialized CNNs for every scenario is still difficult, either with human-based methods or hardware-aware NAS, since such methods need to *repeat* the network design process and

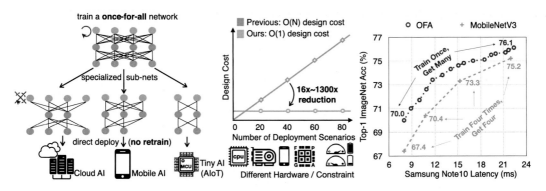

FIGURE 4.22 Left: a single once-for-all network is trained to support versatile architectural configurations including depth, width, kernel size, and resolution. Given a deployment scenario, a specialized subnetwork is directly selected from the once-for-all network without training. Middle: this approach reduces the cost of specialized deep learning deployment from O(N) to O(1). Right: once-for-all network followed by model selection can derive many accuracy-latency trade-offs by training only once, compared to conventional methods that require repeated training (Cai et al., 2020a).

retrain the designed network from scratch for *each* case. Their total cost grows linearly as the number of deployment scenarios increases, which will result in excessive energy consumption and CO_2 emission (Strubell et al., 2019). It makes them unable to handle the vast number of hardware devices (23.14 billion IoT devices till 2018[7]) and highly dynamic deployment environments (different battery conditions, different latency requirements, etc.).

To tackle this challenge, one promising solution is to build a once-for-all (OFA) network (Cai et al., 2020a; Yu et al., 2020) that can be directly deployed under diverse architectural configurations, amortizing the training cost. The inference is performed by selecting only part of the OFA network. It flexibly supports different depths, widths, kernel sizes, and resolutions without retraining. An example of OFA is illustrated in Fig. 4.22 (left). Specifically, the model training stage is decoupled from the neural architecture search stage. In the model training stage, the focus is to improve the accuracy of all subnetworks derived by selecting different parts of the OFA network. A subset of subnetworks is sampled in the model specialization stage to train an accuracy predictor and latency predictors. Given the target hardware and constraint, a predictor-guided architecture search (Liu et al., 2018) is conducted to get a specialized subnetwork, and the cost is negligible.[8] As such, the total cost of specialized neural network design is reduced from O(N) to O(1) (Fig. 4.22 middle).

Table 4.3 reports the comparison between OFA and state-of-the-art hardware-aware NAS methods on the mobile phone (Pixel1). The cost of OFA is *constant* while others are *linear* to the number of deployment scenarios (N). With $N = 40$, the total CO_2 emissions of OFA are 16 × less than ProxylessNAS, 19× less than FBNet, and 1,300 × less than MnasNet.

Fig. 4.23 summarizes the results of OFA under different MACs and Pixel1 latency constraints. An interesting observation is that training the searched neural architectures from

[7] https://www.statista.com/statistics/471264/iot-number-of-connected-devices-worldwide/.
[8] https://github.com/mit-han-lab/once-for-all/blob/master/tutorial/ofa.ipynb.

TABLE 4.3　Summarized results on Pixel1 phone (Cai et al., 2020a). The first group corresponds to human-designed CNN models. The second group corresponds to conventional NAS. The third group corresponds to hardware-aware NAS. The final group corresponds to OFA. "#75" indicates that the specialized subnetworks are fine-tuned for 75 epochs after grabbing weights from the OFA network. "CO_2e" denotes CO_2 emission which is calculated based on (Strubell et al., 2019). AWS cost is calculated based on the price of Amazon AWS on-demand P3.16x large instances.

Network	ImageNet Top1 (%)	MACs	Mobile latency	Search cost (GPU hours)	Training cost (GPU hours)	Total cost ($N = 40$)		
						GPU hours	CO_2e (lbs)	AWS cost
MobileNetV2 (Sandler et al., 2018)	72.0	300M	66 ms	0	150N	6k	1.7k	$18.4k
MobileNetV2 #1200	73.5	300M	66 ms	0	1200N	48k	13.6k	$146.9k
NASNet-A (Zoph et al., 2017)	74.0	564M	–	48,000N	–	1920k	544.5k	$5875.2k
DARTS (Liu et al., 2018b)	73.1	595M	–	96N	250N	14k	4.0k	$42.8k
MnasNet (Tan et al., 2018)	74.0	317M	70 ms	40,000N	–	1600k	453.8k	$4896.0k
FBNet-C (Wu et al., 2019)	74.9	375M	–	216N	360N	23k	6.5k	$70.4k
ProxylessNAS (Cai et al., 2019)	74.6	320M	71 ms	200N	300N	20k	5.7k	$61.2k
SinglePathNAS (Guo et al., 2019)	74.7	328M	–	288 + 24N	384N	17k	4.8k	$52.0k
AutoSlim (Yu and Autoslim, 2019)	74.2	305M	63 ms	180	300N	12k	3.4k	$36.7k
MobileNetV3-Large (Howard et al., 2019)	75.2	219M	58 ms	–	180N	7.2k	1.8k	$22.2k
OFA	76.0	230M	58 ms	40	1200	1.2k	0.34k	$3.7k
OFA #75	76.9	230M	58 ms	40	1200 + 75N	4.2k	1.2k	$13.0k
OFA$_{Large}$ #75	80.0	595M	–	40	1200 + 75N	4.2k	1.2k	$13.0k

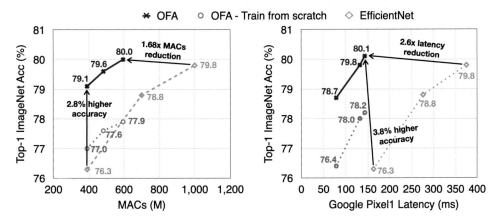

FIGURE 4.23 Training the search neural architectures from scratch cannot achieve the same accuracy as OFA (Cai et al., 2020a).

scratch cannot reach the same level of accuracy as OFA, suggesting that not only neural architectures but also pretrained weights contribute to the superior performances of OFA.

4.3 Conclusion

Over the past few years, deep neural networks have achieved unprecedented success in the field of artificial intelligence; however, their superior performance comes at the cost of high computational complexity. This limits their applications on many edge devices, where the hardware resources are tightly constrained by the form factor, battery, and heat dissipation.

This chapter offers a systematic overview of efficient deep learning to enable both researchers and practitioners to get started in this field quickly. We first introduce various model compression approaches that have become the industry standards, such as pruning, factorization, quantization, and efficient model design. To reduce the design cost of these handcrafted solutions, we then describe many recent efforts on neural architecture search, automated pruning, and quantization, which can outperform the manual design with minimal human efforts. Finally, we describe the once-for-all technique to efficiently handle many hardware platforms and efficiency constraints without repeating the costly search and retraining phases.

References

Baker, Bowen, Gupta, Otkrist, Naik, Nikhil, Raskar, Ramesh, 2016. Designing neural network architectures using reinforcement learning. arXiv preprint. arXiv:1611.02167.

Banner, Ron, Nahshan, Yury, Soudry, Daniel, 2019. Post training 4-bit quantization of convolutional networks for rapid-deployment. In: Advances in Neural Information Processing Systems, pp. 7950–7958.

Bengio, Yoshua, Léonard, Nicholas, Courville, Aaron, 2013. Estimating or propagating gradients through stochastic neurons for conditional computation. arXiv preprint. arXiv:1308.3432.

Brock, Andrew, Lim, Theodore, Ritchie, James M., Weston, Nick, 2017. Smash: one-shot model architecture search through hypernetworks. arXiv preprint. arXiv:1708.05344.

Buciluă, Cristian, Caruana, Rich, Niculescu-Mizil, Alexandru, 2006. Model compression. In: Proceedings of the 12th ACM SIGKDD International Conference on Knowledge Discovery and Data Mining, pp. 535–541.

Cai, Han, Chen, Tianyao, Zhang, Weinan, Yu, Yong, Wang, Jun, 2018a. Efficient architecture search by network transformation. In: AAAI.

Cai , Han, Yang, Jiacheng, Zhang, Weinan, Han, Song, Yu, Yong, 2018b. Path-level network transformation for efficient architecture search. In: ICML.

Cai, Han, Gan, Chuang, Wang, Tianzhe, Zhang, Zhekai, Han, Song, 2020a. Once for all: train one network and specialize it for efficient deployment. In: International Conference on Learning Representations.

Cai, Han, Zhu, Ligeng, ProxylessNAS, Song Han, 2019. Direct neural architecture search on target task and hardware. In: International Conference on Learning Representations.

Cai, Yaohui, Yao, Zhewei, Dong, Zhen, Gholami, Amir, Mahoney, Michael W., Zeroq, Kurt Keutzer, 2020b. A novel zero shot quantization framework. In: Proceedings of the IEEE/CVF Conference on Computer Vision and Pattern Recognition, pp. 13169–13178.

Chen, Guobin, Choi, Wongun, Yu, Xiang, Han, Tony, Chandraker, Manmohan, 2017. Learning efficient object detection models with knowledge distillation. In: Advances in Neural Information Processing Systems, pp. 742–751.

Chen, Liang-Chieh, Collins, Maxwell, Zhu, Yukun, Papandreou, George, Zoph, Barret, Schroff, Florian, Adam, Hartwig, Shlens, Jon, 2018. Searching for efficient multi-scale architectures for dense image prediction. In: Advances in Neural Information Processing Systems, pp. 8699–8710.

Chen, Yu-Hsin, Krishna, Tushar, Emer, Joel S., Eyeriss, Vivienne Sze, 2016. An energy-efficient reconfigurable accelerator for deep convolutional neural networks. IEEE Journal of Solid-State Circuits.

Chen, Yu-Hsin, Yang, Tien-Ju, Emer, Joel, Sze, Vivienne, 2019a. Eyeriss v2: a flexible accelerator for emerging deep neural networks on mobile devices. IEEE Journal on Emerging and Selected Topics in Circuits and Systems 9 (2), 292–308.

Chen, Yukang, Yang, Tong, Zhang, Xiangyu, Meng, Gaofeng, Xiao, Xinyu, Detnas, Jian Sun, 2019b. Backbone search for object detection. In: Advances in Neural Information Processing Systems, pp. 6642–6652.

Cheong, Robin, Daniel, Robel, 2019. Transformers. Zip: Compressing transformers with pruning and quantization. Technical report. Stanford University, Stanford, California.

Courbariaux, Matthieu, Bengio, Yoshua, 2016. Binarynet: training deep neural networks with weights and activations constrained to+ 1. arXiv:1602.02830.

Courbariaux, Matthieu, Bengio, Yoshua, Binaryconnect, Jean-Pierre David, 2015. Training deep neural networks with binary weights during propagations. In: NIPS.

Deng, Jia, Dong, Wei, Socher, Richard, Li, Li-Jia, Li, Kai, Li Imagenet, Fei-Fei, 2009. A large-scale hierarchical image database. In: IEEE Conference on Computer Vision and Pattern Recognition, 2009. CVPR 2009. IEEE, pp. 248–255.

Deng, Lei, Li, Guoqi, Han, Song, Shi, Luping, Xie, Yuan, 2020. Model compression and hardware acceleration for neural networks: a comprehensive survey. Proceedings of the IEEE 108 (4), 485–532.

Denton, Emily L., Zaremba, Wojciech, Bruna, Joan, LeCun, Yann, Fergus, Rob, 2014. Exploiting linear structure within convolutional networks for efficient evaluation. In: Advances in Neural Information Processing Systems, pp. 1269–1277.

Elsken, Thomas, Metzen, Jan Hendrik, Hutter, Frank, 2018. Efficient multi-objective neural architecture search via Lamarckian evolution. arXiv preprint. arXiv:1804.09081.

Engelbrecht, Andries Petrus , 2001. A new pruning heuristic based on variance analysis of sensitivity information. IEEE Transactions on Neural Networks 12 (6), 1386–1399.

Frankle, Jonathan, Carbin, Michael, 2018. The lottery ticket hypothesis: finding sparse, trainable neural networks. arXiv preprint. arXiv:1803.03635.

Frankle, Jonathan, Dziugaite, Gintare Karolina, Roy, Daniel, Carbin, Michael, 2020. Linear mode connectivity and the lottery ticket hypothesis. In: International Conference on Machine Learning. PMLR, pp. 3259–3269.

Ghiasi, Golnaz, Lin, Tsung-Yi, Le Nas-fpn, Quoc V., 2019. Learning scalable feature pyramid architecture for object detection. In: Proceedings of the IEEE Conference on Computer Vision and Pattern Recognition, pp. 7036–7045.

Giles, C. Lee, Omlin, Christian W., 1994. Pruning recurrent neural networks for improved generalization performance. IEEE Transactions on Neural Networks 5 (5), 848–851.

Girshick, Ross, 2015. Fast r-cnn. In: Proceedings of the IEEE International Conference on Computer Vision, pp. 1440–1448.

Golub, Gene H., Van Loan, Charles F., 1996. Matrix Computations. Johns Hopkins University Press, Baltimore and London.

Gong, Yunchao, Liu, Liu, Yang, Ming, Bourdev, Lubomir, 2014. Compressing deep convolutional networks using vector quantization. arXiv preprint. arXiv:1412.6115.

Guo, Yiwen, Yao, Anbang, Chen, Yurong, 2016. Dynamic network surgery for efficient dnns. In: Advances in Neural Information Processing Systems, pp. 1379–1387.

Guo, Zichao, Zhang, Xiangyu, Mu, Haoyuan, Heng, Wen, Liu, Zechun, Wei, Yichen, Sun, Jian, 2019. Single path one-shot neural architecture search with uniform sampling. arXiv preprint. arXiv:1904.00420.

Han, Song, 2017. Efficient methods and hardware for deep learning.

Han, Song, Kang, Junlong, Mao, Huizi, Hu, Yiming, Li, Xin, Li, Yubin, Xie, Dongliang, Luo, Hong, Yao, Song, Wang, Yu, et al., 2017. Ese: efficient speech recognition engine with sparse lstm on fpga. In: Proceedings of the 2017 ACM/SIGDA International Symposium on Field-Programmable Gate Arrays. ACM, pp. 75–84.

Han, Song, Liu, Xingyu, Mao, Huizi, Pu, Jing, Pedram, Ardavan, Horowitz, Mark A., Dally, William J., 2016. Eie: efficient inference engine on compressed deep neural network. In: Proceedings of the 43rd International Symposium on Computer Architecture. IEEE Press, pp. 243–254.

Han, Song, Mao, Huizi, Dally, William J., 2015a. Deep compression: compressing deep neural networks with pruning, trained quantization and Huffman coding. arXiv preprint. arXiv:1510.00149.

Han, Song, Pool, Jeff, Tran, John, Dally, William, 2015b. Learning both weights and connections for efficient neural network. In: Advances in Neural Information Processing Systems, pp. 1135–1143.

Hassibi, Babak, Stork, David G., 1993. Second Order Derivatives for Network Pruning: Optimal Brain Surgeon. Morgan Kaufmann.

He, Kaiming, Zhang, Xiangyu, Ren, Shaoqing, Sun, Jian, 2016. Deep residual learning for image recognition. In: Proceedings of the IEEE Conference on Computer Vision and Pattern Recognition, pp. 770–778.

He, Yihui, Lin, Ji, Liu, Zhijian, Wang, Hanrui, Li, Li-Jia, Amc, Song Han, 2018. Automl for model compression and acceleration on mobile devices. In: Proceedings of the European Conference on Computer Vision (ECCV), pp. 784–800.

He, Yihui, Zhang, Xiangyu, Sun, Jian, 2017. Channel pruning for accelerating very deep neural networks. In: Proceedings of the IEEE Conference on Computer Vision and Pattern Recognition, pp. 1389–1397.

Hinton, Geoffrey, Vinyals, Oriol, Dean, Jeff, 2015. Distilling the knowledge in a neural network. arXiv preprint. arXiv:1503.02531.

Howard, Andrew, Sandler, Mark, Chu, Grace, Chen, Liang-Chieh, Chen, Bo, Tan, Mingxing, Wang, Weijun, Zhu, Yukun, Pang, Ruoming, Vasudevan, Vijay, et al., 2019. Searching for mobilenetv3. In: ICCV 2019.

Howard, Andrew G., Zhu, Menglong, Chen, Bo, Kalenichenko, Dmitry, Wang, Weijun, Weyand, Tobias, Andreetto, Marco, Mobilenets, Hartwig Adam, 2017. Efficient convolutional neural networks for mobile vision applications. arXiv preprint. arXiv:1704.04861.

Iandola, Forrest N., Moskewicz, Matthew W., Ashraf, Khalid, Han, Song, Dally, William J., Squeezenet, Kurt Keutzer, 2016. Alexnet-level accuracy with 50x fewer parameters and< 1mb model size. arXiv preprint. arXiv:1602.07360.

Ioffe, Sergey, Szegedy, Christian, 2015. Batch normalization: accelerating deep network training by reducing internal covariate shift. arXiv preprint. arXiv:1502.03167.

Jacob, Benoit, Kligys, Skirmantas, Chen, Bo, Zhu, Menglong, Tang, Matthew, Howard, Andrew, Adam, Hartwig, Kalenichenko, Dmitry, 2017. Quantization and training of neural networks for efficient integer-arithmetic-only inference. arXiv preprint. arXiv:1712.05877.

Jaderberg, Max, Vedaldi, Andrea, Zisserman, Andrew, 2014. Speeding up convolutional neural networks with low rank expansions. arXiv preprint. arXiv:1405.3866.

Judd, Patrick, Albericio, Jorge, Hetherington, Tayler, Aamodt, Tor M., Stripes, Andreas Moshovos, 2016. Bit-serial deep neural network computing. In: MICRO.

Kim, Yong-Deok, Park, Eunhyeok, Yoo, Sungjoo, Choi, Taelim, Yang, Lu, Shin, Dongjun, 2015. Compression of deep convolutional neural networks for fast and low power mobile applications. arXiv preprint. arXiv:1511.06530.

Krizhevsky, Alex, Hinton, Geoffrey, 2009. Learning multiple layers of features from tiny images.

Krizhevsky, Alex, Sutskever, Ilya, Hinton, Geoffrey E., 2012. Imagenet classification with deep convolutional neural networks. In: Advances in Neural Information Processing Systems, pp. 1097–1105.

Lebedev, Vadim, Ganin, Yaroslav, Rakhuba, Maksim, Oseledets, Ivan, Lempitsky, Victor, 2014. Speeding-up convolutional neural networks using fine-tuned cp-decomposition. arXiv preprint. arXiv:1412.6553.

Lebedev, Vadim, Lempitsky, Victor, 2016. Fast convnets using group-wise brain damage. In: Proceedings of the IEEE Conference on Computer Vision and Pattern Recognition, pp. 2554–2564.

LeCun, Yann, Cortes, Corinna, Burges, Christopher JC, 2010. Mnist handwritten digit database. AT&T Labs [Online]. Available: http://yann.lecun.com/exdb/mnist.

LeCun, Yann, Denker, John S., Solla, Sara A., Howard, Richard E., Jackel, Lawrence D., 1989. Optimal brain damage. In: NIPs, vol. 2, pp. 598–605.

Li, Fengfu, Zhang, Bo, Liu, Bin, 2016a. Ternary weight networks. arXiv preprint. arXiv:1605.04711.

Li, Hao, Kadav, Asim, Durdanovic, Igor, Samet, Hanan, Graf, Hans Peter, 2016b. Pruning filters for efficient convnets. arXiv preprint. arXiv:1608.08710.

Li, Muyang, Lin, Ji, Ding, Yaoyao, Liu, Zhijian, Zhu, Jun-Yan, Han, Song, 2020. Gan compression: efficient architectures for interactive conditional gans. In: Proceedings of the IEEE/CVF Conference on Computer Vision and Pattern Recognition, pp. 5284–5294.

Lillicrap, Timothy P., Hunt, Jonathan J., Pritzel, Alexander, Heess, Nicolas, Erez, Tom, Tassa, Yuval, Silver, David, Wierstra, Daan, 2015. Continuous control with deep reinforcement learning. arXiv preprint. arXiv:1509.02971.

Lin, Ji, Rao, Yongming, Lu, Jiwen, 2017. Runtime neural pruning. In: NeurIPS.

Lin, Zhouhan, Courbariaux, Matthieu, Memisevic, Roland, Bengio, Yoshua, 2015. Neural networks with few multiplications. arXiv preprint. arXiv:1510.03009.

Liu, Chenxi, Chen, Liang-Chieh, Schroff, Florian, Adam, Hartwig, Hua, Wei, Yuille, Alan L., Li Auto-deeplab, Fei-Fei, 2019. Hierarchical neural architecture search for semantic image segmentation. In: Proceedings of the IEEE Conference on Computer Vision and Pattern Recognition, pp. 82–92.

Liu, Chenxi, Zoph, Barret, Neumann, Maxim, Shlens Wei Hua, Jonathon, Li, Li-Jia, Fei-Fei, Li, Yuille, Alan, Huang, Jonathan, Murphy, Kevin, 2018. Progressive neural architecture search. In: Proceedings of the European Conference on Computer Vision (ECCV), pp. 19–34.

Liu, Chenxi, Zoph, Barret, Shlens Wei Hua, Jonathon, Li, Li-Jia, Fei-Fei, Li, Yuille, Alan, Huang, Jonathan, Murphy, Kevin, 2017. Progressive neural architecture search. arXiv preprint. arXiv:1712.00559.

Liu, Hanxiao, Simonyan, Karen, Vinyals, Oriol, Fernando, Chrisantha, Kavukcuoglu, Koray, 2018a. Hierarchical representations for efficient architecture search. In: ICLR.

Liu, Hanxiao, Simonyan, Karen, Darts, Yiming Yang, 2018b. Differentiable architecture search. arXiv preprint. arXiv:1806.09055.

Liu, Yifan, Chen, Ke, Liu, Chris, Qin, Zengchang, Luo, Zhenbo, Wang, Jingdong, 2019a. Structured knowledge distillation for semantic segmentation. In: Proceedings of the IEEE Conference on Computer Vision and Pattern Recognition, pp. 2604–2613.

Liu, Zechun, Mu, Haoyuan, Zhang, Xiangyu, Guo, Zichao, Yang, Xin, Cheng, Kwang-Ting, Metapruning, Jian Sun, 2019b. Meta learning for automatic neural network channel pruning. In: Proceedings of the IEEE International Conference on Computer Vision, pp. 3296–3305.

Ma, Ningning, Zhang, Xiangyu, Zheng, Hai-Tao, Sun, Jian, 2018. Shufflenet v2: practical guidelines for efficient cnn architecture design. In: ECCV.

Ma, Xiaolong, Guo Wei Niu, Fu-Ming, Lin, Xue, Tang, Jian, Ma, Kaisheng, Ren, Bin, Pconv, Yanzhi Wang, 2020. The missing but desirable sparsity in dnn weight pruning for real-time execution on mobile devices. In: AAAI, pp. 5117–5124.

Mao, Huizi, Han, Song, Pool, Jeff, Li, Wenshuo, Liu, Xingyu, Wang, Yu, Dally, William J., 2017. Exploring the granularity of sparsity in convolutional neural networks. In: Proceedings of the IEEE Conference on Computer Vision and Pattern Recognition Workshops, pp. 13–20.

Molchanov, Pavlo, Tyree, Stephen, Karras, Tero, Aila, Timo, Kautz, Jan, 2016. Pruning convolutional neural networks for resource efficient transfer learning. CoRR. arXiv:1611.06440 [abs].

Molchanov, Pavlo, Tyree, Stephen, Karras, Tero, Aila, Timo, Kautz, Jan, 2017. Pruning convolutional neural networks for resource efficient transfer learning. In: International Conference on Learning Representations.

Nagel, Markus, van Baalen, Mart, Blankevoort, Tijmen, Welling, Max, 2019. Data-free quantization through weight equalization and bias correction. In: Proceedings of the IEEE International Conference on Computer Vision, pp. 1325–1334.

Niu, Wei, Ma, Xiaolong, Lin, Sheng, Wang, Shihao, Qian, Xuehai, Lin, Xue, Wang, Yanzhi, Patdnn, Bin Ren, 2020. Achieving real-time dnn execution on mobile devices with pattern-based weight pruning. In: Proceedings of the Twenty-Fifth International Conference on Architectural Support for Programming Languages and Operating Systems, pp. 907–922.

Pham, Hieu, Guan, Melody Y., Zoph, Barret, Le, Quoc V., Dean, Jeff, 2018. Efficient neural architecture search via parameter sharing. In: ICML.

Ramachandran, Prajit, Zoph, Barret, Le, Quoc V., 2017. Searching for activation functions. arXiv preprint. arXiv: 1710.05941.

Rastegari, Mohammad, Ordonez, Vicente, Redmon, Joseph, Xnor-net, Ali Farhadi, 2016. Imagenet classification using binary convolutional neural networks. In: European Conference on Computer Vision. Springer, pp. 525–542.

Real, Esteban, Aggarwal, Alok, Huang, Yanping, Le, Quoc V., 2018. Regularized evolution for image classifier architecture search. arXiv preprint. arXiv:1802.01548.

Romero, Adriana, Ballas, Nicolas, Ebrahimi Kahou, Samira, Chassang, Antoine, Gatta, Carlo, Bengio, Yoshua, 2014. Fitnets: hints for thin deep nets. arXiv preprint. arXiv:1412.6550.

Sandler, Mark, Howard, Andrew, Zhu, Menglong, Zhmoginov, Andrey, Chen, Liang-Chieh, 2018. Mobilenetv2: inverted residuals and linear bottlenecks. In: Proceedings of the IEEE Conference on Computer Vision and Pattern Recognition, pp. 4510–4520.

Sanh, Victor, Debut, Lysandre, Chaumond, Julien, Wolf, Thomas, 2019. Distilbert, a distilled version of bert: smaller, faster, cheaper and lighter. arXiv preprint arXiv:1910.01108.

Sharify, Sayeh, Delmas Lascorz, Alberto, Siu, Kevin, Judd, Patrick, Loom, Andreas Moshovos, 2018. Exploiting weight and activation precisions to accelerate convolutional neural networks. In: DAC.

Sharma, Hardik, Park, Jongse, Suda, Naveen, Lai, Liangzhen, Chau, Benson, Chandra, Vikas, Esmaeilzadeh, Hadi, 2018. Bit fusion: bit-level dynamically composable architecture for accelerating deep neural networks. In: Proceedings of the 45th Annual International Symposium on Computer Architecture. IEEE Press, pp. 764–775.

Simonyan, Karen, Zisserman, Andrew, 2014. Very deep convolutional networks for large-scale image recognition. arXiv preprint. arXiv:1409.1556.

Srinivas, Suraj, Babu, R. Venkatesh, 2015. Data-free parameter pruning for deep neural networks. arXiv preprint. arXiv:1507.06149.

Strubell, Emma, Ganesh, Ananya, McCallum, Andrew, 2019. Energy and policy considerations for deep learning in nlp. In: ACL.

Szegedy, Christian, Liu, Wei, Jia, Yangqing, Sermanet, Pierre, Reed, Scott, Anguelov, Dragomir, Erhan, Dumitru, Vanhoucke, Vincent, Rabinovich, Andrew, 2015. Going deeper with convolutions. In: Proceedings of the IEEE Conference on Computer Vision and Pattern Recognition, pp. 1–9.

Szegedy, Christian, Vanhoucke, Vincent, Ioffe, Sergey, Shlens, Jon, Wojna, Zbigniew, 2016. Rethinking the inception architecture for computer vision. In: CVPR.

Tan, Mingxing, Chen, Bo, Pang, Ruoming, Vasudevan, Vijay, Le Mnasnet, Quoc V., 2018. Platform-aware neural architecture search for mobile. arXiv preprint. arXiv:1807.11626.

Tan, Mingxing, Pang, Ruoming, Le Efficientdet, Quoc V., 2020a. Scalable and efficient object detection. In: Proceedings of the IEEE/CVF Conference on Computer Vision and Pattern Recognition, pp. 10781–10790.

Tan, Zhanhong, Song, Jiebo, Ma, Xiaolong, Tan, Sia-Huat, Chen, Hongyang, Miao, Yuanqing, Wu, Yifu, Ye, Shaokai, Wang, Yanzhi, Li, Dehui, et al., 2020b. Pcnn: pattern-based fine-grained regular pruning towards optimizing cnn accelerators. arXiv preprint. arXiv:2002.04997.

Umuroglu, Yaman, Rasnayake, Lahiru, Bismo, Magnus Sjalander, 2018. A scalable bit-serial matrix multiplication overlay for reconfigurable computing. In: FPL.

Vanhoucke, Vincent, Senior, Andrew, Mao, Mark Z., 2011. Improving the Speed of Neural Networks on Cpus. In: Proc. Deep Learning and Unsupervised Feature Learning NIPS Workshop. In: Citeseer, vol. 1, p. 4.

Wang, Kuan, Liu, Zhijian, Lin, Yujun, Lin, Ji, Haq, Song Han, 2018. Hardware-aware automated quantization. arXiv preprint. arXiv:1811.08886.

Wen, Wei, Wu, Chunpeng, Wang, Yandan, Chen, Yiran, Li, Hai, 2016. Learning structured sparsity in deep neural networks. In: Advances in Neural Information Processing Systems, pp. 2074–2082.

Wu, Bichen, Dai, Xiaoliang, Zhang, Peizhao, Wang, Yanghan, Sun, Fei, Wu, Yiming, Tian, Yuandong, Vajda, Peter, Jia, Yangqing, Fbnet, Kurt Keutzer, 2019. Hardware-aware efficient convnet design via differentiable neural architecture search. In: Proceedings of the IEEE Conference on Computer Vision and Pattern Recognition, pp. 10734–10742.

Wu, Jiaxiang, Leng, Cong, Wang, Yuhang, Hu, Qinghao, Cheng, Jian, 2016. Quantized convolutional neural networks for mobile devices. In: Proceedings of the IEEE Conference on Computer Vision and Pattern Recognition, pp. 4820–4828.

Xue, Jian, Li, Jinyu, Gong, Yifan, 2013. Restructuring of deep neural network acoustic models with singular value decomposition. In: Interspeech, pp. 2365–2369.

Yang, Tien-Ju, Howard, Andrew, Chen, Bo, Zhang, Xiao, Go, Alec, Sze, Vivienne, Netadapt, Hartwig Adam, 2018. Platform-aware neural network adaptation for mobile applications. arXiv preprint. arXiv:1804.03230.

Yu, Jiahui, Autoslim, Thomas Huang, 2019. Towards one-shot architecture search for channel numbers. arXiv preprint. arXiv:1903.11728.

Yu, Jiahui, Jin, Pengchong, Liu, Hanxiao, Bender, Gabriel, Kindermans, Pieter-Jan, Tan, Mingxing, Huang, Thomas, Song, Xiaodan, Pang, Ruoming, le, Quoc, 2020. Bignas: scaling up neural architecture search with big single-stage models. arXiv preprint. arXiv:2003.11142.

Yu, Jiecao, Lukefahr, Andrew, Palframan, David, Dasika, Ganesh, Das, Reetuparna, Scalpel, Scott Mahlke, 2017. Customizing dnn pruning to the underlying hardware parallelism. ACM SIGARCH Computer Architecture News 45 (2), 548–560.

Zagoruyko, Sergey, Komodakis, Nikos, 2016. Paying more attention to attention: improving the performance of convolutional neural networks via attention transfer. arXiv preprint. arXiv:1612.03928.

Zhang, Chen, Li, Peng, Sun, Guangyu, Guan, Yijin, Xiao, Bingjun, Cong, Jason, 2015. Optimizing fpga-based accelerator design for deep convolutional neural networks. In: Proceedings of the 2015 ACM/SIGDA International Symposium on Field-Programmable Gate Arrays, pp. 161–170.

Zhang, Shijin, Du, Zidong, Zhang, Lei, Lan, Huiying, Liu, Shaoli, Li, Ling, Guo, Qi, Chen, Tianshi, Cambricon-x, Yunji Chen, 2016a. An accelerator for sparse neural networks. In: 2016 49th Annual IEEE/ACM International Symposium on Microarchitecture (MICRO). IEEE, pp. 1–12.

Zhang, Xiangyu, Zou, Jianhua, He, Kaiming, Sun, Jian, 2016b. Accelerating very deep convolutional networks for classification and detection. IEEE Transactions on Pattern Analysis and Machine Intelligence 38 (10), 1943–1955.

Zhang, Xiangyu, Zhou, Xinyu, Lin, Mengxiao, Jian Sun, 2017. An extremely efficient convolutional neural network for mobile devices. arXiv preprint. arXiv:1707.01083.

Zhong, Zhao, Yan, Junjie, Liu, Cheng-Lin, 2017. Practical network blocks design with q-learning. arXiv preprint. arXiv:1708.05552.

Zhou, Shuchang, Wu, Yuxin, Ni, Zekun, Zhou, Xinyu, Wen, He, Dorefa-net, Yuheng Zou, 2016. Training low bitwidth convolutional neural networks with low bitwidth gradients. arXiv preprint. arXiv:1606.06160.

Zhu, Chenzhuo, Han, Song, Mao, Huizi, Dally, William J., 2016. Trained ternary quantization. arXiv preprint. arXiv:1612.01064.

Zoph, Barret, Le, Quoc V., 2016. Neural architecture search with reinforcement learning. arXiv preprint. arXiv:1611.01578.

Zoph, Barret, Vasudevan, Vijay, Shlens, Jonathon, Le, Quoc V., 2017. Learning transferable architectures for scalable image recognition. arXiv preprint. arXiv:1707.07012.

5

Deep conditional image generation
Towards controllable visual pattern modeling

Gang Hua[a] and Dongdong Chen[b]

[a]Wormpex AI Research, Bellevue, WA, United States [b]Microsoft Cloud & AI, Redmond, WA, United States

CHAPTER POINTS

- A visual pattern is a discernible visual regularity, which repeats in a predictable manner.

- Pattern recognition, discovery and synthesis are three fundamental tasks in visual pattern analysis.

- Pattern synthesis is the most comprehensive visual modeling task.

- Learning disentangled representation is the key to more controllable visual pattern synthesis.

- Unsupervised learning of disentangled representations requires appropriate inductive biases.

- More controllable pattern synthesis leads to more explainable pattern analysis.

5.1 Introduction

Visual perception is a challenging task, since natural scenes are composed by an overwhelming number of visual patterns, which often follow either a stochastic, or deterministic process, or a combination of both. By definition, a visual pattern is a *discernible* visual regularity in the world, whose compositional elements as a whole, *repeat* in a predictable manner. There are three fundamental tasks in visual pattern analysis, i.e., pattern recognition, pattern discovery, and pattern synthesis. Discerning different visual patterns corresponds to the task of pattern recognition.[1] The fact that visual patterns repeat themselves in the world is the

---

[1] Here we refer to pattern recognition in a narrow sense of differentiating different visual patterns. It may also be used in a wide sense to cover all pattern analysis tasks.

191

Copyright © 2022 Elsevier Inc. All rights reserved.

FIGURE 5.1 Examples to show that to recognize an image, we may only need a handful of discriminative visual features (a.k.a. atomic visual patterns). Even by only looking at the visual characteristics in local image regions, we can well recognize the semantic categories of the images.

FIGURE 5.2 Examples of unsupervised object discovery from a video sequence, using the technologies proposed in Zhao et al. (2013). The figures are by courtesy of Zhao et al. (2013).

basis for various pattern discovery methods. Pattern synthesis refers to the task of generating new instance of a visual pattern. This entails a process of modeling and characterizing the underneath processes which govern (and hence can predict) the variations of the visual pattern.

From the modeling and learning perspective, the minimal information needed to fulfill the above three fundamental tasks are different. Specifically, for pattern recognition, we only need to identify the most distinctive visual features[2] of a pattern to differentiate it from other visual patterns. In other words, successfully performing pattern recognition (e.g., classification) tasks does not require the feature representations to be comprehensive in describing every single detail of the pattern. This is one of the reasons that most state-of-the-art pattern recognition methods, if not all of them, adopted a discriminative modeling approach. To better illustrate this, we show several example images in Fig. 5.1. Obviously, even only by looking at visual characteristics from some local regions in the image, we can already reliably recognize the semantic category of these images, even for the image with artistic style.

Modeling patterns for the task of visual pattern discovery is a little more demanding, as the task is designed to discover and localize repetitively occurred visual patterns from a collection of images and videos, without a priori information about them (Yuan, 2011; Zhao et al., 2013). The unsupervised nature of the pattern discovery tasks necessitates a modeling approach that can establish relatively generic representations and a metric space, where perceptual grouping could be conducted to identify meaningful semantic visual patterns.

---

[2] Indeed, a visual feature can also be regarded as an atomic visual pattern.

Some existing works have also attacked the problem of pattern discovery in a weakly supervised problem (Liu et al., 2010). Often only a limited number of examples are provided, sometimes even just at frame level indicating a frame contains a pattern or not without specifying where the pattern is located. Beyond modeling the visual characteristics of the visual patterns, spatial, temporal and/or spatiotemporal contextual information were proven to be effective in helping with the visual pattern discovery tasks. An effective model for visual pattern discovery necessitates a feature representation that seeks a trade-off between a comprehensive description of a pattern and yet being sufficiently discriminative, such that the patterns can be effectively grouped and differentiated from the background. Therefore, existing approaches for pattern discovery could either be generative (Zhao et al., 2013) or discriminative (Weng et al., 2018), or a combination of both.

Pattern synthesis requires a comprehensive modeling of the target visual patterns, as ultimately the model needs to capture every single details along with the variations of the visual patterns. A learning-based approach to the problem of pattern synthesis often takes a generative modeling approach. While there has been tremendous success made in generative modeling, either using traditional statistical methods (Zhu et al., 1998; Van de Wouwer et al., 1999; Zhu, 2003; Guo et al., 2003), or more recent deep learning approaches (Kingma and Welling, 2014; Goodfellow et al., 2014), the generation processes (or sampling from the models) is often driven by a random process, where controlling how the generated visual samples look like is difficult, if not impossible.

In this chapter, we focus our discussion on how we may achieve more controllable visual pattern synthesis by using deep learning-based conditional image generation. Here controllable means that there is a way we may specifically set the value of a certain parameter or subset of parameters to control the variation of the generated samples of the visual patterns along certain semantic, physical and/or geometric dimensions (a.k.a., factors), such as facial expression, color and poses. We achieve such controllability through deep conditional image generation with an encoder-decoder structure, where a probabilistic space is defined over a disentangled vector embedding to drive the generation of samples of target visual patterns. Learning such a disentangled representation is challenging, which remains an active research.

While many previous works have claimed to learn disentangled representation in a unsupervised way, Locatello et al. (2019) manifest that unsupervised learning of disentangled representation is theoretically impossible without inductive biases on both the models and the datasets. We will devote our discussions to how such inductive biases may be introduced in a supervised, semisupervised and selfsupervised fashion in real computer vision applications to learn disentangled representation for controllable visual pattern synthesis. The applications we explore include image/video style transfer, texts to image generation and face synthesis. Nevertheless, we hope that the insights learnt from these explorations would be transformative in solving other problems in different applications as well.

The remainder of the chapter will be organized as follows: in Section 5.2, we present a brief review of the historical development of visual pattern learning. Then the fundamentals of traditional statistical learning based generative models and deep generative models are elaborated in Section 5.3 and Section 5.4 respectively. Next, in Section 5.5, we will introduce how to leverage deep generative models for visual pattern learning and synthesis under the conditional image generation framework. Accordingly, in Section 5.6, we will use three case studies with different supervision levels to illustrate how inductive bias could be introduced to learn disentangled representations for controllable visual pattern synthesis, along with analysis on results, putting them along their literature. Finally, we draw conclusions with potential future explorations in Section 5.7.

5.2 Visual pattern learning: a brief review

The early research of visual pattern learning can trace back several decades. And the main stream adopts the Bayesian framework and has developed many explicit models for visual pattern modeling. Among them, the pioneering works (Grenander, 1976; Cooper, 1979; Fu, 1982) may be the earliest to use statistical models for visual pattern modeling. In the late 1980s and early 1990s, image models grow popular and indispensable. Early models assume local and piece-wise smoothness in natural images and many different works have been proposed. For instance, physically-based models are proposed in the works (Blake and Zisserman, 1987; Terzopoulos, 1983), and the regularization theory is used in Poggio et al. (1985). In the work (Mumford and Shah, 1989), this problem is formulated as an energy function and solved by energy minimization.

Then the above concepts soon start convergence to the statistical descriptive models because of two influential works. One is Markov random field (MRF) modeling (Besag, 1974; Cross and Jain, 1983). It considers texture pattern to follow a stochastic, possibly periodic, two-dimensional image field, and explores Markov random fields as texture models. A texture model is defined as a mathematical procedure capable of producing and describing a texture image. Another one is the work (Geman and Geman, 1984), which makes an analogy between images and statistical mechanics systems and formulates pattern modeling as a Gibbs distribution and sampling problem under the Bayesian framework. More specifically, the pixel values and the presence and orientation of edges are regarded as states of atoms or molecules in a lattice-like physical system. And the assignment of an energy function in the physical system determines a Gibbs distribution. Because of the Gibbs distribution, MRF equivalence, it can be also viewed as a MRF image model. However, these works have two limitations: 1) Markov random field models are based on pair cliques, so it cannot characterize natural images very well. 2) Gibbs sampling is very time-consuming, making it less applicable in real systems. There are also some other probability models proposed for visual pattern representation learning, such as deformable templates for face (Yuille, 1991), eyes (Xie et al., 1994) and objects (Shackleton, 1994). Compared to homogeneous MRF models, deformable templates are inhomogeneous.

To tackle the computational complexity of the above descriptive models, generative models emerge and postulate hidden variables as the causes for the complicated dependencies in raw signals. A simple illustration figure is shown in Fig. 5.3. Take human as an example, it often has of one head (denoted as "h"), one body (denoted as "b"), two arms and two legs (denoted as "a_l, a_r, l_l, l_r"). The descriptive model considers the joint distribution of five pieces $p(h, b, a_l, a_r, l_l, l_r)$ without understanding the hidden concept of "human". In contrast, generative models regard the five pieces to be conditional dependent under a hidden variable h for the human, then formulate them with a conditional probability model $p(h, b, a_l, a_r, l_l, l_r | d)$.

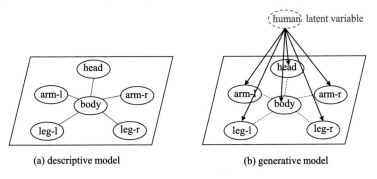

(a) descriptive model   (b) generative model

FIGURE 5.3 A simple illustration of difference between descriptive models and generative models in terms of the desk object. Compared to descriptive models, generative models introduce hidden variables to model the strong dependency within the observed images.

Typical generative models include sparse coding (Roweis and Ghahramani, 1999; Hoyer and Hyvärinen, 2002; Manat and Zhang, 1993), wavelet image representation (Do and Vetterli, 2003; Lu et al., 1992), principle component analysis (PCA) (Kambhatla and Leen, 1997; Kong et al., 2005), independent component analysis (ICA) (Hyvärinen, 1999; Hyvärinen and Oja, 2000) and the random college model (Lee et al., 2001). Such models assume an image can be represented by a series of bases. In this way, the representation dimension can be greatly reduced by projecting the original image space to the hidden space. Therefore, the computation cost becomes less intensive. In the literature, generative models are often inseparable from descriptive models, as the hidden variables are often characterized by a descriptive model. For instance, the sparse coding scheme is a two-layer generative model and assumes that the image bases are independent and identically distributed hidden variables. And the hidden layer in the hidden Markov models for speech and motion modeling is a Markov chain.

Recently, benefiting from large scale datasets and high-end computation devices, deep learning has achieved great success in many artificial intelligence tasks, like visual recognition (Szegedy et al., 2015; He et al., 2016) and object detection (Girshick, 2015; Ren et al., 2015). Similarly, many deep neural networks based generative models have been proposed, including pixel CNN (Van den Oord et al., 2016), variational auto-encoder (VAE) (Doersch, 2016), and generative adversarial networks (Goodfellow et al., 2014). Essentially, in order to generate high-quality images, such deep models must learn to memorize the visual patterns and underlying structures in their weight space. To some extent, better generation quality is equivalent to better pattern learning. In the following parts, we will briefly introduce the fundamentals of classical generative models and deep generative models.

5.3 Classical generative models

From the mathematical perspective (Hua, 2020), classic generative models are statistical models which target modeling the joint probability distribution $p(X, \mathbf{z}|\Theta)$, where X is an observed multivariate variable, $\mathbf{z}$ is the aforementioned hidden variable, and Θ are the parameters of the model to optimize. The hidden variable Z could represent different confounders. For example, if Z are the class labels, generative models can also be used for classification tasks. But the modeling objectives of generative models are different from *discriminative models*, which directly model the conditional distribution $p(Z|X, \Theta)$.

As it is the case for any statistical models, *learning* and *inference* are two fundamental problems to be addressed in generative models. Learning refers to the process of estimating the parameters of these generative models from the data. When the data is complete where X and Z are both observed as a sample pair, parameter estimation is often formalized as a standard Maximum Likelihood estimation problem. A more common setting in real world applications is to estimate the parameters from incomplete data, where only X is observed and the target variable Z is not observed in the data sample. This is a problem of maximum likelihood with incomplete data. Such a problem is often approached by the seminal Expectation-Maximization (EM) algorithm (Dempster et al., 1977), where an E-step and a M-step are iteratively conducted to maximize the looklihood of the incomplete data. From the perspective of optimization, such an iterative process can be regarded as a surrogate optimization process.

The E-Step, by definition, computes the expectation of the data likelihood over the distribution of the hidden or target variables. This is done by first conducting an *inference* step, i.e., calculating the posterior probability $p(Z|X, \Theta_c)$ given the current parameter setting Θ_c. Then we have

$$E(\Theta|\Theta_c) = \int_Z p(X, Z|\Theta)p(Z|X, \Theta_c)dZ. \tag{5.1}$$

Then the M-Step maximizes $E(\Theta|\Theta_c)$ to obtain updated parameters

$$\Theta_c^{new} = \arg\max_\Theta E(\Theta|\Theta_c). \tag{5.2}$$

These two steps iterate until convergence. This iteration represents a surrogate maximization process of $\mathcal{L}(X|\Theta)$, which is monotonically nondecreasing. Since it is obviously upper bounded, the process is guaranteed to converge. Ghahramani (Ghahramani and Beal, 2001) derived a Bayesian EM algorithm by leveraging probabilistic variational analysis. Indeed, the M-Step does not need to fully solve the maximization problem. Instead it only needs to find a new parameter Θ_c^{new}, which has higher $E(\Theta_c^{new}|\Theta_c)$ than $E(\Theta_c|\Theta_c)$. This is also referred to the generalized EM algorithm.

The inference of the posterior probability $p(Z|X, \Theta_c)$ would have a closed-form solution under limited cases. For example, when the prior distribution $p(Z)$ and likelihood distribution $p(X|Z)$ are conjugated, then the posterior would take the same form as the prior. Such conjugate priors are common when the distributions are confined to the conjugate-exponential family (Ghahramani and Beal, 2001). However, for more general distributions, it is often intractable to calculate the posterior in closed-form.

When closed-form inference is intractable, one can often resort to numerical solution, such as using Markov chain Monte Carlo (MCMC) methods, e.g., the Gibbs sampling, to produce samples from this distribution and then compute the integral in Eq. (5.1) numerically. Hinton et al. (Hinton, 2002) presented a method, namely contrastive divergence to use one step sampling to replace the full MCMC sampling. It could significantly speed up the learning process with a certain guarantee of convergence.

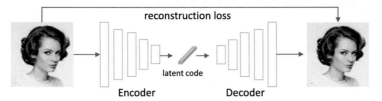

FIGURE 5.4 The simple illustration of the typical deep generative model "auto-encoder", which first encodes the input image into a hidden latent code, then reconstruct the original image from the latent code with the decoder.

5.4 Deep generative models

Following a similar methodology, deep generative models also aim to model the distribution of X with the hidden variables Z. But different from the above traditional generative methods which often use a handcrafted model like wavelet and sparse coding, deep neural networks are utilized instead, whose learning capacity is demonstrated to be much more powerful.

Auto-encoder

We start our discussion with the vanilla auto-encoders (Wang et al., 2016), though strictly speaking, they are not typical generative models, because it facilitates the understanding of other deep generative models. As shown in Fig. 5.4, it usually consists of two parts: an encoder and a decoder part. Both of them are stacked by a series of fully connected or convolutional layers. The encoder continually reduces the dimensionality to a smaller latent representation z (code), and the decoder then reconstructs the input image from the latent representation to the original resolution symmetrically. Need to note that the goal of auto-encoder is not simply to reconstruct the original images with a trivial identity function, but to learn the underlying visual patterns so that we can even generate some new unseen images from the learned latent space. To achieve this goal, a lot of auto-encoder variants have been proposed. The first typical one is the undercomplete auto-encoder (Zhai et al., 2018), which requires the dimension of latent code to be significantly smaller than that of the input. In this way, the latent code should be informative enough to represent the input image, otherwise the reconstruction loss will be very large. The second popular variant is denoising auto-encoder (Vincent et al., 2008), which deliberately adds some noises into the training dataset so that the auto-encoder needs to learn how to denoise. Instead of changing the training dataset, the third variant sparse auto-encoder (Ng et al., 2011) imposes a sparsity constraint on its loss and encourages the number of active units in the code layer to be minimal and learn a sparse representation of the dataset.

Variational auto-encoder

Though the above auto-encoders can successfully map the training image to an embedding space, there is no explicit probabilistic model associated with the embedding space. Therefore, randomly sampling from this embedding space can not guarantee that we meaningfully sample from the image space when feeding the sampled code to the decoder. To address this problem, variational auto-encoder (VAE) (Doersch, 2016) forces the learned latent distribution to follow a Gaussian distribution $P(z)$. Then the goal of VAE is to maximize the probability (maximum likelihood) of each X in the training set under the entire generative process:

$$P(X) = \int P(X|z; \theta)P(z)dz \qquad (5.3)$$

Solving the above integral over z is a nontrivial problem. In practice, for most z, $P(X|z)$ will be nearly zero and hence contribute almost nothing to the estimation $P(x)$. Therefore, the key idea behind the VAE is to attempt to sample values of z that are likely to have produced X and compute $P(x)$ just from those. To this end, it introduces a new function $Q(z|X)$ that takes a value of X and produces a distribution over z values that are likely to generate X. Considering the space of z values under Q may be much smaller than that under the prior $P(z)$, computing $E_{z \sim Q} P(X|z)$ is relatively easier.

As one of the corner-stone of variational Bayesian methods, the KL divergence between $P(z|X)$ and $Q(z)$ is first studied to derive the final optimization objective.

$$\mathcal{D}\big[Q(z) \| P(z|X)\big] = E_{z \sim Q}\big[\log Q(z) - \log P(z|X)\big]. \qquad (5.4)$$

By applying Bayes rule to $P(z|X)$, both $P(X)$ and $P(X|z)$ can be taken into this equation:

$$\mathcal{D}\big[Q(z) \| P(z|X)\big] = E_{z \sim Q}\big[\log Q(z) - \log P(X|z) - \log P(z)\big] + \log P(X). \qquad (5.5)$$

Since $\log P(X)$ does not depend on z, it can be taken out of the expectation. By reorganizing the terms on both sides, we can get:

$$\log P(X) - \mathcal{D}\big[Q(z) \| P(z|X)\big] = E_{z \sim Q}\big[\log P(X|z)\big] - \mathcal{D}\big[Q(z) \| P(z)\big]. \qquad (5.6)$$

In fact, Q can be any distribution, not just a distribution which does a good job mapping X to the z's that can produce X. As $P(X)$ is the final inferring goal, it makes sense to construct a Q which depends on X and makes $\mathcal{D}[Q(z) \| P(z|X)]$ small:

$$\log P(X) - \mathcal{D}\big[Q(z|X) \| P(z|X)\big] = E_{z \sim Q}\big[\log P(X|z)\big] - \mathcal{D}\big[Q(z|X) \| P(z)\big]. \qquad (5.7)$$

This equation is the core of VAE. Specifically, the left side consists of the objective $\log P(X)$ we want to optimize plus an error term, which makes Q produce z's that can reproduce a given X. This error term will become small if Q is modeled with a high-capacity deep network. And the right hand side is something that can be optimized via stochastic gradient descent given the right choice of Q. In fact, the right hand side of Eq. (5.7) takes a form which looks like an autoencoder, i.e., Q is "encoding" X into z, and P reconstructs X from z.

To optimize the right side with stochastic gradient descent, the usual form of $Q(z|X)$ is set to $Q(z|X) = \mathcal{N}(z|\mu(X; \vartheta), \Sigma(X; \vartheta))$, where μ and Σ are arbitrary deterministic functions with parameters ϑ that can be learned from data. In practice, μ and Σ are again implemented via neural networks, and Σ is constrained to be a diagonal matrix. And the gradient of the first term on the right side is estimated by sampling different values of X from the dataset D and one important "reparameterization trick" as shown in Fig. 5.5.

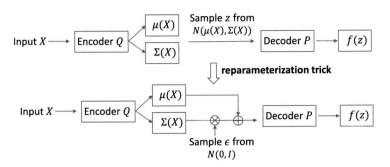

FIGURE 5.5 The reparameterization trick used in variational auto-encoder to approximate the stochastic gradient during training.

FIGURE 5.6 The adversarial training of generative adversarial networks: the discriminator network tries to classify the input image to be real or generated by the generator network, while the generator network tries to generate more realistic images to fool the discriminator network.

At the test time, if we want to generate new samples, VAE randomly samples $z \sim \mathcal{N}(0, I)$ and input it into the decoder. That is, the encoder including the multiplication and addition operations are all removed. For more theoretical and derivation details, we encourage the readers to read the tutorial (Doersch, 2016).

Generative adversarial network

In recent years, another influential work is the generative adversarial networks (GAN) proposed in the pioneering work (Goodfellow et al., 2014). Compared to previous generative models, its key idea is introducing an auxiliary discriminator network to help the learning of the target generative network. Another key idea is the adversary training strategy as shown in Fig. 5.6, i.e., the discriminator network learns to discriminate whether a sample is from the real data distribution or generated by the generative network, while the generative network tries to generate more realistic images to fool the discriminator network. Therefore, these two networks compete with each other in a zero-sum game.

Mathematically, this can be viewed as a min-max optimization problem with the value function $V(G, D)$:

$$\min_{G} \max_{D} V(G, D) = \mathbf{E}_{x \sim p_{data}(x)}\big[\log D(x)\big] + \mathbf{E}_{z \sim p_z(z)}\big[\log\big(1 - D\big(G(z)\big)\big)\big] \tag{5.8}$$

where $D(x)$ represents the probability that x comes from the real data distribution rather than the generated data distribution $G(z)$. Similar to VAE, $p_z(z)$ is a prior distribution to be sampled from, which can be specified as a normal distribution. During training and inference, a random z will be sampled and fed into G to generate an image. In practice, G and D are

FIGURE 5.7 The ideal training evolution of generative adversarial networks. The green line is the real data distribution, the blue line is the data distribution of the generative network, and the red dashed line is the discrimination boundary of the discriminator network. In the ideal case (last column) where the generative network can perfectly match the real distribution, the discriminator network cannot distinguish the real and generated data distribution.

trained in an alternative way, i.e., training G while fixing D, then training D while fixing G. After training, the discriminator network will be discarded and only G will be used for generating new images.

Essentially, G is trying to match the real data distribution during the learning process as shown in Fig. 5.7. In the ideal case where G can exactly simulate the real data distribution, D will be unable to discriminate a generated image to be real/fake, i.e., it will predict an image to be real/fake with a random guessing probability 0.5.

Despite the great success of GANs, training a good GAN model is shown to be not that easy. To improve the quality and stability, plenty of variants (Mirza and Osindero, 2014; Chen et al., 2016; Metz et al., 2016; Arjovsky et al., 2017; Bao et al., 2017) have also been proposed. A more comprehensive summarization can be found in Wang et al. (2019).

5.5 Deep conditional image generation

The above generative models are for unconditional image generation, i.e., the new images are generated by random codes sampled from the prior distribution without other conditional requirements. In other words, visual pattern synthesis is conducted in an uncontrolled way. In contrast, we focus on how to achieve more controllable visual pattern synthesis in this chapter under the conditional image generation framework.

Different from unconditional image generation, conditional image generation (Isola et al., 2017; Zhu et al., 2017; Chen et al., 2017a, 2020a) imposes extra input conditions into the generation process. It has a broad scope and covers a large variety of computer vision problems, such as edge-to-image translation (Isola et al., 2017), style transfer (Gatys et al., 2015; Chen et al., 2018), image colorization (He et al., 2018), controllable image processing (Fan et al., 2018, 2019; Chen et al., 2020a), image restoration (Wan et al., 2020a,b), semantic synthesis (Tan et al., 2020b) and face synthesis and editing (Tan et al., 2020a).

As shown in Fig. 5.8, given the input condition like text description, attribute vectors and images, the conditional generative models aim to generate an output image that satisfies the conditional requirements in a controllable way. But similar to unconditional image generation, learning and modeling the intrinsic visual patterns is still the essential factor that determines the generation quality. By more efficient modeling of the visual patterns, conditional image generation is essentially a process of constrained pattern sampling and recompositing.

"painting of a woman" **or** *curve hair, profile face,* **or** *painting style woman* → conditional generative model →

FIGURE 5.8 Illustration of typical conditional image generation framework. Given input conditions like text description, attributes or images, the conditional generative model needs to generate an output image that satisfies the conditional requirement.

However, compositing natural images involves a lot of visual confounders, such as pose, lighting and shape. In order to encourage the generative model to learn the underlying pattern better, "disentangling" in a hidden embedding space is a key design principle and widely used in many existing conditional image generation methods (Yan et al., 2016; Chen et al., 2017b; Bao et al., 2018; Ma et al., 2018). But achieving disentanglement is a nontrivial problem. In fact, Locatello et al. (2019) show that it is theoretically impossible to achieve unsupervised disentanglement without inductive biases on both the models and the datasets. Therefore, it often requires some dedicated network design and training recipe in a supervised, semisupervised or selfsupervised way.

5.6 Disentanglement for controllable synthesis

We present three cases studies on how we may introduce inductive bias in learning disentangled representations in deep conditional image generation for controllable visual pattern synthesis. The applications we explore include style transfer (Section 5.6.1), vision-language image generation (Section 5.6.2) and identity preserving face synthesis (Section 5.6.3).

5.6.1 Disentangle visual content and style

Style transfer (Gatys et al., 2015; Johnson et al., 2016; Chen et al., 2020c) is a typical conditional image generation task. As shown in Fig. 5.9, it aims to migrate the style from one style image to another content image while keeping the original semantic structure of the content image. The core problem behind this task is modeling the visual patterns of the style image and separating the content and style of the content image. And then style transfer is to resample the learned style pattern under the content structure constraint.

Since the pioneering work of Gatys et al. (2015), style transfer adopting convolutional neural networks (CNN) has ignited a renewed interest in both academia and industry. In the work (Gatys et al., 2015), Gatys et al. innovatively apply a pretrained CNN network to decompose an image into content components and style components. Specifically, they regard the correlation matrices (Gram-matrices) of feature responses in different layers as the hierarchical style representation. By further representing the content structure of one image with its high-level feature response, they model style transfer as an optimization problem, i.e., search for an image that has similar feature Gram-matrices as the style image and similar high-level content feature as the content image. This optimization based method can produce

| content image | style image | stylized image |

FIGURE 5.9 Illustration of style transfer. It aims to render one content image with the visual style from another style image while keeping the original content structure. The figures are by courtesy of Chen et al. (2020b).

very impressive stylization results and is much better than traditional methods. However, it is very time-consuming due to the optimization process, thus imposing a big limitation on real applications.

To speedup, many feed-forward based methods (Johnson et al., 2016; Ulyanov et al., 2016) have been proposed to use a feed-forward network to approximate the above optimization procedure. And the stylization results can be obtained by directly feeding the content image into the feed-forward network, hence it is faster than the optimization based methods by several magnitudes. However, such feed-forward networks are trained in a black-box way, and the content and style (visual patterns) components are highly coupled in the learned networks. This not only prevents us from learning an explicit representation for either style or content, but also makes such networks only able to capture a specific style at a time.

Therefore, Chen et al. (2017b) designed a new disentangled network structure to learn an explicit representation for each style in a supervised way, which naturally supports multiple-style transfer. It is motivated by the "texton" concept in classical texture synthesis and we propose to use a series of filter banks to represent different style images. All the channels in one filter bank can be regarded as the bases of style elements within one style image, such as texture pattern and strokes. And the stylization process is then conducted by convolving the corresponding filter banks with the content feature maps, which is analogous to the convolution operation between the texton and Delta function in the image space for texture synthesis (as shown in Fig. 5.11).

Two-branch design

The detailed disentanglement framework is illustrated in Fig. 5.10. Basically, it consists of three parts: one shared encoder $\mathcal{E}$, the stylebank layer $\mathcal{K}$ and one shared decoder $\mathcal{D}$. To force the network to decouple the content and style in an explicit way, we build two learning branches: the reconstruction branch $\mathcal{E} \rightarrow \mathcal{D}$ and the stylizing branch $\mathcal{E} \rightarrow \mathcal{K} \rightarrow \mathcal{D}$. Given an input content image I, it is first transformed into the feature space (F) by using the encoder subnetwork. Then for the reconstruction branch, F is directly passed to $\mathcal{D}$ to produce an image $(O = \mathcal{D}(F))$ that should be as close as the input I. In parallel, when transferring the style i to I, we convolve the corresponding filter bank $\mathcal{K}_i$ with F and then feed the transformed feature $\widetilde{F}_i$ ($\widetilde{F}_i = F \otimes \mathcal{K}_i$) into $\mathcal{D}$ to produce the stylization result $O_i = \mathcal{D}(\widetilde{F}_i)$. The above two branches are trained in an alternative way, and different loss functions are designed accordingly. In details, the simple MSE (Mean Square Error) loss between the input image I and O

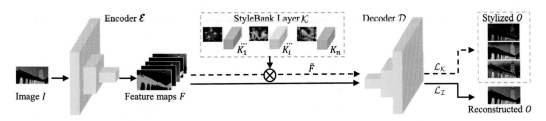

FIGURE 5.10 The disentangled style transfer framework proposed in Chen et al. (2017b), which consists of one reconstruction branch (below) and one stylizing branch (top). The content is designed to be encoded in the encoder and decoder part, and the style is represented by a set of filterbanks in the Stylebank layer. The figures are by courtesy of Chen et al. (2017b).

FIGURE 5.11 The process of texture synthesis can be viewed as the convolution operation between the texton image and a delta function in the image space, which motivates us to perform style transfer in the feature space by convolving the content feature with the corresponding style filterbank.

is adopted as the identity mapping loss $\mathcal{L}_\mathcal{I}$ for the reconstruction branch:

$$\mathcal{L}_\mathcal{I}(I, O) = \|O - I\|^2. \tag{5.9}$$

For the stylizing branch, following the objective function in Johnson et al. (2016), a content loss $\mathcal{L}_c$, a style loss $\mathcal{L}_s$ and a variation regularization loss $\mathcal{L}_{tv}(O_i)$ are used as the stylization loss $\mathcal{L}_\mathcal{K}$:

$$\mathcal{L}_\mathcal{K}(I, S_i, O_i) = \alpha \mathcal{L}_c(O_i, I) + \beta \mathcal{L}_s(O_i, S_i) + \gamma \mathcal{L}_{tv}(O_i) \tag{5.10}$$

where I, S_i, O_i are the input content image, style image and stylization result (for the i-th style) respectively. $\mathcal{L}_{tv}(O_i)$ is a total variation regularizer used in Johnson et al. (2016) to encourage smoothness. And $\mathcal{L}_c$ and $\mathcal{L}_s$ use the same definition as in Gatys et al. (2015):

$$\mathcal{L}_c(O, I) = \sum_{l \in \{l_c\}} \left\| F^l(O) - F^l(I) \right\|^2$$

$$\mathcal{L}_s(O, S) = \sum_{l \in \{l_s\}} \left\| G\left(F^l(O)\right) - G\left(F^l(S)\right) \right\|^2 \tag{5.11}$$

$$\text{where} \quad G(X) = X X^T$$

Here F^l and G are the l-th layer feature map of a pretrained VGG-16 network and the corresponding Gram matrix computed from F. $\{l_c\}$, $\{l_s\}$ are VGG-16 layers used to compute the content loss and the style loss. Since the reconstruction branch is designed to recover the original content image, it guarantees no style information is absorbed into the encoder $\mathcal{E}$ and

content:

style:

result:

FIGURE 5.12 The stylization results of multiple different styles, which are trained in one single network simultaneously. It shows that different styles are well decoupled (one stylization result only consists of its own style patterns) and learned into their corresponding filterbanks. The figures are by courtesy of Chen et al. (2017b).

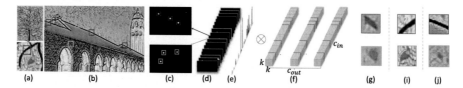

FIGURE 5.13 Style elements reconstruction for two representative patches in an examplar stylization image. The figures are by courtesy of Chen et al. (2017b).

decoder $\mathcal{D}$. Simultaneously, to achieve the stylization goal in the stylizing branch, all the style information is forced to be learned into the intermediate stylebank layer. In this way, the content and style are explicitly decoupled.

Multistyle transfer results

Thanks to the above decoupled two-branch design, one style is encoded in one specific set of convolutional filters and multiple styles can be learned in one network simultaneously. This is more friendly to real applications than previous single-style methods that usually train one independent network for each style. During inference, to apply one specific style, the corresponding filterbank is chosen and applied. As shown in Fig. 5.12, different styles are decoupled very well and the corresponding stylization results only consist of their own style patterns.

Style elements reconstruction

To better understand how the stylebank layer represents the visual pattern of each style image, the style elements from a learned filter bank are reconstructed in an examplar stylization image shown in Fig. 5.13. Specifically, two kinds of patches in the stylization result are selected: stroke patch (red box) and texture patch (green box).

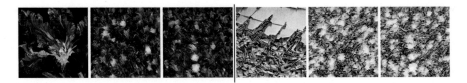

FIGURE 5.14 Visualizing the learned style elements: for each case, the left one is the style image, and the right two are the stylization results by feeding two different noise images. The figures are by courtesy of Chen et al. (2020c).

First, all other regions except the corresponding positions of the selected patches are masked out as shown in (c, d), and the feature distribution of these patches are visualized in (e). It can be observed that such feature responses are sparsely distributed and some peak responses occur in individual channels.

Then, only nonzero feature channels and the corresponding filter bank channels are considered in the feature transformation operation. The transformed features are finally passed to the decoder to obtain the reconstructed style elements (g), which are visually similar to the original style patches in (i) and the stylization patch in (j). By such analysis, we can hypothesize that the different weighted combinations of the filter bank channels can constitute the diverse style elements in one style image, which are similar to the dictionaries/bases in the representation learning literature.

To further study how many different style elements are learned for each style image, a large noise image is used to approximate the content patch distribution and let the network render different content patches with different style elements. It can be seen from Fig. 5.14 that the stylization network has learned a set of representative style elements for each style image.

Importance of the two-branch design

To demonstrate no style information absorbed in the auto-encoder and the importance of the two-branch design, another ablation experiment is conducted by removing the reconstruction branch during the training stage. And given one input image, only the encoder and decoder are used to output the reconstruction result. As shown in Fig. 5.15, without the reconstruction branch during the training stage, the decoded image (middle) fails to reconstruct the original input image (left) and seems to carry some style information. By comparison, when the reconstruction branch is used, the decoded image reconstructs the input image perfectly and has very close appearance to the input image. That is to say, the proposed two-branch design forces all the content information to be only covered in the encoder and decoder part, and the style information to be learned in the intermediate stylebank layer.

Benefits of disentanglement

The disentanglement of content and style can bring many additional advantages. The first is fast incremental learning. Specifically, to enable a new style, we do not need to retrain the whole network, which is often time-consuming. Instead, given a learned multistyle network, we only need to retrain a new filter bank for the newly added style while keeping the encoder and decoder part fixed. This process converges very fast since only the new style bank part needs to be trained. Empirically, it often only takes several minutes, which is tens of times faster than retraining the whole network. As shown in Fig. 5.16, incremental learning can

FIGURE 5.15 The reconstruction result of the encoder and decoder with (right) and without (middle) the reconstruction branch during the training stage. Obviously, involving the reconstruction branch helps guarantee no style information to be absorbed in the encoder and decoder part and almost perfect reconstruction. The figures are by courtesy of Chen et al. (2017b).

FIGURE 5.16 The disentanglement enables fast incremental learning for new styles. For each case, the left and right are the stylization results of incremental training and fresh training respectively. The figures are by courtesy of Chen et al. (2017b).

FIGURE 5.17 Stylization results by linear combination of two style filter banks. By adopting different fusing weights, we change the proportion of each style accordingly. The figures are by courtesy of Chen et al. (2017b).

obtain comparable stylization results to those from fresh training, which retrains the whole network with the new styles.

The second advantage is enabling style fusion in two different ways: linear fusion and region-specific fusion. For linear fusion, since different styles are encoded into different filter banks $\{K_i^1, ..., K_i^m\}$, we can linearly fuse multiple styles by linearly fusing filter banks in the *StyleBank* layer. Then the fused filter bank is used to convolve with the content feature F:

$$\widetilde{F} = \left(\sum_{i=1}^m w_i * K_i \right) \otimes F \quad \sum_{i=1}^m w_i = 1, \quad (5.12)$$

where m is the number of styles, K_i is the filter bank of style i. $\widetilde{F}$ is then fed to the decoder to get the final stylization result. Fig. 5.17 shows such linear fusion results of two different styles with variant fusion weight w_i.

For region-specific style fusion, suppose that the image is decomposed into n disjoint regions in the feature space, and M_i denotes every region mask, then the feature maps can be

FIGURE 5.18 Region-specific style fusion results with two paintings by applying different filterbanks onto different image regions. The left two paintings are from Picasso and Van Goph respectively, while the right two are both from Van Goph. The figures are by courtesy of Chen et al. (2020c).

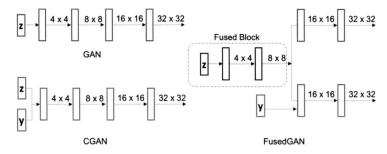

FIGURE 5.19 The simple illustration figure to show the general idea of FusedGAN, which fuses an unconditional GAN and a conditional GAN and trains them in a semisupervised way. The figures are by courtesy of Bodla et al. (2018).

described as $F = \sum_{i=1}^{m}(M_i \times F)$ and region-specific style fusion can be formulated as:

$$\widetilde{F} = \sum_{i=1}^{m} K_i \otimes (M_i \times F), \tag{5.13}$$

Fig. 5.18 shows such two region-specific style fusion results. The left case borrows styles from two famous paintings of Picasso and Van Goph, while the right two styles are both from Van Goph.

5.6.2 Disentangle structure and style

Different from the above style transfer task, where the semantic content structure and style elements are directly specified by the input content and style image, it is more difficult to achieve fine-grained disentanglement in general image synthesis. For example, in the classical text-to-image method StackGAN (Zhang et al., 2017), we can control only over the styles by feeding the text descriptions but cannot achieve the fine-grained controllability upon both structure and styles. Here, the structures mainly denote the semantic shape and posture, and the styles denote the fine-grained visual patterns/textures.

To address the above limitation, Bodla et al. (2018) design a new cascaded generative model FusedGAN to achieve the disentanglement of structure and style in a semisupervised way. The general idea of FusedGAN is shown in Fig. 5.19, which basically combines an unconditional GAN to produce the structure prior and another conditional GAN to match the

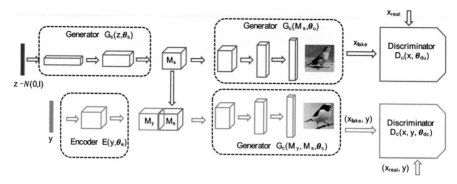

FIGURE 5.20 The disentangled learning framework for FusedGAN, where blue and orange blocks are the unconditional and conditional image generation pipelines respectively. The figures are by courtesy of Bodla et al. (2018).

style condition from the text description. The underlying motivation is that, when human beings want to paint a picture about one object like a bird, the most intuitive way is to first sketch the outline of the bird with a specific posture and shape (structure), then add the fine-grained texture details (style).

Disentanglement by sharing the structure prior

Based on the above motivation, an end-to-end disentangled learning framework is designed in Fig. 5.20, where the blue and orange blocks correspond to the unconditional and conditional image generation branches respectively. In details, the unconditional branch consists of a generator network G_1 and a discriminator network D_u, which are trained in a similar adversarial way as other GAN models. In order to provide the structure prior for the conditional generation branch, G_1 is split into two modules: G_s and G_u. G_s takes a random noise vector z as the input, and generates a structure prior M_s after a series of convolution and unsampling operations. Then the structure prior M_s is fed into G_u to generate the final image after another series of convolution and upsampling operations.

Different from the traditional conditional generation framework, which often takes one condition and one random noise vector as inputs, the conditional generator network G_c in FusedGAN instead takes the structure prior M_s and the condition vector M_y as inputs. That is to say, the unconditional generation branch and the conditional generation branch share the same structure prior M_s. To encourage the disentanglement of the structure and style, the unconditional and the conditional generation branch adopt different training datasets, i.e., an unlabeled dataset for G_1 and a dataset labeled with conditional descriptions for G_c. According to different tasks, the condition vector M_y may have different formats. For example, in the text-to-image task, the original text description is first encoded into an embedding representation y and then fed into an encoder E to generate a new condition tensor M_y as shown in Fig. 5.20. Note that, in order to generate diverse results, conditional augmentation is performed to sample latent vector $\hat{c}$ from an independent Gaussian distribution $N(\mu(y), \Sigma(y))$ around the text embedding, and $\hat{c}$ is then spatially repeated to match the spatial dimension of M_s to produce M_y.

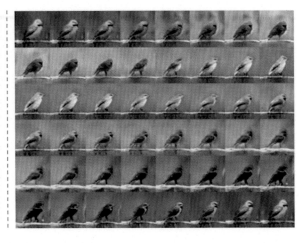

This bird has a bright yellow body, with brown on its crown and wings.

This bird is completely red with black wings and pointy beak.

This bird has wings that are brown and has a white body.

A bird with bright yellow belly, and colors of orange on it tail and back.

A small colorful bird that contains bright blue feathers covering most of its body except for on its black tail.

Style

Structure

FIGURE 5.21 Synthesized results of FusedGAN. The left part are the synthesis results of different styles (rows) and structures (column), and the right part are the style interpolation results of different styles under a fixed structure. The figures are by courtesy of Bodla et al. (2018).

To guide the learning of the unconditional and conditional generation branch, the discriminator D_u takes the generated image x_{uf} of G_u or the real image x_r as input and discriminate whether it is real or fake, while the discriminator D_c takes the generated image x_{cf} of G_c or the real image x_r and the corresponding condition as inputs to ensure that G_c generates images that match the condition. In the training stage, these two pipelines are trained end-to-end in an alternative way. And the model parameters are updated by optimizing the combined GAN and CGAN objectives, *i.e.*,

$$\mathcal{L}_{G_u} = \log D_u\big(G_u(z)\big), \quad \mathcal{L}_{D_u} = \log D_u(x), \quad \mathcal{L}_{D_c} = \log D_c(x, y),$$
$$\mathcal{L}_{G_c} = \log D_c\big(G_c(M_y, M_s), y\big) + \lambda D_{KL}\big(N\big(\mu(y), \Sigma(y)\big)\|N(0, I)\big)$$

(5.14)

where z is the sampled noise vector from a normal distribution $N(0, I)$. To summarize, the key disentanglement design in this work is sharing the structure prior but training the two branches with different objectives.

During inference, to generate a conditional image, a random sample z is first drawn and passed through G_s to generate the structure prior M_s. M_s then takes two paths, one through the generator G_u to produce an unconditional image x_{uf}. In the second path, the input text description is fed into the encoder E and draws a sample from the Gaussian around the text embedding. The output of E and M_s are concatenated together and passed through G_c to generate the conditional image x_{cf}. That is to say, in one inference step, two images are synthesized: x_{cf} the conditional image and x_{uf} the unconditional image, a byproduct of the model which helps to analyze and better understand the proposed model and the results.

Synthesis results

To verify whether the structure and style are well decoupled, some synthesis results are presented in Fig. 5.21. In the left part, the last row displays the unconditional generation

FIGURE 5.22 Synthesized results of various styles by FusedGAN via varying different amount of fine-grained details. The figures are by courtesy of Bodla et al. (2018).

results, and other rows show the conditional generation results by using the same structure prior. It can be seen that M_s is able to successfully capture and transfer significant amount of information about the bird structure into the conditional synthesis results of various text descriptions. By using a constant structure prior, it is also easy to conduct style interpolation between various styles. To achieve this, two text-samples t_1 and t_2 are taken into E to draw two samples from their respective Gaussian distributions. Then eight uniform samples are picked by linear interpolation between them, such that the first sample corresponds to t_1 and last one corresponds to t_2. As shown in the right part of Fig. 5.21, very smooth style interpolation effects can be achieved in a controllable way.

Fig. 5.22 further shows the synthesized results of different styles by varying different amount of fine-grained details. In details, a particular texture description is first fed into E and five samples are then drawn from the Gaussian distribution around the text embedding. By using the same structure prior, each text description can drive synthesis results of different fine-grained details. For example, for the second row in the left part, though all the birds are red with black wing, they have varying amount of black on their wings and the length of the tail.

5.6.3 Disentangle identity and attributes

The third representative work (Bao et al., 2018) is about the typical conditional image generation task "identity-preserving face synthesis". Given one input image I_{id} of a certain subject identity and one input image I_a to extract the attributes, this task aims to generate a new high-quality face image I' that has the same identity as I_{id} and the same attributes as I_a. Here, the attributes include but are not limited to pose, emotion, skin color, illustration and background. This is a very challenging task, especially when the face identity is not presented in the training set, and the underlying challenge is how to disentangle the identity-related visual patterns (e.g., nose and eye shape) and attribute-related visual patterns (e.g., skin color, mouth shape under different emotions).

To solve this problem, many related methods have been proposed, such as TP-GAN (Huang et al., 2017) and FF-GAN (Yin et al., 2017) that can synthesize the frontal view of a given face image, and DR-GAN (Tran et al., 2017) that can synthesize different face poses. However, they often rely on full annotation of attributes and the supported attribute types are also limited. By comparison, another work CVAEGAN (Bao et al., 2017) can support a variety of attribute changes, but it is not able to synthesize a face with an identity outside the training dataset.

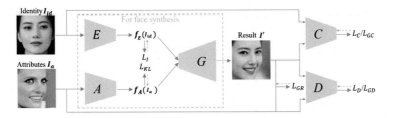

FIGURE 5.23 The overall framework for identity-preserving face synthesis, which disentangles identity and attributes from the input face images. The new face I' is generated by recombining the identity information of I_{id} and the attribute information of I_a. The figures are by courtesy of Bao et al. (2018).

To enable a large variety of attribute changes and open-set identities, Bao et al. (2018) design a new disentangled face synthesis framework as shown in Fig. 5.23. It contains five subnetworks: the identity encoder network E, the attributes encoder network A, the generative synthesis network G, the auxiliary classification network C and discriminator network D. The function of the identity encoder network E and the attribute network A is to extract the identity vector $f_E(I_{id})$ from I_{id} and the attribute vector $f_A(I_a)$ from I_a respectively. By recombining $f_E(I_{id})$ and $f_A(I_a)$, the network G then generates a new image I' that follows the identity of I_{id} and attributes of I_a. The auxiliary network C and D are only used during training. Specifically, C is used to ensure that I' have the same identity as I_{id} and D encourages G to generate higher quality images by distinguishing the generated image and real images in an adversarial training manner.

Disentanglement

Though the above framework is designed in a disentangled way, training it is not a trivial problem. Because most existing face datasets only have the identity annotation but no attribute annotation. In fact, annotating some attributes precisely is sometimes even impossible, such as the illustration and the background. This work enforces the disentanglement by extracting the identity representation in a supervised way and attribute representation in an unsupervised way.

In details, to extract the identity representation, E is formulated as a face recognition network and trained via softmax loss on a labeled face dataset $\{I_i, y_i\}$, where y_i is the identity label of image I_i. During training, to discriminate different individuals, E is encouraged to learn similar feature representation for images with the same identity and dissimilar feature representation for images with different identities. Therefore, the feature response of the last pooling layer of E is adopted as the identity vector of I_{id}.

To obtain the attribute representation, a simple and effective training strategy is designed by leveraging a reconstruction loss and a KL divergence loss. Specifically, during training, I_{id} and I_a are randomly selected to be the same image or different images. In both cases, the result image I' is required to reconstruct the attribute image I_a, but with different loss weight.

Formally, the reconstruction loss is

$$\mathcal{L}_{GR} = \begin{cases} \dfrac{1}{2}\|I_a - I'\|^2 & \text{if } I_s = I_a, \\ \dfrac{\lambda}{2}\|I_a - I'\|^2 & \text{otherwise,} \end{cases} \tag{5.15}$$

where λ is the reconstruction loss weight for the case $I_s \neq I_a$. Specifically, when the identity image I_{id} is the same as the attribute image I_a, the synthesized image I' must be the same as I_{id} or I_a. Since there are many face images I_a for each identity in the training set, their identity vector is almost the same for these images and the only possible difference would be the attribute vector. Therefore, by requiring the reconstructed images to be same as these different face images, it forces the attribute encoder network A to learn different attribute representations accordingly. When the identity image I_{id} and the attribute image I_a are different, though it is difficult to accurately predict what the reconstructed result should look like, we can expect the reconstruction to be approximately similar to the attribute image I^a, such as the background, overall illumination, and pose. Therefore, a raw pixel reconstruction loss with a relatively small weight ($\lambda = 0.1$) is adopted to maintain the attributes.

Besides the above reconstruction loss, a KL divergence loss is further used to regularize the attribute vector with an appropriate prior $P(z) \sim N(0, 1)$. It constrains the attribute vector not to contain much identity information and helps the attribute encoder network learn better representation. Given the input attribute face image, the network A will output the mean μ and covariance of the latent vector. Then the KL divergence loss is defined as:

$$\mathcal{L}_{KL} = \frac{1}{2}\left(\mu^T \mu + sum\big(\exp(\epsilon) - \epsilon - 1\big)\right). \tag{5.16}$$

During training, the reparameterization trick is used to sample the attribute vector by using $z = \mu + r \odot exp(\epsilon)$, where $r \sim N(0, I)$ is a random vector and $\odot$ represents the element-wise multiplication.

Asymmetric training

After extracting the identity vector $f_I(I_{id})$ and attribute vector $f_A(I_a)$, they are concatenated in the latent space ($z' = [f_I(I_{id}), f_A(I_a)]$) and then fed into the network G to synthesize a new face image. Similar to general GANs, the generative network G plays a two-player minmax game with the discriminator network D, i.e., D tries to distinguish real training data from synthesized data while G tries to fool the network D. Concretely, network D tries to minimize the loss function

$$\mathcal{L}_D = -\mathbb{E}_{I_a \sim P_r}\big[\log D(I_a)\big] - \mathbb{E}_{z \sim P_z}\big[\log(1 - D(G(z)))\big]. \tag{5.17}$$

However, if the network G directly attempts to maximize $\mathcal{L}_D$ as the traditional GAN, the training process will be unstable. This is because the distributions of "real" and "fake" images may not overlap with each other in practice, especially at the early training stage. Therefore, the discriminator network D can separate them perfectly and cause gradient vanishing. To address this problem, the pairwise feature matching loss is used for the generator as in

CVAE-GAN. In details, assuming $f_D(\cdot)$ to be the features of the intermediate layers of D, the pairwise feature matching loss is defined as:

$$\mathcal{L}_{GD} = \frac{1}{2}\|f_D(I') - f_D(I_a)\|_2^2. \tag{5.18}$$

That is to say, it encourages the features of the real and fake images as close as possible. By default, the output feature of the last fully connected layer of D is utilized as f_D.

Similarly, in order to achieve the identity-preserving goal (I' has the same identity as I_{id}), a similar pairwise feature matching loss is used to encourage I' and I_{id} to have similar feature representations in the face classification network C:

$$\mathcal{L}_{GC} = \frac{1}{2}\|f_C(I') - f_C(I_{id})\|_2^2. \tag{5.19}$$

Here the input of the last FC layer of network C is used as the feature f_C. In practice, network C and network E share the parameters and are initialized by a pretrained face classification network to speedup the convergence.

Unsupervised learning strategy

Synthesizing face images of identities that do not appear in the training set is challenging, which requires the generative network to cover both intraperson and interperson variations. Considering existing public datasets with labeled identities often has a limited size and does not contain extreme poses or illuminations, one million face images with large variation and diversity are collected from flicker and Google. Then an unsupervised training process is conducted to help the learnt generator generalize to unseen identities. In details, the collected unlabeled images can be used either as the identity image I_{id} or the attribute image I_a. When used as the attribute image I_a, the whole training process remains unchanged. When used as the identity image I_{id}, since it does not have a class label, they are not involved in the learning of E and C (fixed). Empirically, we find these unlabeled data can increase intraclass and interclass variation of the face distributions, hence improving the diversity of the synthesized faces, such as larger changes in poses and expressions.

Synthesis results

To demonstrate the effectiveness of the above disentangled face synthesis framework, Fig. 5.24 and Fig. 5.25 show the synthesis results that use the images whose identities appear and do not appear in the training set respectively. It can be seen that the trained network can disentangle the identity and attributes components and learn corresponding visual patterns very well, for both the close-set and open-set settings.

The disentanglement also enables continuous attribute change in the generated images by tuning the latent vector, which is called "attribute morphing". Specifically, given a pair of images I_{a1} and I_{a2}, the attribute network A is first used to extract their attribute vector z_{a1} and z_{a2} respectively, and then a series of attribute vectors z can be obtained by linear interpolation, i.e., $z = \alpha z_{a1} + (1 - \alpha)z_{a2}, \alpha \in [0, 1]$. Fig. 5.26 presents the results of face attribute morphing, which can gradually change the pose, emotion, or lighting by selecting a proper pair of attribute images.

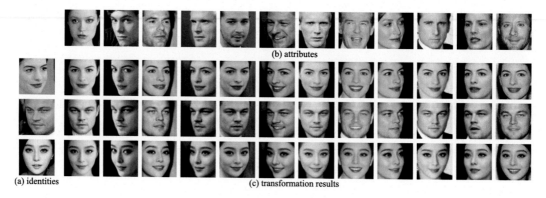

(b) attributes

(a) identities   (c) transformation results

FIGURE 5.24 Identity-preserving face synthesis results by using the images whose identities appear in the training dataset. It can be seen that the proposed method can disentangle the identity-related and attribute-related visual patterns well and then recomposite them in the final results. The figures are by courtesy of Bao et al. (2018).

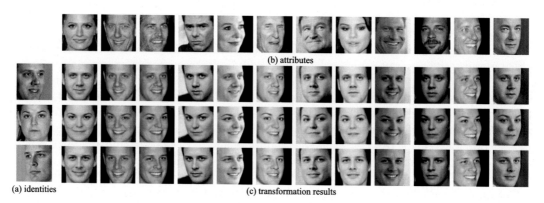

(b) attributes

(a) identities   (c) transformation results

FIGURE 5.25 Identity-preserving face synthesis results by using the images whose identities do not appear in the training dataset, which demonstrates the strong generalization ability of the learned identity disentanglement. The figures are by courtesy of Bao et al. (2018).

Application of synthesis

Deep network based face verification systems have been widely used in surveillance and access control. Given two faces, a pretrained face classification DNN model will be first used to extract their features. Then the two faces are regarded as the same identity if the feature distance is smaller than a threshold. However, recent research shows deep neural networks are vulnerable to adversarial examples, which fool the network to some specific targets by adding some imperceptible perturbations onto the clean images. In details, supposing two faces I_1 and I_2 have different identities, we can find imperceptible perturbations r, such that $I_1 + r$ will be regarded as the same person as I_2 by using the above face verification system.

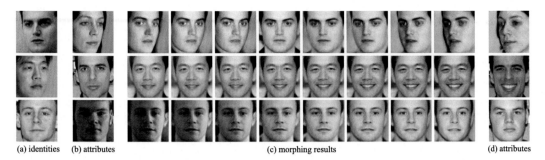

(a) identities   (b) attributes   (c) morphing results   (d) attributes

FIGURE 5.26 Face morphing results using unseen identities between two attributes in terms of pose, emotion, and lighting change respectively. The figures are by courtesy of Bao et al. (2018).

(a)   (b)   (c)   (d)   (e)   (f)

FIGURE 5.27 Application to adversarial example detection in face verification systems. (a) is the source image, (b) is the adversarial example which aims to mislead the verification network to the identity shown in (c). (d), (e) and (f) are the reconstruction results from our framework. It shows that although the adversarial example shares a similar appearance with the source image, their reconstruction results have different appearances. The figures are by courtesy of Bao et al. (2018).

TABLE 5.1   Adversarial examples detection accuracy at different thresholds.

Threshold	1.0	0.8	0.6	0.4
acc	76.73%	82.58%	87.18%	92.41%

This process is often formulated as an optimization problem:

$$\min \|r\|_2^2$$
$$s.t. \|f_C(I_1 + r) - f_C(I_2)\|_2^2 < \tau, \tag{5.20}$$

where f_C is the extracted feature from the pretrained network, and τ is the predefined threshold.

In Fig. 5.27, (a) and (c) are the two inputs I_1 and I_2, and (b) is the generated adversarial example $I_1 + r$. Since the adversarial examples have similar identity features with other faces, if we reconstruct the image from the feature using the proposed framework, it will generate an image of the other person in (e). Obviously, the adversarial example and its reconstruction clearly have different identities. Based on this observation, the above synthesis framework can be used to identify the adversarial samples by comparing the identity of the original faces and the reconstruction results.

Taking the LFW dataset as an example, for each of the 3000 pairs of different identities, two adversarial examples are generated by performing adversarial attacks with each other, thus producing total 6000 adversarial examples. Here, four different thresholds τ are used: $[0.4, 0.6, 0.8, 1]$. At the same time, we have 6000 source images and their reconstruction. Then the LBP feature of the input and its reconstruction image is extracted and concatenated together. Finally, a linear SVM is trained as a binary classifier. The results are shown in Table 5.1, 92.41% detection accuracy can be obtained if the feature distance is required to be smaller than 0.4.

5.7 Conclusion and discussions

Learning and modeling visual patterns is a fundamental building block for visual intelligence. Through the unconditional or conditional image generation framework, deep generative models attempts to recover the lower dimensional structure of the target visual models in an embedding space. In this chapter, we discussed how to leverage deep generative models to achieve more controllable visual pattern synthesis via conditional image generation. And we argue that the key to achieve such controllable pattern synthesis is disentanglement of the visual representation, where different controlling factors are encouraged to be separated in the hidden embedding space. Then three different case studies are used to illustrate how to achieve the disentanglement for pattern synthesis in unsupervised or weakly supervised setting, by introducing inductive bias, from the network design and training strategy perspectives.

In classical generative models, often the various confounding factors are explicitly modeled as interacting random processes. This kind of explicit modeling has not been well explored in deep generative models. It may be beneficial to explore how we may introduce such explicit and structured representations in learning deep generative models, which may lead to more explainable deep models.

References

Arjovsky, M., Chintala, S., Bottou, L., 2017. Wasserstein gan. arXiv preprint. arXiv:1701.07875.
Bao, J., Chen, D., Wen, F., Li, H., Hua, G., 2017. Cvae-gan: fine-grained image generation through asymmetric training. In: Proceedings of the IEEE International Conference on Computer Vision, pp. 2745–2754.
Bao, J., Chen, D., Wen, F., Li, H., Hua, G., 2018. Towards open-set identity preserving face synthesis. In: Proceedings of the IEEE Conference on Computer Vision and Pattern Recognition, pp. 6713–6722.
Besag, J., 1974. Spatial interaction and the statistical analysis of lattice systems. Journal of the Royal Statistical Society, Series B, Methodological 36, 192–225.
Blake, A., Zisserman, A., 1987. Visual Reconstruction. MIT Press.
Bodla, N., Hua, G., Chellappa, R., 2018. Semi-supervised fusedgan for conditional image generation. In: Proceedings of the European Conference on Computer Vision (ECCV), pp. 669–683.
Chen, D., Liao, J., Yuan, L., Yu, N., Hua, G., 2017a. Coherent online video style transfer. In: Proceedings of the IEEE International Conference on Computer Vision, pp. 1105–1114.
Chen, D., Yuan, L., Liao, J., Yu, N., Hua, G., 2017b. Stylebank: an explicit representation for neural image style transfer. In: Proceedings of the IEEE Conference on Computer Vision and Pattern Recognition, pp. 1897–1906.
Chen, D., Yuan, L., Liao, J., Yu, N., Hua, G., 2018. Stereoscopic neural style transfer. In: Proceedings of the IEEE Conference on Computer Vision and Pattern Recognition, pp. 6654–6663.

Chen, D., Fan, Q., Liao, J., Aviles-Rivero, A., Yuan, L., Yu, N., Hua, G., 2020a. Controllable image processing via adaptive filterbank pyramid. IEEE Transactions on Image Processing 29, 8043–8054.

Chen, D., Yuan, L., Hua, G., 2020b. Deep style transfer. Computer Vision: A Reference Guide, 1–8.

Chen, D., Yuan, L., Liao, J., Yu, N., Hua, G., 2020c. Explicit filterbank learning for neural image style transfer and image processing. IEEE Transactions on Pattern Analysis and Machine Intelligence.

Chen, X., Duan, Y., Houthooft, R., Schulman, J., Sutskever, I., Abbeel, P., 2016. Infogan: interpretable representation learning by information maximizing generative adversarial nets. In: Advances in Neural Information Processing Systems, pp. 2172–2180.

Cooper, D.B., 1979. Maximum likelihood estimation of Markov-process blob boundaries in noisy images. IEEE Transactions on Pattern Analysis and Machine Intelligence, 372–384.

Cross, G.R., Jain, A.K., 1983. Markov random field texture models. IEEE Transactions on Pattern Analysis and Machine Intelligence, 25–39.

Dempster, A.P., Laird, N.M., Rubin, D.B., 1977. Maximum likelihood from incomplete data via the em algorithm. Journal of the Royal Statistical Society, Series B, Methodological 39, 1–38.

Do, M.N., Vetterli, M., 2003. The finite ridgelet transform for image representation. IEEE Transactions on Image Processing 12, 16–28.

Doersch, C., 2016. Tutorial on variational autoencoders. arXiv preprint. arXiv:1606.05908.

Fan, Q., Chen, D., Yuan, L., Hua, G., Yu, N., Chen, B., 2018. Decouple learning for parameterized image operators. In: Proceedings of the European Conference on Computer Vision (ECCV), pp. 442–458.

Fan, Q., Chen, D., Yuan, L., Hua, G., Yu, N., Chen, B., 2019. A general decoupled learning framework for parameterized image operators. IEEE Transactions on Pattern Analysis and Machine Intelligence.

Fu, K.S., 1982. Syntactic Pattern Recognition. Prentice-Hall.

Gatys, L.A., Ecker, A.S., Bethge, M., 2015. A neural algorithm of artistic style. arXiv preprint. arXiv:1508.06576.

Geman, S., Geman, D., 1984. Stochastic relaxation, Gibbs distributions, and the Bayesian restoration of images. IEEE Transactions on Pattern Analysis and Machine Intelligence, 721–741.

Ghahramani, Z., Beal, M.J., 2001. Graphical models and variational methods. In: Graphical Models and Variational Methods. In: Neural Information Processing Series. MIT Press, Cambridge, MA.

Girshick, R., 2015. Fast r-cnn. In: Proceedings of the IEEE International Conference on Computer Vision, pp. 1440–1448.

Goodfellow, I., Pouget-Abadie, J., Mirza, M., Xu, B., Warde-Farley, D., Ozair, S., Courville, A., Bengio, Y., 2014. Generative adversarial nets. In: Advances in Neural Information Processing Systems, pp. 2672–2680.

Grenander, U., 1976, 1976–1981. Lectures in pattern theory i, ii and iii.

Guo, C.E., Zhu, S.C., Wu, Y.N., 2003. Modeling visual patterns by integrating descriptive and generative methods. International Journal of Computer Vision 53, 5–29.

He, K., Zhang, X., Ren, S., Sun, J., 2016. Deep residual learning for image recognition. In: Proceedings of the IEEE Conference on Computer Vision and Pattern Recognition, pp. 770–778.

He, M., Chen, D., Liao, J., Sander, P.V., Yuan, L., 2018. Deep exemplar-based colorization. ACM Transactions on Graphics (TOG) 37, 1–16.

Hinton, G.E., 2002. Training products of experts by minimizing contrastive divergence. Neural Computation 14, 1771–1800.

Hoyer, P.O., Hyvärinen, A., 2002. A multi-layer sparse coding network learns contour coding from natural images. Vision Research 42, 1593–1605.

Hua, G., 2020. Deep generative models. In: Computer Vision: a Reference Guide. Springer.

Huang, R., Zhang, S., Li, T., He, R., 2017. Beyond face rotation: global and local perception gan for photorealistic and identity preserving frontal view synthesis. In: Proceedings of the IEEE International Conference on Computer Vision, pp. 2439–2448.

Hyvärinen, A., 1999. Survey on independent component analysis.

Hyvärinen, A., Oja, E., 2000. Independent component analysis: algorithms and applications. Neural Networks 13, 411–430.

Isola, P., Zhu, J.Y., Zhou, T., Efros, A.A., 2017. Image-to-image translation with conditional adversarial networks. In: Proceedings of the IEEE Conference on Computer Vision and Pattern Recognition, pp. 1125–1134.

Johnson, J., Alahi, A., Fei-Fei, L., 2016. Perceptual losses for real-time style transfer and super-resolution. In: European Conference on Computer Vision. Springer, pp. 694–711.

Kambhatla, N., Leen, T.K., 1997. Dimension reduction by local principal component analysis. Neural Computation 9, 1493–1516.

Kingma, D.P., Welling, M., 2014. Auto-encoding variational Bayes. In: Bengio, Y., LeCun, Y. (Eds.), Interactional Conference on Learning Representation. http://dblp.uni-trier.de/db/conf/iclr/iclr2014.html#KingmaW13.

Kong, H., Wang, L., Teoh, E.K., Li, X., Wang, J.G., Venkateswarlu, R., 2005. Generalized 2d principal component analysis for face image representation and recognition. Neural Networks 18, 585–594.

Lee, A.B., Mumford, D., Huang, J., 2001. Occlusion models for natural images: a statistical study of a scale-invariant dead leaves model. International Journal of Computer Vision 41, 35–59.

Liu, D., Hua, G., Chen, T., 2010. A hierarchical visual model for video object summarization. IEEE Transactions on Pattern Analysis and Machine Intelligence 32, 2178–2190.

Locatello, F., Bauer, S., Lucic, M., Rätsch, G., Gelly, S., Schölkopf, B., Bachem, O., 2019. Challenging common assumptions in the unsupervised learning of disentangled representations. In: Proc. of the 36th International Conference on Machine Learning. Long Beach, CA.

Lu, X., Katz, A., Kanterakis, E., Li, Y., Zhang, Y., Caviris, N., 1992. Image analysis via optical wavelet transform. Optics Communications 92, 337–345.

Ma, L., Sun, Q., Georgoulis, S., Van Gool, L., Schiele, B., Fritz, M., 2018. Disentangled person image generation. In: Proceedings of the IEEE Conference on Computer Vision and Pattern Recognition, pp. 99–108.

Manat, S., Zhang, Z., 1993. Matching pursuit in a time-frequency dictionary. IEEE Transactions on Signal Processing 12, 3397–3451.

Metz, L., Poole, B., Pfau, D., Sohl-Dickstein, J., 2016. Unrolled generative adversarial networks. arXiv preprint. arXiv:1611.02163.

Mirza, M., Osindero, S., 2014. Conditional generative adversarial nets. arXiv preprint. arXiv:1411.1784.

Mumford, D.B., Shah, J., 1989. Optimal approximations by piecewise smooth functions and associated variational problems. Communications on Pure and Applied Mathematics.

Ng, A., et al., 2011. Sparse Autoencoder. CS294A. Lecture Notes, vol. 72, pp. 1–19.

Poggio, T., Torre, V., Koch, C., 1985. Computational vision and regularization theory. Nature 317, 314–319.

Ren, S., He, K., Girshick, R., Sun, J., 2015. Faster r-cnn: towards real-time object detection with region proposal networks. In: Advances in Neural Information Processing Systems, pp. 91–99.

Roweis, S., Ghahramani, Z., 1999. A unifying review of linear Gaussian models. Neural Computation 11, 305–345.

Shackleton, M., 1994. Learned deformable templates for object recognition. In: IEE Colloquium on Genetic Algorithms in Image Processing and Vision, IET, p. 7.

Szegedy, C., Liu, W., Jia, Y., Sermanet, P., Reed, S., Anguelov, D., Erhan, D., Vanhoucke, V., Rabinovich, A., 2015. Going deeper with convolutions. In: Proceedings of the IEEE Conference on Computer Vision and Pattern Recognition, pp. 1–9.

Tan, Z., Chai, M., Chen, D., Liao, J., Chu, Q., Yuan, L., Tulyakov, S., Yu, N., 2020a. Michigan: multi-input-conditioned hair image generation for portrait editing. ACM Transactions on Graphics (TOG) 39, 95.

Tan, Z., Chen, D., Chu, Q., Chai, M., Liao, J., He, M., Yuan, L., Hua, G., Yu, N., 2020b. Semantic image synthesis via efficient class-adaptive normalization. arXiv preprint. arXiv:2012.04644.

Terzopoulos, D., 1983. Multilevel computational processes for visual surface reconstruction. Computer Vision, Graphics, and Image Processing 24, 52–96.

Tran, L., Yin, X., Liu, X., 2017. Disentangled representation learning gan for pose-invariant face recognition. In: Proceedings of the IEEE Conference on Computer Vision and Pattern Recognition, pp. 1415–1424.

Ulyanov, D., Lebedev, V., Vedaldi, A., Lempitsky, V.S., 2016. Texture networks: feed-forward synthesis of textures and stylized images. In: ICML, p. 4.

Van de Wouwer, G., Scheunders, P., Van Dyck, D., 1999. Statistical texture characterization from discrete wavelet representations. IEEE Transactions on Image Processing 8, 592–598.

Van den Oord, A., Kalchbrenner, N., Espeholt, L., Vinyals, O., Graves, A., et al., 2016. Conditional image generation with pixelcnn decoders. In: Advances in Neural Information Processing Systems, pp. 4790–4798.

Vincent, P., Larochelle, H., Bengio, Y., Manzagol, P.A., 2008. Extracting and composing robust features with denoising autoencoders. In: Proceedings of the 25th International Conference on Machine Learning, pp. 1096–1103.

Wan, Z., Zhang, B., Chen, D., Zhang, P., Chen, D., Liao, J., Wen, F., 2020a. Bringing old photos back to life. In: Proceedings of the IEEE/CVF Conference on Computer Vision and Pattern Recognition, pp. 2747–2757.

Wan, Z., Zhang, B., Chen, D., Zhang, P., Chen, D., Liao, J., Wen, F., 2020b. Old photo restoration via deep latent space translation. arXiv preprint. arXiv:2009.07047.

Wang, Y., Yao, H., Zhao, S., 2016. Auto-encoder based dimensionality reduction. Neurocomputing 184, 232–242.

Wang, Z., She, Q., Ward, T.E., 2019. Generative adversarial networks in computer vision: a survey and taxonomy. arXiv preprint. arXiv:1906.01529.

Weng, J., Weng, C., Yuan, J., Liu, Z., 2018. Discriminative spatio-temporal pattern discovery for 3d action recognition. IEEE Transactions on Circuits and Systems for Video Technology 29, 1077–1089.

Xie, X., Sudhakar, R., Zhuang, H., 1994. On improving eye feature extraction using deformable templates. Pattern Recognition 27, 791–799.

Yan, X., Yang, J., Sohn, K., Lee, H., 2016. Attribute2image: conditional image generation from visual attributes. In: European Conference on Computer Vision. Springer, pp. 776–791.

Yin, X., Yu, X., Sohn, K., Liu, X., Chandraker, M., 2017. Towards large-pose face frontalization in the wild. In: Proceedings of the IEEE International Conference on Computer Vision, pp. 3990–3999.

Yuan, J., 2011. Discovering Visual Patterns in Image and Video Data: Concepts, Algorithms, Experiments Paperback. VDM Verlag Dr. Müller.

Yuille, A.L., 1991. Deformable templates for face recognition. Journal of Cognitive Neuroscience 3, 59–70.

Zhai, J., Zhang, S., Chen, J., He, Q., 2018. Autoencoder and its various variants. In: 2018 IEEE International Conference on Systems, Man, and Cybernetics (SMC). IEEE, pp. 415–419.

Zhang, H., Xu, T., Li, H., Zhang, S., Wang, X., Huang, X., Metaxas, D.N., 2017. Stackgan: text to photo-realistic image synthesis with stacked generative adversarial networks. In: Proceedings of the IEEE International Conference on Computer Vision, pp. 5907–5915.

Zhao, G., Yuan, J., Hua, G., 2013. Topical video object discovery from key frames by modeling word co-occurrence prior. In: Proc. IEEE Conf. on Computer Vision and Pattern Recognition (CVPR'2013). Portland, OR.

Zhu, J.Y., Park, T., Isola, P., Efros, A.A., 2017. Unpaired image-to-image translation using cycle-consistent adversarial networks. In: Proceedings of the IEEE International Conference on Computer Vision, pp. 2223–2232.

Zhu, S.C., 2003. Statistical modeling and conceptualization of visual patterns. IEEE Transactions on Pattern Analysis and Machine Intelligence 25, 691–712.

Zhu, S.C., Wu, Y., Mumford, D., 1998. Filters, random fields and maximum entropy (frame): towards a unified theory for texture modeling. International Journal of Computer Vision 27, 107–126.

Deep face recognition using full and partial face images

Hassan Ugail

Centre for Visual Computing, University of Bradford, Bradford, United Kingdom

CHAPTER POINTS

- Methods and practical techniques for deep learning based model definition, training, and testing.

- Examples of using deep learning based models for full and partial face recognition.

- Highlights of the present state of the art in face recognition as well as the challenges in deep learning assisted face recognition.

6.1 Introduction

Face recognition is one of the most exciting and prominent applications of deep learning applied in the domain of visual computing. Not only can it showcase the power of deep learning but also it can highlight some of the concerns of using a methodology whereby a black-box approach is utilized for solving problems whose solutions have real-life consequences.

Computer-based face recognition is still riddled with many challenges compared to the face recognition ability of humans. For a human, seeing someone for a brief moment is often enough to learn sufficient about the face of that individual (Young and Burton, 2018). This is because the brain memorizes important details relating to the person. In fact, it seems that when a familiar face is presented within various contexts, the brain compares the 'before' and 'after' images, without the use of any significant new information. Conversely, the variability of the appearance of a face has a greater effect on the ability of a machine to identify the person from the face.

There are many machine learning algorithms within computer vision specific to face-related functions. These algorithms are either unsupervised techniques, which are not ex-

Copyright © 2022 Elsevier Inc. All rights reserved.

plicitly programmed, or supervised, which are based on the idea that someone can select a portion of data with known labels (provided by the operator) or fed to an algorithm as a training set. For example, Principal Component Analysis (PCA), was introduced in 1905 and is an unsupervised machine learning model widely used for reducing dimensionality in complex data and image compression (Jolliffe, 2002). It has also been a common technique for coding faces in a dataset into the "face space". In 1991, the Eigenface algorithm was applied to construct face recognition algorithms (Turk and Pentland, 1991). Eigenfaces extract the most important details in the data in the form of eigenvectors corresponding to the largest eigenvalues, representing the variations in the face space. Essentially, in the area of face recognition, the concept of an average face is interesting, and in some ways, it appears to be at the core of human face recognition. Attempts to use the average face as a tool for computer-based face recognition are not uncommon in the published literature, e.g., authors (Elmahmudi and Ugail, 2019a). Also, commonly used is the Local Binary Patterns (LBP) technique, which is a simple yet powerful approach to texture classification. This divides an image into different regions to extract features from each region separately. These features are subsequently used for aiding classification in face recognition tasks (Kas et al., 2020).

In conventional computing, a given algorithm is a group of explicitly programmed commands utilized by a machine to work out a problem. Machine learning approaches permit algorithms to be trained by using input data and to utilize statistical analysis to infer values that fall within a particular domain. For this reason, machine learning enables computers to build models from example data in order to automate a decision-making process, based on data input.

Deep learning (LeCun et al., 2015) is a mechanism by which a machine algorithm can learn by example. It aims to obtain an optimal configuration to a model so that the desired outputs can be obtained from a set of input data. It has been widely used for solving challenging problems in image processing and analysis. As a result, deep learning has probably become the standard de facto technique for use in modern face recognition systems.

Prior to deep learning, a high proportion of face recognition algorithms have utilized image processing techniques such as filtering, histograms and feature coding, containing only two or three layers of computation. There was a significant drawback in such methods, mainly because they can only solve one aspect of the face recognition problem at the expense of the others. For example, while a Gabor filtering method (Dora et al., 2017) can enhance face recognition under varying lighting conditions, that same method can perform poorly in the presence of facial expressions and pose variations. As a result, researchers struggled to come up with a coherent and integral method to solve the bulk of the issues that have plagued face recognition.

However, much of that changed in 2012, when it was decisively demonstrated that deep learning can take care of many of the problems faced by computer-assisted face recognition at the time. At that stage, the AlexNet deep learning model won the ImageNet competition, demonstrating that it was consistently ahead of the game on image recognition. Since then, deep learning has been on an upward trajectory and has stayed ahead of the curve in showing that it can match human performance in face recognition. For example, in 2014, DeepFace on the Labelled Faces in the Wild (LFW) dataset has shown, it can achieve human-level accuracy, the exact figures being 97.35% for DeepFace and 97.533% for humans (Taigman et al., 2014).

6.1.1 Deep learning models

Deep learning seeks to emulate and learn complex arrangements in datasets using multiple learning layers of processing units (neurons). This is motivated by the structure and functions of the human nervous system. Deep learning models learn patterns in very complex data via a series of repeated or multiple artificial neurons by manipulating the parameters associated with the neuron to produce the desired output. Thus, a deep learning model is composed of a neural network with multiple hidden layers. This is intended to emulate human decision making when solving complex recognition problems. In practical terms, deep learning usually take the form of Convolutional Neural Networks (CNNs) (LeCun et al., 2015).

6.1.1.1 *The structure of a CNN*

Essentially, a CNN is a collection of individual perceptrons (artificial neurons) forming a network of neurons connected together to achieve parallel signal processing across a wide network structure. Adjustable weights are used to control the interaction between the perceptron units. A key component of a CNN is the hidden layers which are placed between the input and output of the network. The hidden layers help in assigning nonlinear weights to the input and direct them to the output. i.e., they help to apply nonlinear mathematical functions; to specific parts of the network to produce a desired final output. For example, in the case of a face recognition task, one hidden layer can be tuned to identify the colors in the input image, another can be tuned to identify the physical features, and so on. Together, the network can then recognize and classify a face in the input image.

A typical CNN is comprised of multiple layers that fall into three broad categories. They are the convolution layers (CONV), subsampling layers (POOL), and fully connected layers (FC). Usually, a combination of these layers is arranged in a specific manner with the sole goal of transforming the input of the network into a useful representation that gives an output prediction, as shown in Fig. 6.1.

The convolution layer derives its name from the mathematical (convolution) operator. This layer computes the scalar product between the weights of neurons and a small region of the input volume. The neurons are arranged as a stack of 2-dimensional filters or kernels that extend the depth of the input volume. Hence, they are 3D structured. During the forward pass, each kernel is convolved across the width and height of the input volume to produce a 2D feature map, as shown in Fig. 6.2. Thus, these feature maps are the outputs of the convolution operation at each spatial operation. In comparison to the feed-forward neural network, these filters represent neurons which activate when they come across visual features such as edges. As discussed earlier, CNNs use local connectivity to reduce complexity. Hence each neuron is connected to a local region whose spatial dimension is defined by the filter size known as the receptive field of the neuron and its depth is equal to the depth of the input volume.

Hence, for a $256 \times 256 \times 3$ input image, if the receptive field is 3×3, each neuron in the CONV layer will have a total of $3 \times 3 \times 3 = 27$ connections, and 1 bias parameter. Obviously, the connectivity is spatially local but full along the input depth. Subsequently, the size of the feature map (i.e., the output) is computed using three hyperparameters – depth, zero padding, and stride. Depth refers to the number of filters deployed. The greater the number of filters, the greater the amount of information retrieved, since each filter learns to look for a specific feature. Stride defines a pattern used to slide the filter across the input, i.e., $S = 1$

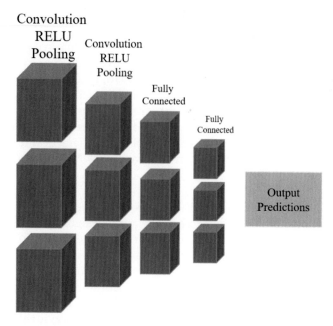

FIGURE 6.1 The structure of a CNN. It comprises of multiple layers which include convolution layers, pooling layers, and fully connected layers.

means that the filter has to move one pixel at a time along with the input. Zero padding defines the number of zero pixels placed around the input volume in order to keep the spatial size of the output volume constant. One can compute the spatial size of the output using:

$$\frac{I - F + 2P}{S} + 1 = 0, \tag{6.1}$$

where I is the spatial size of the input, F is the filter size, P is the number of zero paddings and S is the stride (Dumoulin and Visin, 2018). Hence, if applied to input images of size $224 \times 224 \times 3$ and assuming the neurons have a receptive field of 3×3 in size, depth $K = 64$, a single stride $S = 1$, and zero padding $P = 1$, one obtains $(224 - 3 + 2)/1 + 1 = 224$. This means that the output volume of this particular CONV layer will have size $224 \times 224 \times 64$. Consequently, there will be $224 \times 224 \times 64 = 3211264$ neurons, each having $3 \times 3 \times 3 = 27$ weights and 1 bias. Interestingly, rather than having $3,211,264 \times 27$ weights and $32,112,64$ biases, the concept of weight sharing makes all the neurons on one slice share the same weight and bias. Hence, the number of weights and biases is drastically reduced to $1,728$ and 64, respectively.

As mentioned earlier, in neural networks, the activation function plays a significant role in introducing nonlinearity to the output of a neuron. Introducing this nonlinearity makes the neural network a universal function approximator, thereby, giving it the ability to cope with various types of relationships. The most effective and commonly used activation function for CNNs is the rectified linear unit (ReLU). This involves the element-wise application of

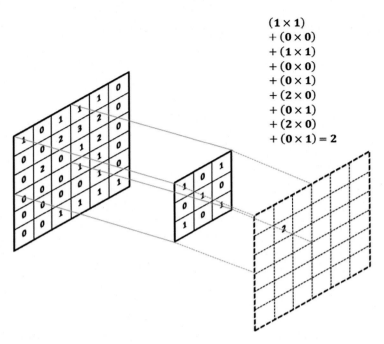

$$(1 \times 1)$$
$$+ (0 \times 0)$$
$$+ (1 \times 1)$$
$$+ (0 \times 0)$$
$$+ (0 \times 1)$$
$$+ (2 \times 0)$$
$$+ (0 \times 1)$$
$$+ (2 \times 0)$$
$$+ (0 \times 1) = 2$$

FIGURE 6.2 **Description of convolution operation.** During the forward pass, each kernel is convolved across the width and height of the input volume to produce a 2D feature map.

a zero threshold function, $f(x) = \max(0, x)$, where x is the input of the neuron. Compared to other activation functions, CNNs with ReLUs train several times faster. This is because the ReLU function and its gradient can be efficiently computed. Note, the activation layer does not introduce additional parameters to its input. Furthermore, it does not change the dimension of the input. In the architecture, activation layers are placed after every CONV layer. Additionally, networks with more than one fully connected (FC) layer also deploy it – with the exception of the last FC layer.

Pooling layers are usually inserted between successive CONV layers. Their primary function is to consistently reduce the number of parameters and consequently decrease the computational complexity of the network by reducing the spatial size of the feature maps. Hence, they summarize the output of the neighboring neurons. For every 2D slice of the feature map, the most common type of pooling operation (Gholamalinezhad and Khosravi, 2020), called Max Pooling, usually takes the maximum of each 2×2 region, thus discarding 75% of the activations as shown in Fig. 6.3.

Thus, the pooling operation does not introduce new parameters. Rather, it leads to shrinkage of the first and second dimensions of the feature map. The operation takes two parameters, the stride S and the spatial dimension F. Hence the pooling operation reduces a feature map from $W_1 \times H_1 \times D$ to $W_2 \times H_2 \times D$ dimensions. Here W_2 and H_2 are computed using:

$$W_2 = \frac{W_1 - F}{S + 1}, \qquad H_2 = \frac{H_1 - F}{S + 1}. \tag{6.2}$$

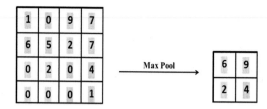

FIGURE 6.3 Pooling layer in action. Pooling layers help reduce the number of parameters and consequently decrease the computational complexity of the network.

Interestingly, this operation introduces translational invariance with respect to elastic distortions. The fully connected layer has neurons that have full connection to the activation of the previous layer and unlike the CONV and POOL layers, the FC layer is 2D. They are typically configured to output the predicted label/classes of the network. Hence the FC is usually the last layer of the network. In the work that won the 2012 ImageNet Large Scale Visual Recognition Competition (ILSVRC) (Krizhevsky et al., 2012), 3 FC layers were used, and since then this has been a typical arrangement. Intuitively, flattening the 3D feature maps at the end of the computation gives us an avenue for interpreting the learned spatially invariant features.

The most usual arrangement used by researchers starts with the image input layer and ends with an FC (decision) layer, in between these two are repeated stacks of CONV-ReLU layers followed by POOL layers, then a few FC-ReLU layers. This layered pattern can be described mathematically as:

$$INPUT \implies NM(CONV \implies ReLU) \implies POOL \implies K(FC \implies ReLU) \implies FC, \quad (6.3)$$

where N, M, and K are positive real-valued parameters. Usually, the number of CONV-ReLU layers that appear before POOL is within the range $0 < N < 4$, and the combinations of variables M and K are usually greater than 1.

6.1.1.2 Methods of training CNNs

Generally, there are three ways of deploying CNNs – training a network from scratch, fine-tuning an existing model, or using off-the-shelf CNN features. The latter two approaches are referred to as transfer learning. Since training CNNs from scratch, using the backpropagation algorithm involves the automatic learning of millions of parameters, this approach requires an enormous amount of data. More so, the data-hungry nature of CNNs consequently demands large computational power. Furthermore, the procedure involves the adjustment of several hyperparameters. Thus, an entire network is rarely trained from scratch.

Fine tuning involves transferring the weights of the first n layers learned from a base network to a target network, and then continuing the backpropagation using the new dataset. Hence, the target network is trained using the new dataset for a specific task, usually different from that of the base network. Fine tuning is normally used when the new dataset is moderately large (tens to hundreds of thousands) and very different from the dataset used to train the base network. Using the weights of the old network to initialize helps the backpropagation algorithm, thereby leading to relatively fast automatic learning of more specific features.

In situations where the dataset is rather small, say a few hundreds, even fine tuning the weights may result in overfitting. However, since CNNs efficiently learn generic image features, it is possible to directly use a trained network as a fixed feature extractor. Hence, features from new data are extracted by projecting them on to activations of a specific layer of the pretrained network. After that, the learned representations are fed into simple classifiers to solve the task at hand. This approach, known as off-the-shelf feature extraction, has been used by researchers to achieve promising results (Weiss et al., 2016; Day and Khoshgoftaar, 2017).

Similarly, data augmentation is the simplest way to combat the overfitting problem encountered by deep neural networks. This technique works by artificially enlarging the size of the dataset through various methods such as changing the image orientation by flipping (which produces the mirrored image) and rotating the original images, which subsequently results in a new image. This ensures that the learning algorithm infers features from data with different orientations.

6.1.1.3 *Datasets for deep face recognition experimentation*

There are a number of face datasets that can be utilized for training and testing the deep face recognition models. Here, we provide details of some of them.

The LFW Dataset: The Labelled Faces in the Wild (LFW) (Huang et al., 2008) is a large dataset of face pictures, which is designed for testing the capability of face recognition in simulated uncontrolled scenarios. All the images have been collected from the Internet and consist of a spectrum of variations in expression, pose, age, illumination, and resolution. The LFW database contains images of 5749 subjects with a combined total of some 13000 images. The images themselves in the dataset have variable and significant background clutter.

The YouTube Faces Database: The YouTube Faces DB (He et al., 2018), is composed of face videos with varying lighting, pose and age conditions. The database is specifically designed to study and analyze face recognition algorithms in videos. It contains over 3000 videos of over 1500 individuals. The videos have been downloaded from YouTube. The database has been put together by closely following the LFW as a benchmark.

The FEI Dataset (Thomaz and Giraldi, 2010) contains 200 images of Brazilian students and staff with an equal number of males and females. For each subject, there are 14 images, the total number of images in the dataset being 2800. The resolution of the images is 640 pixels by 480 pixels, and all the images are in color taken against a homogeneous white background. The subjects are between the ages of 19 and 40 years and the dataset contains images displaying variations in facial expressions and pose.

6.2 Components of deep face recognition

There are essentially three parts to a modern deep learning based face recognition system. They are face detection by which a face is successfully identified within the image, face processing by which the face is cropped and often normalized, and face recognition by which a deep learning algorithm is used to classify or match the face. This process is described in Fig. 6.4.

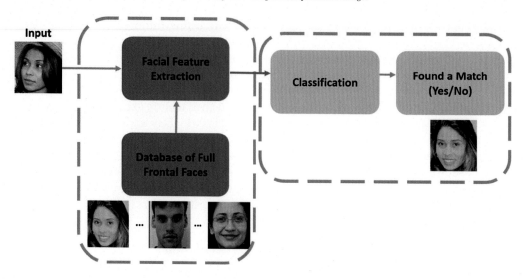

FIGURE 6.4 **Components of deep face recognition.**

Though deep learning can be efficiently utilized to solve most of the object recognition problems, for face recognition problems, we still have to resort to a face processing step in order to extract the face from the scene to ensure that the problems of facial pose, illumination, facial expressions, and occlusion are minimized.

Thus, in symbolic terms, a deep face recognition system can be described as:

$$M\big[F(I_i), F(I_j)\big], \tag{6.4}$$

where I_i and I_j are two face images that are compared, F are the deep features from the CNN model, and M defines the matching criteria. Following, a usually very long process of training with massive datasets and also with the supervision of appropriate loss functions, the optimal layers from which features are to be extracted and compared are determined. The matching process itself can be undertaken using distance measures such as Euclidean distance, Cosine Similarity (CS) and Support Vector Machines (SVMs).

Thus, today, most of the face recognition in practice is undertaken with the aid of a deep learning model, and as mentioned earlier, there are many to choose from. In what follows, we discuss some examples to further explain the process of deep learning-based face recognition and the challenges one needs to be vigilant about.

The prominent part of any deep learning-based face recognition system is the deep features that are derived from a trained CNN model (Liu et al., 2017). A number of model architectures are available for a user to choose from. These include the AlexNet (Krizhevsky et al., 2012), GoogleNet (Szegedy et al., 2015), ResNet (He et al., 2015), and VGGNet (Parkhi et al., 2015).

6.2.1 An example of a trained CNN model for face recognition

As discussed in the previous chapters, generally speaking, there are several ways one can deploy CNNs. These include training a network from scratch or fine-tuning an existing model or using off-the-shelf CNN features from a pretrained model. The latter is referred to as transfer learning.

It is important to highlight that training a CNN from scratch requires an enormous amount of data, which is often a challenging task, e.g., it took millions of faces and hundreds of hours of computing time to train the FaceNet model (Schroff et al., 2015) from scratch. On the other hand, fine-tuning involves transferring the weights of the first few layers learned from a base network to a target network. The target network can then be trained using a new dataset.

A good example of a pretrained CNN model within the context of face recognition is the VGG-F model (Chatfield et al., 2016), which was developed by the Oxford Visual Geometry Group. This model has been trained on a large dataset of 2.6 million face images of more than 2.6 thousand individuals. The architecture of VGG-F consists of 38 layers starting from the input layer up to the output layer. As a fixed criterion, the input is normally a color image of 224×224 dimensions and, as the preprocessing step, an average is normally computed from the input image.

In general, the VGG-F contains thirteen convolutional layers, each layer having a special set of hybrid parameters. Each group of convolutional layers contains 5 Max-Pooling layers and 15 Rectified Linear Units (ReLUs). After these, there are three FC layers, namely FC6, FC7, and FC8. The first two have 4096 channels, while FC8 has 2622 channels which are used to classify the 2622 identities. The last layer is a Softmax classifier whose function is to provide the probability of an image belonging to given class. The architecture of the VGG-F is represented in Fig. 6.5.

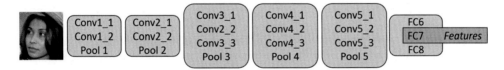

FIGURE 6.5 **Architecture of the VGG-F model.** The model contains 13 convolutional layers, each layer having a special set of hybrid parameters.

Here, we show how the pretrained VGG-F model for feature extraction can be utilized for facial feature coding and how the cosine similarity measure or linear SVM measures can be used for classification for efficient face recognition from partial faces.

6.2.1.1 Feature extraction

Given an input image, X_0, it can be represented as a tensor $X_0 \in R^H W D$, where H is the image height, W is the width, and D represents the color channels. A pretrained layer L of the CNN can be expressed as a series of functions, $g_L = f_1 \longrightarrow f_2 \longrightarrow ... \longrightarrow f_L$.

Let $X_1, X_1, ..., X_n$ be the outputs of each layer in the network. Then, the output of the ith intermediate layer can be computed from the function f_i and the learned weights w_i such that $X_i = f_i(X_{(i-1)} : w_i)$.

As we know, CNNs learn features through the training stage and use such features to classify images later. Each convolutional (conv) layer learns different features. For example, one layer may learn about entities such as edges and colors of an image while further complex features may be learnt in the deeper layers. For example, a result of the conv layer involves numerous $2D$ arrays which are called channels. In the VGG-F, there are 37 layers, 13 of which are convolutions, and the remaining layers are mixed between the ReLU, pooling, fully connected, and the Softmax. However, after applying the conv5_3 layer with 512 filters of size 3×3, the features can be extracted for classification. By examining the activations of that layer, one can obtain the main features, as shown in Fig. 6.6, where a sample of the features is presented.

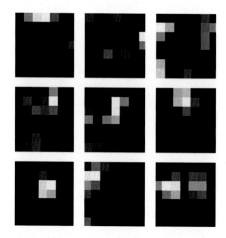

FIGURE 6.6 Feature visualization from a VGG-F layer. Visualization obtained from the conv5_3 layer for an input of a face image.

To decide the best layer within the VGG-F model to utilize for facial feature extractions, one must conduct a number of trial and error experiments. Tests of this nature suggest that layers 34 through to 37 are typically most effective, e.g., Elmahmudi and Ugail (2019b). Often, the best results are derived from the layer 34. It is noteworthy that this layer is the fully connected layer and is placed at the end of the CNN, which means the extracted features represent the whole face.

The features from layer 34 are the results that arise from the fully connected layer FC7 after applying ReLU6, which gives a vector of 4096 dimensions. The reason for suggesting that layer 34 is the best layer is inferred as a result of undertaking a number of face recognition tests using the full frontal face image for both training and testing, which shows that recognition rate can reach up to 100%.

6.2.1.2 Feature classification

One of the objectives of the classification is to build a brief model of the distribution of class labels in terms of the predicted features. There are several techniques for such a classification, namely decision trees, k-nearest neighbors (kNN) and SVM.

The SVM is a supervised machine learning algorithm, which can be used for both binary classification and multiclassification problems. The SVM focuses on identifying the "margins" via hyperplanes to separate the data into classes. Maximizing the margin reduces the upper bound on the expected generalization error by creating the largest possible distance between the separating hyperplanes. It is clear that the SVM is geared to solve binary classification problems. It is common for researchers to use the linear SVM to solve the multiclass classification problem based on One-vs.-One (OVO) approach, also known as pairwise classification. The OVO decomposition constructs $\frac{n(n-1)}{2}$ binary classifiers for n classes. Then, for a final decision, the Error Correcting Codes (ECC) combination approach decides how the various classifiers can be combined.

Consider that we have a training dataset (x_i, y_i), we can use the linear SVM such that:

$$\min \frac{1}{2}|w|^2 + C \sum_i^N \max(0, 1 - y_i w^T x_i), \quad w \in R^d, \tag{6.5}$$

where w is a weight vector, N is a number of classes and C is a trade-off parameter between the error and the margin.

Furthermore, one can also utilize the Cosine Similarity (CS) for classification. The CS is a measure between two nonzero vectors. It uses the inner product space to measure the cosine of the angle between those two vectors. The Euclidean dot product formula can be used to compute the cosine similarity such that:

$$a.b = |a||b|\cos\theta, \tag{6.6}$$

where a and b are two vectors, and θ is an angle between them. By using the magnitude or length, which is the same as the Euclidean norm or the Euclidean length of vector $x = [x_1, x_2, x_3, ..., x_n]$, the similarity S is computed using the formulation:

$$|x| = \sqrt{x_1^2 + x_2^2 + ... + x_n^2}, \tag{6.7}$$

$$S = \cos\theta = \frac{A.B}{|A||B|} = \frac{\sum_i^N A_i B_i}{\sqrt{\sum_i^N A_i^2}\sqrt{\sum_i^N B_i^2}}, \tag{6.8}$$

where A and B are two vectors.

For classification one can compute the CS to find the minimum "distance" between the test face image, $test_{im}$ and training images, $training_{im}^n$ by using Eq. (6.9), such that:

$$M_{cs} = \min\left(CS\left(test_{im}, training_{im}^n\right)\right), \tag{6.9}$$

where im is an image number, n is total images in the training set.

6.3 Face recognition using full face images

As far as deep learning is concerned, a well trained CNN model often provides excellent recognition for full frontal faces. In this section, we show how a typical pretrained model can be utilized for the task of face matching and verification.

The FaceNet model (Schroff et al., 2015) is a pretrained deep learning architecture inspired by GoogLeNet models for efficient face recognition. For the task of face recognition, it uses pictures of a list of people in a dataset along with data from a new person or people to be recognized. A key element of the FaceNet architecture is the generation of an embedding of a given dimension from a face image of predefined size. The input image is fed through a deep CNN architecture which has a fully connected layer at the end. This results in an embedding of 128 features which may or may not be visually understandable to a human. Then, for the recognition task, the network can calculate the distance between the individual features of each of the embeddings. Metrics such as the squared error or the absolute error can be utilized to compute the distance between the embeddings.

FIGURE 6.7 **The general architecture of FaceNet model.** A key aspect of the FaceNet model is the ability to classify a given face image using an embedding of 128 deep features.

Common FaceNet models use two types of architecture. They are the Zeiler and Fergus architecture (Zeiler and Fergus, 2014) and the GoogLeNet style Inception model (Szegedy et al., 2015). The essential idea in the training of the FaceNet is the "triplet loss" to capture the similarities and differences between classes of faces in a 128-dimensional embedding. Hence, given an embedding $E(x)$ from an image to a feature set R^n, FaceNet looks into the squared L_2 distance between face images, this value being small for images of the same identity and large for different identities.

Fig. 6.7 shows the general architecture of the FaceNet model. An important element of the model is the triplet loss function. Often in common deep learning models, the loss function tries to map all faces of the same identity to a single point in R^n. The triplet loss function attempts to discriminate each pair of faces from one person to all the others, thus enforcing strong discrimination between faces. Thus, the triplet loss function chosen in this model ensures that an image of a specific person is closer to all other images of that person than any other image in the dataset. This idea is illustrated in Fig. 6.8. The training process assumes that we pick a random image – the anchor – from the dataset. We would want to ensure that the distance of that image from another image of the same identity – the positive image – is closer to that of the images not belonging to the same person.

The FaceNet model was trained using around 100 million face samples with over 8 million identities. Through experimentation, it has been found that the optimal embedding for a faces is of dimensionality 128, meaning that a given face is classified using 128 feature points extracted from it. Experiments were also conducted on varying the training data, and it was found that after a certain point the number of training samples does not add value to the level of accuracy, i.e., while 10s of millions of samples can improve the accuracy, adding 100s of millions to the sample has diminishing returns in the accuracy of recognition obtained.

6.3.1 Similarity matching using the FaceNet model

One of the main benefits of the FaceNet model is that it can achieve very high classification accuracy using a simpler embedding comprising of only 128 features. Experiments done on faces on both the LFW dataset and the YouTube Faces DB show the recognition is very high:

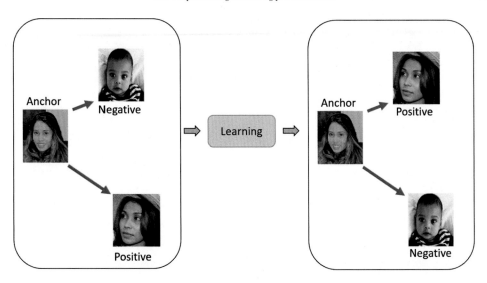

FIGURE 6.8 **Description of how triplet loss based training is utilized by FaceNet.** The triplet loss function attempts to discriminate each pair of faces from one person to all the others, thus enforcing strong discrimination between faces.

i.e., on the LFW dataset, the recognition accuracy of 98.87% is achieved while on the YouTube Faces DB a recognition accuracy of 95.18% is achieved. It is also crucial to highlight that both the datasets contain faces taken in varying lighting, pose, occlusion and age conditions. Despite this, the performance on recognition tasks by FaceNet is impressive.

Fig. 6.9 describes how the FaceNet model can be used to see the degree of similarity between faces. In this example, the 128 deep features for each face are computed using the FaceNet model. Then the cosine similarity measure described earlier can be used to compute the distance between the deep face features for each face against the face image in the center. The results show that the degree of accuracy for face matching, in this case, appears to be excellent. Note that in experiments involving the FaceNet model, a similarity match of 75% or higher between two faces is usually considered to be an identity match.

6.4 Deep face recognition using partial face data

Based on the many works that have been undertaken in the field of face recognition using deep learning, it is clear that many of the state of the art algorithms provide human-level accuracy for face recognition when the query images are full frontal. For example, the FaceNet model discussed above provides an impressive level of accuracy for face recognition using frontal facial images. However, in many practical scenarios, the full face may not be available either as a probe or as a comparison image in a dataset. Here, we discuss how deep learning based methods can be taken further forward in that such models can be trained to successfully recognize faces even with partial information. Achieving this would provide the potential for deep learning models to surpass human-level face recognition.

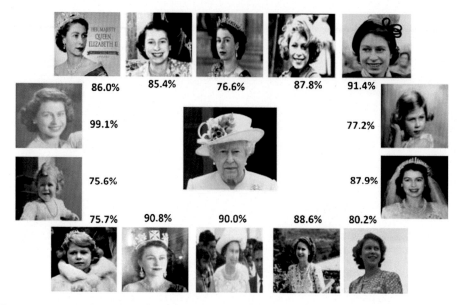

FIGURE 6.9 **An example of similarity matching.** The cosine similarity measure is utilized for matching faces of the Queen using the 128 deep features of the FaceNet model. The central image is used for matching.

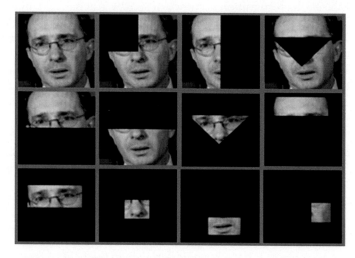

FIGURE 6.10 **An example of partial faces from the LFW dataset.** For experiments on partial face recognition one can consider half face, 3/4 face and key parts of the face such as the eyes, nose, mouth, and forehead.

In this section, we discuss how deep face recognition using partial faces can be undertaken. The work described here is predominately based on the deep learning based face recognition work reported in Elmahmudi and Ugail (2019b).

The approach here is to utilize features extracted from a pretrained model and using a standard classifier to see how the various parts of the face (see Fig. 6.10), are embedded within the

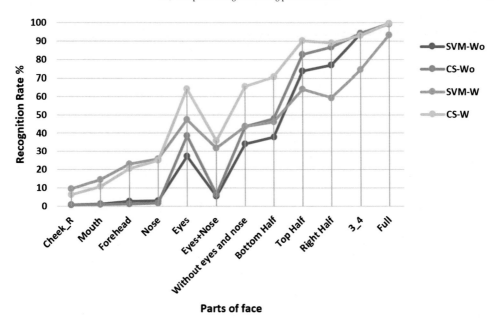

FIGURE 6.11 **Results for partial face recognition using VGG-F features.** Recognition rates (%) using images from the LFW dataset, based on parts of the face using both the SVM and CS classifiers.

model during the training phase. This can then be utilized for face recognition and matching tasks. In this particular case, we have utilized the standard VGG-F model for feature training and extraction. The various parts of the face considered here include the eyes, nose, mouth, top and bottom half of the face, left half of the face, 3/4 face, and the full face. Thus, all the extracted features from the VGG-F model can be passed onto both the classifiers, in this case, SVM and CS. Faces within the VGG-F can be enrolled with (W) or without (Wo) various parts of the face to see the variations in results.

For example, the classification capacity of SVM and CS can be verified using faces with and without parts in the training (SVM-Wo and CS-Wo) and "with" parts in training (SVM-W and CS-W). In order to investigate the recognition rates for each facial part, each classifier can be applied separately. In the case of "without" facial parts, it is clear that, in general, CS-Wo outperforms SVM-Wo for most of the regions of the face. The results of these experiments are shown in Figs. 6.11 and 6.12. From these figures, we can observe that the recognition rates for the right cheek, mouth, forehead, and the nose are low, with about 1% for both the classifiers. In contrast, the rate of recognition increases significantly for facial parts such as the eyes, and reach 40% using CS-Wo. We also notice that as we increase the proportion of the face, the recognition rate also improved significantly, with the best recognition rate of near 100% for the 3/4 face and full face. It is also noteworthy that for all the tests carried out in these experiments, the CS measures appeared to outperform the SVM measures.

From the results presented in Figs. 6.11 and 6.12 for the partial facial experiments using the FEI dataset, the highest recognition rate (as far as partial face is concerned) reported is for the 3/4 face using SVM-Wo. Under this experimental condition, the training set did not consist

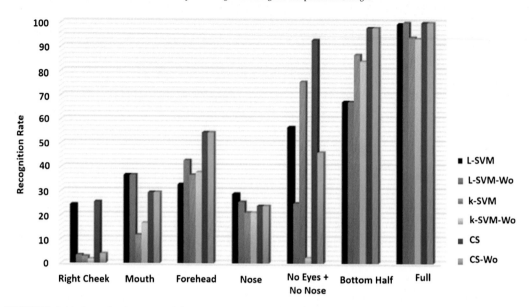

FIGURE 6.12 Results for partial face recognition using VGG-F features. The results of the recognition task (%) when some parts of the face (cheeks and facial-part with no eyes and no nose) are removed from the training sets. The tested classifiers include linear SVM, kernel SVM and CS.

of various facial parts. Also, in the case of CS-Wo, the right half of the face, the top half and the 3/4 face produces high recognition rates. However, the worst recognition observations are for the smaller and perhaps, less significant parts of the face, such as cheek, mouth and nose. When applying the same methodology to the uncontrolled LFW dataset and training with larger proportions of the face, a slight decrease in the recognition rates is observed when compared with the FEI dataset, which was between 76% to 99% for SVM-Wo and 83% to 99% for the CS-W classifier. According to the results obtained for smaller regions of the face, the worst recognition rate is observed for the cheeks, mouth, forehead and nose. However, the eyes do appear to hold more information.

Similarly, when the individual parts of the face are added to the training sets, the output results in a dramatic improvement to the rate of recognition. For example, the recognition rate for the right cheek improved from 0% to 15% when using the FEI dataset. It is also noteworthy that the eyes still have the highest recognition rate when considering other individual parts of the face using the FEI and LFW datasets, although the combined eyes and nose features report around 90% recognition when using the controlled FEI dataset. However, in the case of the uncontrolled LFW dataset, this percentage drops slightly. Furthermore, we notice that in general, better recognition results are achieved by using the CS measure.

Thus, an important point to highlight here in this case is that the CS measure, in general, appears to be a better classifier, compared to both Linear and NonLinear SVMs. SVMs require complete retraining when new data is added, which subsequently introduces computational issues. However, in the case of the CS classifier, this is not an issue. Nevertheless, in the testing stage, the CS classifier is more computationally intensive, but given the greater degree of accuracy, it makes better logical sense to employ the CS classifier over SVMs.

6.5 Specific model training for full and partial faces

In this section, we discuss how one can train specific deep learning models for efficient face recognition using full or partial faces. Here we show how an experimental framework can be built to train specific CNNs corresponding to a specific parts of the face. In designing systems of this nature, it is important to bear in mind the level of training data utilized, the model complexities and the amount of training time required to create a given model.

Let's suppose we are building a deep learning model that is fine tuned for identifying individuals simply from the images of their eyes. We will start with the base VGG-16 architecture (Parkhi et al., 2015) which is similar to VGG-F discussed earlier. This model consists of 13 convolutional layers (CONVs) with the same filters of size 3 × 3. These layers themselves are divided as follows. The first two layers have depths of size 64. There are two layers which have a depth of size 128, three layers with a depth of size 256 and six layers with a depth of size 512. Following these layers, there are three FC layers, of which two layers contain 4096 units while the last layer has 1000 units. The structure of VGG-16 is thought to be particularly neat in that it can efficiently manage the number of hyperparameters. Specifically, the arrangement of layers has been well thought out so that the number of hyperparameters is minimized. This arrangement is structured as follows. The size of the CONV filters is of 3 × 3 and there are 2 × 2 masks for padding and max-pool layers with stride 2. This suggests that for training for a specific part of the face we should adopt a two-stage strategy. In the first stage, the first model will be constructed and trained on a dataset of eye images only. In the second stage, the generated model will be used for building the comprehensive model by utilizing the same structure as the VGG network using fine tuning (Yosinski et al., 2014). Fig. 6.13 shows a flow diagram for training the model.

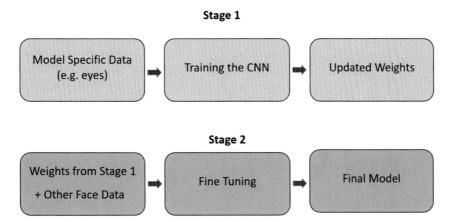

FIGURE 6.13 **Proposed procedure for model training for specific parts of the face.** The training process can be composed of two stages where weights for the specific part of the face are generated which are then used to produce the final shape of the model.

The process of fine tuning for machine learning algorithms is a procedure whereby the chosen CNN model, already trained for a given job and/or data type, is utilized for performing

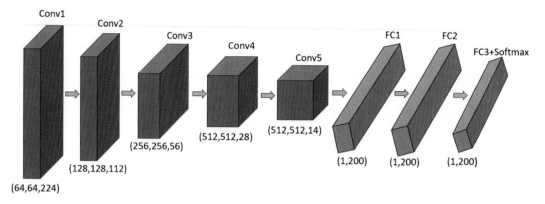

FIGURE 6.14 Proposed CNN model architecture for face recognition using specific parts of the face. This model is derived from the standard VGG-16.

a similar task. This is done by replacing an output layer, which was originally trained for recognizing previous classes, with one which can recognize a new number of classes for a new task. An advantage of using fine tuning is the reduction of computational time during the training phase. In fact, the first layers already have the capability of dealing with a new task, and training will utilize the last layers, without the need for training from scratch. Another advantage is improved performance because the pretrained models are typically trained on large datasets.

6.5.1 Suggested architecture of the model

As discussed above, the proposed model for training for specific parts of the face is inspired by the VGG-16 model (Parkhi et al., 2015) and requires an input image with a prefixed size (224, 224, 3). The model has 5 convolutional blocks, each containing a number of convolutional layers followed by nonlinear activation functions, as shown in Fig. 6.14.

As can be inferred from Fig. 6.14, the proposed CNN model for partial faces is architecturally similar to the VVG-F model with 5 convolutional layers and 3 FC layers. These convolutional layers are designed to integrate various max pooling layers to create efficient feature maps. A crucial aspect the proposed arrangement of convolutional and FC layers is to reduce the number of trainable parameters (as defined by Eq. (6.1)) to keep the length of training time reasonable while keeping the accuracy of the network within the acceptable limits.

6.5.2 Training phase

Training a CNN model is a procedure to minimize the variations between the ground truth labels and the predicted outputs from within a training dataset. This is achieved by locating learnable parameters (kernels and weights) within the convolutional and fully connected layers. In the model suggested here, the weights of filters are initialized by using a Gaussian and standard deviation (Bishop, 2006) and the biases are initialized to zero. To evaluate the performance of the model through feed-forward propagation, the most common loss function for a multi class classification problem, called cross-entropy function, can be used (Busoniu et al., 2011). The value of the loss function identifies how well or poorly a model performs after every iteration of the optimization. In addition, based on the error gradient in the current state of the model, all learnable parameters will be updated during the optimization, e.g., using gradient descent along with backpropagation. The learning rate is another hyperparameter which has the role of controlling how fast the CNN model learns using data presented to it. The value of this parameter is positive and in the range 0 to 1.

The last two hyperparameters in the model signify the number of epochs the model must pass through – which indicates the number of times the learning method sees the training dataset. It also signifies the batch size required to complete an epoch – the suggested batch size is 64. As for actual training, it is useful to follow a two-stage process. In the first stage, the model is trained using around 20 epochs with a batch size of 64. After that, the resulting weights from the model are saved. These weights are then utilized to initialize the weights for the second part of the training. During phase two, around 50 epochs of training are undertaken. These epochs are divided into ten parts with 5 epochs in each part. Thus, the weights from the previous training are used to initialize the new weights and train the model for five epochs. The resulting new weights will be used for the next training run. This procedure is continued until the selected training loss and validation criteria are reached. During the process, training loss of a small value (i.e., approximately < 0.02) and a higher validation accuracy (i.e., $> 85\%$) must be aimed for.

As far as training data is concerned, sufficient face images with sufficient identities must be available. A typical figure suggested is $70,000$ images corresponding to a specific part of the face with 200 identities. Such a dataset can be split into two groups, i.e., 70% for training and 30% validation. Finally, the training set is used to train the model, and the loss is computed by forward propagation while updating the learnable parameters through backpropagation.

6.6 Discussion and conclusions

In this chapter, we have broadly discussed the present state of the art in face recognition using deep learning methods and techniques. We have shown how deep learning methods can be utilized for identity and facial similarity matching – using both full frontal faces and partial facial data. We have shown how off-the-shelf features from well trained deep learning models can be used to efficiently craft accurate face recognition systems. We have also shown how to train specific models based on the various model architectures, trainable layers and weights from well known deep models that are used for image processing and analysis.

For example, we have explored the question that surrounds the idea of face recognition using facial features. We have explored them by showing key examples to test the performance of machine learning using the full, and parts of, the face in recognition tasks. In particular, we have shown how deep face recognition performs when parts, such as the eyes, mouth, nose and cheek are presented as learning and recognition cues. We have also shown that face recognition systems can follow a number of approaches. For instance, we have shown how to implement state of the art CNN along with the pretrained models (such as the VGG-F) through which key facial features can be extracted. Well known classifiers such as the Cosine Similarity measure and Linear Support Vector machines can then be utilized to test the recognition rates. Similarly, training specific models taking into consideration specific parts of the face such as the mouth, nose, eyes, forehead and their combinations for designing efficient face recognition systems from partial faces can be applied.

It is clear that, with the advances in deep learning methods and techniques, face recognition has immensely benefited (Guo and Zhang, 2019). In this sense, many of the problems previously thought unsolvable are now considered to be straightforward. For example, given two well lit frontal photos of an individual, confirming an identity match between them, using deep facial features, is now considered to be a solved problem. However, in general, the subject of face recognition is still a current issue, with a number of challenging and unsolved problems. For example, the issue of computer-based face recognition using partial facial data as probes is still largely an unexplored area with key research challenges. Given that humans and computers perform face recognition and authentication inherently differently, it is intriguing to understand how a computer will react to various parts of the face when they are presented for the challenge of face recognition.

Computer assisted face recognition, starting from the 1960s, has come a long way. However, there are many challenges and hurdles still to be overcome. These challenges include, for example, recognizing faces in poor illumination, pose variations, partial faces, inverted faces, aged faces and faces from long distances. The question of accuracy on facial similarity measures using deep learning techniques is largely unanswered. For example, the question of similarity between faces related through kinship, siblings and identical twins has not yet been answered satisfactorily. Furthermore, the question of bias in training data and how deep learning systems should be made transparent and explainable are also pressing questions that require to be answered satisfactorily.

References

Bishop, C.M., 2006. Pattern Recognition and Machine Learning. Springer, Berlin.

Busoniu, L., Ernst, D., De Schutter, B., Babuska, R., 2011. Cross-entropy optimization of control policies with adaptive basis functions. IEEE Transactions on Systems, Man, and Cybernetics, Part B 41 (1), 196–209.

Chatfield, K., Simonyan, K., Vedaldi, A., Zisserman, A., 2016. Return of the devil in the details: delving deep into convolutional nets. In: Proceedings of the British Machine Vision Conference (BMVC).

Day, O., Khoshgoftaar, T.M., 2017. A survey on heterogeneous transfer learning. Journal of Big Data 4 (1).

Dora, L., Agrawal, S., Panda, R., Abraham, A., 2017. An evolutionary single Gabor kernel based filter approach to face recognition. Engineering Applications of Artificial Intelligence 62, 286–301.

Dumoulin, V., Visin, F., 2018. A guide to convolution arithmetic for deep learning. arXiv:1603.07285v2.

Elmahmudi, A., Ugail, H., 2019a. The biharmonic eigenface. Signal, Image and Video Processing 3, 1639–1647.

Elmahmudi, A., Ugail, H., 2019b. Deep face recognition using imperfect facial data. Future Generations Computer Systems 41 (99), 213–225.

Gholamalinezhad, H., Khosravi, H., 2020. Pooling methods in deep neural networks, a review. arXiv:2009.07485.

Guo, G., Zhang, N., 2019. A survey on deep learning based face recognition. Computer Vision and Image Understanding 189, 102905.

He, K., Zhang, X., Ren, S., Sun, J., 2015. Deep residual learning for image recognition. In: Proceedings of the British Machine Vision Conference (BMVC).

He, L., Li, H., Zhang, Q., Sun, Z., 2018. Dynamic feature learning for partial face recognition. In: 2018 IEEE/CVF Conference on Computer Vision and Pattern Recognition, pp. 7054–7063.

Huang, G.B., Ramesh, M., Berg, T., Learned-miller, E., 2008. Labeled faces in the wild: a database for studying face recognition in unconstrained environments. In: Workshop on Faces in Real-Life Images: Detection, Alignment, and Recognition, pp. 1–11.

Jolliffe, I.T., 2002. Principal Component Analysis. Springer, New York.

Kas, M., El-merabet, Y., Ruichek, Y., Messoussi, R., 2020. A comprehensive comparative study of handcrafted methods for face recognition LBP-like and non LBP operators. Multimedia Tools and Applications 79, 375–413.

Krizhevsky, A., Sutskever, I., Hinton, G.E., 2012. ImageNet classification with deep convolutional neural networks. In: NIPS.

LeCun, Y., Bengio, Y., Hinton, G., 2015. Deep learning. Nature 521, 436–444.

Liu, W., Wang, Z., Liu, X., Zeng, N., Liu, Y., Alsaadi, F.E., 2017. A survey of deep neural network architectures and their applications. Neurocomputing 234, 11–26.

Parkhi, O.M., Vedaldi, A., Zisserman, A., 2015. Deep face recognition. In: IEEE CVPR.

Schroff, F., Kalenichenko, D., Philbin, J, 2015. FaceNet: A unified embedding for face recognition and clustering. In: IEEE CVPR.

Szegedy, C., Liu, W., Jia, Y., Sermanet, P., Reed, S., Anguelov, D., Erhan, D., Vanhoucke, V., Rabinovich, A., 2015. Going deeper with convolutions. In: IEEE CVPR.

Taigman, Y., Yang, M., Ranzato, M., Wolf, L., 2014. DeepFace: closing the gap to human-level performance in face verification. In: 2014 IEEE Conference on Computer Vision and Pattern Recognition, pp. 701–1708.

Thomaz, C.E., Giraldi, G.A., 2010. A new ranking method for principal components analysis and its application to face image analysis. Image and Vision Computing 28 (6), 902–913.

Turk, M., Pentland, A., 1991. Eigenfaces for recognition. Journal of Cognitive Neuroscience 13, 71–86.

Weiss, K., Khoshgoftaar, T.M., Wang, D., 2016. A survey of transfer learning. Journal of Big Data 3 (1).

Yosinski, J., Clune, J., Bengio, Y., Lipson, H., 2014. How transferable are features in deep neural networks? In: Proceedings of the Advances in Neural Information Processing Systems.

Young, A.W., Burton, A.M., 2018. Are we face experts. Trends in Cognitive Sciences 22, 100–110.

Zeiler, M.D., Fergus, R., 2014. Visualizing and understanding convolutional networks. In: Computer Vision – ECCV 2014.

Biographies

Hassan Ugail is the director of the Centre for Visual Computing in the Faculty of Engineering and Informatics at the University of Bradford, UK. He has a first class BSc Honors degree in Mathematics from King's College London and a PhD in the field of geometric design from the School of Mathematics at the University of Leeds. Professor Ugail's research interests include computer based geometric and functional design, imaging and machine learning.

Unsupervised domain adaptation using shallow and deep representations

Yogesh Balaji[a], *Hien Nguyen*[b], *and Rama Chellappa*[c]

[a]Department of Computer Science and UMACS, University of Maryland, College Park, MD, United States [b]Department of Electrical and Computer Engineering, University of Houston, Houston, TX, United States [c]Departments of Electrical and Computer Engineering and Biomedical Engineering, Johns Hopkins University, Baltimore, MD, United States

CHAPTER POINTS

- Unsupervised domain adaptation using shallow and deep features.
- Interpolation between source and target domains using manifolds and dictionaries.
- Generative adversarial networks for mediating the domain shift.

7.1 Introduction

With the availability of web-scale data obtained from different devices and varied acquisition conditions, we are often faced with scenarios where the data used to train a classifier has some properties that are different from what is presented during testing. Instances like this arise very naturally in object recognition applications where the training and testing data are captured under different lighting conditions, in speech processing where a speaker model trained in a noise-free indoor setting needs to be deployed in more realistic outdoor environments, or in multimedia indexing where tagged Flickr photos or YouTube videos are readily available from which a user would want to automatically index his/her own photo/video collection harvested with a consumer camera. Domain adaptation (DA) refers to the class of techniques aimed to learn representations from a low-resource dataset by transferring knowledge from a related, yet shifted, resource-rich dataset. Shifts between datasets can exist in the form of changes in illumination, lighting, texture patterns, or any such biases inherent to the

Advanced Methods and Deep Learning in Computer Vision
https://doi.org/10.1016/B978-0-12-822109-9.00016-3

Copyright © 2022 Elsevier Inc. All rights reserved.

dataset. To transfer knowledge across domain-shifted data distributions, a domain-invariant feature representation is sought that effectively phases out biases contributing to the domain shift.

Transfer learning (TL) (Pan et al., 2010) is a related approach that addresses the data set bias issue. The primary difference between TL and DA is due to what properties of data are preserved across training and testing conditions. While TL deals with the case where the conditional distribution of data labels is changing (i.e., tasks in the two domains are different) while the marginal distribution of data is preserved, DA addresses the opposite scenario (Daume and Marcu, 2006), where the data distributions vary between the two domains but the task remains the same. Although we often see practitioners applying TL and DA methods interchangeably to both these settings, in this chapter, we will focus on DA since it applies naturally to computer vision applications such as object recognition, where one is interested in preserving the identity of objects across variations in viewpoint, illumination, among others.

There are two broad categories of DA techniques depending on whether the target domain test data has either partial labels (semisupervised) or is completely unlabeled (unsupervised). While semisupervised DA often uses correspondence from labeled target data to learn domain transformation (Daume and Marcu, 2006; Saenko et al., 2010), unsupervised DA uses strategies that assume (i) certain class of transformations between the domains (Wang and Mahadevan, 2009), (ii) the availability of discriminative features that are common to or invariant across both domains (namely 'domain invariants') (Blitzer et al., 2008; Mansour et al., 2009), or (iii) a latent space where the difference in distribution of source and target data is minimal (Blitzer et al., 2011). In addition to adapting between a single source and a single target domain, there have been studies on multidomain adaptation (e.g. Mansour et al., 2009) that consider more than one domain in source and/or target. While some of these approaches pursue 'representation adaptation' by learning a domain shifting transformation, others (Duan et al., 2009; 2012) advocate a 'classifier'-oriented approach that attempts to obtain target classifiers by manipulating or reoptimizing classifiers trained on the source domain. In the rest of the chapter, we focus mainly on the unsupervised setting.

Since 2011, a significant amount of research has been done to address the challenge of unsupervised domain adaptation. Methods based differential geometry (Gopalan et al., 2011; 2014; Gong et al., 2012; Ho and Gopalan, 2014), sparse dictionaries (Lu et al., 2015; Nguyen et al., 2012; 2015; Shekhar et al., 2013; Xu et al., 2015) and more recently, Generative Adversarial Networks (GANs) (Sankaranarayanan et al., 2018; Shi and Sha, 2012) have been developed. In this chapter, we present some typical examples from the three approaches mentioned above. For more detailed expositions of domain adaptation research, the reader is referred to Patel et al. (2015).

7.2 Unsupervised domain adaptation using manifolds

In the spirit of adapting 'gradually between extremes', Gopalan et al. (2014) presented an approach for unsupervised DA by constructing a smooth path between source and target domains using intermediate data representations that convey relevant information about the domain shift. Thus, without making assumptions on domain invariant properties, this work presented the first unsupervised DA framework for object recognition by computing an adaptation path of certain statistical characteristics from the source domain(s) to the target domain(s). A special case of this framework, presented in Gopalan et al. (2011), used a linear generative subspace as the representation of a domain. More specifically, principal component analysis (PCA) is applied to each of the domains, followed by representing the subspaces as points on a Grassmann manifold. Then the geodesic between these points is used as a statistically meaningful path to represent the domain shift. By sampling points along the geodesic, intermediate cross-domain data representations are obtained with which a discriminative classifier is trained to perform recognition. This work subsequently studied other special cases of this framework such as domain representation in a high-dimensional Reproducing Kernel Hilbert Space (RKHS) using kernel methods, and a low-dimensional manifold representation using Laplacian Eigenmaps. Interestingly, this framework also accommodates semisupervised and multidomain adaptation settings, and has been further enhanced by simulating fine-grained domains blended with gradually varying proportions of source and target samples, as well as by applying boosting to a pool of intermediate representations yielded by different parameter choices.

Since the publication of Gopalan et al. (2011), there have been other related studies such as Gong et al. (2012); Zheng et al. (2012), that discussed alternative sampling strategies along the geodesic, and Shi and Sha (2012) that proposed an information-theoretic approach for jointly learning domain shift features and classifiers. Multisource adaptation that could accommodate different feature types across domains was addressed in Duan et al. (2012) using a data-dependent regularization mechanism, and robustness to noise or outliers was addressed in a low-rank reconstruction approach by Jhuo et al. (2012). More discussions and comparative evaluations against many of these notable contributions are included in Patel et al. (2015).

7.2.1 Unsupervised domain adaptation using product manifolds

An application of Unsupervised Domain Adaptation (UDA) is unconstrained face recognition which is a very challenging problem due to appearance variations between the probe and gallery images caused by multiple factors such as blur, expression, illumination, pose and resolution. As a result, face classifiers trained with the assumption that the training and testing data are drawn from similar distributions usually have very poor performance, especially when applied to uncontrolled environments. For instance, face recognition algorithms trained on samples from a source domain containing sharp, well-illuminated face images do not perform well when used on a target domain containing blurred, poorly illuminated face images (Vageeswaran et al., 2013). These algorithms' performance further degrades when only a limited number of images per subject is available due to the cost and other challenges in data acquisition.

While there have been several studies addressing prespecified facial variations across source and target domains (Zhao et al., 2003), such as the nine points of light study for illumination (Lee et al., 2005), analyzing domain shifts caused by multiple, unknown factors have not received much attention. Domain adaptation is a recent paradigm for addressing such transformations in a broader setting, where given labeled data from the source domain and few (or no) labeled data from target domain, unsupervised and semisupervised approaches have been devised to account for variations in data across domains (Saenko et al., 2010; Ben-David et al., 2010; Gopalan et al., 2011). Most of these techniques address domain shifts in a statistical sense as models causing variations in data are not known. This limits their application to the particular problem of face recognition where there is a rich literature on models for pose, lighting, blur, expression and aging. As a result, it is important to understand domain shifts with respect to the underlying constraints pertaining to models that generate the observed data. Such an analysis would necessitate the study of geometrical properties of the image space induced by these models.

Many traditional approaches, however, often either ignore the geometric structures of the space or naively treat the space as Euclidean (Lui, 2012). While nonlinear manifold learning algorithms such as ISOMAP (Tenenbaum et al., 2000) or Locally Linear Embedding (LLE) (Roweis and Saul, 2000) offer alternatives, they require large amounts of training data to estimate the underlying nonlinear manifold structure of the data. Such a requirement on data may not always be satisfied in many real-world applications. One possible solution for handling facial variations due to multiple factors is by employing a mathematical framework called multilinear algebra – the algebra of higher-order tensors. As matrices represent linear operators over a vector space, their generalization, tensors, define multilinear operators over a set of vector spaces (Vasilescu and Terzopoulos, 2002). While there have been studies using multilinear algebraic framework for face recognition (Vasilescu and Terzopoulos, 2002; 2007), such approaches ignore the curved geometry of the image space and resort to a Euclidean treatment. Attempts to incorporate nonlinear geometrical structures into the tensor computing framework have been reported in Lui and Beveridge (2010); Park and Savvides (2011a; 2011b), but they again need large training data.

A domain adaptive solution for face recognition using the tensor geometry corresponding to models explaining facial variations, with as few as a single image per subject in the source domain, is presented in Ho and Gopalan (2014). Instead of finding linear transformations representing the shift across domains as in Saenko et al. (2010); Kulis et al. (2011), we propose a model-driven approach to construct a latent domain where multifactor facial variations across the source and target domains can be captured together. One main advantage of such an approach is even if data within the source domain and/or the target domain is heterogeneous, for instance when the domain shift is due to blurring and both source and target data contain a mix of sharp and blurred faces, the process of accounting for domain shift remains unaltered unlike other techniques that expect the domains to be more or less homogeneous (Saenko et al., 2010; Kulis et al., 2011; Gopalan et al., 2011). Furthermore, the proposed method overcomes the data requirement constraint for modeling domain variations by synthesizing multiple face images under different illumination, blur and 2D alignment from a single input image on the source or target domain, and uses them to formulate a multidimensional tensor unlike other methods like (Lui and Beveridge, 2010) that places more stringent data-requirement constraints. The tensor obtained from the set of synthesized images can then be represented on a product manifold by performing Higher-Order Singular Value Decomposition (HOSVD) and mapping each orthogonal factored matrix to a point on a Grassmann manifold. The order of the tensors is the number of factors used in the synthesis process. We then recognize the target domain face labels by performing computations pertaining to the tensor geometry for cases where the source domain either contains only one image per subject or has multiple images per subject. This work also addresses the problem of image set matching which is relevant to video-based face recognition where multiple frames in a video provide evidence related to the facial identity.

Manifold-based approaches appear to have fallen out of favor due to superior performance from methods based on Generative Adversarial Networks.

7.3 Unsupervised domain adaptation using dictionaries

Sparse and redundant signal representations have drawn much interest in vision, signal and image processing (Bruckstein et al., 2009; Elad et al., 2010; Rubinstein et al., 2010; Wright et al., 2010; Bo et al., 2011). This is partly because signals and images of interest can be sparsely represented or compressible given an appropriate dictionary. In particular, we say a signal $y \in R^n$ is sparsely represented by a dictionary $D \in R^{n \times K}$ when it can be well approximated by a linear combination of a few columns of D as $y \approx Dx$, where $x \in R^K$ is the sparse representation vector and D is a dictionary that contains a set of basics (atoms) as its columns. Finding a sparse representation vector entails solving the following optimization problem

$$\hat{x} = \arg \min_{x} \|x\|_0 \text{ s.t.} \|y - Dx\|_2 \leq \epsilon \tag{7.1}$$

where ϵ is an error tolerance, $\|x\|_0$ is the ℓ_0-sparsity measure that counts the number of nonzero elements in the vector x, and $\|y - Dx\|_2$ is the mean squared error resulted from the sparse approximation. Solving (7.1) is NP-hard and can be approximated by various methods (Chen et al., 2001; Patil et al. 1993; Tropp, 2004). Instead of using a predetermined dictionary, one can directly learn a dictionary from the data. Indeed, it has been observed that learning a dictionary directly from the training data rather than using a predetermined dictionary (e.g. wavelet) usually leads to a more compact representation and hence can provide better results in many image processing applications such as restoration and classification (Elad et al., 2010; Rubinstein et al., 2010; Wright et al., 2010; Olshausen and Field, 1996; Mairal et al., 2009, 2011).

Several algorithms have been developed for the task of learning a dictionary. Two of the most well-known algorithms are the method of optimal directions (MOD) (Engan et al., 1999) and the K-SVD algorithm (Aharon et al., 2006). Given a set of N signals $Y = [y_1, y_2, \ldots, y_N]$, the goal of KSVD and MOD algorithms is to find a dictionary D and a sparse matrix X that minimize the following representation error

$$(D, X) = \arg \min_{D,X} \|Y - DX\|_F^2 \text{ s.t.} \|x_i\| \leq T_0, \forall i = 1, \ldots, N \tag{7.2}$$

where x_i represents the ith column of X, $\|A\|_F$ denotes the Frobenius norm of A, and T_0 denotes the sparsity level. Both MOD and KSVD are iterative methods that alternate between sparse-coding and dictionary update steps. First, a dictionary D with ℓ_2-normalized columns is initialized. Then, the main iteration is composed of the following two stages:

- Sparse coding: In this step, D is fixed and the following optimization problem is solved to compute the representation vector x_i for each example y_i

$$\min_{x_i} \|y_i - Dx_i\|_2^2 \text{ s.t.} \|x_i\|_0 \leq T_0, \quad \forall i = 1, \ldots, N \tag{7.3}$$

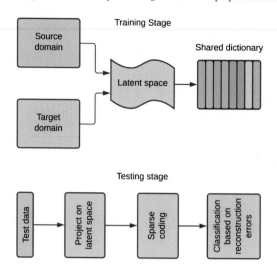

FIGURE 7.1 Overview of generalized domain adaptive dictionary (Shekhar et al., 2013).

- Dictionary update: This is where both MOD and KSVD algorithms differ. The MOD algorithm updates all the atoms simultaneously by solving an optimization problem whose solution is given by $D = YX^+$, where X^+ denotes the Moore-Penrose pseudo-inverse. Even though the MOD algorithm is very effective and usually converges in a few iterations, it suffers from the high complexity of the matrix inversion as discussed in Aharon et al. (2006). In the case of KSVD, the dictionary update is performed atom-by-atom in an efficient way rather than using a matrix inversion. It has been observed that the KSVD algorithm requires fewer iterations to converge than the MOD method.

7.3.1 Generalized domain adaptive dictionary learning

When the target data has a different distribution than the source data, the learned sparse representation may not be optimal. This section investigates if it is possible to optimally represent both source and target by a common dictionary. Specifically, we describe a technique that jointly learns projections of data in the two domains, and a latent dictionary that can succinctly represent both the domains in the projected low-dimensional space as shown in Fig. 7.1. An efficient optimization technique is presented, which can be easily extended to multiple domains. The learned dictionary can then be used for classification. The proposed approach does not require any explicit correspondence between the source and target domains, and shows good results even when there are only a few labels available in the target domain. Various recognition experiments show that the method performs on par or better than competitive methods.

Consider a special case, where we have data from two domains, $Y_1 \in R^{d \times N_1}$ and $Y_2 \in R^{d \times N_2}$. We wish to learn a shared K-atoms dictionary, $D \in R^{n \times K}$ and mappings $P_1 \in R^{n \times d}$, $P_2 \in R^{n \times d}$ onto a common low-dimensional space, which will minimize the representation error in the projected space. Formally, we desire to minimize the following cost function:

$$C_1(D, P_1, P_2, X_1, X_2) = \|P_1Y_1 - DX_1\|_F^2 + \|P_2Y_2 - DX_2\|_F^2 \tag{7.4}$$

subject to sparsity constraints on X_1 and X_2. We further assume that rows of the projection matrices, P_1 and P_2 are orthogonal and normalized to unit-norm. This prevents the solution from becoming degenerate.

Regularization: While bringing the data from two domains to a shared subspace, the transformations should not lose too much information available in the original domains. To facilitate this, we add a PCA-like regularization term that preserves energy in the original signal, given as:

$$C_2(P_1, P_2) = \|Y_1 - P_1^TP_1Y_1\|_F^2 + \|Y_2 - P_2^TP_2Y_2\|_F^2 \tag{7.5}$$

We can rewrite the parameters as follows:

$$\tilde{P} = [P_1, P_2], \tilde{Y} = \begin{pmatrix} Y_1 \\ 0 \end{pmatrix}\begin{pmatrix} 0 \\ Y_2 \end{pmatrix} \quad \text{and} \quad \tilde{X} = [X_1, X_2] \tag{7.6}$$

Using the new notations, the overall optimization can be rewritten as:

$$\{D^*, \tilde{P}^*, \tilde{X}^*\} = \arg \min_{D, \tilde{P}, \tilde{X}} C_1(D, \tilde{P}, \tilde{X}) + \lambda C_2(\tilde{P}) \text{ s.t. } P_iP_i^T = I, i = 1, 2 \text{ and}\|\tilde{x}_j\|_0 \leq T_0, \forall j \tag{7.7}$$

Handling multiple domains: The above formulation can be extended so that it can handle multiple domains. For M domain problem, we simply construct matrices $(\tilde{Y}, \tilde{P}, \tilde{X})$ as:

$$\tilde{P} = [P_1, P_2, \ldots, P_M], \tilde{Y} = \begin{pmatrix} Y_1 & \cdots & 0 \\ \vdots & \ddots & \vdots \\ 0 & \cdots & Y_M \end{pmatrix}, \text{and} \tilde{X} = [X_1, X_2, \ldots, X_M] \tag{7.8}$$

Optimization: We minimize the objective function in Eq. (7.7) by alternating between optimizing $\tilde{P}$ and $(D, \tilde{X})$. Specifically, when $\tilde{P}$ is fixed, the optimization becomes a standard dictionary learning problem where KSVD and MOD algorithms are effective. When $(D, \tilde{X})$ are fixed, we can minimize Eq. (7.7) using manifold optimization techniques (Wen and Yin, 2013). Alternatively, we can derive the optimal form of (P_i, D) and convert the objective function to a simpler form before optimization as done in Shekhar et al. (2013).

Classification: Once the shared dictionary among multiple domains is learned, our approach will carry out the classification using the following procedure. First, we project the test sample to the latent space using the learned transformation P_i, perform sparse coding, then compute the reconstruction error for each domain transformation. The class corresponding to the smallest error will be the final class prediction. The procedure can be written as follows:

$$\hat{x}_{te}^i = \arg \min_x \|P_iy - Dx\|_F^2 \text{ s.t.}\|x\|_0 \leq T_0, \forall i \tag{7.9}$$

$$\text{Output class} = \arg \min_{i=1,\ldots,C} \|P_iy - D\hat{x}_{te}^i\|_2 \tag{7.10}$$

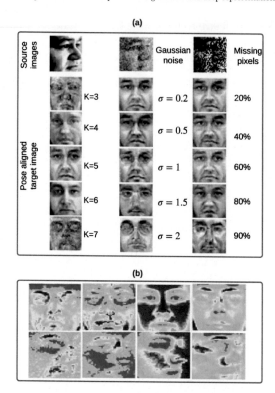

FIGURE 7.2 Pose alignment results (Shekhar et al., 2013) (a) Examples of pose-aligned images using the proposed method. Synthesis in various conditions demonstrates the robustness of the method. (b) First few components of the learned projection matrices for the two poses.

The error term in Eq. (7.10) is called residual error, which is a measure of unfitness of the test sample to a particular class.

Experiments: We use the Multipie dataset (Gross et al., 2010) – a comprehensive face dataset of 337 subjects, having images taken across 15 poses, 20 illuminations, 6 expressions and 4 different sessions. For the purpose of our experiment, we used 129 subjects common to Session 1 and 2. The experiment was done on 5 poses, ranging from frontal to 75°. Frontal faces were taken as the source domain, while different off-frontal poses were taken as target domains. Dictionaries were trained using illuminations {1, 4, 7, 12, 17} from the source and the target poses, in Session 1 per subject. All the illumination images from Session 2, for the target pose, were taken as probe images.

We first consider the problem of pose alignment using the proposed dictionary learning framework. Pose alignment is challenging due to the highly nonlinear changes induced by 3-D rotation of face. Images at the extreme pose of 60° were taken as the target pose. A shared discriminative dictionary was learned using the approach described in this section. Given the probe image, it was projected on the latent subspace and reconstructed using the dictionary. The reconstruction was back-projected onto the source pose domain, to give the aligned image. Fig. 7.2(a) shows the synthesized images for various conditions. We can draw some

TABLE 7.1 Comparison of the proposed method with other algorithms for face recognition across pose (Shekhar et al., 2013).

Method	Probe pose					Average
	15°	30°	45°	60°	75°	
PCA	15.3	5.3	6.5	3.6	2.6	6.7
PLS (Sharma and Jacobs, 2011)	39.3	40.5	41.6	41.1	38.7	40.2
LDA	98.0	94.2	91.7	84.9	79.0	89.5
CCA (Sharma and Jacobs, 2011)	92.1	89.7	88.0	86.1	83.0	83.5
GMLDA (Sharma et al., 2012)	**99.7**	**99.2**	98.6	94.9	95.4	97.6
FDDL (Yang et al., 2011)	96.8	90.6	94.4	91.4	90.5	92.7
SDDL (Shekhar et al., 2013)	98.4	98.2	**98.9**	**99.1**	**98.8**	**98.7**

useful insights about the method from this figure. Firstly, it can be seen that there is an optimal dictionary size, $K = 5$, where the best alignment is achieved. Further, by learning a discriminative dictionary, the identity of the subject is retained. For $K = 7$, the alignment is not good, as the learned dictionary is not able to successfully correlate the two domains when there are more atoms in the dictionary. Dictionary with $K = 3$ has a higher reconstruction error, hence the result is not optimal. We chose $K = 5$ for additional experiments with noisy images. It can be seen from rows 2 and 3 that the proposed method is robust even at high levels of noise and missing pixels. Moreover, denoised and in-painted synthesized images are produced as shown in rows 2 and 3 of Fig. 7.2(a), respectively. This shows the effectiveness of our method. Moreover, the learned projection matrices (Fig. 7.2(b)) show that our method can learn the internal structure of the two domains. As a result, it is able to learn a robust common dictionary.

We also conducted recognition experiment using the set-up described above. Table 7.1 shows that our method compares favorably with other multiview recognition algorithms (Sharma et al., 2012) and gives the best performance on average. The dictionary learning algorithm, FDDL (Yang et al., 2011) is not optimal here as it cannot efficiently represent the nonlinear changes introduced by the pose variation.

7.3.2 Joint hierarchical domain adaptation and feature learning

Complex visual data contain discriminative structures that are difficult to be fully captured by any single feature descriptor. While recent work on domain adaptation focuses on adapting a single hand-crafted feature, it is important to perform adaptation on a hierarchy of features to exploit visual data's richness. This section discusses an approach for domain adaptation based on a sparse and hierarchical network (DASH-N). Our method jointly learns a hierarchy of features together with transformations that rectify the mismatch between different domains. The building block of DASH-N is the latent sparse representation. It employs a dimensionality reduction step to prevent the data dimension from increasing too fast as one traverses deeper into the hierarchy. Experimental results show that this method compares favorably with competing state-of-the-art sparse learning methods. In addition, it is shown that a multilayer DASH-N performs better than a single-layer DASH-N.

Latent sparse representation: From the observation that signals often lie on a low dimensional manifold, previous section performs dictionary learning and sparse coding in a

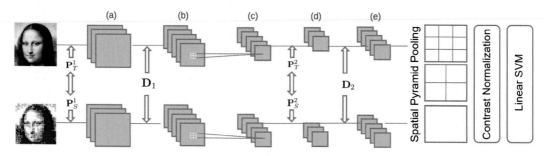

FIGURE 7.3 Overview of DASH-N algorithm (Nguyen et al., 2015). First, images are divided into small overlapping patches. These patches are vectorized while maintaining their spatial arrangements. (a) Performing contrast-normalization and dimensionality reduction using P_S for source images and P_T for target images. The feedbacks between P_S and P_T indicate that these two transformations are learned jointly. (b) Obtaining sparse codes using the common dictionary D_1. (c) Performing max pooling. The process then repeats for layer 2 (d & e), except that the input is the sparse codes from layer 1 instead of pixel intensities. At the final stage, spatial pyramid with max pooling is used to create image descriptors. Classification is done using linear support vector machine at the final layer.

low-dimensional latent space. To facilitate further discussion, we define the latent sparse representation and its corresponding optimization as follows:

$$L(Y, P, D, X) = \|PY - DX\|_F^2 + \alpha \|Y - P^T PY\|_F^2 + \beta \|X\|_1$$
$$\text{s.t. } PP^T = I, \text{and} \|d_i\|_2 = 1, \forall i \in [1, K] \tag{7.11}$$

where $P \in R^{p \times d}$ is a linear transformation that brings the data to a low-dimensional feature space ($p < d$). Note that the dictionary is now in the low-dimensional space $D \in R^{p \times K}$. The first term of the cost function promotes the sparsity of signals in the reduced space. The second term is the amount of energy discarded by the transformation P, or the difference between low-dimensional approximations and the original signals. The minimization of the second term encourages the learned transformation to preserve the useful information present in the original signals. Besides the computational advantage, (Nguyen et al., 2012) shows that this optimization can better recover the underlying sparse representation than traditional dictionary learning methods. This formulation is attractive since it allows the transformation of the data into another domain to handle better different sources of variation such as illumination and geometric articulation.

We propose a method to perform hierarchical domain adaptation jointly with feature learning. Fig. 7.3 shows an overview of the proposed method. The network contains multiple layers, each of which includes 3 sublayers as illustrated in Fig. 7.3. The first sublayer performs contrast-normalization and dimensionality reduction on the input data. Sparse coding is carried out in the second sublayer. In the final sublayer, adjacent features are max-pooled together to produce a new feature. The output from one layer becomes the input to the next layer. For the simplicity of notation, we consider a single source domain. The extension of DASH-N to multiple source domains can be done using the procedure in Eq. (7.8).

Let $Y_S \in R^{d \times N_S}$ and $Y_T \in R^{d \times N_T}$ be the input data at each layer from the source domain and target domain, respectively. Note that there are N_S, d-dimensional samples in the source domain and N_T, d-dimensional samples in the target domain. Note that we assume source and target data have the same dimension d for simplicity of our discussion. However, our formulation can also accommodate the scenario where data dimensions are different across domains. Given Y_S and Y_T, in each layer of DASH-N, we learn a joint latent sparse representation by minimizing the following cost function with respect to (P_S, P_T, D, X_S, X_T):

$$L(Y_S, P_S, D, X_S, \alpha, \beta) + \lambda L(Y_T, P_T, D, X_T, \alpha, \beta) \text{ s.t. } P_S P_S^T = P_T P_T^T = I, \|d_i\|_2 = 1, \forall i \in [1, K] \tag{7.12}$$

where (α, β, λ) are the nonnegative constants, $D \in R^{p \times K}$ is the common dictionary, $P_S \in R^{p \times d}$ and $P_T \in R^{p \times d}$ are the transformations to the latent domain, $X_S \in R^{K \times N_S}$ and $X_T \in R^{K \times N_T}$ are the sparse codes of the source and the target domains, respectively. As can be seen from the above formulation, two domains are forced to share a common dictionary in the latent domain. Together with the sparsity constraint, the common D provides a coupling effect that promotes the discovery of common structures between the two domains. For simplicity, in what follows, we provide a detailed discussion on a two-layer DASH-N network. Extension of DASH-N to multiple layers is straightforward.

Layer 1: We perform dense sampling on each training image to get a set of overlapping patches. These patches are then contrast normalized. If f is a vector corresponding to a patch, then the contrast-normalization can be written as follows:

$$\hat{f} = \frac{f}{\sqrt{\|f\|^2 + \epsilon}} \tag{7.13}$$

where ϵ is a small constant. We set the value of ϵ equal to 0.1 as it is found to work well in our experiments. To make the computation more efficient, only a random subset of patches from each image is used for learning the latent sparse representation. We found that setting this number to 150 for images of maximum size of 150x150 provides a good trade-off between accuracy and computational efficiency. After learning the dictionary D_1 and the transformations (P_S^1, P_T^1) the sparse codes (X_S^1, X_T^1) are computed for all sampled patches by solving the following optimization problem:

$$\min_{X_*^1} \left\| P_*^1 Y_*^1 - D_1 X_*^1 \right\|_2^2 + \beta_1 \left\| X_*^1 \right\|_1 \tag{7.14}$$

where $*$ indicates that the above problem can either correspond to source data or target data. Each column of Y_1 is the vectorized pixel values inside a patch. A fast implementation of the LARS algorithm is used for solving this optimization problem (Mairal et al., 2009). Spatial max pooling is used to aggregate the sparse codes over each 4x4 neighborhood as this pooling method is particularly well-suited for the separation of sparse features (Boureau, 2012; Boureau et al., 2010).

Layer 2: In this layer, we perform similar computations except that the input is the sparse codes from layer 1 instead of image pixels. The features obtained from the previous layer are aggregated by concatenation over each 4x4 neighborhood and contrast-normalized. This results in a new representation that is more robust to occlusion and illumination. Similar to layer 1, we also randomly sample 150 normalized feature vectors $\hat{f}$ from each image for training. ℓ_1 optimization is again employed to compute the sparse codes of the normalized features $\hat{f}$. At the end of layer 2, the sparse codes are then aggregated using max-pooling in a multilevel patch decomposition (i.e. spatial pyramid max pooling). At level 0 of the spatial pyramid, a single feature vector is obtained by performing max-pooling over the whole image. At level 1, the image is divided into four quadrants and max-pooling is applied to each quadrant, yielding 4 feature vectors. Similarly, for level 2, we obtain 9 feature vectors, and so on. In this section, max-pooling using a three-level spatial pyramid is used. As a result, the final feature vector returned by the second layer for each image is a result of concatenating 14 feature vectors from the spatial pyramid.

Optimization: This section describes the procedure for optimization the objective function in Eq. (7.12). First, let us define

$$K_S = Y_S^T Y_S, K_T = Y_T^T Y_T, K = \begin{pmatrix} K_S & 0 \\ 0 & \sqrt{\lambda}K_T \end{pmatrix}, K_S = V_S \Lambda_S V_S^T, K_T = V_T \Lambda_T V_T^T \qquad (7.15)$$

to be the Gram matrix of source data, target data, and their block diagonal concatenation, respectively. It can be shown that the optimal solution of (7.12) takes the following form:

$$D = [A_S^T K_S, \sqrt{\lambda} A_T^T K_T]B, P_S = (Y_S A_S)^T, P_T = (Y_T A_T)^T \qquad (7.16)$$

For some $A_S \in \mathbb{R}^{N_S \times p}$, $A_T \in \mathbb{R}^{N_T \times p}$, and $B \in \mathbb{R}^{(N_S+N_T) \times K}$. We can substitute these forms into the objective function in Eq. (7.12) and optimize it with respect to (A_S, A_T, B) instead. Notice that rows of each transformation live in the column subspace of the data from its own domain. In contrast, columns of the dictionary are jointly created by the data of both source and target. When (B, X) are fixed, we can solve for (A_S, A_T) by first solving the following constrained optimization:

$$\min_{G} \operatorname{tr}(G^T H G) \text{ s.t. } G_S^T G_S = G_T^T G_T = I, G = [G_S, \sqrt{\lambda}G_T] \qquad (7.17)$$

where

$$H = \Lambda^{0.5} V^T K ((I - BX)(I - BX)^T - \alpha I) K V \Lambda^{0.5} \qquad (7.18)$$

Then the solutions of (A_S, A_T) are given by

$$A_S = V_S \Lambda_S^{-0.5} G_S, A_T = V_T \Lambda_T^{-0.5} G_T \qquad (7.19)$$

When (A_S, A_T) are fixed, we can solve for (B, X) using standard sparse coding procedure:

$$\|Z - DX\|_F^2 + \beta(\|X_S\|_1 + \lambda\|X_T\|_1) \text{ s.t. } Z = [A_S K_S, \sqrt{\lambda}A_T K_T], X = [X_S, \sqrt{\lambda}X_T], B = Z^+ D \quad (7.20)$$

Here Z^+ denotes the Moore-Penrose pseudo-inverse of the matrix Z.

Experiments. The proposed algorithm is evaluated in the context of object recognition using a recent domain adaptation dataset (Saenko et al., 2010), containing 31 classes, with the addition of images from the Caltech-256 dataset (Griffin et al., 2007). Domain shifts are caused by variations in factors such as pose, lighting, resolution, etc., between images in different domains. Furthermore, to better assess the ability to adapt to a wide range of domains, experimental results are also reported on new images obtained by performing halftoning and

TABLE 7.2 Recognition rates of different approaches on four domains (C: *Caltech*, A: *Amazon*, D: *DSLR*, W: *Webcam*). 10 common classes are used. Red color denotes the best recognition rates. Blue color denotes the second best recognition rates.

Method	C→ A	C→ D	A→ C	A→ W	W→ C	W→ A	D→ A	D→W
Metric (Saenko et al., 2010)	33.7	35.0	27.3	36.0	21.7	32.3	30.3	55.6
SGF (Gopalan et al., 2011)	40.2	36.6	37.7	37.9	29.2	38.2	39.2	69.5
GFK (PLS+PCA) (Gong et al., 2012)	46.1	55.0	39.6	56.9	32.8	46.2	46.2	80.2
SDDL (Shekhar et al., 2013)	49.5	76.7	27.4	72.0	29.7	49.4	48.9	72.6
HMP (Manjunath and Chellappa, 1993)	67.7	70.2	51.7	70.0	46.8	61.5	64.7	76.0
DASH-N (1ˢᵗlayer) (Nguyen et al., 2015)	60.3	79.6	52.2	74.1	45.3	68.7	65.9	76.3
DASH-N (Nguyen et al., 2015)	71.6	81.4	54.6	75.5	50.2	70.4	68.9	77.1

edge detection algorithms on images from the datasets in Saenko et al. (2010); Griffin et al. (2007).

The recognition results of different algorithms on 8 pairs of source-target domains are shown in Table 7.2. It can be seen that DASH-N outperforms all compared methods in 7 out of 8 pairs of source-target domains. For pairs such as Caltech-Amazon, Webcam-Amazon, or DSLR-Amazon, we achieve more than 20% improvements over the next best algorithm without feature learning used in the comparison (from 49:5% to 71:6%, 49:4% to 70:4%, and 48:9% to 68:9%, respectively). It is worth noting that while we employ a generative approach for learning the feature, our method consistently achieves better performance than (Shekhar et al., 2013), which uses discriminative training together with nonlinear kernels. It is also clear from the table that the multilayer DASH-N outperforms the single-layer DASH-N. In the case of adapting from Caltech to Amazon, the performance gain, based on a combination of features obtained from both layers rather than just features from the first layer, is more than 10% (from 60:3% to 71:6%).

7.3.3 Incremental dictionary learning for unsupervised domain adaptation

This section discusses an incremental dictionary learning method where some target data called supportive samples are selected to assist adaptation, as shown in Fig. 7.4. The incremental nature of this approach enables users to choose different numbers of supportive samples according to their budget, or until when the performance is satisfactory. Supportive samples are close to the source domain and have two properties: first, their predicted class labels are reliable and can be used for building more discriminative classification models; second, they act as a bridge to connect the two domains and reduce the domain mismatch. Theoretical analysis shows that both properties are important for adaptation, enabling the idea of adding supportive samples to the source domain. A stopping criterion is designed to guarantee that the domain mismatch decreases monotonically during adaptation. Experimental results on several widely used visual datasets show that the proposed approach performs better than many state-of-the-art sparse learning methods.

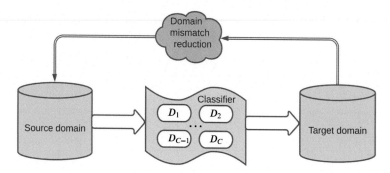

FIGURE 7.4 Overview of incremental dictionary learning for domain adaptation. The source dictionaries are adapted to the target domain using a set of supportive samples. The iterative procedure guarantees that the domain mismatch decreases monotonically.

Given the dictionary $D^{(k)}$, we want to select a subset of target samples as supportive samples. We have two constraints on this selection. First, the supportive samples selected in the previous iterations should be excluded as we want to add new data for adaptation. Second, we select equal number of supportive samples for each class to ensure class balance during adaptation (Gong et al., 2013). With these two constraints, we select the most confident samples that minimize the reconstruction error when represented by $D^{(k)}$. Then we update the augmented source domain by adding supportive samples and retrain the dictionary. After that, the stopping criterion is checked to see whether adding new supportive samples will reduce the domain dissimilarity. An overview of the proposed approach is shown in Fig. 7.4. The algorithm consists of the following main components:

Confidence matrix update: We use $X_S = X^{(0)}$, and X_T to denote the data from source and target domains. Let $L = [1, \ldots, C]$ represent the existing label set. Let $D^{(0)} = [D_1^{(0)} \mid \cdots \mid D_C^{(0)}]$ denote the original dictionary trained on source domain where $D_j^{(0)}$ denote the subdictionary that corresponds to class j. Let $P \in R^{N_t \times C}$ denote a confidence matrix whose elements $p_{ij} \in [0, 1]$ represents the probability that a target sample x_i^t belongs to class j. In the $(k + 1)^{th}$ iteration, we update the confidence matrix $P^{(k+1)}$ using the current class-specific dictionaries $D^{(k)} = [D_1^{(k)} \mid \cdots \mid D_C^{(k)}]$.

$$P_{ij}^{(k+1)} = \begin{cases} \dfrac{2^{-0.5}\, \sigma\, \exp\left(-e_{ij}^{(k+1)}/2\sigma^2\right)}{\sum_{l=1}^{C} 2^{-0.5}\, \sigma\, \exp\left(-e_{il}^{(k+1)}/2\sigma^2\right)} & \text{if } j = \text{argmax}_l\, p_{il}^{(k+1)} \\ 0 & \text{otherwise} \end{cases} \tag{7.21}$$

where σ^2 is the normalization parameter and e_{ij} denotes the reconstruction error of target sample x_i^t using $D_j^{(k)}$:

$$e_{ij}^{(k+1)} = \left\| x_i^t - D_j^{(k)} \cdot Z_{ij}^{(k+1)} \right\|_2^2 \tag{7.22}$$

Here, $Z_{ij}^{(k+1)}$ is the sparse code. We put a constraint $p_{ij}^{(k+1)} \neq 0$ only when j is the most likely class that sample i belongs to. This constraint guarantees that a sample cannot be selected as a supportive sample for multiple classes.

Supportive sample selection: We select new supportive samples using $W^{(k+1)}$ by solving the following optimization:

$$W_j^{(k+1)} = \arg\max_{W_j} \; \mathrm{tr}\big(W_j P_j^{(k+1)}\big) \tag{7.23}$$

$$\text{s.t. } W_j^{(k+1)} \cdot \sum_{l=1}^{k} W_j^{(l)} = 0, \; \big\|W_j^{(k+1)}\big\|_0 = Q, \quad j = 1, \ldots, C \tag{7.24}$$

where $W_j \in R^{N_t \times N_t}$ are diagonal matrices containing j^{th}-column of W on the diagonal. That is, $W_j = \mathrm{diag}([w_{1j}, w_{2j}, \ldots])$. Similarly, $P_j = \mathrm{diag}([p_{1j}, p_{2j}, \ldots])$. Q is the number of supportive samples for each class. The objective function Eq. (7.23) maximizes the confidence of the selected supportive samples. The first constraint requires that the supportive samples in the $(k+1)^{th}$ iteration are disjoint from the previously chosen ones which ensures that we keep adding new supportive samples to the source domain. The second constraint ensures that the number of supportive samples for each class is balanced. The solution to Eq. (7.23) is to find the corresponding Q supportive samples that maximize the confidence with the constraint that old supportive samples are excluded.

Augmented source domain update: After selecting the supportive samples, we update the augmented source data by adding weighted supportive samples to the previous source data:

$$X_j^{(k+1)} = \big[X_j^{(k)} \mid X^t W_j^{(k+1)} P_j^{(k+1)}\big], \quad k = 1, \ldots, C \tag{7.25}$$

Since the supportive samples' labels may have errors, each selected supportive sample is weighted by its confidence. The weights indicate the reliability of the supportive samples' labels and highly confident supportive samples will contribute more to the model.

Dictionary update: The dictionary is updated by solving the following optimization problem:

$$D_j^{(k+1)} = \arg\min_{D_j, Z_j} \big\|X_j^{(k+1)} - D_j Z_j\big\|_F^2 + \lambda \|Z_j\|_1, \quad j = 1, \ldots, C \tag{7.26}$$

We solve Eq. (7.26) using the online dictionary learning method (Mairal et al., 2009). The dictionary obtained in the previous iteration is used as the initial dictionary in the next iteration. In this way, the computational cost is relatively low.

Stopping criterion: One trivial stopping criterion is to stop when there are no new supportive samples for one of the classes. But our goal is to guarantee that the adaptation process monotonically reduces the domain divergence. In this way, the classification error bound in the target domain will decrease as stated in Ben-David et al. (2010); Smetana et al. (2009). So, we design in the next section a domain similarity measure and we only perform adaptation when the domain similarity increases after each iteration. It can be shown that when adding supportive samples to the source domain, the similarity between the source and target domains will increase. Readers are referred to Lu et al. (2015) for more details on proving this property.

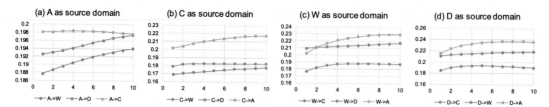

FIGURE 7.5 The change in domain similarity when the supportive samples are added to the source domain. In our experiments, we only continue our adaptation as long as the similarity value goes up, which is represented by the solid lines before the slash symbols. A: Amazon, C: Caltech, W:Webcam, D: Dslr.

Experiments. We use Office+Caltech dataset containing images from four domains: Amazon (A), Webcam (W), DSLR (D), and Caltech (C). This leads to a total of 12 domain pairs for testing. 10 common classes are selected in all domains. For each class, A, C, D and W have about 100, 100, 15 and 30 images, respectively. We follow the protocol used in Gong et al. (2013) to generate the source and target domain data. DeCAF features (Donahue et al., 2014) are used in our experiment. We compare two nonadaptation (NA) methods, and five state-of-the-art unsupervised DA methods: SVM and Dictionary Learning Based Classification (DLC) are the two NA methods, Subspace Interpolation via Dictionary Learning (SIDL) (Ni et al., 2013), Geodesic Flow Kernel (GFK) (Gong et al., 2012), Transfer Joint Matching (TJM) (Long et al., 2014), Landmarks (Gong et al., 2012) and DA-NBNN (Tommasi and Caputo, 2013) are the unsupervised DA methods. DLC is implemented using the online dictionary learning method as Mairal et al. (2009) and is also used as the initial dictionary in the proposed approach.

We also compare in change of domain similarity in Fig. 7.5 with classification results in Table 7.3 and find that the accuracy likely increases when the domain similarity value continues to go up as more supportive samples are added to the source domain.

We showed that dictionary learning is an effective approach in addressing domain shifts under unsupervised setting. The general idea is to project data representations from multiple domains to the same latent space where their distributions are more similar. Instead of doing this transformation on one semantic level, we also showed the benefits of hierarchical dictionary learning for gradual adaptation on multiple semantic levels. Finally, we demonstrate the benefits of incrementally adding selective support samples to the source domain are guaranteed to increase the domain similarity, which generally leads to better classification performance.

Dictionary-based approaches appear to have fallen out of favor due to superior performance from methods based on Generative Adversarial Networks.

7.4 Unsupervised domain adaptation using deep networks

Deep neural networks are a powerful class of machine learning models for extracting meaningful representations from images. Despite their success in achieving state-of-the-art performance in several visual recognition tasks (Ren et al., 2015; He et al., 2016; 2017), neural networks suffer from domain shift i.e., the performance of neural networks drops signifi-

TABLE 7.3 Recognition accuracies on 12 pairs of cross-domain unsupervised object recognition. A: Amazon, C: Caltech, W: Webcam, D: Dslr.

	Method	A→C	A→D	A→W	C→A	C→D	C→W	W→A	W→D	W→C	D→A	D→C	D→W
NA	SVM	85.0	87.9	79.0	91.4	89.8	80.0	75.7	99.4	72.0	87.1	78.8	98.6
	DLC	85.3	82.1	75.6	91.3	87.9	78.6	78.4	98.7	76.0	88.1	81.6	99.3
	GFK (Gong et al., 2012)	77.3	84.7	81.0	88.5	86.0	80.3	81.8	100	73.9	85.8	76.0	97.3
	SIDL (Ni et al., 2013)	84.5	81.5	74.2	90.9	89.8	78.3	75.1	100	71.1	87.9	80.1	99.3
	TJM (Long et al., 2014)	80.1	84.7	75.2	89.0	85.3	76.9	84.8	100	78.0	87.4	77.4	98.6
	DA-NBNN (Tommasi and Caputo, 2013)	83.4	80.9	76.6	89.6	87.9	80.3	88.0	100	82.4	91.3	86.1	98.0
DA	Landmarks (Gong et al., 2013)	84.7	86.0	82.4	92.4	92.3	84.1	84.0	98.7	71.7	77.0	74.4	95.2
	Online dictionary (Lu et al., 2015)	86.7	92.4	88.5	93.3	88.5	95.6	92.8	100	88.7	93.1	89.1	99.3

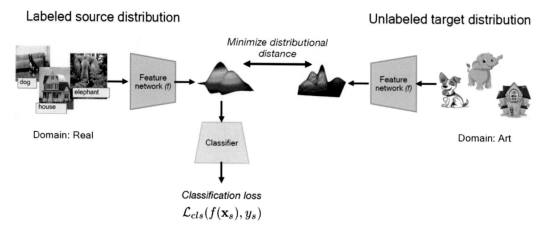

FIGURE 7.6 Domain adaptation using deep networks. Source and target feature spaces are aligned using a distributional distance minimization objective.

cantly when the test distribution is different from the training distribution. To alleviate the issue of domain shift, additional loss functions are used in the training objective to prevent the source and target feature spaces from drifting.

The framework for deep domain adaptation is shown in Fig. 7.6. The source and the target feature distributions are first obtained by passing the corresponding images through a feature network F. Their feature distributions look dissimilar due to the domain shift. During domain adaptation, distributional distance between these feature spaces is minimized, while simultaneously training a classifier model on the labeled source data.

Several discriminative and generative approaches have been proposed for the domain adaptation problem (Ganin et al., 2016; Long et al., 2016; 2017, Hoffman et al., 2018; Sankaranarayanan et al., 2018b). In discriminative approaches, an additional loss function is typically used along with the classification objective to prevent the feature space drift. These functions typically take the form of a distributional distance minimization between source and target feature spaces (Ganin et al., 2016; Long et al., 2016; 2017). In generative approaches, a generative model (typically a Generative Adversarial Network) is trained to model the distribution of source and target images. This knowledge about the learnt distributions is then used to induce domain invariance during training (Hoffman et al., 2018; Sankaranarayanan et al., 2018b).

7.4.1 Discriminative approaches for domain adaptation

Let F denote a deep neural network for extracting feature representations from images. The network F is typically a convolutional network that acts on input images and returns a vector as output. Let C denote a classification network that takes the feature representation as input and returns the classification logits. Let (x^s, y^s) denote input-output pairs of source domain and (x^t) denote the target inputs.

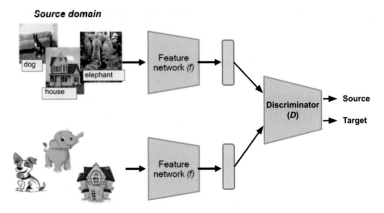

FIGURE 7.7 Domain adversarial training.

To train a classifier model in the source domain, we minimize a cross-entropy loss given by

$$L_{cls} = E\left[-y^s \log\left(C\left(F\left(x^s\right)\right)\right)\right] \qquad (7.27)$$

The resulting classifier would perform poorly on target domain due to the domain shift. The feature space drift is minimized using a domain adversarial loss that measures how different the source and target feature representations are. To implement the domain adversarial loss, we utilize an additional network called a discriminator (D) as shown in Fig. 7.7 The discriminator network takes as input the feature representations obtained from the F network and predicts if the features come from the source or the target domain. If there is sufficient mismatch between source and target features, the discriminator will obtain low loss value. The feature network is then trained adversarially to maximize the discriminator's loss. The training will converge when the discriminator fails at its task i.e., when both source and target feature distributions are indistinguishable.

To accomplish the above objective, the discriminator network maximizes the following loss function:

$$L_{disc} = \log\left[D\left(F\left(x^s\right)\right)\right] + \log\left[1 - D\left(F\left(x^t\right)\right)\right] \qquad (7.28)$$

During training, the models are optimized using a combination of classification loss and domain adversarial loss:

$$\min_{F, C} \max_{D} \left[L_{cls} + \lambda L_{disc}\right] \qquad (7.29)$$

The term λ controls the weight given to the domain adversarial term in the objective. It is a hyper-parameter that needs to be tuned while training the algorithm. The resulting algorithm is called *Domain Adversarial Training* (Ganin et al., 2016).

TABLE 7.4   Domain adaptation performance (in %) on Office-31 dataset. A: Amazon, D: DSLR, W: Webcam.

Method	$A \rightarrow W$	$D \rightarrow W$	$W \rightarrow D$
Source only	64.2	96.1	97.8
DDC (Tzeng et al., 2014)	61.8	95.0	98.5
DAN (Long et al., 2015)	68.5	96.0	99.0
Domain Adversarial (Ganin et al., 2016)	73.0	96.4	99.2

The networks F, C and D are typically implemented as neural networks. In particular, the F network is a deep convolutional network (such as Resnet), and C and D networks are multilayer perceptrons. In Ganin et al. (2016), the adversarial loss is implemented using a gradient reversal layer in which the gradients coming from the discriminator network with L_{disc} loss is modulated with a $-\lambda$ factor before updating the feature network. Note that the feature network is also updated using the classification loss. All networks are optimized using stochastic gradient descent.

The performance of domain adversarial adaptation in comparison with other deep adaptation methods are shown in Table 7.4 for the Office-31 dataset, which has three domains – Amazon (A), DSLR (D) and Webcam (W). The task is to perform classification of 31 object categories. In Table 7.4, we observe that domain adversarial adaptation achieves significant gains in performance compared to source only baseline, i.e., baseline models which are trained only on source domain without any adaptation. Additionally, the model also performs better than Deep Domain Confusion (Tzeng et al., 2014) and Deep Adaptation Networks (Long et al., 2015), two other discriminative adaptation techniques.

Other distance measures: While domain adversarial training is a popular choice for discriminative adaptation, other distance measures can also be used for measuring distributional mismatch between the source and the target distributions. Two such distance measures are Maximum Mean Discrepancy (MMD) and Wasserstein distance. In MMD, distributional distance is computed as distances between mean embeddings represented using a kernel from a Reproducing Kernel Hilbert Space (RKHS). Adaptation is then performed by minimizing the MMD between source and target feature distributions (Long et al., 2015; 2016).

In Wasserstein-based adaptation, Wasserstein distance between source and target feature distributions are used as a choice of distance measure. Dual form of Wasserstein distance is computed using a discriminator function. Adaptation is performed by minimizing the dual of Wasserstein distance between source and target feature maps (Shen et al., 2018). These approaches perform on-par with adversarial adaptation.

Pseudo-labeling approaches: Different from adversarial approaches, in pseudo-labeling / self-training (Saito et al., 2017a; Zou et al., 2018; 2019), pseudo-labels are generated on the unlabeled target samples using the current estimate of the model. Since pseudo-labels are typically noisy, most confident pseudo-labels are picked using some measures of confidence scores. One such measure is consistency across an ensemble of classifier models (Saito et al., 2017a). The mined pseudo-labels are then used to retrain the model to improve the predictive power on the target domain. This process is repeated iteratively until convergence. These techniques can also be used in combination with adversarial adaptation.

Regularization-based approaches: In regularization-based approaches, networks are trained with an additional regularization term along with the cross-entropy loss for the source domain. One common regularization function is entropy minimization (Vu et al., 2019) in which entropy of the target logits are minimized. This enforces target samples to produce high confidence predictions. Other forms of regularization include adversarial dropout regularization (Saito et al., 2017b) and maximum classifier discrepancy (Saito et al., 2018) in which feature network is trained to minimize the discrepancy in logits arising from different classifier models that are trained in combination.

Extension to multisource daptation: In multisource adaptation, the objective is to adapt from multiple source distributions to the target distribution. Multisource adaptation is a very useful setting in practice as real-world datasets typically contain a mixture of multiple latent domains. The domain adversarial training objective can be extended to the multisource setting using a k-way domain classifier as done in Xu et al. (2018). In Yang et al. (2020), a curriculum-based weighting mechanism is used for selecting the best source domain samples to adapt for the given target domain. In cases where domain labels are available, latent domain discovery can be used for mining the unknown domain labels for multiway adversarial adaptation (Mancini et al., 2018).

7.4.2 Generative approaches for domain adaptation

In generative approaches, the objective is to use generative models to estimate source and target distributions. The learnt generative models are then used in the adaptation process to learn domain-invariant representations. Deep neural networks can be a popular choice for learning the generative models due to their high expressive power. Three popular choices of deep generative models are Generative Adversarial Networks (GANs) (Goodfellow et al., 2014), Variational Autoencoders (VAEs) (Kingma and Welling, 2013) and normalizing flows (Papamakarios et al., 2019). GANs have been extensively used for domain adaptation as they have been very successful in generating high fidelity samples.

Generative adversarial networks

Let $\{x_i\}_{i=1:N}$ be samples corresponding to the input distribution. The objective of GANs (Goodfellow et al., 2014) is to train a model that enables generating samples that resemble the input distribution. To do this, GANs train a model G that maps samples from latent space to the space of input images. The latent space is typically modeled using a known tractable distribution such as multivariate isotropic Gaussian. Once the model G is trained, we can synthesize samples by sampling from the latent distribution and passing it through the generative model G.

To train the generator G, we utilize a second model called the discriminator (D) or the critic network. The objective of the discriminator is to discriminate if the samples come from real or generated distribution. This is posed as a binary classification problem, with real samples treated as one class while the generated samples are treated as the second class. The generator model G is then trained to make the discriminator fail at this task. Both models G and D are implemented using deep neural networks as shown in Fig. 7.8. The objective of GANs can

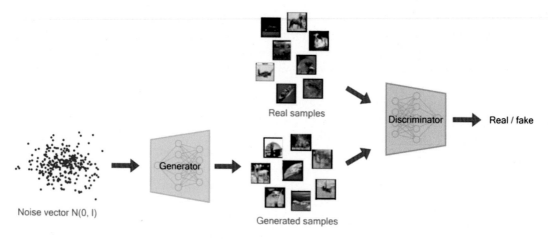

FIGURE 7.8 GAN framework.

then be written as

$$\min_G \max_D \left[E_{x \sim p_{data}} \log D(x) + E_{z \sim p_z} \log \left(1 - D\big(G(z)\big)\right) \right] \qquad (7.30)$$

The term inside the parathesis is the negative of the binary classification loss for classifying samples as being drawn from the real distribution p_{data} or the generated distribution $G(p_z)$. While the discriminator D maximizes the objective, the generator minimizes it leading to a 2-player minimax game. The overall objective is to find the saddle point of this 2-player game. At convergence, the generator network synthesizes realistic samples and discriminator are maximally confused between the real and the generated distribution.

Training GANs are extremely challenging due to the min-max nature of the training objective. Several stabilization tricks are used in practice to make the GANs converge to good solutions (Liu et al., 2020). Some of these tricks include use of Wasserstein (Arjovsky et al., 2017) or hinge-loss based objective (Miyato et al., 2018), using spectral normalization in network architecture (Miyato et al., 2018), regularization techniques such as weight decay (Liu et al., 2020), gradient penalty (Gulrajani et al., 2017) and feature matching losses (Liu et al., 2020).

Conditional image synthesis: In the previous section, we focused on unconditional image synthesis where the objective was simply to generate images that resemble the input distribution. In conditional image synthesis, we are interested in generating samples conditioned on some variables of interest. One example is class-conditional synthesis where we are required to generate samples belonging to one specific class. When the conditioning variable is itself an image, the task is called image-to-image translation. Here, we are interested in transforming an image belonging to one domain to a different one. In conditional image synthesis, the conditioning variable is used as input in addition to the latent vector.

In domain adaptation, image-to-image translation models are a popular choice of generative models. Since, we are interested in unsupervised domain adaptation, we use unpaired image-to-image translation models in which source and target images do not have any corre-

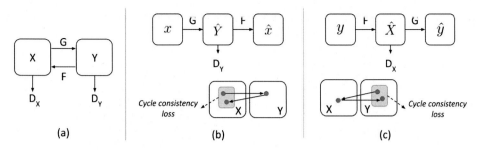

FIGURE 7.9 (a) CycleGAN framework, (b) Adversarial and cycle consistency loss for the forward model, (c) Adversarial and cycle-consistency losses for the reverse model.

spondences. The idea is to translate source domain images to look like target, and then train a classifier on the translated source domain samples using source labels as ground truth. The trained classifier can then be used for making test-time predictions on the target domain.

CycleGAN: CycleGAN is a popular model for unpaired image-to-image translation. Let X and Y denote two domains between which we need to train the image translation models. In CycleGAN, we use two models – a forward model G which translates images from a domain X to domain Y, and a reverse model F which translates images from domain Y to domain X. A discriminator network D_X is used to obtain adversarial loss to ensure that samples produced by F are indistinguishable from real distribution X. Similarly, a second discriminator D_Y encourages samples produced by G to be indistinguishable from the domain Y. Fig. 7.9 shows an overview of CycleGAN framework.

While the discriminator losses encourage realism in generated samples, there is no term in the objective to enforce content preservation. That is, there is no way to prevent a sample from one domain (e.g. Cat) to map to a different semantic class in the other domain (e.g. Dog). To prevent this, a cycle-consistent term is used in the objective. The idea is to ensure that applying both forward and reverse maps to the same sample produces back the input, i.e., $F(G(x)) \approx x$ and $G(F(y)) \approx y$. L1 loss is typically used for cycle consistency.

$$L_{cyc}(G, F) = E_{x \sim p_{data}(x)}\big[\| F(G(x)) - x \|\big] + E_{x \sim p_{data}(y)}\big[\| G(F(y)) - y \|\big] \qquad (7.31)$$

The CycleGAN model is trained using a combination of adversarial losses and cycle-consistency losses.

$$L = L_{adv1} + L_{adv2} + \lambda \, L_{cyc} \qquad (7.32)$$

In Fig. 7.10, we depict the results of training CycleGAN on the following datasets: paintings, Horse to zebra, winter to summer and aerial photos to google maps. We observe that images are translated from one domain to another while preserving the content. The reconstruction F(G(x)) well approximates the input sample x, which shows the importance of cycle consistency term in preserving the content. In addition, the generated samples are realistic, which shows the effectiveness of the adversarial losses.

CycleGAN based domain adaptation: In the previous section, we discussed the Cycle-GAN model and how it can be used for unpaired image-to-image translation. Now, we show how these models can be used for unsupervised domain adaptation (Hoffman et al., 2018).

Input x　　　　Output G(x)　　　Reconstruction
F(G(x))

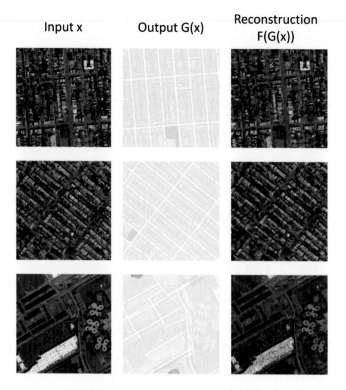

FIGURE 7.10 CycleGAN results. Input images are shown in the left panel, while the translation of the forward model and back-translation is shown in the middle and right panel respectively.

First, a CycleGAN model is trained to translate between the source and the target domains. Then, a task model (classifier / segmentation model) is trained on the translated source images using source labels as ground truth. For the task model to train well, an additional discriminator loss is used on the feature maps of translated source images and true target images. This ensures that the feature distribution of translated source images and target images is aligned. The training framework of this approach, also known as CyCADA (Hoffman et al., 2018), is shown in Fig. 7.11.

Distributional distance minimization with generative models: An alternative approach to generative domain adaptation is by utilizing GANs for guiding distributional distance minimization. Recall from Section 7.4.1 that domain adaptation is posed as a distributional distance minimization problem between source and target feature distributions. Instead of directly performing distance minimization in the feature space, the idea is to project the feature back into pixel-space using GANs and perform distributional distance minimization in this projected image space (Fig. 7.12). The authors of Sankaranarayanan et al. (2018b) called this approach "Generate to Adapt".

Projecting features back into image space has two main benefits: First, the capacity of the discriminator network is effectively increased due to the projection step. Second, the projection step helps preserve semantic content in the generated features. That is, it prevents target

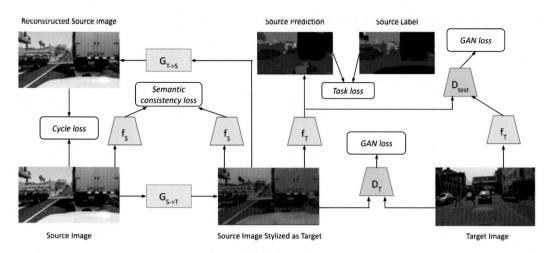

FIGURE 7.11 Framework of CyCADA.

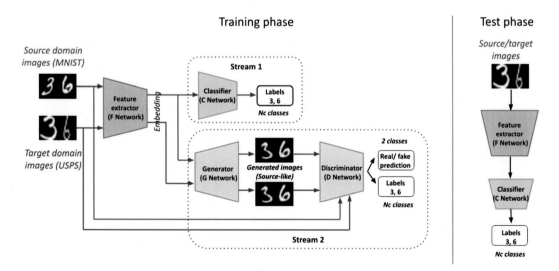

FIGURE 7.12 Framework of Generate to Adapt.

TABLE 7.5 Domain adaptation performance using generative approaches on Digits dataset. Classification accuracy in % is reported. MN – MNIST, US – USPS, SV – SVHN.

Method	$MS \rightarrow US$	$US \rightarrow MN$	$SV \rightarrow MN$
Source only	79.1	57.1	60.3
CoGAN (Liu and Tuzel, 2016)	61.8	95.0	98.5
PixelDA (Bousmalis et al., 2017)	73.0	96.4	99.2
CyCADA (Hoffman et al., 2018)	95.6	96.5	90.4
Generate to Adapt (Sankaranarayanan et al., 2018b)	92.8	95.3	92.4

TABLE 7.6 Semantic segmentation performance on GTA-5 - > Cityscapes adaptation. IoU scores for each category and mean IoU scores (mIoU) are reported.

Method	Road	Side-walk	Bldg.	Wall	Fence	Pole	T. Light	T. Sign	Veg	Ter-rain	Sky	Per-son	Rider	Car	Truck	Bus	Train	M. Bike	Bike	mIoU
Source only	73.5	21.3	72.3	18.9	14.3	12.5	15.1	5.3	77.2	17.4	64.3	43.7	12.8	75.4	24.8	7.8	0.0	4.9	1.8	29.6
Cy-CADA	79.1	33.1	77.9	23.4	17.3	32.1	33.3	31.8	81.5	26.7	69.0	62.8	14.7	74.5	20.9	25.6	6.9	18.8	20.4	39.5
Generate to Adapt	88.0	30.5	78.6	25.2	23.5	16.7	23.5	11.6	78.7	27.2	71.9	51.3	19.5	80.4	19.8	18.3	0.9	20.8	18.4	37.1
Target only	96.5	74.6	86.1	37.1	33.2	30.2	39.7	51.6	87.3	52.6	90.4	60.1	31.7	88.4	54.9	52.3	34.7	33.6	59.1	57.6

TABLE 7.7 A comparison of manifold, dictionary, and deep learning approaches for unsupervised domain adaptation.

Source	Target	Manifolds (2012)	Dictionaries (2015)	Deep features (2017)	Deep features and GANs (2018)
Webcam	Dslr	71.2		99.5	99.8
Dslr	Webcam	68.8	72	98.2	97.9
Amazon	Webcam	55.6	72	62.4	86.5
Amazon	Dslr			64	87.7
Dslr	Amazon		48.9	52	72.8
Webcam	Amazon		49.4	48.4	71.4

features from one class to be mapped to a different class as the reconstruction network provides supervisory signal in preserving the semantic content.

The framework of Generate to Adapt is shown in Fig. 7.12. Features of source and target images are extracted using a feature network F. The obtained features are then passed through two streams: The first stream is a classification branch that is trained using cross-entropy loss on the source domain. The second steam is the distance minimization branch. In this stream, the source and the target features are first inverted back into image space using a generator network G. The generated samples are then passed through a discriminator network that discriminates if the samples come from real (source) or fake (target) domain. In addition, it also performs class label prediction on the source reconstructions using source labels as ground truth. This helps obtain class consistency in the reconstructed samples.

The feature network F is then trained adversarially to make the target reconstructions look like source. This happens only when the source and target feature distributions overlap. Additionally, the signal obtained from stream 1 helps features obtain class consistent predictions.

Results: In Table 7.5, we show the results of generative approaches on cross-domain classification tasks using digits dataset. Three datasets are used for this purpose: MNIST, USPS and SVHN. In each of the adaptation setting, we observe that source only baseline achieves low performance. Both CyCADA and Generate to Adapt achieve significant performance gains compared to the baseline model. In addition, they also outperform CoGAN and PixelDA, two other GAN-based adaptation approaches.

In Table 7.6, we report the performance on cross-domain semantic segmentation task. The source domain is GTA-5, which is a synthetic dataset of street scenes, while the target domain is the real dataset – Cityscapes. We use VGG-based FCN architecture for the feature network (Sankaranarayanan et al., 2018b). Intersection of Union (IoU) scores for each semantic class along with the mean IoU (mIoU) scores is reported. We observe that both CyCADA and Generate to Adapt achieve significant IoU gains compared to the source only baseline model. However, there is a huge gap compared to the target-only model, which is the oracle model trained using true target labels.

Table 7.7 given below provides a comparison among manifold, dictionary, and GAN-based approaches for unsupervised domain adaptation on the Office dataset. GAN-based methods are providing the best performance.

7.5 Summary

In this chapter, we discussed approaches based on differential geometry, sparse representation and deep neural networks for domain adaptation. Two broad classes of techniques were presented: discriminative and generative approaches. In discriminative approaches, we train a classifier model while employing additional losses to make the source and target feature distributions similar. A distributional distance minimization objective is used for this task. In generative approaches, we utilize a generative model to perform domain adaptation. One approach is to train intermediate dictionaries and a cross-domain GAN for mapping samples from source domain to target and training a classifier model on the transformed target images. The second approach takes a distributional distance minimization formulation and uses GANs to guide the distributional distance minimization. All these approaches are validated on cross-domain recognition and semantic segmentation tasks.

References

Aharon, M., Elad, M., Bruckstein, A., 2006. K-SVD: an algorithm for designing overcomplete dictionaries for sparse representation. IEEE Transactions on Signal Processing 54 (11), 4311–4322.

Arjovsky, M., Chintala, S., Bottou, L., 2017. Wasserstein generative adversarial networks. In: International Conference on Machine Learning, pp. 214–223.

Ben-David, S., Blitzer, J., Crammer, K., Kulesza, A., Pereira, F., Vaughan, J.W., 2010. A theory of learning from different domains. Machine Learning 79 (1), 151–175.

Bo, L., Xiaofeng, R., Dieter, F., 2011. Hierarchical matching pursuit for image classification: Architecture and fast algorithms. In: Advances in Neural Information Processing Systems, vol. 24.

Boureau, Y.L., Ponce, J., LeCun, Y., 2010. A theoretical analysis of feature pooling in visual recognition. In: Proceedings of the 27th International Conference on Machine Learning (ICML-10), pp. 111–118.

Bousmalis, K., Silberman, N., Dohan, D., Erhan, D., Krishnan, D., 2017. Unsupervised pixel-level domain adaptation with generative adversarial networks. In: Proceedings of the IEEE Conference on Computer Vision and Pattern Recognition, pp. 3722–3731.

Blitzer, J., Crammer, K., Kulesza, A., Pereira, F., Wortman, J., 2008. Learning bounds for domain adaptation.

Blitzer, J., Kakade, S., Foster, D., 2011. Domain adaptation with coupled subspaces. In: Proceedings of the Fourteenth International Conference on Artificial Intelligence and Statistics, pp. 173–181. JMLR Workshop and Conference Proceedings.

Boureau, Y.L., 2012. Learning hierarchical feature extractors for image recognition. Doctoral dissertation. New York University.

Bruckstein, A.M., Donoho, D.L., Elad, M., 2009. From sparse solutions of systems of equations to sparse modeling of signals and images. SIAM Review 51 (1), 34–81.

Chen, S.S., Donoho, D.L., Saunders, M.A., 2001. Atomic decomposition by basis pursuit. SIAM Review 43 (1), 129–159.

Daume III, H., Marcu, D., 2006. Domain adaptation for statistical classifiers. Journal of Artificial Intelligence Research 26, 101–126.

Donahue, J., Jia, Y., Vinyals, O., Hoffman, J., Zhang, N., Tzeng, E., Decaf, Darrell T., 2014. A deep convolutional activation feature for generic visual recognition. In: International Conference on Machine Learning, pp. 647–655. PMLR.

Duan, L., Tsang, I.W., Xu, D., Chua, T.S., 2009. Domain adaptation from multiple sources via auxiliary classifiers. In: Proceedings of the 26th Annual International Conference on Machine Learning, pp. 289–296.

Duan, L., Xu, D., Exploiting Web, Chang SF., 2012. Images for event recognition in consumer videos: a multiple source domain adaptation approach. In: Proceedings of the IEEE Conference on Computer Vision and Pattern Recognition. Providence, RI, pp. 1338–1345.

Elad, M., Figueiredo, M.A., Ma, Y., 2010. On the role of sparse and redundant representations in image processing. Proceedings of the IEEE 98 (6), 972–982.

Engan, K., Aase, S.O., Husoy, J.H., 1999. Method of optimal directions for frame design. In: 1999 IEEE International Conference on Acoustics, Speech, and Signal Processing. In: Proceedings ICASSP99, vol. 5. IEEE, pp. 2443–2446. (Cat. No. 99CH36258).

Ganin, Y., Ustinova, E., Ajakan, H., Germain, P., Larochelle, H., Laviolette, F., Marchand, M., Lempitsky, V., 2016. Domain-adversarial training of neural networks. Journal of Machine Learning Research 17 (1), 2096–2130.

Gong, B., Grauman, K., Sha, F., 2013. Connecting the dots with landmarks: discriminatively learning domain-invariant features for unsupervised domain adaptation. In: International Conference on Machine Learning, pp. 222–230. PMLR.

Gong, B., Shi, Y., Sha, F., Grauman, K., 2012. Geodesic Flow Kernel for Unsupervised Domain Adaptation. Proceedings of the IEEE Conference on Computer Vision and Pattern Recognition, vol. 16. IEEE, pp. 2066–2073.

Goodfellow, I.J., Pouget-Abadie, J., Mirza, M., Xu, B., Warde-Farley, D., Ozair, S., Courville, A., Bengio, Y., 2014. Generative adversarial networks. arXiv preprint. arXiv:1406.2661.

Gopalan, R., Li, R., Chellappa, R., 2011. Domain adaptation for object recognition: an unsupervised approach. In: 2011 International Conference on Computer Vision. IEEE, pp. 999–1006.

Gopalan, R., Li, R., Chellappa, R., 2014. Unsupervised adaptation across domain shifts by generating intermediate data representations. IEEE Transactions on Pattern Analysis and Machine Intelligence 36, 2288–2302.

Griffin, G., Holub, A., Perona, P., 2007. Caltech-256 object category dataset.

Gross, R., Matthews, I., Cohn, J., Kanade, T., Baker, S., 2010. Multi-pie. Image and Vision Computing 28 (5), 807–813.

Gulrajani, I., Ahmed, F., Arjovsky, M., Dumoulin, V., Courville, A., 2017. ImprovedTraining of Wasserstein gans. arXiv preprint. arXiv:1704.00028.

He, K., Zhang, X., Ren, S., Sun, J., 2016. DeepResidual learning for image recognition. In: Proceedings of the IEEE Conference on Computer Vision and Pattern Recognition, pp. 770–778.

He, K., Gkioxari, G., Dollár, P., Girshick, R., 2017. Mask R-CNN. In: Proceedings of the IEEE International Conference on Computer Vision, pp. 2961–2969.

Ho, H.T., Gopalan, R., 2014. Model-driven domain adaptation on product manifolds for unconstrained face recognition. International Journal of Computer Vision 109, 110–125.

Hoffman, J., Tzeng, E., Park, T., Zhu, J.Y., Isola, P., Saenko, K., Efros, A., Darrell, T., 2018. Cycada: cycle-consistent adversarial domain adaptation. In: International Conference on Machine Learning, pp. 1989–1998.

Jhuo, I.H., Liu, D., Lee, D.T., Chang, S.F., 2012. RobustVisual domain adaptation with low-rank reconstruction. In: Proceedings of the IEEE Conference on Computer Vision and Pattern Recognition. Providence, RI, pp. 2168–2175.

Kingma, D.P., Welling, M., 2013. Auto-encoding variational Bayes. arXiv preprint. arXiv:1312.6114.

Kulis, Brian, Saenko, Kate, Darrell, Trevor, 2011. What you saw is not what you get: Domain adaptation using asymmetric kernel transforms. In: CVPR 2011. IEEE, pp. 1785–1792.

Lee, J., Moghaddam, B., Pfister, H., Machiraju, R., 2005. A bilinear illumination model for robust face recognition. In: Tenth IEEE International Conference on Computer Vision (ICCV'05), vol. 2. IEEE, pp. 1177–1184.

Liu, M.Y., Tuzel, O., 2016. Coupled generative adversarial networks. arXiv preprint. arXiv:1606.07536.

Liu, M.Y., Huang, X., Yu, J., Wang, T.C., Mallya, A., 2020. Generative adversarial networks for image and video synthesis: algorithms and applications. arXiv preprint. arXiv:2008.02793.

Long, M., Wang, J., Ding, G., Sun, J., Yu, P.S., 2014. Transfer joint matching for unsupervised domain adaptation. In: Proceedings of the IEEE Conference on Computer Vision and Pattern Recognition, pp. 1410–1417.

Long, M., Cao, Y., Wang, J., Jordan, M., 2015. Learning transferable features with deep adaptation networks. In: International Conference on Machine Learning, pp. 97–105.

Long, M., Zhu, H., Wang, J., Jordan, M.I., 2016. Unsupervised domain adaptation with residual transfer networks. arXiv preprint. arXiv:1602.04433.

Long, M., Zhu, H., Wang, J., Jordan, M.I., 2017. Deep transfer learning with joint adaptation networks. In: International Conference on Machine Learning, pp. 2208–2217.

Lu, B., Chellappa, R., Nasrabadi, N.M., 2015. Incremental dictionary learning for unsupervised domain adaptation. In: BMVC, pp. 108.1–108.12.

Lui, Y.M., Beveridge, J.R., Kirby, M., 2010. Action classification on product manifolds. In: 2010 IEEE Computer Society Conference on Computer Vision and Pattern Recognition. IEEE, pp. 833–839.

Lui, Y.M., 2012. Human gesture recognition on product manifolds. Journal of Machine Learning Research 13 (1), 3297–3321.

Mairal, J., Bach, F., Ponce, J., Sapiro, G., 2009. Online dictionary learning for sparse coding. In: Proceedings of the 26th Annual International Conference on Machine Learning, pp. 689–696.

Mairal, J., Bach, F., Ponce, J., 2011. Task-driven dictionary learning. IEEE Transactions on Pattern Analysis and Machine Intelligence 34 (4), 791–804.

Mancini, M., Porzi, L., Bulo, S.R., Caputo, B., Ricci, E., 2018. Boosting domain adaptation by discovering latent domains. In: Proceedings of the IEEE Conference on Computer Vision and Pattern Recognition, pp. 3771–3780.

Manjunath, B.S., Chellappa, R., 1993. A unified approach to boundary perception: edges, textures, and illusory contours. IEEE Transactions on Neural Networks 4 (1), 96–108.

Mansour, Y., Mohri, M., Rostamizadeh, A., 2009. Domain adaptation: learning bounds and algorithms. arXiv preprint. arXiv:0902.3430.

Miyato, T., Kataoka, T., Koyama, M., Yoshida, Y., 2018. Spectral normalization for generative adversarial networks. arXiv preprint. arXiv:1802.05957.

Nguyen, H.V., Ho, H.T., Patel, V.M., Chellappa, R., 2015. DASH-n: joint hierarchical domain adaptation and feature learning. IEEE Transactions on Image Processing 24 (12), 5479–5491.

Nguyen, H.V., Patel, V.M., Nasrabadi, N.M., Chellappa, R., 2012. Sparse embedding: a framework for sparsity promoting dimensionality reduction. In: Proceedings of the European Conference on Computer Vision. Springer, Berlin, Heidelberg, pp. 414–427.

Ni, J., Qiu, Q., Chellappa, R., 2013. Subspace interpolation via dictionary learning for unsupervised domain adaptation. In: Proceedings of the IEEE Conference on Computer Vision and Pattern Recognition. Portland, oR, pp. 692–699.

Olshausen, B.A., Field, D.J., 1996. Emergence of simple-cell receptive field properties by learning a sparse code for natural images. Nature 381 (6583), 607–609.

Pan, Weike, Evan, Xiang, Nathan, Liu, Qiang, Yang, 2010. Transfer learning in collaborative filtering for sparsity reduction. In: Proceedings of the AAAI Conference on Artificial Intelligence, vol. 24(1).

Papamakarios, G., Nalisnick, E., Rezende, D.J., Mohamed, S., Lakshminarayanan, B., 2019. Normalizing flows for probabilistic modeling and inference. arXiv preprint. arXiv:1912.02762.

Park, Sung Won, Savvides, Marios, 2011a. The multifactor extension of Grassmann manifolds for face recognition. In: 2011 IEEE International Conference on Automatic Face & Gesture Recognition (FG). IEEE, pp. 464–469.

Park, Sung Won, Savvides, Marios, 2011b. Multifactor analysis based on factor-dependent geometry. In: CVPR 2011. IEEE, pp. 2817–2824.

Patel, V.M., Gopalan, R., Li, R., Chellappa, R., 2015. Visual domain adaptation: a survey of recent advances. IEEE Signal Processing Magazine 32 (3), 53–69.

Patil, Y.C., Rezaiifar, R., Krishnaprasad, P.S., 1993. Orthogonal matching pursuit: recursive function approximation with applications to wavelet decomposition. In: Proceedings of 27th Asilomar Conference on Signals, Systems and Computers. IEEE, pp. 40–44.

Ren, S., He, K., Girshick, R., Sun, J., 2015. Faster R-CNN: towards real-time object detection with region proposal networks. arXiv preprint. arXiv:1506.01497.

Roweis, S.T., Saul, L.K., 2000. Nonlinear dimensionality reduction by locally linear embedding. Science 290 (5500), 2323–2326.

Rubinstein, R., Bruckstein, A.M., Elad, M., 2010. Dictionaries for sparse representation modeling. Proceedings of the IEEE 98 (6), 1045–1057.

Saenko, K., Kulis, B., Fritz, M., Darrell, T., 2010. Adapting visual category models to new domains. In: European Conference on Computer Vision. Springer, Berlin, Heidelberg, pp. 213–226.

Saito, K., Ushiku, Y., Harada, T., 2017a. Asymmetric tri-training for unsupervised domain adaptation. In: International Conference on Machine Learning, pp. 2988–2997.

Saito, K., Ushiku, Y., Harada, T., Saenko, K., 2017b. Adversarial dropout regularization. arXiv preprint. arXiv:1711.01575.

Saito, K., Watanabe, K., Ushiku, Y., Harada, T., 2018. Maximum classifier discrepancy for unsupervised domain adaptation. In: Proceedings of the IEEE Conference on Computer Vision and Pattern Recognition, pp. 3723–3732.

Sankaranarayanan, S., Balaji, Y., Chellappa, R., 2018. Adapting across Domains Using Generative Adversarial Networks. Proceedings of the IEEE Computer Society Conference on Computer Vision and Pattern Recognition (Spotlight paper). Salt Lake City, UT.

Sankaranarayanan, S., Balaji, Y., Chellappa, R., 2008. Learning from Synthetic Data: Semantic Segmentation across Domain Shift. (Spotlight Paper). In: Proceedings of IEEE Computer Society Conference on Computer Vision and Pattern Recognition, Salt Lake City, UT.

Sharma, A., Jacobs, D.W., 2011. Bypassing synthesis: PLS for face recognition with pose, low-resolution and sketch. In: CVPR 2011. IEEE, pp. 593–600.

Sharma, A., Kumar, A., Daume, H., Jacobs, D.W., 2012. Generalized multiview analysis: a discriminative latent space. In: 2012 IEEE Conference on Computer Vision and Pattern Recognition. IEEE, pp. 2160–2167.

Shekhar, S., Patel, V.M., Nguyen, H.V., Chellappa, R., 2013. Generalized domain-adaptive dictionaries. In: Proceedings of the IEEE Conference on Computer Vision and Pattern Recognition, pp. 361–368.

Shen, J., Qu, Y., Zhang, W., Wasserstein, Yu Y., 2018. Distance guided representation learning for domain adaptation. In: Proceedings of the AAAI Conference on Artificial Intelligence, vol. 32(1).

Shi, Y., Sha, F., 2012. Information-theoretical learning of discriminative clusters for unsupervised domain adaptation. In: Proceedings of International Conference on Machine Learning, pp. 1079–1086.

Smetana, Judith G., Villalobos, Myriam, Tasopoulos-Chan, Marina, Gettman, Denise C., Campione-Barr, Nicole, 2009. Early and middle adolescents' disclosure to parents about activities in different domains. Journal of Adolescence 32 (3), 693–713.

Tenenbaum, J.B., De Silva, V., Langford, J.C., 2000. A global geometric framework for nonlinear dimensionality reduction. Science 290 (5500), 2319–2323.

Tommasi, T., Caputo, B., 2013. Frustratingly easy domain adaptation. In: Proceedings of the IEEE International Conference on Computer Vision, pp. 897–904.

Tropp, J.A., 2004. Greed is good: algorithmic results for sparse approximation. IEEE Transactions on Information Theory 50 (10), 2231–2242.

Tzeng, E., Hoffman, J., Zhang, N., Saenko, K., Darrell, T., 2014. Deep domain confusion: maximizing for domain invariance. arXiv preprint. arXiv:1412.3474.

Vageeswaran, P., Mitra, K., Chellappa, R., 2013. Blur and illumination robust face recognition via set-theoretic characterization. IEEE Transactions on Image Processing 22 (4), 1362–1372.

Vu, T.H., Jain, H., Bucher, M., Cord, M., Pérez, P., 2019. Advent: adversarial entropy minimization for domain adaptation in semantic segmentation. In: Proceedings of the IEEE/CVF Conference on Computer Vision and Pattern Recognition, pp. 2517–2526.

Vasilescu, M.A., Terzopoulos, D., 2002. Multilinear analysis of image ensembles: tensorfaces. In: European Conference on Computer Vision. Springer, Berlin, Heidelberg, pp. 447–460.

Vasilescu, M.A., Terzopoulos, D., 2007. Multilinear projection for appearance-based recognition in the tensor framework. In: 2007 IEEE 11th International Conference on Computer Vision. IEEE, pp. 1–8.

Wang, C., Mahadevan, S., 2009. Manifold alignment without correspondence. In: Twenty-First International Joint Conference on Artificial Intelligence, p. 26.

Wen, Z., Yin, W., 2013. A feasible method for optimization with orthogonality constraints. Mathematical Programming 142 (1), 397–434.

Wright, J., Ma, Y., Mairal, J., Sapiro, G., Huang, T.S., Yan, S., 2010. Sparse representation for computer vision and pattern recognition. Proceedings of the IEEE 98 (6), 1031–1044.

Xu, H., Zheng, J., Chellappa, R., 2015. Bridging the domain shift by domain adaptive dictionary learning. In: British Machine Vision Conference 2015. Brighton, UK.

Xu, R., Chen, Z., Zuo, W., Yan, J., Lin, L., 2018. Deep cocktail network: multi-source unsupervised domain adaptation with category shift. In: Proceedings of the IEEE Conference on Computer Vision and Pattern Recognition, pp. 3964–3973.

Yang, M., Zhang, L., Feng, X., Zhang, D., 2011. Fisher discrimination dictionary learning for sparse representation. In: International Conference on Computer Vision, pp. 543–550.

Yang, L., Balaji, Y., Lim, S.N., Shrivastava, A., 2020. Curriculum manager for source selection in multi-source domain adaptation. arXiv preprint. arXiv:2007.01261.

Zhao, W., Chellappa, R., Phillips, P.J., Rosenfeld, A., 2003. Face recognition: a literature survey. ACM Computing Surveys (CSUR) 35 (4), 399–458.

Zheng, J., Liu, M.Y., Chellappa, R., Phillips, J.P., 2012. A Grassmann manifold-based domain adaptation approach. In: Proceedings of the International Conference on Pattern Recognition, pp. 2095–2099.

Zou, Y., Yu, Z., Kumar, B.V., Wang, J., 2018. Unsupervised domain adaptation for semantic segmentation via class-balancedSself-training. In: Proceedings of the European Conference on Computer Vision (ECCV), pp. 289–305.

Zou, Y., Yu, Z., Liu, X., Kumar, B.V., Wang, J., 2019. Confidence regularized self-training. In: Proceedings of the IEEE/CVF International Conference on Computer Vision, pp. 5982–5991.

Biographies

Yogesh Balaji is a PhD candidate in the University of Maryland, College Park. He received his MS in Computer Science from the University of Maryland and B. Tech in Electrical Engineering from Indian Institute of Technology, Madras. His research interests are in computer vision and machine learning, with a focus on domain adaptation and generative modeling.

Hien V. Nguyen is an Assistant Professor in the Department of Electrical and Computer Engineering, University of Houston. He received his Ph.D. from the University of Maryland at College Park (2013), and B.S. from the National University of Singapore (2007). His research interests are at the intersection of machine learning and medicine. He has published 50 peer-reviewed articles, and coauthored 12 U.S. patents. His work on physician-friendly medical diagnosis was featured in the "Great Innovative Ideas" series by the Computing Research Association. He is a senior member of the National Academy of Inventors.

Rama Chellappa is a Bloomberg Distinguished Professor in the Departments of Electrical and Computer Engineering and Biomedical Engineering at Johns Hopkins University (JHU). At JHU, he is also affiliated with MINDS, CIS and CLSP. He holds a nontenure position as a College Park Professor in the ECE department at University of Maryland (UMD). Before coming to JHU, he was a Distinguished University Professor, a Minta Martin Professor of Engineering, a Professor in the ECE department and the University of Maryland Institute Advanced Computer Studies at UMD. His current researcher interests are computer vision, pattern recognition and machine intelligence. He has received numerous research, teaching, service, and mentoring awards from the University of Southern California, University of Maryland, IBM, IEEE Computer and Signal Processing Societies, the IEEE Biometrics Council and the International Association of Pattern Recognition. He has been recognized as an Outstanding Electrical Engineer by the ECE department at Purdue University and as a Distinguished Alumni of the Indian Institute of Science. He is a Fellow of IEEE, IAPR, OSA, AAAS, ACM, AAAI, and NAI and holds eight patents.

8

Domain adaptation and continual learning in semantic segmentation

Umberto Michieli, Marco Toldo, and Pietro Zanuttigh

Department of Information Engineering, University of Padova, Padova, Italy

CHAPTER POINTS

- The domain adaptation task for semantic segmentation is formally introduced and the different levels at which adaptation can be performed are presented.

- Wide review of domain adaptation techniques for semantic segmentation.

- First review of the recent advancements in continual learning for semantic segmentation.

8.1 Introduction

The standard supervised learning setting assumes the availability of a large training set containing data with the same statistical properties as the target data labeled according to the problem at hand. This learning setup has been used to design a huge number of machine learning strategies, from simple linear classifiers to advanced deep learning methods and has allowed to obtain a robust theoretical framework.

When moving to practical applications, however, some limitations of this setup arise. First of all, a large amount of training data for the considered setting and problem is typically not available. While very large generic datasets are publicly available, the acquisition and labeling of a large amount of data for a specific setting is typically too expensive and time consuming for most companies developing machine learning systems. This raised the need for domain adaptation techniques able to transfer the learned knowledge from a generic source dataset to the target data of the problem at hand. These can be either partially supervised, *i.e.*, a small amount of labeled data for the target set is available, or unsupervised, when no labeling information or even no data at all, is available for the target domain.

Advanced Methods and Deep Learning in Computer Vision
https://doi.org/10.1016/B978-0-12-822109-9.00017-5

Copyright © 2022 Elsevier Inc. All rights reserved.

Another common limitation is that the labeling data may only partially or not exactly match the target problem (*i.e.*, the so-called weakly supervised learning problem). Many recent research strategies aim at exploiting labeling data belonging to a different but related problem or to a set of classes different from the target one. Related to this, in many cases the target problem may not even be completely defined at the starting point with new classes or new tasks being added at run-time. Continual learning strategies deal with these settings aiming at progressively learning the new tasks or classes without retraining from scratch the machine learning model.

This general discussion applies to many learning models and target problems, but becomes very relevant when a huge amount of data and a large computational effort for training are needed. In particular, this is the case of image and video understanding problems, which are nowadays typically solved with complex deep learning models. For this reason, transfer learning for this kind of problem has been widely studied and many solutions have been proposed especially in the image classification field, which is the simplest and most classical global-level image understanding problem.

In this chapter, instead, we focus on the more challenging semantic segmentation task where, in spite of an image-level classification, a dense pixel-level labeling is performed. This problem is more demanding but also particularly interesting since the labeling operation is highly time-consuming (much more than in image classification), thus making the construction of large training sets more difficult. Even if a huge number of strategies can be exploited for this task, nowadays most methods exploit deep learning strategies and, in particular, Convolutional Neural Networks (CNNs) with an auto-encoder structure; in this chapter we will focus on this architecture.

We start by dealing with the Unsupervised (or partially supervised) Domain Adaptation (UDA) task. In the standard setting, the task remains the same while there is a shift in the statistical properties of the data when moving from the source to the target domain set. We will formally introduce the problem in Section 8.2.1, considering also more advanced configurations where both the domain and the task can change. The adaptation of the deep learning network to the target domain can be done at different stages of the deep network, *i.e.*, (1) at the *input level* by translating the images to a new domain more similar to the target one, (2) at the *feature level* by constructing a feature space where the various classes are better separated making the description more robust to the change of domain, and finally (3) at the *output level* by enforcing a coherence between the output probability spaces when using data from the two domains. These targets can be achieved using different strategies such as adversarial learning, generative networks for domain translation, self-teaching, entropy minimization, and many others. Section 8.2.3 will present in detail the most successful strategies.

In the second part of this chapter, we will analyze Continual Learning (CL) strategies, which are designed to handle changes of tasks over time. Such changes are typically represented by an expansion of the label set, by adding new labels or by splitting the existing ones into more refined subclasses. One of the main targets is the capability to adapt the network to the new setting using only data concerning the new tasks and without retraining the model from scratch. However, this is highly nontrivial due to the so-called *catastrophic forgetting* phenomenon, *i.e.*, a machine learning model tends to forget knowledge about previous tasks when it learns new ones. We will start by formally introducing this problem in Section 8.3.1 together with the wide range of experimental scenarios that can be considered.

Then, in Section 8.3.3 we will show how it can be tackled by means of knowledge preservation strategies, especially based on the knowledge distillation framework. Other strategies are based on parameter freezing or slowing down the learning in some parts of the network, on trying to regenerate past (and no longer available) data of the previous classes using generative networks or exploiting web-crawled data.

8.1.1 Problem formulation

In this section we provide a formal definition of the Transfer Learning problem and we introduce some notation that we will use in the rest of the chapter. Let us define a domain $\mathcal{D} = \{\mathcal{X}, P(X)\}$, where $\mathcal{X}$ is the space of input data and $P(X)$ is the probability distribution function over that input data. A task $\mathcal{T}$ over the domain $\mathcal{D}$ is a combination of a label space $\mathcal{Y}$ with the predictive function $f(\cdot)$ modeling the conditional probability distribution $P(Y|X)$. Thus, any supervised machine learning problem can be generally attributed to the search for a function $h : \mathcal{X} \to \mathcal{Y}$ which better approximates the unknown underlying $f(\cdot)$, by examining a set of labeled training samples drawn from the joint distribution $P(X, Y)$ over $\mathcal{X} \times \mathcal{Y}$.

Let us assume that the input data domain $\mathcal{D}$ is not unique, *e.g.*, there exist separate source $\mathcal{D}_S$ and target $\mathcal{D}_T$ domains (*e.g.*, in classical UDA) or the domain is split into multiple pieces $\mathcal{D}^{(t)}, t = 1, ..., T_{max}$ available for training at separate times (*e.g.*, in classical CL). Furthermore, over these domains different tasks may need to be solved, *e.g.*, two different tasks $\mathcal{T}_S$ and $\mathcal{T}_T$ could have been respectively chosen for the source and target domains or there can be a sequence of tasks $\mathcal{T}^{(t)}, t = 1, ..., T_{max}$ to be solved at different stages t of the learning process. Then, *transfer learning* is defined as seeking for an improved predictive function $f_T(\cdot)$ over the target domain (or over multiple target domains), relying on useful information extracted from source task $\mathcal{T}_S$ on $\mathcal{D}_S$, in case $\mathcal{D}_S \neq \mathcal{D}_T$ or $\mathcal{T}_S \neq \mathcal{T}_T$.

Notice how domain adaptation and continual learning can be viewed as two special cases of transfer learning; in the first case, the source and target domain are different while the task is the same, while in the second case the macro-domain is the same (but is made available in separate portions) and the task changes.

8.2 Unsupervised domain adaptation

Over the past few years, deep learning has had a groundbreaking impact on the computer vision field. Before the deep learning revolution, semantic segmentation was a very challenging task and even sophisticated algorithms were achieving only mediocre performance; nowadays, with the advent of deep neural networks we can obtain remarkable results, provided that enough computational resources are available. Nonetheless, the potential stored in deep learning models can be fully unleashed only when sufficiently plentiful and carefully labeled training data is available. The complexity enclosed in the millions of learnable parameters of state-of-the-art deep learning models easily leads to overfitting of the training data, rather than to an enhanced model performance, and this has to be counteracted by using huge datasets for training. A major example of the central role played by the availability of large amounts of training data is the ImageNet large-scale dataset (Deng et al., 2009), whose contribution in the early development and expansion of deep neural networks for image classification is certainly very relevant.

Unfortunately, the collection and annotation of data samples is often extremely expensive, time consuming and error-prone, since it requires a large amount of human supervision in the process. The excessive cost may prevent from gathering enough data to address a new task or to move in a new environment, thus posing a serious threat to the remarkable advance brought by deep learning approaches. Therefore, it may be extremely beneficial to rely on previously built datasets, whenever they share similar properties to the target data. In this way, already available samples can be efficiently exploited to address the current task, since they belong to a domain correlated to the target one.

Even though the transferring of information from related domains appears quite appealing and fairly straightforward, in practice the process requires careful handling. Deep neural networks typically lack generalization skills; in other words, even a small change in data distribution between training and test statistical distributions might cause a severe drop of performance. For this reason, the simple application of a pretrained model in a new environment is likely to fail, as domain-specific attributes are usually captured alongside domain-invariant ones, thus preventing an effective knowledge transfer. In this scenario, Domain Adaptation comes in handy, as it allows to handle the statistical gap between source and target representations. The ultimate goal of the adaptation effort is to learn a prediction model on a selected task working optimally on both source and target domains, while supervision largely (or solely) comes from the label-abundant source domain. To this end, an efficient transfer of knowledge learned in the source domain to the target one is crucial to eventually reach an overall good performance. Particularly interesting is the Unsupervised Domain Adaptation (UDA) setting, in which target annotations are totally missing. This is an extremely favorable, yet challenging, scenario, as data from the target domain no longer requires expensive labeling.

Recently, the domain adaptation task has been very actively studied in the context of deep learning applied to visual tasks. While deep convolutional frameworks have proven to be capable of learning visual features useful to solve multiple related problems (*e.g.* image classification, object detection, semantic segmentation), the transferability of those representations typically shows to decrease when moving to deeper network layers (Long et al., 2015).

Early works on domain adaptation for deep networks mainly focused on the image classification task. In many approaches a layer-wise measure of statistical domain discrepancy is jointly estimated and minimized, thus promoting the extraction of domain-invariant feature representations, while discrimination ability is guaranteed by a task-specific loss. Later, adversarial domain adaptation strategies have proven to be extremely successful, in which schemes the domain discrepancy is expressed in the form of a learnable discriminator and its minimization is performed in an adversarial manner. More details on adversarial learning and its use in domain adaptation will be provided in Section 8.2.3.1. This has effectively opened the door to domain adaptation solutions apt to solve the semantic segmentation task, where the inherent higher complexity in terms of network representations needed for pixel-wise classification calls for more advanced solutions.

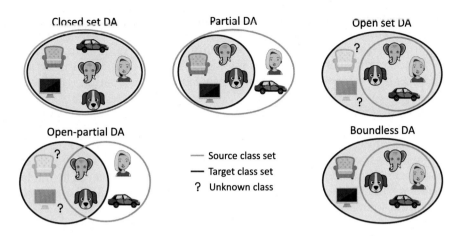

FIGURE 8.1 Different settings for domain adaptation, according to how source and target class sets are related.

8.2.1 Domain adaptation problem formulation

Domain Adaptation (DA) is a special case of transfer learning, called Transductive Transfer Learning, in which the source and target tasks coincide ($\mathcal{T}_S = \mathcal{T}_T$), whereas the discrepancy lies in the domain difference ($\mathcal{D}_S \neq \mathcal{D}_T$). In addition, domain adaptation is commonly intended in a homogeneous fashion, when the domain shift happens at a statistical level ($P(X_S, Y_S) \neq P(X_T, Y_T)$) rather than being due to distinct input spaces ($\mathcal{X}_S$ and $\mathcal{X}_T$ belong to the same semantic domain, *e.g.*, urban scene images) (Wang and Deng, 2018).

Recently, some researches have considered more challenging scenarios than the standard homogeneous DA, allowing for separate sets of semantic classes in the source and target domains ($\mathcal{C}_S$ and $\mathcal{C}_T$). Depending on how the source $\mathcal{C}_S$ and target $\mathcal{C}_T$ sets are related, it is possible to identify multiple DA scenarios (Fig. 8.1):

- *Closed Set DA*: it corresponds to the homogeneous case, where semantic classes are completely shared between source and target domains ($\mathcal{C}_S = \mathcal{C}_T$).
- *Partial DA*: in this setup there exist some source classes that do not appear in the target domain ($\mathcal{C}_S \supset \mathcal{C}_T$).
- *Open Set DA*: conversely to partial DA, here the presence of some target private classes is admitted, for which no training examples in the source domain are available ($\mathcal{C}_S \subset \mathcal{C}_T$).
- *Open-Partial Set DA*: the source and target domains include separate sets of semantic classes (Saito et al., 2020), with a subset of those in common ($\mathcal{C}_S \neq \mathcal{C}_T$, $\mathcal{C}_S \cap \mathcal{C}_T \neq \varnothing$). However, elements belonging to the class subset exclusive to the target domain have only to be acknowledged as not part of the shared classes.
- *Boundless DA*: this setup is very similar to the open set one, but objects of target private classes must be explicitly classified rather than only be associated to a general unknown target class. This setting has been recently introduced (Bucher et al., 2020) and it represents the most ambitious one, since it admits complete unawareness beforehand about semantic content of target data.

In the following sections, the focus will be placed on the *standard* most-diffused closed set adaptation, as, up to now, this is by far the most explored setup.

According to the degree of annotations availability in the target domain, the adaptation problem is subject to a further categorization, ranging from the full or partially annotated *supervised* or *semisupervised* settings to the completely label deprived *unsupervised* scenario. In particular, Unsupervised Domain Adaptation (UDA) will be discussed, as it has recently witnessed an increase in popularity, especially in relation to the semantic segmentation task, and it involves many practical applications. More specifically, it is assumed that a set of labeled source data $\{x_i^s, y_i^s\}$ drawn accordingly to the source joint distribution over $\mathcal{X}_S \times \mathcal{Y}_S$ is provided, paired with a set of unlabeled samples $\{x_i^t\}$, retrieved from a distinct target marginal distribution over $\mathcal{X}_T$. The objective is to discover a predictive function correctly modeling the task input-label relation in the target domain, while knowledge on the chosen task can be extracted only from source labeled samples.

Furthermore, for standard domain adaptation techniques to work, source and target domains should be somehow related, meaning that they should share task-relevant content, while low-level attributes may differ. This scenario is commonly referred to as *one-step* DA, as knowledge transferring happens directly across source and target data without intermediate stages.

8.2.2 Adaptation focus

As previously discussed, behind the performance degradation suffered by deep prediction models applied on new target environments lies the covariate shift phenomenon affecting source and target input data samples. For this reason, most of domain adaptation research builds upon bridging the statistical gap between domain distributions, in order for the prediction model to yield satisfactory results whenever those distributions are matched.

Various strategies have been explored to achieve the statistical matching, which will be thoroughly discussed in Section 8.2.3. A more general categorization of domain adaptation techniques, however, can be inferred, according to where in the employed semantic segmentation model the statistical discrepancy happens to be addressed. In particular, different data representations could be subject to adaptation, from the bare images prior to classification up to intermediate and output network activations (Fig. 8.2). In the following, a description of the main ideas behind adaptation approaches will be provided, grouped by where the adaptation effort is focused.

8.2.2.1 *Input level adaptation*

A first strategy is to perform adaptation at the input level, directly on images before they are fed to the segmentation network (as shown in the leftmost part of Fig. 8.2). The idea is to force data samples from either domain to reach an uniform visual appearance, meaning that they not only have to carry high-level semantic similarity, but their low-level statistical discrepancy should be matched as well. This because low-level domain dependent attributes, even though they do not define the semantic content of the input image, can still be captured by the prediction model, thus leading to incorrect predictions when a domain change alters them. A clear example of this is the synthetic to real adaptation; although it may be quite realistic, synthetic data can mimic real-word properties up to a certain extent. Thus, it is usually

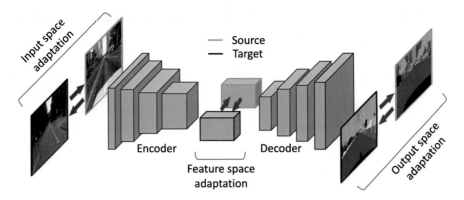

FIGURE 8.2 General scheme of an auto-encoder network for semantic segmentation highlighting the different network stages on which domain adaptation strategies can be applied, from the input image space up to intermediate or output network activations.

possible to find synthetic peculiar traits, however small, which can undermine the efficacy of a model trained on synthetic data in a real-world environment.

The common approach to address domain adaptation at the input level is to map the data to a new image space, where the projected source (or target) samples carry an enhanced perceptual resemblance to target (or source) ones. This is normally achieved with the help of style-transfer techniques, whose objective is turned into matching source and target marginal distributions in the image space. By feeding in input supervised data from the new domain-invariant space to the segmentation network, the predictor should now able to retain consistent results across domains.

An upside of this approach is its complete independence with respect to the segmentation network currently in use that does not require any modification. This, however, comes with a cost, which is that, in its vanilla scheme without any extra regularizing factors, marginal alignment may be performed without the class-conditional distributions being simultaneously matched. In other words, it may be possible to end up with domain invariant representations, which yet lack the semantic coherence with the original data crucial to solve the segmentation task. To get past this problem, multiple solutions have been proposed to achieve semantically consistent image translations, for example through image reconstruction constraints or additional loss components enforcing the coherence of segmentation predictions.

8.2.2.2 Feature level adaptation

An alternative approach is to focus the adaptation on feature representations, pursuing a distribution alignment of network latent embeddings, which are normally retrieved from the encoder output in the commonly employed auto-encoder architecture (even if adaptation at other network stages has also been employed). The primary objective is to build a domain invariant latent space, in which features extracted either from source or target input images observe the same distribution. In the end, learning solely from supervision on source representations should result in a good performance also on the target domain, as shared classification in the adapted latent space should be jointly effective on both source and target representations when distributed alike.

In the context of semantic segmentation, the feature space retains significant complexity due to its high-dimensionality, which is necessary for the prediction model to simultaneously capture global semantic clues, while attaining pixel-level accuracy. In addition, as for the input level adaptation, a semantically unaware alignment of marginal distributions (*e.g.*, standard adversarial adaptation) does not guarantee that the joint input-label distributions are matching, since no information can be derived from unlabeled target samples about the target joint distribution. For these reasons, many feature level adaptation techniques that have been successfully devised for image classification do not easily extend to the dense segmentation task, and in general require careful tuning and further regularization.

8.2.2.3 Output level adaptation

Finally, the last class of domain adaptation techniques exploits a cross-domain distribution alignment over the network output, *i.e.*, typically the output per-class probability space. Not only prediction probability maps have proven to retain sufficient complexity and richness of semantic information, but they also span a low-dimensional space over which statistical alignment happens to be achieved much more effectively, for example by the domain adversarial strategy. In addition, source knowledge can be indirectly translated over the unlabeled target domain by resorting to some form of self-taught supervision extracted from target prediction maps, whose careful introduction in the learning process to support the standard source supervision may result in an effective cross-domain adaptation of the network performance. Source priors derived from label distribution have proven to provide an useful regularization to the learning process as well, since they usually identify high-level semantic properties shared across domains.

8.2.3 Unsupervised domain adaptation techniques

In the following sections the most relevant approaches to address the Unsupervised Domain Adaptation in semantic segmentation will be discussed. For each set of techniques, some works whose proposed adaptation solutions can be associated to that class will be presented. However, it should be stressed that most of the domain adaptation frameworks recently introduced resort to a combination of multiple techniques to improve performance.

8.2.3.1 Domain adversarial adaptation

Adversarial learning has been originally introduced for image generation (Goodfellow et al., 2014). The main objective behind the generative task is to retrieve an unknown probability distribution modeling data from the employed training set. The adversarial strategy has proven extremely effective in solving this problem, since no explicit expression of the underlying target data distribution has to be found, and, more importantly, no specific learning objective to train the generative model is required. The learning process builds upon a minmax game, where a generator network is progressively guided by a discriminator network to produce realistic samples. In Goodfellow et al. (2014), the discriminator is a binary classifier whose goal is to discern between the original training data and the data produced by the generator. The generator is instead a generative model that takes in input random noise (or some

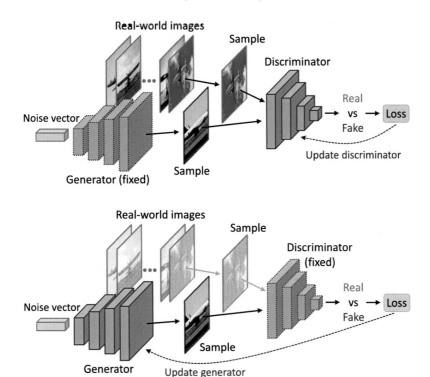

FIGURE 8.3 Training of a generative adversarial network. Update step of the discriminator (top) and of the generator (bottom).

conditioning data in more recent variations of the approach) and produces data (*i.e.*, images in the setting of interest) resembling the ones in the training set. It aims at constantly improving the realism of its output samples to fool the discriminative action of its opponent, and this is achieved by using a loss function whose minimization in turn maximizes the errors of the discriminator. The model is trained by alternating a discriminator training step, aiming at maximizing its accuracy, and a generator optimization phase with the opposite target (see Fig. 8.3). If correctly carried out, the adversarial competition should result in a statistical distribution of generated data that fully matches the training set one, meaning that original and generated data should be statistically indistinguishable. In addition, the discriminator should be able to both capture and express a measure of statistical discrepancy in the form of a structured learnable loss. Therefore, the objective function can be thought as being jointly learned and optimized in the adversarial process, allowing it to adapt to the specific context.

Adversarial learning has been successfully extended to the domain adaptation task. The real-fake discriminator is now turned into a domain classifier that is used to drive the adaptation process. Its discriminative action is, in fact, focused on capturing the statistical discrepancy between representations from separate domains, which is responsible for the performance degradation and thus has to be reduced in order to achieve an effective adaptation.

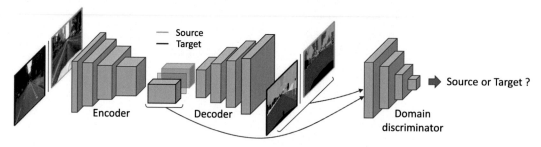

FIGURE 8.4 Graphical representation of the standard adversarial adaptation strategy. A domain discrimination captures the statistical discrepancy between source and target representations (*e.g.*, segmentation network's output or features maps computed from one or the other domain). Its supervisory signal is then exploited to perform domain alignment.

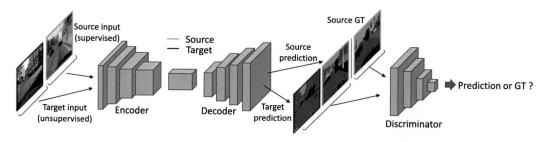

FIGURE 8.5 Graphical representation of an output adversarial adaptation strategy, where domain alignment is performed indirectly by bridging the distribution gap between source annotation maps and network predictions from either source or target domains.

There are two possible targets for the domain classifier. The first is to discriminate between internal or output representations extracted from data in either source or target domains (Fig. 8.4). This allows to introduce additional loss terms enforcing the construction of feature or output spaces that are more domain invariant. Alternatively, it is possible to use the discriminator to distinguish between the output of the network (that can correspond both to inputs from the source and from the target domain) and the ground truth segmentations (that in the unsupervised setting are only present in the source domain). Since in the adversarial model there is no need to have ground truth data matching the provided samples, this allows to use also the target domain images for which no ground truth is available and to enforce that their predicted segmentation maps have statistical properties similar to the ground truth ones (Fig. 8.5). Using these strategies, the standard supervision from the annotated source data is joined by a supervisory signal from the domain discriminator, which pushes the prediction network towards domain invariance, in turn mitigating the intrinsic bias towards the supervised source domain.

In the wake of the success of adversarial domain adaptation for image classification (Ganin and Lempitsky, 2015; Ganin et al., 2016), the adversarial strategy has been introduced also in the context of semantic segmentation to achieve domain alignment over latent feature embeddings (Hoffman et al., 2016). Yet, as previously discussed, the global domain alignment of marginal distributions provided by the vanilla domain adversarial scheme may end up in incorrect semantic knowledge transfer across domains, with class-conditional distributions neglected in the learning process. For this reason, to reach an effective adversarial adaptation when dealing with the semantic segmentation task, additional modules should be embedded in the adaptation pipeline.

A possible solution is to integrate adversarial feature alignment in a generative approach (see Section 8.2.3.2), as done in several works (Li et al., 2019; Hoffman et al., 2018; Chen et al., 2019c; Toldo et al., 2020). Here the goal is to strengthen the image space adaptation, so that the attribute transferring to match visual appearances of images from different domains is extended inside the feature space. An alternative is to perform category-wise adaptation (Chen et al., 2017; Du et al., 2019). The idea is to resort to class-wise adversarial learning, by introducing multiple per-class distinct feature discriminators, that, in principle, should provide a semantically consistent knowledge transfer, which is absent in the standard global adaptation. Finally, with a different perspective it is possible to rely on a reconstruction constraint to enforce domain invariance over latent feature embeddings (Sankaranarayanan et al., 2018; Murez et al., 2018; Zhu et al., 2018). Adversarial learning in this case is applied over the reconstruction image-level space, to guarantee that feature representations can be projected back to either source or target image spaces without distinction.

As previously observed, the semantic segmentation task entails a quite complex feature space, due to the high dimensionality of its representations. Thus, to bypass the complexity encompassed in feature space adaptation, a research direction has been to focus the adaptation effort to the segmentation output space (Tsai et al., 2018; Chen et al., 2018; Chang et al., 2019; Luo et al., 2019; Yang et al., 2020a; Biasetton et al., 2019; Michieli et al., 2020; Spadotto et al., 2020). The low-dimensional output representations, in fact, have been shown to retain enough semantic information for a successful adaptation. In this new output level adversarial scheme, a domain discriminator learns to discover the domain from which segmentation maps are originated. Simultaneously, the segmentation network plays the generative role by providing cross-domain statistically close predictions to fool the discerning action of the domain classifier. While the common solution has been to align source and target output representations (Tsai et al., 2018; Chen et al., 2018; Chang et al., 2019; Luo et al., 2019; Yang et al., 2020a), some works have revisited the standard approach by seeking for an indirect domain alignment (Biasetton et al., 2019; Michieli et al., 2020; Spadotto et al., 2020), by forcing predictions from either domain to be distributed as ground-truth source labels (as depicted in Fig. 8.5).

Recently, new approaches (Vu et al., 2019b,a; Tsai et al., 2019) have been built upon the extraction of meaningful patterns from the segmentation output space. The intention is to provide the domain discriminator with a more functional and significant understanding of source and target representations, allowing it to yield a more effective guidance in the adaptation process. This is done by manually extracting some meaningful information from both source and target data (*e.g.*, entropy maps over segmentation predictions (Vu et al., 2019b,a)) to be fed to the discriminator network. In turn, by focusing on meaningful semantic clues from data representations, the adversarial alignment over activations of the segmentation network should be enhanced.

8.2.3.2 *Generative-based adaptation*

Image-to-image translation is a class of generative techniques whose main objective is to learn a suitable function to project images from one domain to another. In other words, the idea is to discover the joint distribution of image data from separate domains. The image-to-image translation task can be effectively exploited in domain adaptation. What could be achieved is, in fact, the transferring of visual attributes from the target domain to the source one, while preserving source semantic information. By doing so, the covariate shift phenomena at the root of the classifier performance drop is mitigated. In this direction, several works have resorted to an input-level adaptation strategy based on image translation between source and target domains. What all these works have in common, is the search for domain invariance of visual appearance of images from different domains. This ultimately allows to exploit target-like supervision from translated, yet still annotated, source images.

A common approach shared by a multitude of works (Hoffman et al., 2018; Chen et al., 2019c; Toldo et al., 2020; Zhou et al., 2020; Li et al., 2019; Murez et al., 2018; Qin et al., 2019; Li et al., 2018; Yang et al., 2020b; Gong et al., 2019) is to exploit the successful CycleGAN (Zhu et al., 2017) model to perform unsupervised image-to-image translation (Fig. 8.6). The framework proposed by Zhu et al. (2017) concurrently learns in adversarial manner the conditional image translations in both the source-to-target and target-to-source directions. The two adversarial generative modules are further tied by a cycle-consistency constraint, driving each of them to learn the inverse projection of the other. The purpose of the reconstruction requirement is to preserve geometry and layout of the input scene, but in turn does not guarantee that the semantic content of an input image is preserved in the translation.

To tackle this issue, the lack of semantic consistency in the vanilla translation scheme is counteracted by taking advantage of the semantic prediction capability of the segmentation network (Hoffman et al., 2018; Chen et al., 2019c; Toldo et al., 2020; Zhou et al., 2020; Li et al., 2019). In particular, the semantic predictor can be exploited to detect and thus discourage any perturbation of the semantic output which may happen during the translation provided by the CycleGAN's generators. This is generally done by enforcing consistent prediction maps over original and translated versions of the same image. An alternative solution could be, instead, to achieve semantic awareness in the image-to-image translation based adaptation by acting directly over the adversarial translation modules. This, for example, has been performed with a soft gradient-sensitive loss (Li et al., 2018) and a phase consistency constraint (Yang et al., 2020b), both providing a regularizing effect over the standard adversarial learning. Finally, it is possible to focus on the target-to-source translation rather than the more commonly used source-to-target one. This has proved to decrease the bias towards the source domain (Yang et al., 2020a). As opposed to the standard generative adaptation, the image-level domain invariance is in this case achieved inside the source image domain, where pseudo-labeling allows to exploit the translated source-like target images in the supervised training.

As an alternative to the CycleGAN-based adaptation, style transfer techniques have also been explored to achieve domain invariance of low-level image attributes. Behind these methods lies the principle that any image can be disentangled into its content and style representations. While the style of an image is related to low-level domain-specific traits, its content indicates domain-invariant high-level semantic properties. Therefore, joining source content

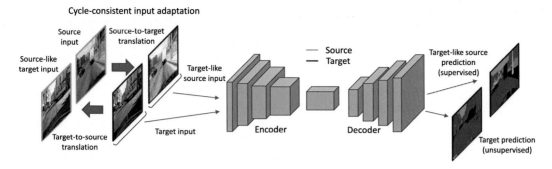

FIGURE 8.6 Overview of the generative-based adaptation approach built upon cycle-consistent image-to-image translation. In particular, source translated input images are exploited as a form of target-like artificial supervision during the learning process.

with target style should provide target-distributed training data, still preserving source semantic annotations. Once more, target supervised training can be performed thanks to the new generated target supervision. A common approach for style transfer is to resort to content and style decomposition in the latent space (Chang et al., 2019; Pizzati et al., 2020). Then, source to target translation becomes combining the extracted source content representations to random target style ones, with the mixed representations to be reprojected into the image space. To avoid the complexity involved in GAN's image generation (specially high resolution images are challenging to be obtained), different types of style transfer techniques have also been explored, ranging from neural or photo-realistic style transfer (Zhang et al., 2018; Dundar et al., 2018) to feature renormalization (Choi et al., 2019; Wu et al., 2019) and low-level frequency spectrum manipulation (Yang and Soatto, 2020).

8.2.3.3 Classifier discrepancy

As mentioned in Section 8.2.3.1, feature level adversarial adaptation in its standard design involves an additional domain classifier, whose discriminative action over feature representations from source and target domains provides an effective supervisory signal to the learning process, pushing the segmentation network towards domain invariance within the latent space it spans. A separate task-specific objective is instead responsible for the prediction network to learn the actual task with source supervision, *i.e.*, the standard cross-entropy loss for semantic segmentation.

Despite being fairly effective, the standard adversarial adaptation lacks semantic awareness (Saito et al., 2018a,b). A proper adversarial alignment, in fact, entails a match of marginal distributions of source and target data, which is typically not followed by a class-conditional statistical alignment as well. This because category-level joint distributions necessarily remain unknown to the domain classifier, as complete lack of supervision in the target domain implies that no information about semantic content of target data is available. The result is that features may be moved near class boundaries, where classification uncertainty could lead to incorrect predictions. Even worse, a class-agnostic transfer of target features might incorrectly align them to source representations of a different semantic class in the domain invariant latent space, which means negative transfer has been introduced.

Aiming at overcoming those issues, Saito et al. (2018a) completely redesign the original domain adversarial approach. In particular, they provide the task-specific dense classifier (*i.e.*, the encoder network) with the discriminative role that was before assigned to an external domain discriminator. By perturbing the classifier using dropout, it is possible to detect where predictions are more uncertain, which happens to be strongly related to the distance of feature representations from decision boundaries. In this novel adversarial scheme, the dense classifier is trained to improve its sensitivity to semantic variations on target representations. By acting against it, the feature extractor (*i.e.*, encoder) aims at providing categorical certainty to the target features it computes. This should effectively remove task-unrelated information enclosed in target representations, which is responsible for highly variable predictions, ultimately pushing them far from decision boundaries.

A downside of the classifier discrepancy strategy in its primary scheme is the inherent noise sensitivity acquired by the decoder (Saito et al., 2018a), which is crucial for it to capture the proximity of target samples to the classification boundaries, yet it negatively affects the accuracy of the whole segmentation network, requiring, in fact, an extra training stage to correctly learn the segmentation task. On this regard, the original scheme could be improved by replacing the dropout strategy to retrieve multiple predictions over the same feature representation with a couple of distinct decoders, which are simultaneously trained to provide correct, yet distinct, dense classifications (Saito et al., 2018b). This should effectively improve the accuracy of the decoder section of the prediction model, at the cost of an extra module to be learned within the training process.

Following a different path, the original adversarial learning scheme could be modified, opting for a non-stochastic virtual dropout mechanism to discover minimum distance dropout masks causing maximum prediction divergence (Lee et al., 2019b). By doing so, the original dropout-based solution to get distinct predictions from a single classifier is retained, while the aforementioned noise susceptibility problem is simultaneously solved.

The cotraining principle based on multiple predictors to estimate the current adaptation performance has further been investigated (Luo et al., 2019). In particular, the detection of inconsistent predictions can be exploited to focus the discriminator effort (now in a standard domain adversarial framework) towards less adapted regions of the input image, *i.e.*, those affected by the highest uncertainty. This has proved to ultimately lead towards a more effective domain alignment of source and target representations.

8.2.3.4 *Self-supervised learning*

Due to the similarity between the two tasks, multiple techniques for unsupervised domain adaptation have been borrowed from the semi-supervised learning (SSL) field. Indeed, UDA can be thought as an extension of the SSL problem, since in both cases part of the training data is unlabeled. However, the original lack of annotations of SSL within the unlabeled training set is joined by a statistical shift in UDA between source and target data, which demands for an extra effort to be addressed.

Self-training

In this direction, a first class of adaptation techniques (Zou et al., 2018, 2019; Li et al., 2019; Biasetton et al., 2019; Michieli et al., 2020; Spadotto et al., 2020; Choi et al., 2019; Chen et al., 2019a; Yang and Soatto, 2020; Zhou et al., 2020) have resorted to self-training. This approach, commonly employed in semi-supervised learning (Grandvalet and Bengio, 2005), revolves around the creation of pseudo-labels from highly confident network predictions inferred on unlabeled target data. A form of self-taught supervision is therefore available on the target domain, to be exploited in conjunction with the standard supervision from source labeled data.

As opposed to other adaptation approaches described in previous sections, such as the most successful adversarial ones, feature-level cross-domain alignment is implicitly pursued through target self-supervised learning, as source supervision is indirectly transferred to the target domain by pseudo-labels. Self-training pushes network probability outputs to reach a peaked distribution, which translates into predictions displaying a more confident behavior. A key issue, however, lies in the self-referential nature of this technique, which could lead to catastrophic error propagation if not properly handled. Over-confident incorrect predictions on uncertain pixels, in fact, may result undetected, since any form of supervision on unlabeled target data is missing. In turn, those prediction mistakes could be reinforced by the self-teaching strategy, causing a progressive deviation from the correct solution. To cope with this issue, most of self-training based adaptation approaches apply some filtering strategies to the pseudo-labeling process, so that prediction errors inherently affecting target segmentation maps are largely discarded.

A common approach towards a self-training based adaptation involves offline techniques for pseudo-label computation (Zou et al., 2018, 2019; Li et al., 2019), with the confidence threshold updated multiple times during the adaptation process by looking at the whole available training set. In particular, an iterative self-training optimization procedure is followed, which alternates steps of task supervised learning on both source original and target artificial annotations and pseudo-labeling to generate a self-taught target supervision. Thus, during an entire training stage, artificial target annotations are kept fixed. This offline strategy allows for a stable learning process, at the price of the extra computational burden due to multiple steps of pseudo-annotation on the whole target dataset.

In a different direction, the self-training strategy could be tied to an output level adversarial adaptation (Biasetton et al., 2019; Michieli et al., 2020; Spadotto et al., 2020). In particular, the output map from a fully convolutional output-level domain discriminator can be regarded as an accurate measure of prediction reliability on target data, thus providing useful information to refine target pseudo-labels. Thus, the quality of artificial annotations is progressively improved throughout the training process, being computed over single batches of target images rather than on the whole dataset, leading to an overall rather effective adaptation.

As an alternative, pseudo-label reliability can be enhanced by resorting to a form of prediction ensembling (Choi et al., 2019; Chen et al., 2019a; Yang and Soatto, 2020; Zhou et al., 2020). For example, it is possible to exploit an additional network to produce self-guidance over the unlabeled samples (Choi et al., 2019; Zhou et al., 2020). This is done by introducing a teacher network in addition to the original one, which plays the role of a student. Then, the teacher model, whose weights are averaged over the student ones from past training steps, is exploited to guide the learning process of the student network, by yielding target predictions the student network is compelled to emulate. The supervisory action provided by the teacher network leads to more accurate target predictions, on top of which less noisy pseudo-labeling can be performed. The result is a more effective self-training adaptation.

Entropy minimization

Adopted from the semisupervised learning field as well, entropy minimization has been recently introduced to UDA (Vu et al., 2019b). The intuition behind this approach is that source predictions are more inclined to show a confident behavior, which is revealed by a low entropy level in probability outputs. Conversely, the segmentation output maps from target inputs are likely to display more uncertainty (high entropy), with the noise pattern widely spread and not just limited to regions close to semantic boundaries. Thus, by mirroring the over-confident source behavior in the uncertain target domain, the segmentation network should, in principle, bridge the performance gap that exists between domains. More precisely, the effect of entropy minimization is to avoid classification boundaries in the latent space crossing high density regions, while, at the same time, target representations are well clustered far from those decision boundaries.

The original entropy minimization strategy (Vu et al., 2019b) works at the pixel-level, with each single spatial unit independently contributing to the overall objective. However, to overcome some inherent limitations of this approach, further arrangements have been introduced (Vu et al., 2019b; Chen et al., 2019a; Yang and Soatto, 2020). A possible solution is to pursue a global distribution alignment of entropy behavior by means of a domain adversarial approach, where the domain discriminator is provided with entropy maps rather than directly with output probability maps as in the standard Scheme 8.2.3.1 (Vu et al., 2019b). By doing so, structural information enclosed in entropy maps is leveraged, thus leading to a more effectively domain statistical adaptation. In addition, it should be remarked that the entropy minimization objective in its original form (Vu et al., 2019b) leads to rapidly exploding gradients when moving from high to low uncertainty regions, which could seriously hinder the learning process. On this regard, a solution could be to modify the standard objective, for example, with a quadratic loss with analogous purposes to the original one, but improved gradient signal properties (Chen et al., 2019a). This strategy, together with category-wise weighting factors to balance the contribution of different semantic classes, has proved to greatly enhance adaptation process.

Entropy minimization has been used together with feature space shaping techniques in a couple of recent works (Toldo et al., 2021; Barbato et al., 2021). Toldo et al. (2021), besides using entropy minimization, forces internal feature representations to be clustered, sparse and orthogonal (if belonging to different classes) in both source and target domains to improve feature-level adaptation. Another recent work (Barbato et al., 2021) further introduces a norm alignment constraint to aid a class-wise feature orthogonality objective in promoting disjoint sets of active feature channels between distinct semantic categories, while driving target embeddings towards the highly confident (*i.e.*, associated with high values of feature norm) source distribution.

8.2.3.5 Multitasking

The last class of adaptation techniques to be discussed regards multitasking (Lee et al., 2019a; Vu et al., 2019a; Chen et al., 2019b; Watanabe et al., 2018). A regularizing action is provided by solving multiple related tasks (*e.g.*, many approaches focus on depth regression) in addition to the semantic segmentation one. The goal is to implicitly extract domain invariant and semantically meaningful representations from images, as they should be more suitable to simultaneously address the related tasks. Depth regression is usually tackled in combination with semantic segmentation, to regularize an input level adaptation process based on source to target image translation (Section 8.2.3.2). Multitask adaptation has also shown to be effectively integrated with other adaptation approaches. For example, it can be exploited to enhance the maximum classifier strategy (Section 8.2.3.3), where instead of a single dense classifier two separate decoder modules are employed to obtain both depth and segmentation maps (Watanabe et al., 2018). It is possible to combine multitasking with an adversarial entropy minimization technique (Section 8.2.3.4) (Vu et al., 2019a), by fusing self-information maps with depth prediction ones before feeding them to a domain discriminator. By doing so, the detection capability of the domain discriminator in identifying domain discrepancy over source and target representations is boosted, thus providing, in the end, a more robust statistical alignment.

8.3 Continual learning

Recently, deep learning techniques have evolved rapidly to tackle a wide variety of tasks considered extremely challenging beforehand, in particular in the computer vision field, where deep learning models are able to achieve human-like performance in many tasks. Deep learning techniques have matured along the way and the transition from academic research to various practical and industrial applications has begun to be successful. However, practical applications soon raised the need for techniques able to improve the learned knowledge over time in order to accomplish new tasks without forgetting previous knowledge. This represents the building paradigm of continual learning in its essence. In other words, when deep learning models are deployed into the real-world, we would like to have the possibility of improving the capability of the models with new experiences or labels, without retraining them from scratch.

In general, the main issue of these computational models is that they are prone to *catastrophic forgetting* (McClelland et al., 1995; McCloskey and Cohen, 1989), *i.e.*, training a model with new information interferes with previously learned knowledge and typically greatly degrades the performance. Deep learning models, in particular, assume that all the data samples are available during the training phase and, therefore, they require that the training is performed on the entire dataset in order to adapt to changes in the data distribution. When trained on sequential tasks with samples progressively available over time, the performance significantly decreases on previously learned tasks as the network parameters are optimized for the new task without accounting for the old ones, if no *ad-hoc* provisions are employed (Kemker et al., 2018; Parisi et al., 2019).

Continual learning (also called incremental learning, lifelong learning or never ending learning), then, is the set of techniques designed to face this challenging scenario in which a sequence of tasks comes in succession. Despite being a long-standing problem in computational models (McCloskey and Cohen, 1989), in deep learning it has been tackled with some successes only recently.

To further prove its relevance, it is worthwhile to consider an analogy between machine learning and human learning. Indeed, humans encounter a continual stream of learning tasks and are able to generalize to similar unseen tasks (Thrun and Pratt, 2012). To understand the brain mechanisms and to translate them into computational models, many connectionist and biologically-inspired attempts have been made throughout the years (Grossberg, 2013; Ditzler et al., 2015).

Recently, the problem has been actively investigated in some image-level visual tasks (*i.e.*, with one or few labels per each image) such as image classification (Rebuffi et al., 2017; Li and Hoiem, 2017; Castro et al., 2018) and object detection (Shmelkov et al., 2017). In dense labeling tasks such as semantic segmentation, the problem has been faced only recently (Ozdemir and Goksel, 2019; Tasar et al., 2019; Michieli and Zanuttigh, 2019, 2021b; Cermelli et al., 2020; Klingner et al., 2020; Douillard et al., 2021) due to the inherent increased complexity.

Before digging into the definition of continual learning and exploring its application to dense labeling tasks, we point out some general reviews on the topic, not directly dealing with semantic segmentation. We refer to (French, 1999) for catastrophic forgetting in connectionist models, while Parisi et al. (2019) is the first review to critically compare recent works about the phenomenon in deep learning models. In De Lange et al. (2019) many approaches are compared into a common framework and in Lesort et al. (2020) the challenges of continual learning are described with a special focus on robotics.

8.3.1 Continual learning problem formulation

Continual learning (CL) could be regarded as a particular case of transfer learning, where the data domain distribution changes at every incremental step and the model should perform well on all the distributions. It is also strongly connected to the domain adaptation problem but, in this case, the focus is devoted toward both the input data and the annotations, whose distributions change over time and their number may be increased as well (*i.e.*, more classes to be distinguished).

On the other hand, we remark that the focus of domain adaptation discussed in the previous section was devoted toward the input domain distributions where, typically, a single domain shift is performed. Hybrid approaches combining UDA and CL are emerging to overcome the need for multiple changes in domains and tasks (Busto and Gall, 2017; Zhuo et al., 2019; Kundu et al., 2020).

Due to the intrinsic variety of challenges in continual learning and their respective difficulties, most of the approaches relax the general setup of continual learning to an easier one of incremental task learning. In the latter scenario, tasks are received one at the time and training is performed on the available training data.

Hence, this represents a mitigation of the true continual learning system, which is more likely to be encountered in practice (De Lange et al., 2019). For instance, in class-incremental learning, the learned model is updated to recognize new classes whilst preserving knowledge about previous ones. An overview of this setup is reported in Fig. 8.7. In more formal terms, we consider the t-th incremental step (with $t = 1, 2, ..., T_{max}$) and we are given the previous model $\mathcal{M}_{t-1}$ and two sets of data $\{\mathcal{X}^{(t)}, \mathcal{Y}^{(t)}\}$ randomly drawn from the distribution $\mathcal{D}^{(t)}$, which is an observation (or subset) of the complete domain $\mathcal{D}$. Here, $\mathcal{X}^{(t)}$ denotes a set of data samples for step t and $\mathcal{Y}^{(t)}$ denotes the corresponding ground truth annotations (*i.e.*, a single label for the image classification problem or a dense labeling map for the semantic

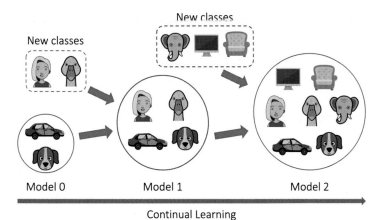

New classes

New classes

Model 0

Model 1

Model 2

Continual Learning

FIGURE 8.7 Graphical representation of the class-incremental continual learning framework. The model is updated to recognize new classes over time without forgetting previously learned ones.

segmentation problem). In the considered class-incremental setup, we assume that each step corresponds to a different learning task.

To mimic what happens in many real world scenarios and to reduce the need for storage or privacy limitations, most of the frameworks do not store any sample of data $\{\mathcal{X}^{(s)}, \mathcal{Y}^{(s)}\}$ for any step s preceding the current step t. Hence, the problem becomes even more challenging as the goal is to control an objective function comprising of all seen tasks without having access to previous samples. More formally, the empirical risk minimization framework translates into the research of the optimal parameters θ^* by optimizing:

$$\operatorname{argmin}_\theta \sum_{t=0}^{T} \mathbb{E}_{(\mathcal{X}^{(t)}, \mathcal{Y}^{(t)})} \left[\mathcal{L}\left(\mathcal{M}_t\left(\mathcal{X}^{(t)}; \theta\right), \mathcal{Y}^{(t)}\right) \right] \tag{8.1}$$

with model's parameters θ, loss function $\mathcal{L}$, T incremental tasks seen so far, and $\mathcal{M}_t$ the model function at step t. We remark, however, that this objective function cannot be optimized directly as old samples may not be present at all or may be very limited (depending on the continual learning scenario, see Section 8.3.2). We refer to the case in which all samples are available from the beginning as *joint training*, representing an upper bound of the performance of a continual learning system (*i.e.*, a single stage of training with all samples).

Furthermore, it is useful to gain insight of the problem in terms of marginal output and input distributions, *i.e.*, $P(\mathcal{Y}^{(t)})$ and $P(\mathcal{X}^{(t)})$ respectively, of a generic step t. In general, task incremental learning considers that $P(\mathcal{Y}^{(t+1)}) \neq P(\mathcal{Y}^{(t)})$ due to $P(\mathcal{X}^{(t+1)}) \neq P(\mathcal{X}^{(t)})$ and that the task output spaces differ over time, *i.e.*, $\{\mathcal{Y}^{(t)}\} \neq \{\mathcal{Y}^{(t+1)}\}$. Continuing our analogy to the UDA scenario, we could bring task incremental learning back to UDA considering $P(\mathcal{Y}^{(t+1)}) \neq P(\mathcal{Y}^{(t)})$ due to $P(\mathcal{X}^{(t+1)}) \neq P(\mathcal{X}^{(t)})$, but $\{\mathcal{Y}^{(t)}\} = \{\mathcal{Y}^{(t+1)}\}$, with the number of incremental steps $T_{max} = 1$ and a typically sudden change in the data domain distribution, while changes are in general more gradual in continual learning (De Lange et al., 2019; Hsu et al., 2018).

Finally, we remark that ideal continual learning setups consider infinite and continuous stream of training data and at each step the system receives some new samples drawn non-i.i.d. from the current distribution $\mathcal{D}^{(t)}$ that could itself experience sudden or gradual changes with no notification. This is what methods developed in the future should aim to tackle and realize.

8.3.2 Continual learning setups in semantic segmentation

Despite being a quite recent field, continual learning in semantic segmentation already comes in different flavors. In particular, existing works differ in the consideration of the domain distributions $\mathcal{D}^{(t)}$ and of the data sampling $\{\mathcal{X}^{(t)}, \mathcal{Y}^{(t)}\}$. The different choices emerge from different target applications. Let us denote with $\mathcal{S}^{(t-1)}$ the previous label set, which is expanded with a set of new classes $\mathcal{C}^{(t)}$ at step t, yielding a new label set $\mathcal{S}^{(t)} = \mathcal{S}^{(t-1)} \cup \mathcal{C}^{(t)}$. As typically assumed in task incremental settings, the sets of new labels discovered at each step are disjoint, except for a special *background* or *void* class which behavior and meaning depends on the selected scenario. There are many possible scenarios and one of the key differences lies in the way the background class is considered, which is typical of many semantic segmentation benchmarks. Previous approaches proposed four main different scenarios:

1. **Sequential masked.** This setup reflects the simplest idea on continual semantic segmentation; *i.e.*, each learning step contains a unique set of images, whose pixels belong either to novel classes or to a void class, which is not predicted by the model and it is masked out from both the results and the training procedure. This setup has been used in Tasar et al. (2019); Klingner et al. (2020).

2. **Sequential.** This setup has been proposed in Michieli and Zanuttigh (2019, 2021b). Each learning step contains a unique set of images, whose pixels belong to classes seen either in the current or in the previous learning steps. At each step, labels for pixels of both novel classes and old ones are present; however, the specific occurrence of a particular old class is highly correlated to the set of classes being added. For example, if the set of all old classes is $\mathcal{S}^{(t-1)} = \{chair, airplane\}$ and the set of classes being added is $\mathcal{C}^{(t)} = \{dining\ table\}$, then it is reasonable to expect that $\{\mathcal{X}^{(t)}, \mathcal{Y}^{(t)}\}$ contains some images with the *chair* class, that typically appears together with the *dining table*, while the class *airplane* is extremely unlikely to occur.

3. **Disjoint.** This setup has been proposed in Cermelli et al. (2020); Michieli and Zanuttigh (2021b). At each learning step, the unique set of images is identical to the sequential setup. The difference with respect to the sequential setup lies in the set of labels. At each step, only labels for pixels of novel classes are present, while the old ones are labeled as background in the segmentation maps (this causes the *background* class to change distribution at each step).

4. **Overlapped.** This setup moves from the work of Shmelkov et al. (2017) for object detection and has been addressed in Cermelli et al. (2020); Douillard et al. (2021); Michieli and Zanuttigh (2021a) for semantic segmentation. In this setup, each training step contains all the images that have at least one pixel of a novel set of classes, with only the classes of the set annotated and the rest set to *background*. Differently from the other settings, in this scenario images may contain pixels of classes that will be learned in the future, but labeled

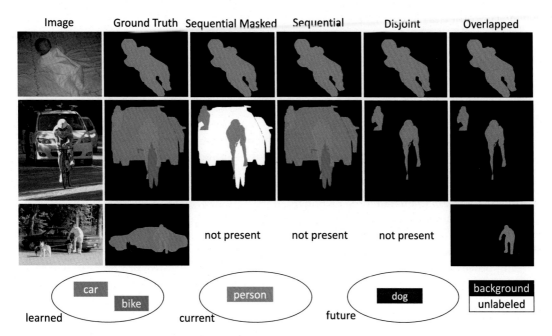

FIGURE 8.8 Overview of the different setups for class-incremental continual learning in semantic segmentation. The black class represents the *background* class and the white one represents the *void/unlabeled*.

as *background* in the current step; for this reason, as in the previous setting, the *background* class changes distribution at every incremental step.

A few examples of the different semantic map annotations are given in Fig. 8.8. Although being subcategories of the same problem they lead to substantially different setups requiring different strategies to be tackled.

This articulated scenario is getting even more articulated as there exist many different ways of sampling the sets $\mathcal{C}^{(t)}$ of unseen classes and of selecting its cardinality $|\mathcal{C}^{(t)}|$, leading to completely different experiments. For instance, let us consider one of the most widely used benchmarks for semantic segmentation, *i.e.*, the Pascal VOC2012 dataset (Everingham et al., 2010), which is composed by 21 semantic classes (*background* included). Regarding the first aspect, one possibility is to sort the classes using a pre-defined order provided by the dataset (*e.g.*, the alphabetical ordering for VOC2012) as done in (Shmelkov et al., 2017; Michieli and Zanuttigh, 2019, 2021b; Cermelli et al., 2020; Douillard et al., 2021; Michieli and Zanuttigh, 2021a). Another possibility is to sort the classes based on their occurrence inside the dataset (Michieli and Zanuttigh, 2021b), to reflect the idea that, in real world application, it is more likely to start from common classes and introduce rarer ones later. With respect to the second aspect, one may add a single class, a batch of classes or multiple classes sequentially one after the other (Michieli and Zanuttigh, 2019, 2021b; Cermelli et al., 2020; Douillard et al., 2021; Michieli and Zanuttigh, 2021a). All these possibilities open up a very variegate picture, which is being explored only recently, hence many research directions remain still unexplored.

8.3.3 Incremental learning techniques

In this section we are going to review the main methods to tackle task-incremental semantic segmentation grouped by employed technique. We will also refer the interested reader to some relevant works in task incremental image classification, this related field being more explored and mature with respect to semantic segmentation.

8.3.3.1 *Knowledge distillation*

The first family of approaches we present is the most commonly employed one thanks to its simplicity and efficacy; *i.e.*, knowledge distillation. This technique was originally proposed by Bucilua et al. (2006) and Hinton et al. (2015) to preserve the output of a complex ensemble of networks when adopting a simpler network for more efficient deployment. The idea was adapted to maintain unchanged the responses of the network on the old tasks whilst updating it with new training samples typically associated to new tasks. This is typically performed applying a constraint (*e.g.*, a loss function) in order to mimic the responses of the previous model in the current one. Its main effect is to act as a powerful regularization term during the learning process of the current classes, often leading to better performance on both previous and current classes (by preserving the capability of recognizing the former set and avoiding the overestimation of the latter set).

Knowledge distillation has been explored in different setups and it is somehow a prerequisite for successful task incremental learning algorithms. In sparse tasks, many algorithms use knowledge distillation in different flavors: Shmelkov et al. (2017) propose an end-to-end learning framework where the representation and the classifier are learned jointly without storing any of the original training samples. Li and Hoiem (2017) distill previous knowledge directly from the last trained model. Dhar et al. (2019) introduce an attention distillation loss as an information preserving penalty for the classifiers' attention maps. Zhou et al. (2019) distill knowledge from all previous model snapshots, of which a pruned version is saved.

These techniques have been found to be extremely effective and reliable also in dense tasks. Ozdemir and Goksel (2019) extend the image classification model of Li and Hoiem (2017) to segmentation simply constructing a knowledge distillation loss as the cross entropy between previous and current model's output probabilities. The authors also devise a strategy to select relevant samples of old data for rehearsal, that improve performances, but violate the assumption used in many scenarios of avoiding previous data storage. Tasar et al. (2019) apply knowledge distillation via cross entropy between previous and current model's output probabilities for each class, as the model predicts binary segmentation maps for each class separately. Michieli and Zanuttigh (2019) evaluate on a standard semantic segmentation benchmark and propose to apply knowledge distillation not only at the output level but also at the intermediate feature space to preserve the geometrical relationships of the extracted features. The work is extended in Michieli and Zanuttigh (2021b) that introduces and compares many knowledge distillation techniques. In particular, distillation on the output layer is enriched by temperature scaling (*i.e.*, rescaling softmax probabilities by a so-called temperature factor) to consider also the uncertainty of the estimations of previous models. Distillation on intermediate feature level is extended to multiple decoding stages and a scheme inspired by Similarity-Preserving Knowledge Distillation (Tung and Mori, 2019) is also proposed. Cermelli et al. (2020) propose a revisited distillation loss on the output level which accounts for the fact that a previous model could have already seen previous classes labeled as *background* (*i.e.*, in the *overlapped* setup). Klingner et al. (2020) propose a masked and weighted distillation loss on the output level to improve the accuracy on small or under-represented classes within the dataset. Finally, Douillard et al. (2021) applies a matching distillation to retain both long-range and short-range statistics at different feature levels and at different scales between the old and current model.

8.3.3.2 Parameter freezing

One of the major achievements of the early connectionist works is that they identified one main strategy to address catastrophic forgetting: *i.e.*, by freezing part of the network weights (Rebuffi et al., 2017). This technique has been applied by a large number of contemporary approaches as a regularization attempt to prevent knowledge degradation caused by upcoming tasks. For instance, Shmelkov et al. (2017) experiment on freezing either all the layers (except for the last) or part of them. Mańdziuk and Shastri (2002) tries to identify and freeze a compact subset of features (nodes) in the hidden layers, that are crucial for the current task, thus preventing forgetting in the future. Similarly, Kirkpatrick et al. (2017) remember old tasks by slowing down the learning process on the relevant weights for those tasks. Jung et al. (2016) try to maintain the performance on old tasks by freezing the final layer and discouraging the change of shared weights in feature extraction layers.

Also in the dense labeling task, parameter freezing has been proposed as a way to prevent forgetting. Michieli and Zanuttigh (2019) propose to freeze all the layers of the encoder in order to preserve unaltered the feature extraction capabilities and only train the decoding parameters. Michieli and Zanuttigh (2021b) exploit the idea of freezing only the first couple of layers of the encoder, to preserve the most task-agnostic part of the feature extractor. However, the choice of which layers to freeze remains an open question and there is an intrinsic trade-off between the capability of efficiently learning new tasks and the preservation of the acquired knowledge. A first attempt of automatic selection of which layers to freeze has been recently introduced by Nguyen et al. (2020) checking the most plastic layers of the network.

8.3.3.3 Geometrical feature-level regularization

The analysis of the latent space organization is becoming crucial towards understanding and improvement of deep neural networks (Bengio et al., 2013; Girshick et al., 2014; Xian et al., 2016; Peng et al., 2019). Recently, some attention has been devoted to latent regularization in continual image classification (Achille et al., 2018; Javed and White, 2019) and in unsupervised domain adaptation (Toldo et al., 2021; Barbato et al., 2021). The key idea of these approaches is to disentangle the intermediate feature space in different ways, to space apart features of different classes. In continual learning this can reduce the overlap when future classes are introduced in the model.

The only work exploiting this idea in dense tasks is Michieli and Zanuttigh (2021a), where the latent space is constrained to reduce forgetting whilst improving the recognition of novel classes. The framework is driven by three main components: first, prototype matching enforces latent space consistency on old classes, constraining the encoder to produce similar latent representation for previously seen classes in the subsequent steps; second, feature sparsification allows to make room in the latent space to accommodate novel classes; third, contrastive learning is employed to cluster features according to their semantics while tearing apart those of different classes.

8.3.3.4 New directions

Other novel ideas have been proposed in the continual learning literature both for dense and sparse tasks. Here, we present and discuss some of the most promising research directions.

Weights Initialization has been employed in Cermelli et al. (2020) to deal with the atypical behavior of the background class in the *disjoint* and *overlapped* scenarios. The authors initialize the classifier's parameters for the novel classes in such a way that the probability of the background is uniformly spread among the novel classes, preventing the model to be biased toward the background class when dealing with unseen classes.

Generative Replay methods train generative models on the current data distribution; afterwards, it is possible to sample data from past experience when learning on new data. By learning on actual data mixed with artificially generated past data, they try to preserve past knowledge while learning the new task. The generative model is generally a GAN (Goodfellow et al., 2014) as in Wu et al. (2018) and Shin et al. (2017) or an auto-encoder as in Draelos et al. (2017) and Kamra et al. (2017). Notice that only the weak classification labeling is available for generated data and some pseudo-labels need to be estimated for segmentation.

Webly-based Learning models (Hou et al., 2018; Modolo and Ferrari, 2017), *i.e.*, models that learn from samples acquired via web searches, could be an extremely powerful tool to retrieve faithful past examples using as queries the label names of the old classes to preserve. Also in this case, only weak classification labels are available and some pseudo-labeling scheme need to be introduced.

8.4 Conclusion

Semantic image segmentation is an active research field aiming at detailed and accurate scene understanding. Being a dense labeling task, it brings additional complexity with respect to classical image classification problems. To solve this task many deep learning models have been proposed, however such architectures require large labeled training datasets and show bad adaptation capabilities to unseen domains or unseen tasks. To tackle these two aspects, a large research activity has been conducted on both domain adaptation and continual learning in the past few years. In this chapter, we discuss domain adaptation in its unsupervised form (*i.e.*, when no labeled data of the target domain is used for training) applied to the dense semantic segmentation task. Then, we showed that continual learning can be achieved in this task in various possible ways.

The algorithms developed so far are able to greatly alleviate the degradation even if they still need to acquire greater maturity before being applied in delicate real-world scenarios (such as in autonomous driving). Indeed, there still remain many challenges to address as regards both the adaptation of complex deep learning architectures across different tasks and domains and their ability to learn new concepts over time without forgetting previous knowledge.

In addition, many ancillary closely related problems are emerging in response to the practical application to real-world use cases. Future directions should include, for instance, the generalization from or to multiple data distributions, open-set UDA (*i.e.*, recognizing in the target domain classes never seen in the source one), and continual UDA (*i.e.*, the continual adaptation process to unseen domains and unseen tasks).

Acknowledgment

Our work was in part supported by the Italian Minister for Education (MIUR) under the "Departments of Excellence" initiative (Law 232/2016).

References

Achille, A., Eccles, T., Matthey, L., Burgess, C., Watters, N., Lerchner, A., Higgins, I., 2018. Life-long disentangled representation learning with cross-domain latent homologies. In: Neural Information Processing Systems (NeurIPS).

Barbato, F., Toldo, M., Michieli, U., Zanuttigh, P., 2021. Latent space regularization for unsupervised domain adaptation in semantic segmentation. In: Proceedings of the IEEE Conference on Computer Vision and Pattern Recognition Workshops (CVPRW).

Bengio, Y., Courville, A., Vincent, P., 2013. Representation learning: a review and new perspectives. In: IEEE Transactions on Pattern Analysis and Machine Intelligence (TPAMI). IEEE, pp. 1798–1828.

Biasetton, M., Michieli, U., Agresti, G., Zanuttigh, P., 2019. Unsupervised domain adaptation for semantic segmentation of urban scenes. In: Proceedings of the IEEE Conference on Computer Vision and Pattern Recognition Workshops (CVPRW).

Bucher, M., Vu, T.H., Cord, M., Pérez, P., 2020. Buda: boundless unsupervised domain adaptation in semantic segmentation. arXiv preprint. arXiv:2004.01130.

Bucilua, C., Caruana, R., Niculescu-Mizil, A., 2006. Model compression. In: Proc. of the 12th ACM SIGKDD International Conference on Knowledge Discovery and Data Mining, pp. 535–541.

Busto, P.P., Gall, J., 2017. Open set domain adaptation. In: Proceedings of the International Conference on Computer Vision (ICCV), pp. 754–763.

Castro, F.M., Marín-Jiménez, M.J., Guil, N., Schmid, C., Alahari, K., 2018. End-to-end incremental learning. In: Proceedings of the European Conference on Computer Vision (ECCV), pp. 233–248.

Cermelli, F., Mancini, M., Bulò, S.R., Ricci, E., Caputo, B., 2020. Modeling the background for incremental learning in semantic segmentation. In: Proceedings of the IEEE Conference on Computer Vision and Pattern Recognition (CVPR).

Chang, W., Wang, H., Peng, W., Chiu, W., 2019. All about structure: adapting structural information across domains for boosting semantic segmentation. In: Proceedings of the IEEE Conference on Computer Vision and Pattern Recognition (CVPR), pp. 1900–1909.

Chen, Y., Li, W., Chen, X., Van Gool, L., 2018. Learning semantic segmentation from synthetic data: a geometrically guided input-output adaptation approach. arXiv preprint. arXiv:1812.05040.

Chen, M., Xue, H., Cai, D., 2019a. Domain adaptation for semantic segmentation with maximum squares loss. In: Proceedings of the International Conference on Computer Vision (ICCV).

Chen, Y., Li, W., Chen, X., Gool, L.V., 2019b. Learning semantic segmentation from synthetic data: a geometrically guided input-output adaptation approach. In: Proceedings of the IEEE Conference on Computer Vision and Pattern Recognition (CVPR), pp. 1841–1850.

Chen, Y.C., Lin, Y.Y., Yang, M.H., Huang, J.B., 2019c. Crdoco: pixel-level domain transfer with cross-domain consistency. In: Proceedings of the IEEE Conference on Computer Vision and Pattern Recognition (CVPR).

Chen, Y.H., Chen, W.Y., Chen, Y.T., Tsai, B.C., Frank Wang, Y.C., Sun, M., 2017. No more discrimination: cross city adaptation of road scene segmenters. In: Proceedings of the International Conference on Computer Vision (ICCV), pp. 1992–2001.

Choi, J., Kim, T., Kim, C., 2019. Self-ensembling with gan-based data augmentation for domain adaptation in semantic segmentation. In: Proceedings of the International Conference on Computer Vision (ICCV), pp. 6830–6840.

De Lange, M., Aljundi, R., Masana, M., Parisot, S., Jia, X., Leonardis, A., Slabaugh, G., Tuytelaars, T., 2019. Continual learning: a comparative study on how to defy forgetting in classification tasks. arXiv preprint. arXiv:1909.08383.

Deng, J., Dong, W., Socher, R., Li, L., Li, K., Li, F., 2009. Imagenet: a large-scale hierarchical image database. In: Proceedings of the IEEE Conference on Computer Vision and Pattern Recognition (CVPR), pp. 248–255.

Dhar, P., Singh, R.V., Peng, K.C., Wu, Z., Chellappa, R., 2019. Learning without memorizing. In: Proceedings of the IEEE Conference on Computer Vision and Pattern Recognition (CVPR), pp. 5138–5146.

Ditzler, G., Roveri, M., Alippi, C., Polikar, R., 2015. Learning in nonstationary environments: a survey. IEEE Computational Intelligence Magazine 10, 12–25.

Douillard, A., Chen, Y., Dapogny, A., Cord, M., 2021. Plop: learning without forgetting for continual semantic segmentation. In: Proceedings of the IEEE Conference on Computer Vision and Pattern Recognition (CVPR).

Draelos, T.J., Miner, N.E., Lamb, C.C., Cox, J.A., Vineyard, C.M., Carlson, K.D., Severa, W.M., James, C.D., Aimone, J.B., 2017. Neurogenesis deep learning: extending deep networks to accommodate new classes. In: 2017 International Joint Conference on Neural Networks (IJCNN). IEEE, pp. 526–533.

Du, L., Tan, J., Yang, H., Feng, J., Xue, X., Zheng, Q., Ye, X., Zhang, X., 2019. SSF-DAN: separated semantic feature based domain adaptation network for semantic segmentation. In: Proceedings of the International Conference on Computer Vision (ICCV).

Dundar, A., Liu, M., Wang, T., Zedlewski, J., Kautz, J., 2018. Domain stylization: a strong, simple baseline for synthetic to real image domain adaptation. arXiv preprint. arXiv:1807.09384.

Everingham, M., Van Gool, L., Williams, C.K., Winn, J., Zisserman, A., 2010. The Pascal visual object classes (VOC) challenge. International Journal of Computer Vision 88, 303–338.

French, R.M., 1999. Catastrophic forgetting in connectionist networks. Trends in Cognitive Sciences 3, 128–135.

Ganin, Y., Lempitsky, V., 2015. Unsupervised domain adaptation by backpropagation. In: Proceedings of the International Conference on Machine Learning (ICML), pp. 1180–1189.

Ganin, Y., Ustinova, E., Ajakan, H., Germain, P., Larochelle, H., Laviolette, F., Marchand, M., Lempitsky, V., 2016. Domain-adversarial training of neural networks. Journal of Machine Learning Research 17, 2096–2130.

Girshick, R., Donahue, J., Darrell, T., Malik, J., 2014. Rich feature hierarchies for accurate object detection and semantic segmentation. In: Proceedings of the IEEE Conference on Computer Vision and Pattern Recognition (CVPR), pp. 580–587.

Gong, R., Li, W., Chen, Y., Gool, L.V., 2019. DLOW: domain flow for adaptation and generalization. In: Proceedings of the IEEE Conference on Computer Vision and Pattern Recognition (CVPR), pp. 2477–2486.

Goodfellow, I., Pouget-Abadie, J., Mirza, M., Xu, B., Warde-Farley, D., Ozair, S., Courville, A., Bengio, Y., 2014. Generative adversarial nets. In: Neural Information Processing Systems (NeurIPS), pp. 2672–2680.

Grandvalet, Y., Bengio, Y., 2005. Semi-supervised learning by entropy minimization. In: Actes de CAP 05, Conférence francophone sur l'apprentissage automatique, pp. 281–296.

Grossberg, S., 2013. Adaptive resonance theory: how a brain learns to consciously attend, learn, and recognize a changing world. Neural Networks 37, 1–47.

Hinton, G., Vinyals, O., Dean, J., 2015. Distilling the knowledge in a neural network. arXiv preprint. arXiv:1503.02531.

Hoffman, J., Tzeng, E., Park, T., Zhu, J.Y., Isola, P., Saenko, K., Efros, A., Darrell, T., 2018. Cycada: cycle-consistent adversarial domain adaptation. In: Proceedings of the International Conference on Machine Learning (ICML).

Hoffman, J., Wang, D., Yu, F., Darrell, T., 2016. FCNs in the wild: Pixel-level adversarial and constraint-based adaptation. arXiv preprint. arXiv:1612.02649.

Hou, Q., Cheng, M.M., Liu, J., Torr, P.H., 2018. Webseg: learning semantic segmentation from web searches. arXiv preprint. arXiv:1803.09859.

Hsu, Y.C., Liu, Y.C., Ramasamy, A., Kira, Z., 2018. Re-evaluating continual learning scenarios: a categorization and case for strong baselines. arXiv preprint. arXiv:1810.12488.

Javed, K., White, M., 2019. Meta-learning representations for continual learning. In: Neural Information Processing Systems (NeurIPS).

Jung, H., Ju, J., Jung, M., Kim, J., 2016. Less-forgetting learning in deep neural networks. arXiv preprint. arXiv:1607.00122.

Kamra, N., Gupta, U., Liu, Y., 2017. Deep generative dual memory network for continual learning. arXiv preprint. arXiv:1710.10368.

Kemker, R., McClure, M., Abitino, A., Hayes, T.L., Kanan, C., 2018. Measuring catastrophic forgetting in neural networks. In: Thirty-Second AAAI Conference on Artificial Intelligence.

Kirkpatrick, J., Pascanu, R., Rabinowitz, N., Veness, J., Desjardins, G., Rusu, A.A., Milan, K., Quan, J., Ramalho, T., Grabska-Barwinska, A., et al., 2017. Overcoming catastrophic forgetting in neural networks. Proceedings of the National Academy of Sciences (PNAS) 114, 3521–3526.

Klingner, M., Bär, A., Donn, P., Fingscheidt, T., 2020. Class-incremental learning for semantic segmentation re-using neither old data nor old labels. In: IEEE International Conference on Intelligent Transportation Systems (ITSC).

Kundu, J.N., Venkatesh, R.M., Venkat, N., Revanur, A., Babu, R.V., 2020. Class-incremental domain adaptation. In: Proceedings of the European Conference on Computer Vision (ECCV).

Lee, K., Ros, G., Li, J., Gaidon, A., 2019a. SPIGAN: privileged adversarial learning from simulation. In: International Conference on Learning Representations (ICLR).

Lee, S., Kim, D., Kim, N., Jeong, S.G., 2019b. Drop to adapt: learning discriminative features for unsupervised domain adaptation. In: Proceedings of the International Conference on Computer Vision (ICCV), pp. 91–100.

Lesort, T., Lomonaco, V., Stoian, A., Maltoni, D., Filliat, D., Díaz-Rodríguez, N., 2020. Continual learning for robotics: definition, framework, learning strategies, opportunities and challenges. Information Fusion 58, 52–68.

Li, P., Liang, X., Jia, D., Xing, E.P., 2018. Semantic-aware grad-gan for virtual-to-real urban scene adaption. In: Proceedings of British Machine Vision Conference (BMVC).

Li, Y., Yuan, L., Vasconcelos, N., 2019. Bidirectional learning for domain adaptation of semantic segmentation. In: Proceedings of the IEEE Conference on Computer Vision and Pattern Recognition (CVPR).

Li, Z., Hoiem, D., 2017. Learning without forgetting. IEEE Transactions on Pattern Analysis and Machine Intelligence (TPAMI) 40, 2935–2947.

Long, M., Cao, Y., Wang, J., Jordan, M., 2015. Learning transferable features with deep adaptation networks. In: Proceedings of the International Conference on Machine Learning (ICML), pp. 97–105.

Luo, Y., Zheng, L., Guan, T., Yu, J., Yang, Y., 2019. Taking a closer look at domain shift: category-level adversaries for semantics consistent domain adaptation. In: Proceedings of the IEEE Conference on Computer Vision and Pattern Recognition (CVPR).

Mańdziuk, J., Shastri, L., 2002. Incremental class learning approach and its application to handwritten digit recognition. Information Sciences 141, 193–217.

McClelland, J.L., McNaughton, B.L., O'Reilly, R.C., 1995. Why there are complementary learning systems in the hippocampus and neocortex: insights from the successes and failures of connectionist models of learning and memory. Psychological Review 102, 419.

McCloskey, M., Cohen, N.J., 1989. Catastrophic interference in connectionist networks: the sequential learning problem. In: Psychology of Learning and Motivation, vol. 24. Elsevier, pp. 109–165.

Michieli, U., Biasetton, M., Agresti, G., Zanuttigh, P., 2020. Adversarial learning and self-teaching techniques for domain adaptation in semantic segmentation. IEEE Transaction on Intelligent Vehicles.

Michieli, U., Zanuttigh, P., 2019. Incremental learning techniques for semantic segmentation. In: Proceedings of the International Conference on Computer Vision Workshops (ICCVW).

Michieli, U., Zanuttigh, P., 2021a. Continual semantic segmentation via repulsion-attraction of sparse and disentangled latent representations. In: Proceedings of the IEEE Conference on Computer Vision and Pattern Recognition (CVPR).

Michieli, U., Zanuttigh, P., 2021b. Knowledge distillation for incremental learning in semantic segmentation. Computer Vision and Image Understanding 205, 103167.

Modolo, D., Ferrari, V., 2017. Learning semantic part-based models from Google images. IEEE Transactions on Pattern Analysis and Machine Intelligence (TPAMI) 40, 1502–1509.

Murez, Z., Kolouri, S., Kriegman, D.J., Ramamoorthi, R., Kim, K., 2018. Image to image translation for domain adaptation. In: Proceedings of the IEEE Conference on Computer Vision and Pattern Recognition (CVPR).

Nguyen, G., Chen, S., Do, T., Jun, T.J., Choi, H.J., Kim, D., 2020. Dissecting catastrophic forgetting in continual learning by deep visualization. arXiv preprint. arXiv:2001.01578.

Ozdemir, F., Goksel, O., 2019. Extending pretrained segmentation networks with additional anatomical structures. International Journal of Computer Assisted Radiology and Surgery 14, 1187–1195.

Parisi, G.I., Kemker, R., Part, J.L., Kanan, C., Wermter, S., 2019. Continual lifelong learning with neural networks: a review. Neural Networks.

Peng, X., Huang, Z., Sun, X., Saenko, K., 2019. Domain agnostic learning with disentangled representations. In: Proceedings of the International Conference on Machine Learning (ICML), PMLR, pp. 5102–5112.

Pizzati, F., Charette, R.d., Zaccaria, M., Cerri, P., 2020. Domain bridge for unpaired image-to-image translation and unsupervised domain adaptation. In: Proceedings of the Winter Conference on Applications of Computer Vision (WACV), pp. 2990–2998.

Qin, C., Wang, L., Zhang, Y., Fu, Y., 2019. Generatively inferential co-training for unsupervised domain adaptation. In: Proceedings of the International Conference on Computer Vision Workshops (ICCVW), pp. 1055–1064.

Rebuffi, S.A., Kolesnikov, A., Sperl, G., Lampert, C.H., 2017. Icarl: incremental classifier and representation learning. In: Proceedings of the IEEE Conference on Computer Vision and Pattern Recognition (CVPR), pp. 2001–2010.

Saito, K., Kim, D., Sclaroff, S., Saenko, K., 2020. Universal domain adaptation through self supervision. In: Neural Information Processing Systems (NeurIPS).

Saito, K., Ushiku, Y., Harada, T., Saenko, K., 2018a. Adversarial dropout regularization. In: International Conference on Learning Representations (ICLR).

Saito, K., Watanabe, K., Ushiku, Y., Harada, T., 2018b. Maximum classifier discrepancy for unsupervised domain adaptation. In: Proceedings of the IEEE Conference on Computer Vision and Pattern Recognition (CVPR), pp. 3723–3732.

Sankaranarayanan, S., Balaji, Y., Jain, A., Nam Lim, S., Chellappa, R., 2018. Learning from synthetic data: addressing domain shift for semantic segmentation. In: Proceedings of the IEEE Conference on Computer Vision and Pattern Recognition (CVPR), pp. 3752–3761.

Shin, H., Lee, J.K., Kim, J., Kim, J., 2017. Continual learning with deep generative replay. In: Neural Information Processing Systems (NeurIPS), pp. 2990–2999.

Shmelkov, K., Schmid, C., Alahari, K., 2017. Incremental learning of object detectors without catastrophic forgetting. In: Proceedings of the International Conference on Computer Vision (ICCV), pp. 3400–3409.

Spadotto, T., Toldo, M., Michieli, U., Zanuttigh, P., 2020. Unsupervised domain adaptation with multiple domain discriminators and adaptive self-training. In: Proceedings of the IEEE International Conference on Pattern Recognition (ICPR).

Tasar, O., Tarabalka, Y., Alliez, P., 2019. Incremental learning for semantic segmentation of large-scale remote sensing data. IEEE Journal of Selected Topics in Applied Earth Observations and Remote Sensing 12, 3524–3537.

Thrun, S., Pratt, L., 2012. Learning to Learn. Springer Science & Business Media.

Toldo, M., Michieli, U., Agresti, G., Zanuttigh, P., 2020. Unsupervised domain adaptation for mobile semantic segmentation based on cycle consistency and feature alignment. Image and Vision Computing.

Toldo, M., Michieli, U., Zanuttigh, P., 2021. Unsupervised domain adaptation in semantic segmentation via orthogonal and clustered embeddings. In: Proceedings of the Winter Conference on Applications of Computer Vision (WACV).

Tsai, Y.H., Hung, W.C., Schulter, S., Sohn, K., Yang, M.H., Chandraker, M., 2018. Learning to adapt structured output space for semantic segmentation. In: Proceedings of the IEEE Conference on Computer Vision and Pattern Recognition (CVPR), pp. 7472–7481.

Tsai, Y.H., Sohn, K., Schulter, S., Chandraker, M., 2019. Domain adaptation for structured output via discriminative patch representations. In: Proceedings of the International Conference on Computer Vision (ICCV), pp. 1456–1465.

Tung, F., Mori, G., 2019. Similarity-preserving knowledge distillation. In: Proceedings of the International Conference on Computer Vision (ICCV), pp. 1365–1374.

Vu, T., Jain, H., Bucher, M., Cord, M., Pérez, P., 2019a. DADA: depth-aware domain adaptation in semantic segmentation. In: Proceedings of the International Conference on Computer Vision (ICCV), pp. 7363–7372.

Vu, T.H., Jain, H., Bucher, M., Cord, M., Pérez, P., 2019b. Advent: adversarial entropy minimization for domain adaptation in semantic segmentation. In: Proceedings of the IEEE Conference on Computer Vision and Pattern Recognition (CVPR), pp. 2517–2526.

Wang, M., Deng, W., 2018. Deep visual domain adaptation: a survey. Neurocomputing 312, 135–153.

Watanabe, K., Saito, K., Ushiku, Y., Harada, T., 2018. Multichannel semantic segmentation with unsupervised domain adaptation. In: Proceedings of the European Conference on Computer Vision (ECCV).

Wu, Y., Chen, Y., Wang, L., Ye, Y., Liu, Z., Guo, Y., Zhang, Z., Fu, Y., 2018. Incremental classifier learning with generative adversarial networks. arXiv preprint. arXiv:1802.00853.

Wu, Z., Wang, X., Gonzalez, J., Goldstein, T., Davis, L., 2019. ACE: adapting to changing environments for semantic segmentation. In: Proceedings of the International Conference on Computer Vision (ICCV), pp. 2121–2130.

Xian, Y., Akata, Z., Sharma, G., Nguyen, Q., Hein, M., Schiele, B., 2016. Latent embeddings for zero-shot classification. In: Proceedings of the IEEE Conference on Computer Vision and Pattern Recognition (CVPR), pp. 69–77.

Yang, J., An, W., Wang, S., Zhu, X., Yan, C., Huang, J., 2020a. Label-driven reconstruction for domain adaptation in semantic segmentation. arXiv preprint. arXiv:2003.04614.

Yang, Y., Lao, D., Sundaramoorthi, G., Soatto, S., 2020b. Phase consistent ecological domain adaptation. arXiv preprint. arXiv:2004.04923.

Yang, Y., Soatto, S., 2020. FDA: Fourier domain adaptation for semantic segmentation. arXiv preprint. arXiv:2004.05498.

Zhang, Y., Qiu, Z., Yao, T., Liu, D., Mei, T., 2018. Fully convolutional adaptation networks for semantic segmentation. In: Proceedings of the IEEE Conference on Computer Vision and Pattern Recognition (CVPR), pp. 6810–6818.

Zhou, P., Mai, L., Zhang, J., Xu, N., Wu, Z., Davis, L.S., 2019. M2kd: multi-model and multi-level knowledge distillation for incremental learning. arXiv preprint. arXiv:1904.01769.

Zhou, Q., Feng, Z., Cheng, G., Tan, X., Shi, J., Ma, L., 2020. Uncertainty-aware consistency regularization for cross-domain semantic segmentation. arXiv preprint. arXiv:2004.08878.

Zhu, J., Park, T., Isola, P., Efros, A.A., 2017. Unpaired image-to-image translation using cycle-consistent adversarial networks. In: Proceedings of the International Conference on Computer Vision (ICCV).

Zhu, X., Zhou, H., Yang, C., Shi, J., Lin, D., 2018. Penalizing top performers: conservative loss for semantic segmentation adaptation. In: Proceedings of the European Conference on Computer Vision (ECCV), pp. 568–583.

Zhuo, J., Wang, S., Cui, S., Huang, Q., 2019. Unsupervised open domain recognition by semantic discrepancy minimization. In: Proceedings of the IEEE Conference on Computer Vision and Pattern Recognition (CVPR), pp. 750–759.

Zou, Y., Yu, Z., Liu, X., Kumar, B.V., Wang, J., 2019. Confidence regularized self-training. In: Proceedings of the International Conference on Computer Vision (ICCV), pp. 5982–5991.

Zou, Y., Yu, Z., Vijaya Kumar, B., Wang, J., 2018. Unsupervised domain adaptation for semantic segmentation via class-balanced self-training. In: Proceedings of the European Conference on Computer Vision (ECCV), pp. 289–305.

Biographies

Umberto Michieli received the M.Sc. degree in Telecommunication Engineering from the University of Padova in 2018. He is currently a final-year Ph.D. student at the same University. In 2018, he spent 6 months as a Visiting Researcher with the Technische Universität Dresden. In 2020 he interned as Research Engineer for 8 months at Samsung Research UK. His research focuses on transfer learning techniques for semantic segmentation, in particular on domain adaptation and on incremental learning.

Marco Toldo received the M.Sc. degree in ICT for Internet and Multimedia in 2019 at the University of Padova. At present, he is doing his Ph.D. at the Department of Information Engineering of the same university. In 2021 he interned as Research Engineer for 7 months at Samsung Research UK. His research interests involve domain adaptation and continual learning applied to computer vision.

Pietro Zanuttigh received the M.Sc. degree in Computer Engineering and the Ph.D. degree from the University of Padova, in 2003 and 2007, respectively. He is currently an Associate Professor with the Department of Information Engineering of the same university. His research interests include semantic understanding of images and 3D data, domain adaptation and incremental learning for visual data, 3D data processing with a special focus on ToF sensors, multiple sensors fusion and hand gesture recognition.

9

Visual tracking
Tracking in scenes containing multiple moving objects

Michael Felsberg[a,b]

[a]Computer Vision Laboratory, Department of Electrical Engineering, Linköping University, Linköping, Sweden [b]School of Engineering, University of KwaZulu-Natal, Durban, South Africa

---

CHAPTER POINTS

- Visual object tracking.
- Discriminative tracking approaches.
- Correlation filters.
- Deep features.

- Deep learning of tracking.
- Video object segmentation.
- Discriminative segmentation.

9.1 Introduction

Tracking is a highly ambiguous term, and even for visual tracking, many different interpretations exist. Therefore, this chapter starts with a thorough definition of the problem addressed to delineate this chapter from other interpretations and to make all assumptions explicit.

9.1.1 Problem definition

In this chapter, we will consider the problem of *generic visual tracking*, in the sense as defined in the visual object tracking (VOT) challenges 2013–2020 (Kristan et al., 2013, 2015a,b, 2016, 2017, 2019,b, 2020), which also coincides with the online tracking benchmark (OTB) definition (Wu et al., 2013). The tracking is performed in the image domain and no prior knowledge about object classes is available: generic means class agnostic.

Copyright © 2022 Elsevier Inc. All rights reserved.

The problem is formulated in the 2D image plane, in contrast to 3D tracking approaches (Garon and Lalonde, 2017). Also, only a single object (target) is considered in contrast to the multiple object tracking case (Dendorfer et al., 2020) that requires association of targets through the sequence. However, focusing on tracking a single object does not imply that no other moving objects, socalled distractors, exist in the sequences. The tracking task thus requires not to mix up the target with any of the distractors even if they partly occlude the target. It is assumed that the target is at least partly visible throughout the sequence such that no redetection is required. The sequence might originate from a moving camera, which excludes simple background modeling (Stauffer and Grimson, 2000) for detecting the target.

The tracking target is provided by a single annotation in the first frame of an image sequence in the form of a bounding-box (or a segmentation mask – see Section 9.5). The task is then to predict the bounding-box (or segmentation mask) that contains the same object in all subsequent frames. Bounding-boxes can be defined in many different coordinates; the exact format does not matter as long as it is consistently used. In the VOT challenge, the upper left corner (x, y) and the width w and height h are used (origin in the upper left). Thus, we get the formal task definition:

input: a video (image sequence) and one annotated bounding-box (x_0, y_0, w_0, h_0) for the initial frame in the sequence

output: prediction of bounding-boxes (x_k, y_k, w_k, h_k) for all frames $k > 0$

requirement: the prediction of the bounding-box for frame K may only use data from frames $k \leq K$.

This task definition formulates a one-shot learning problem, as one annotated training sample for the present sequence is provided. The goal of this learning problem is to train a detector that predicts the best fitting bounding-box in the subsequent frame, also called tracking-by-detection. Typically, the best fitting bounding-box is classified as the target, all others as background. However, it is also possible to use regression methods to determine the bounding-box parameters.

9.1.2 Challenges in tracking

The main difficulty in tracking by detection is that the object of interest might change appearance in the course of the sequence: It might

- rotate in the image plane
- change scale by moving in depth
- change aspect by out-of-plane rotations / changes in viewpoint
- be articulated
- suffer from motion blur
- undergo changes of illumination
- be partly occluded.

These changes might make the detector fail and lose the target, particularly in the presence of cluttered background and distractors. Therefore, modern tracking approaches usually adapt the detector model using its own predictions from previous frames; see Fig. 9.1.

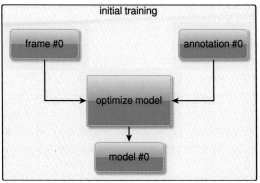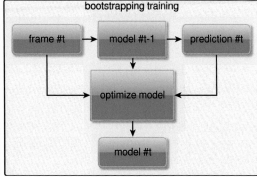

FIGURE 9.1 **Adaptive tracking model.** Block diagram of an adaptive tracking model. After initial training (left), the model adapts using its own predictions as training samples (right).

This adaptive process establishes a bootstrapping procedure, i.e., a recursive self-improvement that proceeds without external input other than the single initial annotation. One key problem of this process is to balance the previous model against the update information. If the model is too rigid, it will lose the target under appearance changes, if the model is too flexible, it will drift away from the target (Matthews et al., 2004; Wang et al., 2016).

9.1.3 Motivation of the setting

The drift problem is largely reduced if the class of the object of interest is known. However, as pointed out in Section 9.1.1, we aim at generic, i.e. class agnostic, models. The main motivation for this choice is the open-world assumption, which, in contrast to the closed-world assumption, does not infer the incorrectness of a statement from its absence (Reiter, 1978), or, translated to detection, does not infer the nonexistence of an object from the absence of a corresponding detector response.

There are two issues with the closed-world assumption in context of visual detectors:

- The visual appearance of objects is characterized by a high intra-class variability, e.g., cars and bicycles in traffic scenarios.
- The set of objects is continuously growing, e.g., e-bikes and hover-boards appear in traffic scenarios.

If an object is not detected due to the former reason, it means that the detector failed to model the variety of appearance of the object. This might be caused by limitations in the training set or by infinite possible variations of appearance. If an object is not detected due to the latter reason, it means that the object class is not represented in the training set at all. However, just because the data set fails to cover a certain class, does not mean that this class is irrelevant to the detector. Designing a system based on the premise that classes not present at training time are neglected is ethically problematic, e.g. in traffic safety applications.

Traffic safety systems are often based on other sensors than vision: Lidar, radar, ultrasound, etc. However, visual sensors will always be of central interest as they establish an approach to human-centered sensing. This happens in several ways: Systems should adapt to the environment that is shaped for humans – and since vision is humans' dominating sense involved in about 80% of our perception, learning, cognition, and activities (Ripley and Politzer, 2010), they need visual sensors. There are many examples for such environments, for instance, in the traffic scenario the highway code defines many signs and symbols that guide the driver visually. Also systems that share their workspace with humans for interaction and collaboration need visual capabilities. The appearance is a strong cue for behavior as well as intentions and humans use this actively by communicating in terms of, e.g., gestures. Finally, in order to be capable to predict human actions, their perception needs to be predicted as well – which is mostly based on vision.

9.1.4 Historical development

Many of the older approaches to visual tracking go back to problems within augmented reality, where visualization and sensing are combined into the same coordinate system to generate a mixed reality environment. Tracking is important here for two purposes: the relative head pose in the scene (Neumann et al., 1999) and the pose of moving objects (Neumann and Park, 1998).

Some major steps in the history of visual tracking as well as key works of the author are

1981: Lucas-Kanade algorithm (Lucas and Kanade, 1981)
1984: Phase-only matched filter (Horner and Gianino, 1984)
1994: "Tracking is a solved problem" (from the presentation of Shi and Tomasi, 1994)
1998: Phase only matched filters and equivariances (Felsberg, 1998)
2004: "Lucas-Kanade 20 Years On: A Unifying Framework" (Baker and Matthews, 2004)
2007: MATRIS project: L1-tracker and covariance (Skoglund and Felsberg, 2007)
2013: New tracking challenges OTB (Wu et al., 2013), VOT (Kristan et al., 2013), and a generative holistic tracker (Felsberg, 2013)
2014: Discriminative tracker (DSST) winner in VOT2014 (Danelljan et al., 2014a; Kristan et al., 2015a)
2018: Discriminative holistic trackers dominate VOT2018 with 75% (Kristan et al., 2019)
2020: Siamese trackers and segmentation methods gain importance (Kristan et al., 2020).

In the subsequent sections, several major developments during this period of 40 years are described in some detail, with a particular focus on discriminative and learning-based methods.

9.2 Template-based methods

Before looking deeper into the learning-based methods, we will briefly discuss relevant methods that build entirely on the initial annotated bounding-box, often called the template.

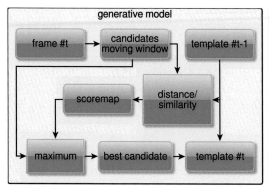
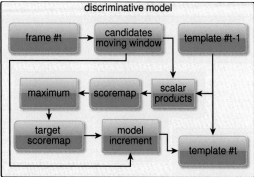

FIGURE 9.2 Template-based methods. Block diagrams of template-based method; generative (left), see Section 9.2.3, and discriminative (right), see Sections 9.2.4 and 9.3.1.

9.2.1 The basics

In general, we can divide the template-based methods into generative and discriminative methods. These methods are conceptually sketched in the block diagrams in Fig. 9.2.

The generative approach is the most frequently used template-based approach. Here, the model is basically the prototypical image patch, usually the template itself. The localization of the object in subsequent frames is obtained by matching the model to candidate positions in the frames. The matching requires a distance measure between image patches, for instance the L2-norm (least-squares) or the L1-norm of the pixel-wise difference.

The search can be done exhaustively, which works for highly efficient L1-methods (Skoglund and Felsberg, 2006), using heuristics to reduce the search window (e.g. dynamic models), or by means of iterative methods. The most dominating approach during the initial years was based on the L2-norm and the iterative approach, the method by Lucas and Kanade (1981). Methods based on this approach have been extensively studied in the review article by Baker and Matthews (2004) and we will restrict the discussion to the basic case in the end of Section 9.9.

Instead, we will focus more on the discriminative approach, where the model is dual to the initial template. In its most basic case, duality is understood in terms of the scalar product in a vector space. Let us assume d vectors v_k that span a d-dimensional vector space. These vectors are not assumed to be normalized or orthogonal. The dual basis is spanned by the d vectors $\tilde{v}_k$, such that

$$\langle v_k | \tilde{v}_l \rangle = \begin{cases} 1 & \text{if } k = l \\ 0 & \text{if } k \neq l \end{cases} \qquad \text{for } k, l \in \{1, \ldots, d\}, \tag{9.1}$$

where $\langle | \rangle$ denotes the scalar product. Note that in the case of an orthonormal basis $\{e_k\}_{k=1\ldots d}$ the basis vectors are dual to themselves $\tilde{e}_k = e_k$.

If we now rewrite vector $v_k = \sum_{l=1}^{d} a_{kl} e_l$ with coefficients a_{kl}, we immediately see that the $d \times d$ matrix A with coefficients a_{kl} has rank d and its inverse exists. Without loss of generality we can assume $\{e_k\}_{k=1\ldots d}$ to be the canonical basis of $\mathbb{R}^d$ and thus v_k, $k = 1, \ldots, d$ are the rows of A. If we now apply a cyclic shift to the rows of A and multiply with A^{-1}, we will no longer get the identity matrix, but a cyclic shift of the identity matrix: we have successfully "tracked" our basis vector v_k.

This concept can be applied to any vector space, including the one of functions or signals. Thus, the idea of the phase-only matched filter (Horner and Gianino, 1984) is to use the dual of the template for determining a distinctive score. This score attends its maximum if the bounding-box is located at the template and will be close to zero in the vicinity.

9.2.2 Performance measures

In the process of tracking, we locate the bounding-box based on some score function; we will get back to segmentation masks further down in this section. To measure how successful the localization was, the predicted bounding-box needs to be compared to the ground truth bounding-box. In both, the VOT-challenge (Kristan et al., 2013) and the OTB (Wu et al., 2013), the intersection-overunion (Everingham et al., 2010), or historically more correct, Jaccard index $\mathcal{J}$ (Jaccard, 1912) has been chosen:

$$\mathcal{J} = \frac{|R_G \cap R_P|}{|R_G \cup R_P|} = \frac{|R_G \cap R_P|}{|R_G| + |R_P| - |R_G \cap R_P|} = \left(\frac{|R_G| + |R_P|}{|R_G \cap R_P|} - 1\right)^{-1}, \quad (9.2)$$

where R_G is the ground truth region, R_P is the predicted region, and $|R|$ is the area of the region R. If the two regions do not overlap we obtain $\mathcal{J} = 0$ and if the two regions perfectly coincide we obtain $\mathcal{J} = 1$. If half of the predicted region overlaps with half of the ground truth region (a typical performance of many trackers) we obtain $\mathcal{J} = \frac{1}{3}$. Note that the Jaccard index is a biased measure that tends to prefer systematic overestimation of the bounding-box size (Häger et al., 2018).

The Jaccard index measures the tracking accuracy per frame and this perframe accuracy can be accumulated in several ways. OTB suggests to calculate two curves, the precision curve and the success curve, see Fig. 9.3. Both are calculated similar to the ROC-curve by setting thresholds and computing the ratio of frames that pass that threshold. The precision curve is obtained by thresholding the box-center distance from 0 to a maximum distance and the success curve is obtained by thresholding the Jaccard index from 0 to 1. The integral under the curve is then an integral measure of performace. Alternatively, particular thresholds can be used, e.g. 20 pixels distance or 50% overlap (Everingham et al., 2010).

In either case, a complete tracking failure leads to a decrease of the performance measure and cannot be distinguished from systematically poor accuracy: accuracy and robustness are strongly correlated under these measures (Kristan et al., 2016b). Therefore, the VOT challenge uses two measures that attempt to decorrelate these two effects: the accuracy measure and the robustness measure. The former measures the average Jaccard index if tracking succeeds and the latter measures the frequency of tracking failures (Kristan et al., 2013). This conditional treatment requires restarts of tracking, either by detecting failures and restarting the tracking using the ground truth, or by starting multiple tracking attempts from several anchor points in the sequence, about every 50th frame (Kristan et al., 2020). The performance is accumulated per sequence length and then averaged, where the performance is weighted by the respective sequence length.

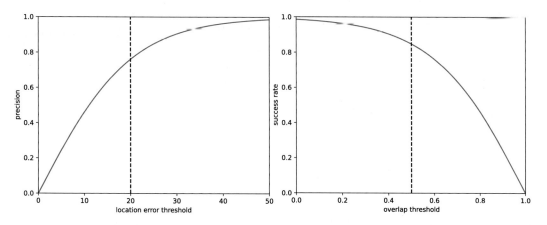

FIGURE 9.3 **Tracking evaluation in OTB.** Precision plot (left) and success plot (right). The ranking can be done by the precision at the location error threshold 20, the success rate at the overlap threshold 0.5, or by the area under the curve (AUC) of the success plot.

The separate treatment of accuracy and robustness places tracking methods into a two-dimensional plot, commonly with increasing robustness on the horizontal axis and increasing accuracy on the vertical axis. Depending on the tracking application, engineers can choose a more robust or more accurate method. However, the average scores do not disclose whether a method has average accuracy throughout the track or initially high accuracy that decays rapidly. For many applications it would be interesting to know what accuracy can be expected after a certain sequence length. This is achieved by means of the expected average overlap curve, measuring the average overlap as a function of the sequence length. The final score of the VOT challenge is obtained by integrating this curve over the range of typical sequence lengths (Kristan et al., 2016).

In its most recent edition (Kristan et al., 2020), the VOT challenge also offers to compute the accuracy measures on segmentation masks instead of bounding boxes. This became necessary as tracking methods achieved a level of accuracy similar to that of the annotations by the bounding-box. In contrast to the Jaccard index for bounding-boxes, which can be computed as a function of the bounding-box parameters, the Jaccard index of segmentation masks requires counting the pixels in the masks and in their intersection.

9.2.3 Normalized cross correlation

One of the oldest methods to predict the position of the bounding-box is template matching. The matching is performed by means of a generative model m, which is obtained from the patch inside the bounding-box in the initial frame. During matching, the model m is compared to candidate patches p in each subsequent frame. The respectively most similar patch determines the position of the bounding-box in that frame, see Fig. 9.2 left.

The model might be the raw image patch in the initial bounding-box, but it is more common to assume that the model m has a zero DC-component, i.e., to subtract the mean of the template. The absolute intensity in the image often depends on other factors than the object, e.g. illumination, shadows, exposure time, etc., and removing the DC component makes the matching more robust under these variations. In photometric cases, e.g. thermal infrared sequences, m might be chosen to include an absolute DC-level, though.

To further improve robustness, also the intensity dynamics is commonly normalized by means of the variances in the template m and the candidate patch p:

$$\sigma_m^2 = \frac{1}{|R|} \sum_{x,y} m(x, y)^2 \tag{9.3}$$

$$\sigma_p^2 = \frac{1}{|R|} \sum_{x,y} p(x, y)^2 - \left(\frac{1}{|R|} \sum_{x,y} p(x, y) \right)^2, \tag{9.4}$$

where $|R|$ is the region area of the patch.

The candidate patches are commonly chosen by a sliding window over the next frame f or a part of it $p_{x,y}(r, s) = f(x + r, y + s)$. The matching score is computed by the scalar product between the template and the patch. This combination of sliding window and scalar product results in the correlation of m and f

$$c(x, y) = \langle p_{x,y} | m \rangle = \sum_{r,s} p_{x,y}(r, s) m(r, s) = \tag{9.5}$$

$$= \sum_{r,s} f(x + r, y + s) m(r, s) \overset{\text{def}}{=} (f \star m)(x, y). \tag{9.6}$$

Note that p does not require any DC-compensation, because m is DC-free and thus the scalar product does not contain any contribution from the mean of p.

The subsequent division by the product of standard deviations results in the normalized cross correlation (NCC)

$$c_n(x, y) = \sigma_m^{-1} \sigma_{p_{x,y}}^{-1} c(x, y), \tag{9.7}$$

where $\sigma_{p_{x,y}}$ is still computed in the sliding window, which makes it a function of (x, y).

The correlation $c(x, y)$ itself, i.e. without the normalization by $\sigma_{p_{x,y}}$, can be efficiently computed in the Fourier domain (Bracewell, 1995)

$$C(u, v) \propto F(u, v) \circ \bar{M}(u, v), \tag{9.8}$$

where capital letters denote the Fourier transforms of the respective signals in lower-case letters and $\bar{M}$ is complex conjugate of M. The frequency coordinates are denoted by (u, v) and $\circ$ is the point-wise product. The position $(\tilde{x}, \tilde{y})$ with the maximum normalized cross correlation (9.7) is then used as a prediction of the bounding-box position.

Normalized cross correlation is not only used with a fixed template, but also with an adaptive template. The only change to the procedure above is that the template from the first frame is only used to initialize the model m. After the first frame, the located bounding-box is used to update the model with the patch in that bounding-box:

$$m \leftarrow (1 - \lambda)m + \lambda p_{\tilde{x}, \tilde{y}}, \tag{9.9}$$

where $\lambda \in (0, 1)$ denotes the update factor. If this factor is chosen too large, the model will suffer from drift (Matthews et al., 2004; Wang et al., 2016), and if it is chosen too small, the model will not adapt sufficiently to appearance changes.

Note that in both the static and the adaptive case, the solution of (9.7) is identical to the least-squares problem of the normalized patches:

$$\min_{x,y} \|\sigma_{p_{x,y}}^{-1} p_{x,y} - \sigma_m^{-1} m\|^2 = \underbrace{\|\sigma_{p_{x,y}}^{-1} p_{x,y}\|^2 + \|\sigma_m^{-1} m\|^2}_{\text{constant}} - 2 \max_{x,y} c_n(x, y). \tag{9.10}$$

If the least-squares problem is solved iteratively by gradient descent, we obtain the Lucas-Kanade method KLT (Lucas and Kanade, 1981). This method is completely local, i.e., it starts from the previous location (or a location predicted by some dynamic model) and locates the closest local minima. In contrast, the normalized cross correlation locates the global maximum in the whole frame f.

9.2.4 Phase-only matched filter

If we reconsider the computation in the Fourier domain (9.8), one could get the idea to choose M as

$$M(u, v) = \frac{F(u, v)}{|F|^2(u, v)} \tag{9.11}$$

such that

$$C(u, v) = \frac{F(u, v) \circ \bar{F}(u, v)}{|F|^2(u, v)} = 1 \quad \text{and} \quad c(x, y) = \delta(x, y). \tag{9.12}$$

A shift of f with (x_0, y_0) leads to a modulation in the Fourier domain (shift theorem) and thus a shifted Dirac in the spatial domain. Thus, we made a transition to a discriminative model as the filter does not represent the appearance of the template and the output score map aims to be 1 for the correct displacement and 0 for incorrect displacements.

To obtain an ideal score map requires that the whole image is noise-free and shifted together with the patch containing the target. In practice, this is hardly the case and the estimated score map is replaced with a target score map located at the maximum score to compute the new model, see Fig. 9.2 right. The target score map is no longer a Dirac because the power spectrum $|F|^2(u, v)$ changes between the initial frame and the subsequent ones. For symmetry reasons, the denominator in (9.11) needs to be changed to $|F||F'|$ where F' is the Fourier transform of the current frame. The resulting filter m is strictly speaking nonlinear and is also known as the symmetric phase-only matched filter (SPOMF) (Chen et al., 1994).

The name phase-only matched filter (POMF) goes back to (Horner and Gianino, 1984) where the nonlinearity is avoided by omitting $|F'|$ in the denominator. Thus the effective matching is a correlation of the new frame with a model m, that has a constant amplitude spectrum

$$M(u, v) = \frac{F(u, v)}{|F|(u, v)}. \tag{9.13}$$

This filter will no longer result in a Dirac response (unless the frame has a constant amplitude spectrum). In terms of a discriminative filter, one can consider the POMF being regularized by the amplitude spectrum of the current frame. Each Fourier coefficient of $C(u, v)$ is effectively multiplied by the corresponding coefficient of the magnitude spectrum $|F|(u, v)$. Thus, the output is not a Dirac, but a smooth response where the shape is obtained from inverse Fourier transform of the magnitude spectrum of f'. This regularization is made explicit in the MOSSE filter, see Section 9.3.1 below.

9.3 Online-learning-based methods

This section takes some ideas from the previous section, the regularization of the response and the incremental model update, to define the concepts of discriminative correlation filters.

9.3.1 The MOSSE filter

Revisiting the POMF, we observed that it comes with an implicit regularization of the response. If this regularization is made explicit in terms of a target response function c, we arrive at the concept of the Minimum Output Sum of Squared Error (MOSSE) filter (Bolme et al., 2010):

$$\min_m \sum_{x,y} \left(\sum_{r,s} p_{x,y}(r, s)m(r, s) - c(x, y) \right)^2 = \min_m \| f \star m - c\|^2. \tag{9.14}$$

Note that although we use the same notation f as for the whole frame, we will only use a local search window. Thus, we obtain an approach that is neither purely local, like the KLT, nor fully global, like the NCC. Typically the search window is two to three times the size of the bounding-box in both dimensions.

The MOSSE filter is computed in closed form using the equivalent formulation in the Fourier domain:

$$\tilde{m} = \arg\min_m \| f \star m - c\|^2 \tag{9.15}$$

$$= \arg\min_m \| F \circ \overline{\mathcal{F}\{m\}} - C\|^2 \tag{9.16}$$

$$= \mathcal{F}^{-1}\left\{ \arg\min_M \| F \circ \bar{M} - C\|^2 \right\}. \tag{9.17}$$

This final expression has the solution (for a derivation, see the appendix in Bolme et al. (2010))

$$\tilde{m} = \mathcal{F}^{-1}\left\{ \frac{\bar{C} \circ F}{\bar{F} \circ F} \right\}. \tag{9.18}$$

Note that the equivalence is only obtained for infinite domains. In practice, the extracted patch is of finite size and the Fourier transform leads to an implicit periodic repetition. In order to reduce the effects of discontinuities at the patch border, a Hann-window (cosine-window in Bolme et al. (2010)) is applied to the patch.

Note further the similarity to (9.11) except for C. As explained for the POMF, we effectively apply a point-wise regularizing function $C(u, v)$ and if we choose $C = |F'|$, we obtain the POMF. For the MOSSE filter, however, the most common choice for the target score map c is a Gaussian function (not normalized) at (x_0, y_0) with fixed width

$$c(x, y) = \exp\left(-\frac{(x - x_0)^2 + (y - y_0)^2}{2\sigma^2}\right). \tag{9.19}$$

This regularized matched filter is usually updated during a sequence of frames, similar to the update of the NCC (9.9), see also Fig. 9.1. In each step, the target score map $c(x, y)$ is located at the maximum score from the previous model, see Fig. 9.2 right.

The MOSSE update equation is obtained by first considering the minimization problem for multiple annotated frames (f_t, c_t):

$$\min_m \sum_t \|f_t \star m - c_t\|^2. \tag{9.20}$$

In the Fourier domain this gives due to linearity

$$\tilde{m} = \arg\min_m \sum_t \|f_t \star m - c_t\|^2 \tag{9.21}$$

$$= \mathcal{F}^{-1}\left\{\arg\min_M \sum_t \|F_t \circ \bar{M} - C_t\|^2\right\}. \tag{9.22}$$

With the solution

$$\tilde{m} = \mathcal{F}^{-1}\left\{\frac{\sum_t \bar{C}_t \circ F_t}{\sum_t \bar{F}_t \circ F_t}\right\}. \tag{9.23}$$

If we now assume that the pairs (f_t, c_t) arrive incrementally, we only need to keep numerator and denominator separated to sequentialize the update:

$$A_t = \bar{C}_t \circ F_t + A_{t-1}, \qquad A_0 = 0 \tag{9.24}$$

and

$$B_t = \bar{F}_t \circ F_t + B_{t-1}, \qquad B_0 = 0 \tag{9.25}$$

such that the filter after time step t is obtained as

$$\tilde{m}_t = \mathcal{F}^{-1}\left\{\frac{A_t}{B_t}\right\}. \tag{9.26}$$

However, for a growing number of frames, the expressions for A_t and B_t are subject to unbounded growth, which we want to avoid for numerical reasons. Also, we would prefer the model to successively reduce the influence of very old samples. Both effects are achieved by introducing a forgetting factor $1 - \eta$, where $0 < \eta < 1$, into the update equations:

$$A_t = \eta \bar{C}_t \circ F_t + (1 - \eta)A_{t-1}, \qquad A_1 = \bar{C}_1 \circ F_1 \tag{9.27}$$

$$B_t = \eta \bar{F}_t \circ F_t + (1 - \eta) B_{t-1}, \qquad B_1 = \bar{F}_1 \circ F_1. \tag{9.28}$$

Note that similar to the updated NCC, the filter with time index t is applied to frame $t + 1$.

In addition to unbounded growth in the expressions, a denominator close to zero can cause numerical issues. This problem has been addressed by regularization of the filter coefficients with penalizing their L2-norm (Henriques et al., 2012). This leads to the following ridge-regression problem:

$$\min_{M} \sum_t \|F_t \circ \bar{M} - C_t\|^2 + \lambda \|M\|^2. \tag{9.29}$$

The solution is a minor modification of (9.23):

$$\tilde{m} = \mathcal{F}^{-1} \left\{ \frac{\sum_t \bar{C}_t \circ F_t}{\lambda + \sum_t \bar{F}_t \circ F_t} \right\} \tag{9.30}$$

and for the incremental case we only change (9.26) to

$$\tilde{m}_t = \mathcal{F}^{-1} \left\{ \frac{A_t}{\lambda + B_t} \right\} \tag{9.31}$$

and keep (9.27) and (9.28) unchanged.

9.3.2 Discriminative correlation filters

The MOSSE filter defined in the previous section acts directly on the image data. The derivations also assume that the image data is scalar-valued (single-channel) and the generalization to color images (multichannel) is nontrivial. Furthermore, we might want to use multiple features extracted from the image instead of, or in addition to, the pure image data. Also in this case, multichannel input data needs to be processed, which leads to a generalization of the (regularized) MOSSE filter: the discriminative correlation filter (DCF).

Note that for the definition of the DCF, we change notation to symbols commonly used in machine learning. Compared to the signal processing inspired notation in the MOSSE filter, we apply the following mappings: $f \mapsto x$ for the input data, $m \mapsto f$ for the filter, and $c \mapsto y$ for the target outputs (Danelljan et al., 2015). For input data with d channels, we reformulate (9.29) as the minimization of the objective

$$\varepsilon_t(f) = \sum_{k=1}^{t} \alpha_k \|S_f(x_k) - y_k\|^2 + \lambda \sum_{l=1}^{d} \|f^l\|^2 \tag{9.32}$$

where we define the score function

$$S_f(x) = \sum_{l=1}^{d} x^l \star f^l \tag{9.33}$$

as the multichannel correlation between input and the filters. Note that the samples (x_k, y_k) are weighted by $\alpha_k \geq 0$, which is a further generalization compared to the previously introduced forgetting factor $(1 - \eta)$.

The solution to the minimization of (9.32) is obtained by plugging in the score function (9.33) and Parseval's theorem such that

$$\min_{f^n} \; \varepsilon_t(f) = \min_{f^n} \sum_{k=1}^{t} \alpha_k \left\| \sum_{l=1}^{d} x_k^l \star f^l - y_k \right\|^2 + \lambda \sum_{l=1}^{d} \| f^l \|^2 \tag{9.34}$$

$$= \min_{F^n} \sum_{k=1}^{t} \alpha_k \left\| \sum_{l=1}^{d} X_k^l \circ \bar{F}^l - Y_k \right\|^2 + \lambda \sum_{l=1}^{d} \| F^l \|^2 \tag{9.35}$$

for all $n = 1 \ldots d$. The necessary condition for a minimum is that the partial derivatives with respect to all components of F^n and for all n vanish

$$0 = \frac{\partial \varepsilon_t}{\partial F^n} = \sum_{k=1}^{t} 2\alpha_k \left(\sum_{l=1}^{d} X_k^l \circ \bar{F}^l - Y_k \right) \bar{X}_k^n + 2\lambda \bar{F}^n \quad \text{for all } u, v, n \tag{9.36}$$

$$0 = \mathbf{X}^* \text{diag}(\alpha) \mathbf{X} \bar{\mathbf{F}} - \mathbf{X}^* \text{diag}(\alpha) \mathbf{Y} + \lambda \bar{\mathbf{F}} \tag{9.37}$$

$$\bar{\mathbf{F}} = \left(\mathbf{X}^* \text{diag}(\alpha) \mathbf{X} + \lambda \mathbf{I} \right)^{-1} \mathbf{X}^* \text{diag}(\alpha) \mathbf{Y} \tag{9.38}$$

where the second equality is a reformulation in matrix notation. The matrix $\mathbf{X}$ collects the Fourier coefficients $X_k^l(u, v)$, $l = 1 \ldots d$ as row vectors, $\mathbf{X}^*$ is its conjugate transpose, $\text{diag}(\alpha)$ is the diagonal matrix with entries α_k, $\bar{\mathbf{F}}$ collects the Fourier coefficients $\bar{F}^l(u, v)$, $l = 1 \ldots d$ as a column vector, and $\mathbf{Y}$ is the column vector containing all Y_k. The original formulation can be found in Danelljan et al. (2015).

The solution (9.38) has several special cases. The regularized MOSSE filter with incremental update is the special case for $d = 1$ and α_k establishing a geometric sequence $\eta(1 - \eta)^{t-k}$. Another special case is obtained if we only consider one single sample. In that case, $\mathbf{X}$ is a row vector and $\mathbf{X}\mathbf{X}^*$ is a scalar, so that

$$\left(\mathbf{X}^* \mathbf{X} + \lambda \mathbf{I} \right)^{-1} \mathbf{X}^* Y = \frac{\mathbf{X}^* Y}{\mathbf{X}\mathbf{X}^* + \lambda} \tag{9.39}$$

and

$$\tilde{F}^n = \frac{\bar{Y} \circ X^n}{\lambda + \sum_{l=1}^{d} \bar{X}^l \circ X^l}, \qquad n = 1 \ldots d. \tag{9.40}$$

If we revisit this solution for multiple samples, the inversion trick in (9.39) only applies if $\mathbf{X}^* \text{diag}(\alpha) \mathbf{X}$ is of rank one. Without this trick, an efficient incremental update becomes impossible as we would have to invert a $d \times d$ matrix for each coefficient and each frame. There exists a generalization of the inversion trick, known as the Sherman-Morrison-Woodbury formula, and it has been applied to multichannel DCFs (Lukežič et al., 2018), but empirical results seem to indicate that a rank-one approximation works sufficiently well (Danelljan et al., 2017). In this appoximation, we simply reuse the modified update Eqs. (9.27) and (9.28)

$$A_t^n = \eta \bar{Y}_t \circ X_t^n + (1 - \eta) A_{t-1}^n, \qquad A_1^n = \bar{Y}_1 \circ X_1^n, \qquad n = 1 \ldots d \qquad (9.41)$$

$$B_t = \eta \sum_{l=1}^d \bar{X}_t^l \circ X_t^l + (1 - \eta) B_{t-1}, \qquad B_1 = \sum_{l=1}^d \bar{X}_1^l \circ X_1^l. \qquad (9.42)$$

Similar to (9.31), we compute the (vector-valued) filter as

$$\tilde{f}_t^n = \mathcal{F}^{-1} \left\{ \frac{A_t^n}{\lambda + B_t} \right\}, \qquad n = 1 \ldots d. \qquad (9.43)$$

9.3.3 Suitable features for DCFs

With the transition from single channel MOSSE filters to DCFs, we now have the freedom to choose which features are used for the input feature map x. There are three major trade-offs to consider regarding the feature selection: the first one regarding d, the number of channels in x, the second one regarding the spatial resolution or the number of Fourier coefficients, and the third one regarding the size of the temporal window in case that other α_k than those from a geometric sequence are to be used.

The number of free parameters of the DCF scales with the size of the feature map, d and the spatial resolution, as can be seen in (9.36). A larger number of parameters requires more computations, as does a larger time-window, the inner dimension in (9.38). Furthermore, more free parameters also require stronger regularization, as the number of original input dimensions, i.e., the input image frame, is independent of the size of the feature map x.

Whereas the size of the temporal window and the spatial resolution will be reconsidered in later sections of this chapter, we will focus on the number of channels in this section. The focus here is on hand-crafted features: learned features and deep features will also be considered later. Thus the problem addressed in this section is: select as few hand-crafted dense features as possible to obtain a high-performance DCF tracker.

Since the DCF approach is based on tracking by detection, we ground the feature selection on experience from the object detection literature. Whereas pure image values might be useful in special cases, e.g. photometric data such as thermal IR, the structure and shape in images is usually much better represented using histograms of oriented gradients (HOG) as proposed for person detection (Dalal and Triggs, 2005). HOG features are a special choice to represent orientation data (Felsberg, 2018) as are dense SIFT features (Lowe, 2004).

A majority of recent tracking methods using hand-crafted features rely on HOG features, and we are not aware of any systematic evaluations of alternative settings, e.g. based on generalized formulations (Felsberg, 2018). Also, features more commonly used for texture analysis, such as binary local patterns (Pietikäinen and Zhao, 2015), are rarely seen in tracking methods, presumably due to the relative low share of textured objects in benchmark datasets. Thus we also stick to HOG features for the purpose of this chapter and refer for details to the original work, alternatively (Felsberg, 2018).

All features discussed so far ignore the color information in the image frames. Again, from the object detection literature, we learn that color names (CNs) are particularly useful

(Khan et al., 2012). The color name descriptor is computed by a learned soft-assignment from Lab color-space to eleven color names in English language (Van De Weijer et al., 2009). The learning is performed on weakly labeled Google images using probabilistic latent semantic analysis (pLSA-bg):

$$P(w|d) = P(w|z)P(z|d), \tag{9.44}$$

where d denotes the color distribution in Lab space, w is one of the eleven color names, and z is a latent variable. The color names are represented by an 11-D soft-assignment vector, i.e., all components are between 0 and 1 and the vector is normalized.

CN features have successfully been used in various tracking methods and have shown to be superior to many other color representations, in particular pure RGB-values (Danelljan et al., 2014b). Pure RGB-values are however commonly used in combination with shallow and deep features from deep networks, see Section 9.4.1. CN features are relatively sparse and going from three to eleven dimensions leads to a significant growth of free parameters. Based on previous arguments, feature compression methods based on, e.g., online PCA have proven to be very useful (Danelljan et al., 2014b) and lead to powerful DCFs based on HOG and CN features.

9.3.4 Scale space tracking

The DCF trackers introduced above still suffer from one fundamental weakness: if the target object moves in depth, e.g. closer to the camera, it will change its size in the image plane. Thus, the bounding-box size should be modified and not only its position in the current frame: we want to track the object at (continuously) varying scale. Obviously, the appearance of the object also changes with scale or depth, but there is a close relation between scale and appearance, which allows us to connect the appearance models across scale: scale space (Koenderink, 1984).

The key idea of scale space is to introduce a third (scale) axis to the image that controls the amount of blur, e.g. caused by the limited depth-of-field of a camera (Felsberg et al., 2005). Blurring is usually modeled by lowpass-filtering, i.e. the reduction of high frequency components. At some scale s_0, frequencies over half the Nyquist frequency $\pi/2$ are sufficiently attenuated to allow down-sampling with a factor 2 without significant aliasing. At scale $2s_0$ we can then down-sample the image with a factor 4 and so on, eventually building a scale pyramid, see Fig. 9.4, left.

The useful side-effect of this pyramid is that the reduced spatial resolution in the pyramid corresponds to the reduced size in the image plane due to the increased distance from the object to the camera. Thus, we can simulate the effect of increasing depth by moving to larger scales in the pyramid and in this way, we can include scale estimation in the tracking (Danelljan et al., 2017). For this process, we have three options:

1. A multiresolution translation filter is applied. All relevant scales are computed in parallel and the best score is selected.
2. A joint scale space filter is applied. The DCF is computed in 3D instead of 2D and the third coordinate determines the object size.
3. Discriminative scale space tracking is applied. A 1D DCF is applied along the scale dimension to determine the object size.

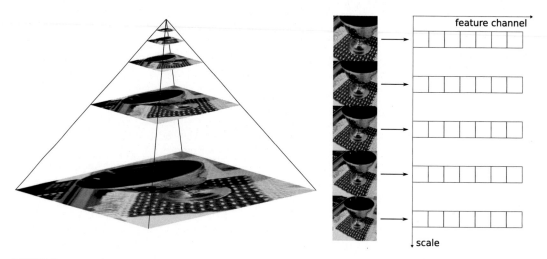

FIGURE 9.4 Scale space tracking. Scale pyramid (left) and scale tracking (right). The scale axis corresponds to the spatial coordinates in the translation case.

As shown in a comparison, the first two approaches are not only slower than the third one, but do not even improve accuracy (Danelljan et al., 2017). Thus we will only describe the third one in this section.

The basic assumption is that scale changes slower than position. Consequently, we can separate position and scale tracking by first estimating the translation at constant scale with a DCF as previously introduced, and secondly estimating the change of scale. The scales that are supposed to produce low scores have to be generated explicitly by the scale pyramid of the region inside the bounding box, see Fig. 9.4 right, in contrast to the translation case where the displaced windows are obtained by shifted bounding-boxes. The corresponding feature vectors are then arranged to form a feature map over the scale coordinate and using a Hann-window.

The pseudo-code for discriminative scale space tracking reads:

- Input:
    - Image I_t
    - Previous target position p_{t-1} and scale s_{t-1}
    - Translation model $A_{t-1,\text{trans}}$, $B_{t-1,\text{trans}}$
    - Scale model $A_{t-1,\text{scale}}$, $B_{t-1,\text{scale}}$
- Output:
    - Estimated target position p_t and scale s_t
    - Updated translation model $A_{t,\text{trans}}$, $B_{t,\text{trans}}$
    - Updated scale model $A_{t,\text{scale}}$, $B_{t,\text{scale}}$
- Translation estimation:
    1. Extract sample $x_{t,\text{trans}}$ from I_t at p_{t-1} and s_{t-1}
    2. Compute correlation scores $y_{t,\text{trans}}$ using (9.33) with $f_{t-1,\text{trans}}$ according to (9.43)
    3. Set p_t, to the target position that maximizes $y_{t,\text{trans}}$

- Scale estimation:
  4. Extract sample $z_{t,\text{scale}}$ from I_t at p_t and s_{t-1}
  5. Compute correlation scores $y_{t,\text{scale}}$ using (9.33) with $f_{t-1,\text{scale}}$ according to (9.43)
  6. Set s_t to the target scale that maximizes $y_{t,\text{scale}}$
- Model update:
  7. Extract samples $x_{t,\text{trans}}$ and $x_{t,\text{scale}}$ from I_t at p_t and s_t
  8. Update the translation model $A_{t,\text{trans}}$, $B_{t,\text{trans}}$ using (9.41) and (9.42)
  9. Update the scale model $A_{t,\text{scale}}$, $B_{t,\text{scale}}$ using (9.41) and (9.42).

Note: The scaling of the window content makes the problem with the periodic extension apparent, as the Hann window is not scaled in this process.

9.3.5 Spatial and temporal weighting

Despite the use of a Hann-window, the implicit periodic extension of the patch caused by the Fourier transform reduces the discriminative power of the filter. Features that are characteristic for the object of interest reappear in the vicinity of the patch and generate side-lobes in the score function. A naive attempt to counter those side-lobes using an increased search windows fails, because a large part of the considered patch now consists of background pixels. The filter tends to associate those background pixels with the peak in the target function, i.e., it learns high coefficients at these locations, and the tracker will stick to the background instead of following with the object (Danelljan et al., 2015).

To counter this effect, adding a spatial regularization to (9.32), which penalizes filter coefficients outside the bounding-box, has been proposed (Danelljan et al., 2015):

$$\varepsilon_t(f) = \sum_{k=1}^{t} \alpha_k \|S_f(x_k) - y_k\|^2 + \sum_{l=1}^{d} \|w \cdot f^l\|^2. \tag{9.45}$$

However, the required point-wise multiplication of w and f in the spatial domain becomes a convolution in the Fourier domain and the major advantage of DCFs, the efficient computation in the Fourier domain, is lost. One option is to alternate between spatial and Fourier domain (Galoogahi et al., 2015), the other is to choose a weight function w that has only a few nonzero Fourier coefficients. In the latter case, the convolution in the Fourier domain is no problem as the computational effort is only marginally increased; see Danelljan et al. (2015).

The number of computations is greatly reduced by replacing the complex Hadamard (point-wise) products in (9.38) with real matrix products. We start from the system of equations

$$(\mathbf{X}^*\text{diag}(\alpha)\mathbf{X} + \lambda \mathbf{I})\bar{\mathbf{F}} = \mathbf{X}^*\text{diag}(\alpha)\mathbf{Y} \tag{9.46}$$

and map $\mathbf{X}$ to $\mathbf{D}$, $\bar{F}$ to $\tilde{F}$, and Y to $\tilde{Y}$, the respective real representations. The Fourier transform of a real-valued function is Hermitian symmetric, i.e., $P(-u, -v) = \bar{P}(u, v)$, and the isometric mapping

$$\big(P(u, v), P(-u, -v)\big) \to \big((P(u, v) + P(-u, -v))/\sqrt{2}, (P(u, v) - P(-u, -v))/i\sqrt{2}\big) \tag{9.47}$$

leads to the same solution, but using real instead of complex matrices. If we further generalize λI as $W^T W$, representing the sparse convolution induced by the spatial weights, we obtain

$$\left(\sum_{k=1}^{t}\alpha_k \mathbf{D}_k^T \mathbf{D}_k + \mathbf{W}^T\mathbf{W}\right)\tilde{\mathbf{F}} = \sum_{k=1}^{t}\alpha_k \mathbf{D}_k^T \tilde{Y}_k. \tag{9.48}$$

We solve this system using the Gauss-Seidel method, i.e., we decompose the LHS into a lower triangular and strictly upper triangular part

$$\left(\sum_{k=1}^{t}\alpha_k \mathbf{D}_k^T \mathbf{D}_k + \mathbf{W}^T\mathbf{W}\right)\tilde{\mathbf{F}} = (\mathbf{L}_t + \mathbf{U}_t)\tilde{\mathbf{F}} = \sum_{k=1}^{t}\alpha_k \mathbf{D}_k^T \tilde{Y}_k \tag{9.49}$$

and compute the solution by iterations

$$\mathbf{L}_t\tilde{\mathbf{F}}^{(j)} = \sum_{k=1}^{t}\alpha_k \mathbf{D}_k^T \tilde{Y}_k - \mathbf{U}_t\tilde{\mathbf{F}}^{(j-1)}. \tag{9.50}$$

In particular, for w with few nonzero Fourier coefficients, this method converges rapidly and results in filters with small coefficients in the vicinity. Note further that this method avoids the approximative solution in (9.42) and the restriction of α_k to a geometric sequence.

This gained freedom to choose α_k has been used to systematically down-weight samples that result from the tracker temporarily drifting off the target (Danelljan et al., 2016a). The weights α_k are set by redeterminating the importance of each frame during later samples. This procedure helps in ambiguous cases (e.g. partial occlusions) and allows us to use all available information.

The weights are initialized from prior information, e.g. how old the sample is. Typically prior sample weights ρ_k induced by a forgetting factor $1 - \eta$, see (9.9), are used. We then optimize over the joint loss

$$\min_{f,\alpha} \quad \varepsilon_t(f,\alpha) + \frac{1}{\mu}\sum_{k=1}^{t}\frac{\alpha_k^2}{\rho_k} \tag{9.51}$$

$$\text{s.t.} \quad \alpha_k \geq 0, \; k = 1, \ldots, t \tag{9.52}$$

$$\sum_{k=1}^{t}\alpha_k = 1 \tag{9.53}$$

by repeatedly updating f using the Gauss-Seidel iteration (9.50) and α using quadratic programming.

An even more efficient way to make optimal use of all available frames will be introduced in section 9.4.2, based on deep features.

9.4 Deep learning-based methods

DCF trackers are powerful methods if the previously introduced concepts of regularization, scale-adaptation, and multichannel score functions are applied. However, the bottleneck for further improvement is the predetermined layer of hand-crafted features. In this section, the transition to deep features, adaptive features, and finally end-to-end learning of DCFs is described.

9.4.1 Deep features in DCFs

The selection of hand-crafted features described in Section 9.3.3 has been inspired by results from object detection. When looking at the development of deep networks and relevant backbones, major progress was first achieved on image classification. Early deep network-based object detection methods often used region proposals (Girshick et al., 2016), which are less suitable as features for a DCF. Therefore, early attempts to integrate deep appearance features into DCFs (Danelljan et al., 2016b) applied backbones from image classification, e.g. imagenet-vgg-m-2048 / CNN-M-2048 (Chatfield et al., 2014), rather than detection networks.

When considering a backbone network for generating a feature map as input to a DCF, the obvious question is which layer to use. Candidates are the convolutional layers, e.g. layers one to five in imagenet-vgg-m-2048. The shallowest layers are characterized by high spatial resolution and few channels (e.g. 109×109, 96 channels for layer one in imagenet-vgg-m-2048) and the deepest layers by low spatial resolution and many channels (e.g. 13×13, 512 channels for layer five in imagenet-vgg-m-2048). In general, deeper layers trade off a higher number of channels for a lower spatial resolution, a necessary consequence from the spatio-featural uncertainty relation (Felsberg, 2009).

Whereas more feature channels enable more discriminative power, the reduced spatial resolution might lead to degraded accuracy. In terms of the VOT-criteria on robustness and accuracy, it is assumed that a DCF based on shallow features will be characterized by high accuracy and low robustness, whereas a DCF based on deep features will be characterized by low accuracy and high robustness. This is also implicitly confirmed by the experiments with imagenet-vgg-m-2048, where layer one and five are local maxima of OTB-50 overlap precision (Danelljan et al., 2016b), a measure that integrates accuracy and robustness (Wu et al., 2013).

In order to make use of both the high spatial accuracy of shallow features and the robustness of deep features, one would like to fuse both feature maps. However, the different spatial resolutions will require either to upsample the deep features, which is computationally infeasible, or to downsample the shallow features, which will diminish the positive effect regarding accuracy. The ideal case would be if both or all feature maps were continuous functions and no sampling was required at all. But how to represent a continuous feature map in a digital computer?

The solution comes with isometric transformations from continuous signals to discrete spectra such as the Fourier series or other series expansions (Danelljan et al., 2016): the C-COT tracker. To keep the presentation simple, we will focus on Fourier series with a finite number of coefficients, but the method generalizes to all isometric transformations. We start by revisiting the calculation of the score map using correlations of the feature map and a set of filters (9.33). If we assume this correlation being performed in the continuous domain, between a continuous feature map and a set of continuous filters

$$S_f(x) = \sum_{l=1}^{d} f^l \star J_l\{x^l\}, \tag{9.54}$$

where $J_l\{x^l\}$ is the hypothetical mapping of the discrete feature map x^l to the continuous domain, we can easily fuse feature maps with different spatial resolutions.

If we now move the whole setting to the Fourier domain, which we will do anyway in order to compute the efficient solutions (9.43), the continuous correlation will become a pointwise product of discrete coefficients. A periodic continuous function has infinitely many discrete Fourier coefficients, but we also know that truncating the sequence of coefficients will lead to an optimal approximation in the L2-sense (note that this is a specific property of the Fourier series though). Approximating the score map in L2-optimal way perfectly fits the MOSSE-idea of minimal L2-error of the output.

In conclusion this means that using the solution (9.50) (or a more efficient formulation using the conjugate gradient method), we already have all tools at hand to fuse multiresolution feature maps – in contrast to the ADMM-based approach (Galoogahi et al., 2015) that switches between spatial and Fourier domain. In addition to the fusion of layers one and five and the raw input data (224 × 224, RGB), the continuous model enables subpixel localization during inference *and* learning, which improves results, even beyond upsampling to the highest resolution. Note in this context that the target function y is known analytically, see (9.19), so the Fourier coefficients can be computed accurately. Note further that this approach can also be used for feature point tracking as an alternative to the KLT (Lucas and Kanade, 1981).

Once moving to multiple layer appearance features, one can integrate other modalities than appearance features. A previously ignored feature type in DCF-trackers is motion features, which can now be fused with appearance features (Danelljan et al., 2019a). The appearance features applied in this work are taken from the imagenet-vgg-verydeep-16 / ConvNet C network (Simonyan and Zisserman, 2015), with 13 convolutional layers, and the deep motion features (Chéron et al., 2015), computed from a three-channel input (optical flow vector and its magnitude) and consisting of five convolutional layers. For tracking, fusing appearance layers four (128 channels, stride 2) and 13 (512 channels, stride 16) and motion layer five (384 channels, stride 16) gives the best results (Danelljan et al., 2019a).

9.4.2 Adaptive deep features

The approach to integrate multiple feature maps for DCF-trackers leads to a significant improvement of the results compared to earlier methods, both using hand-crafted features and single layer deep features, but leads also to many parameters in the filter model. The large number of parameters is problematic in two regards: First, the computational effort scales with the number of parameters. Secondly, the amount of training data is limited and the number of parameters easily grows beyond the dimensionality of the input, leading to overfitting.

More concretely, it has been found that the C-COT tracker using the imagenet-vgg-m-2048 features has as many as 800,000 parameters in online learning (Danelljan et al., 2017). To address the resulting complexity and training data scarcity issues, the ECO tracker, based on efficient correlation operators, suggests to discriminatively learn a mapping to a lower-dimensional feature space (Danelljan et al., 2017). This is achieved by jointly minimizing the classification error, leading to 80% reduction in the number of model parameters.

The key idea is to introduce some projection operator P that maps the d continuous feature channels in (9.54) to $c \ll d$ dimensions before computing the score. Note that this operator is unaffected by the Fourier transform. Including a regularizer for the L2-norm of all coefficients of the operator P, we therefore reformulate (9.45) as

$$\varepsilon_t(f) = \sum_{k=1}^{t} \alpha_k \left\| \sum_{l=1}^{c} f^l \star p_l^\mathsf{T} J\{x_k\} - y_k \right\|^2 + \sum_{l=1}^{c} \|w \cdot f^l\|^2 + \lambda \|P\|_\mathrm{F}^2, \tag{9.55}$$

where p_l are the column vectors in P and the last term is the Frobenius norm. This loss is minimized using the Gauss-Newton method, derived by linearization of the residuals. For further details, the reader is referred to the original publication (Danelljan et al., 2017).

Even though the ratio of parameters and amount of training data is improved, the required diversity in the training data will still require a fairly large sample set, which leads to a significant computational burden. Furthermore, the memory size is limited, which leads to a limited time horizon for the online learning, and discarding old samples leads to overfitting to the recent appearance.

In order to avoid this overfitting, the sum over all samples from $k = 1$ to t in (9.55) is replaced with the expectation over the joint distribution $p(x, y)$:

$$\varepsilon_t(f) = \mathbb{E}_{p(x,y)} \left\{ \left\| \sum_{l=1}^{c} f^l \star p_l^\mathsf{T} J\{x\} - y \right\|^2 \right\} + \sum_{l=1}^{c} \|w \cdot f^l\|^2 + \lambda \|P\|_\mathrm{F}^2. \tag{9.56}$$

Using the original sum implicitly means to assume that the t samples are drawn from $p(x, y)$ and are sufficiently many to approximate $p(x, y)$ accurately, which might both be wrong assumptions.

In order to achieve a compact and diverse representation of the training data, the ECO-tracker models $p(x, y)$ as a mixture of Gaussian models. The combination of the two improvements over C-COT does not only improve speed, but also improves performance by 13.3% (Danelljan et al., 2017).

The approach applied in the C-COT and ECO trackers is not limited to imagenet-vgg-m-2048 features, but somehow surprisingly, using more powerful backbones such as GoogLeNet (Szegedy et al., 2015) or ResNet-50 (He et al., 2016) does not improve the performance further (Bhat et al., 2018). To efficiently unveil the power of deep tracking, the feature maps must not be combined at the level of the tracker input as done in (9.54). Instead, it is more efficient to train deep and shallow models independently, and with different widths of the target function, and to apply a weighted fusion of the separately trained trackers.

As observed in section 9.4.1, predictions based on a deep feature map, $\hat{y}_\mathrm{d} = S_f(x_\mathrm{d})$, are characterized by high robustness, but inferior localization, and predictions based on a shallow feature map, $\hat{y}_\mathrm{s} = S_f(x_\mathrm{s})$, are characterized by accurate localization, but easily drop the target under appearance changes. The two predictions can be combined in various ways, but already the adaptively weighted sum

$$\hat{y}_\beta(p) = \beta_d \hat{y}_d(p) + \beta_s \hat{y}_s(p) \tag{9.57}$$

results in a significant improvement over the ECO tracker. In each time step, the adaptive weights are computed by solving the quadratic programming problem

$$\min_{\xi,\beta} \quad -\xi + \mu(\beta_d^2 + \beta_s^2) \tag{9.58}$$

$$\text{s.t.} \quad \beta_d + \beta_s = 1, \quad \beta_d \geq 0, \quad \beta_s \geq 0 \tag{9.59}$$

$$\hat{y}_\beta(p^*) - \xi \Delta(p^* - p) \geq \hat{y}_\beta(p), \quad \forall p, \tag{9.60}$$

where p^* is a candidate target state (e.g. position) that is sampled from local maxima in $\hat{y}_d$ and $\hat{y}_s$, μ is a regularization parameter, and $\Delta(p) = 1 - \exp(-4|p|^2/s)$ with s being the target size.

9.4.3 End-to-end learning DCFs

The approaches described in the previous section introduce feature adaptation in DCF-trackers with a deep backbone. The backbone is trained offline on image classification data, the tracker is trained online using the current sequence, and the adaptation procedure is an engineered optimization problem. If we want to aim at a learning approach to connect the backbone with the tracker instead, we need to make use of tracking-specific training data. The main difference between adaptation and learning is the data used for the optimization: In contrast to adaptation, learning attempts to generalize from training data to the test data during inference.

The key problem addressed by the adaptive combination of score maps in (9.57) is to generate uni-modal score functions with a maximum centered inside the bounding box. To combine the score functions from shallow and deep features is a well-working heuristic for this purpose, but a machine learning approach would directly address the objective itself. Thus the task is to learn, using the tracking training data, a predictor for the accuracy measure, the Jaccard index. Instead of picking all local maxima and optimizing the weights, the Jaccard index prediction allows to pick the candidate bounding box with the largest overlap.

This idea of overlap maximization is used in the ATOM-tracker (Danelljan et al., 2019b), which combines an online learned classifier (similar to a DCF) with a Jaccard index regressor that is trained on annotated tracking data, see Fig. 9.5. Obviously, the regressor network has to take into account the specific target given in the initial (reference) frame. This is achieved by modulating the coefficient vector of the current frame with the corresponding vector from the reference frame before feeding it into the predictor.

The two branches, for the reference frame and current frame, are very similar in structure. The main difference is that the test branch for the current frame does not have access to a bounding box, but uses the initial estimate from the DCF-like classifier instead. The whole regressor is trained end-to-end, where the classifier is not trained using the back-propagated loss, but using the MOSSE criterion for the target score function. Thus the classifier acts like

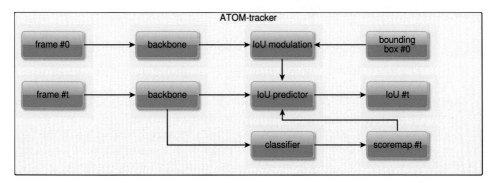

FIGURE 9.5 ATOM-tracker. The ATOM-tracker consists of a classification branch and a regression branch for the Jaccard index (intersection over union).

a fixed input to the regressor network, which has a two-branch architecture that resembles a Siamese tracker (Zhu et al., 2018). For the detailed network model of the Jaccard index regressor we refer to the paper by Danelljan et al. (2019b) or the code repository (https://github.com/visionml/pytracking).

The classifier is basically similar to the one in the ECO tracker, with a first layer that reduces the number of feature channels to 64 (projection operator) and a second layer with a 4 × 4 kernel. Due to the limited size of the spatial kernel, the online optimization is no longer performed in the Fourier domain but directly in the spatial domain. One advantage of this change is that a nonlinear activation function can be added to the output of the convolution, in this case a parametric exponential linear unit.

In the further development of the ATOM-tracker, which can be seen as a merger of DCF-like approaches and Siamese trackers, e.g. Zhu et al. (2018), the MOSSE-objective is dropped and the loss is learned from data instead (Bhat et al., 2019). This method, DiMP, can be considered as the next step of discriminative learning, where the discriminative power is no longer identified with the L2-distance to the target score function, but with some data-dependent distance function. This also increases the flexibility regarding the representation of the tracking output: instead of the parametric bounding-box one can easily move on to segmentation masks.

9.5 The transition from tracking to segmentation

The parametric bounding-box model implicitly enforces the formulation of the target function in terms of a Gaussian function centered at the ground truth position and scale. Besides the bias mentioned in Section 9.2.2, the bounding-box also suffers from its inherent inaccuracy for objects with a shape that deviates from a rectangle. This is already a problem during annotation, where overall accuracy is improved by annotating segmentation masks and automatic fitting of bounding-boxes (Vojíř and Matas, 2017; Kristan et al., 2016).

9.5.1 Video object segmentation

Since 2020 the VOT challenge evaluates the Jaccard index on ground-truth segmentation masks instead of bounding-boxes (Kristan et al., 2020). This means that visual object tracking became similar to video object segmentation (VOS) for the single instance case (Perazzi et al., 2016). The task in VOS is to classify each pixel in a frame of a video either as background or as part of the target object. VOS methods are trained offline on annotated video sequences and evaluated on test sequences with partly new (unseen) object classes.

Similar to the VOT challenge, the evaluation of VOS methods makes use of the Jaccard index, see Section 9.2.2. However, two predictions might have exactly the same Jaccard index but completely different shapes, which can only be addressed by means of a contour measure. In the DAVIS-challenge (Perazzi et al., 2016), the $\mathcal{F}$-measure of the contour is used for this purpose, defined as

$$\mathcal{F} = 2\frac{P \cdot R}{P + R}, \tag{9.61}$$

where P is the precision and R is the recall. For contours, precision is computed as the rate of predicted contour pixels that are truly contour pixels and recall is computed as the rate of true contour pixels that are predicted contour pixels. Commonly the $\mathcal{F}$-measure is computed approximately using morphological operators (Perazzi et al., 2016).

In the semisupervised case (Perazzi et al., 2016), the first frame of the test sequence is annotated with a segmentation mask, similar to the VOT2020 challenge. The VOS method thus needs to adapt to that single annotated sample, solving a one-shot learning problem. In the unsupervised case (Perazzi et al., 2016), no annotation is provided, but it is assumed that the target object is implicitly determined by its motion compared to the background. Finally, there is also a multiinstance semisupervised case, where multiple objects are annotated in the initial frame (Perazzi et al., 2017). This problem is also closely related to the video instance segmentation (VIS) problem, where, similar to detection in still images, all known objects need to be segmented throughout the sequence (Yang et al., 2019).

For the remainder of this chapter, we will focus on the semisupervised VOS problem for the single instance case, as this is closest to the VOT problem with segmentation mask annotation. In a certain way, one can consider this transition to segmentation masks as the end-point of bounding-box tracking, similar to the ATOM-tracker terminating the series of Fourier-based DCFs, and the DiMP tracker terminating the series of MOSSE-based DCFs. Before looking into discriminative VOS methods, we first consider a generative approach to clarify the difference in the case of VOS methods.

9.5.2 A generative VOS method

Even though the VOS field is quite young, many different deep learning approaches have emerged. We do not intend to give a review of deep learning methods for VOS here and refer to recent review papers on this topic, e.g. Yao et al. (2019). Many approaches, however, perform extensive fine-tuning of the network using the first frame annotation of the test sequences. This is not appropriate for practical application as it means that videos can never be processed on the fly, but instead with a delay of at best some minutes.

One approach that avoids this brute-force online fine-tuning is based on a generative appearance model (AGAME) (Johnander et al., 2019). This approach estimates instead the parameters of a Gaussian mixture model (GMM) and draws deep features from

$$p(\mathbf{x}) = \sum_{k=1}^{K} p(z = k) p(\mathbf{x}|z = k), \tag{9.62}$$

where $p(\mathbf{x}|z = k) = \mathcal{N}(\mathbf{x}|\boldsymbol{\mu}_k, \boldsymbol{\Sigma}_k)$. A uniform prior $1/K$ over components k is assumed and the components are either from the background or the foreground. Each of these two classes is divided into a main mode and a mode of hard cases, i.e., we obtain four components in total.

The GMM parameters θ, i.e., means $\boldsymbol{\mu}_k$ and variances $\boldsymbol{\Sigma}_k$, are initialized from the initial frame. During the bootstrapping, i.e., the subsequent frames I^i, when only estimates of the class memberships are available, soft-assignments for the background, object, and the respective hard cases,

$$\alpha_0^i = 1 - \hat{y}(I^i, \theta^{i-1}, \Phi) \tag{9.63}$$

$$\alpha_1^i = \hat{y}(I^i, \theta^{i-1}, \Phi) \tag{9.64}$$

$$\alpha_2^i = \max(0, \alpha_0^i - p(z^i = 0|\mathbf{x}^i, \boldsymbol{\mu}_0^i, \boldsymbol{\Sigma}_0^i)) \tag{9.65}$$

$$\alpha_3^i = \max(0, \alpha_1^i - p(z^i = 1|\mathbf{x}^i, \boldsymbol{\mu}_1^i, \boldsymbol{\Sigma}_1^i)), \tag{9.66}$$

are used for the updates. Here, Φ denotes the parameters of the fusion and prediction networks.

The coarse predictions $\hat{y}$ are computed fusing the mask propagation output and the component score (log-probabilities $\ln p(z = k) p(\mathbf{x}|z = k)$)

$$s_k^i = -\frac{\ln |\boldsymbol{\Sigma}_k^{i-1}| + (\mathbf{x}^i - \boldsymbol{\mu}_k^{i-1})^{\mathsf{T}} (\boldsymbol{\Sigma}_k^{i-1})^{-1} (\mathbf{x}^i - \boldsymbol{\mu}_k^{i-1})}{2}, \tag{9.67}$$

resulting in the assignments for components 0 and 1. The likelihoods in the cases 2 and 3 are computed using the soft-max of the respective scores. The four soft-assignments are used to compute mean- and variance-updates as weighted first and second moments. The updates are then fed into a moving average similar to (9.9).

During inference, the predictions are refined by an upsampling module to achieve full resolution. The level of performance of the AGAME algorithm renders it useful for the semi-automatic annotation in the RGBT challenge of VOT2019 after additional consistency checks (Berg et al., 2019).

9.5.3 A discriminative VOS method

The likelihoods in cases 2 and 3 in the previous section "provide a discriminative mask encoding" (Johnander et al., 2019), which immediately leads to the question whether one can replace the GMM with a discriminative model, a dedicated target model similar to the one in the ATOM-tracker. Similar to the GMM in AGAME, this target model is supposed to adapt to the target appearance by a suitable update in order to maximize foreground-background separability. Focusing on the discriminative power instead of the whole distribution is expected to lead to a more light-weight and faster algorithm compared to AGAME.

This is confirmed in the discriminative VOS approach FRTM (Robinson et al., 2020), which also replaces the mask propagation used in AGAME with spatio-temporal consistency. The target model resembling the ATOM-tracker

$$\mathbf{s} = D(\mathbf{x}; \mathbf{w}) = \mathbf{w}_2 * (\mathbf{w}_1 * \mathbf{x}) \tag{9.68}$$

is trained online using a weighted MOSSE-criterion

$$\mathcal{L}_D(\mathbf{w}) = \sum_k \gamma_k \left\| \mathbf{v}_k \circ \left(\mathbf{y}_k - U\left(D(\mathbf{x}_k)\right) \right) \right\|^2 + \sum_j \lambda_j \|\mathbf{w}_j\|^2, \tag{9.69}$$

where v_k is a balancing weight to increase the importance of small target objects, U performs bilinear upsampling, and γ_k are weights for controlling the impact of different samples in the dataset.

The online training uses, similar to tracking approaches, a temporal window. Data from this window is kept in a memory M.

The pseudo-code for the target model learning reads

1. Initialize memory M, train the discriminative target model D
2. Extract features x from next frame
3. Apply the target model and generate the coarse score map s
4. Enhance s to target mask y with the refinement network
5. Update M with a new sample (x, y, γ)
6. Reoptimize the discriminative target model with M every eighth frame
7. Return to step 2.

The refinement model used in step 4 is trained offline during the VOS training phase. The roles of the coarse score map s and the refined mask y are similar to AGAME, but AGAME uses a coarse prediction for the model update whereas FRTM uses the refined prediction for training, employing bilinear upsampling. FRTM becomes highly efficient by those changes and enables real-time VOS algorithms.

Similar to the transition from ATOM to DiMP, where the loss is generalized, the VOS target model can also be optimized with respect to a representation other than the coarse segmentation prediction. Assume that the target model (9.68) produces some feature map instead of the coarse segmentation map and that the refined segmentation is obtained by a decoder instead of an upsampling module. In this case, the loss for the target model can be computed in the feature map domain by encoding the segmentation mask and computing the weighted L2-error between the encoded segmentation and the target model output (Bhat et al., 2020).

This approach to learn what to learn connects VOS learning to metric learning. Other interesting questions are the generalization to unsupervised VOS, VOS with multiple instances, and VIS.

9.6 Conclusions

The development from simple template matching to video segmentation described in this chapter is one of the cornerstones for many applications in computer vision: video analysis, visual surveillance, remote sensing, augmented reality (AR), visual robotic control, and autonomous vehicles are just a few major application areas of the developed methods. Most of these applications require real-time processing, robust models, and accurate predictions.

For instance, AR requires video real-time processing with low latency in order to reduce AR-sickness. Marker-less methods for AR rely heavily on visual tracking (Chandaria et al., 2007). The alignment of camera pose and virtual pose can also be used to control unmanned areal vehicles. For instance, virtual leashing can be implemented using a DCF tracker on RGB footage from a drone (Häger et al., 2016).

The input data discussed in this chapter was mostly restricted to RGB sequences, but similar methods can be applied to other spectral ranges or depth data as well. For instance, safety systems in trains based on thermal infrared cameras can mitigate consequences of collisions with large animals, people, and obstacles on the track (Berg et al., 2015). In this work, the anomaly detection on the track is performed using a MOSSE-based filter. Also, car safety systems that, e.g., detect pedestrians with IR cameras (Källhammer et al., 2007) and RGB cameras benefit from the progress in combined RGB and thermal tracking (Kristan et al., 2019b).

The methodology of DCFs is not only applicable to tracking and anomaly detection, but can be generalized further in the spirit of scale-space tracking. For instance, visual navigation of a vessel in coastal regions can achieve GPS-accuracy (Grelsson et al., 2020). Here, the efficiency of Fourier-transform DCF-implementations is used to match a (segment of a) horizon line with a huge number of model horizons sampled from a digital elevation model – in real-time.

With the shift of VOS and VIS methods to real-time processing, new application areas and new functionalities in existing application areas arise. For instance, AR systems can exploit the alignment of segmented objects to further improve accuracy and latency. Autonomous cars and advanced driver assistance systems already make frequent use of frame-wise semantic segmentation, but the performance might further increase with powerful VIS algorithms. Vessel navigation needs no longer relies on a separate horizon extraction, but can integrate the horizon analysis into the matching.

Obviously, potential misuse and ethically problematic applications also benefit from the technical development: Massive surveillance of people by authorities or companies is already existing cases. Similarly, VOS methods can be used to falsify videos or to produce deep fakes. Finally, many military applications could be thought of – in particular using IR-images. The openness of research and availability of open-source implementations is the key to scientific progress, but unfortunately also to misuse and spurious applications.

Acknowledgment

Large parts of this chapter describe methods and results developed during the past seven years at the Computer Vision Laboratory in Linköping in collaboration with Fahad Khan and several PhD and Master students, primarily to mention Martin Danelljan, Gustav Häger, Andreas Robinson, Joakim Johnander, Felix Järemo-Lawin, Ghoutam Bhat, Amanda Berg, and Susanna Gladh.

The research work covered in this chapter has been partially supported by KAW through the Wallenberg AI, Autonomous Systems and Software Program WASP, the Swedish Research Council through projects EMC2, NCNN, and ELLIIT, the SSF through the projects CUAS and Symbicloud, and by LiU through CENIIT.

References

Baker, S., Matthews, I., 2004. Lucas-Kanade 20 years on: a unifying framework. International Journal of Computer Vision 56 (3), 221–255.

Berg, A., Öfjäll, K., Ahlberg, J., Felsberg, M., 2015. Detecting rails and obstacles using a train-mounted thermal camera. In: Lecture Notes in Computer Science (Including Subseries Lecture Notes in Artificial Intelligence and Lecture Notes in Bioinformatics). In: LNCS, vol. 9127, pp. 492–503.

Berg, A., Johnander, J., Durand De Gevigney, F., Ahlberg, J., Felberg, M., 2019. Semi-automatic annotation of objects in visual-thermal video. In: Proceedings – 2019 International Conference on Computer Vision Workshop, ICCVW 2019, pp. 2242–2251.

Bhat, G., Johnander, J., Danelljan, M., Khan, F.S., Felsberg, M., 2018. Unveiling the power of deep tracking. In: Lecture Notes in Computer Science (Including Subseries Lecture Notes in Artificial Intelligence and Lecture Notes in Bioinformatics). In: LNCS, vol. 11206, pp. 493–509.

Bhat, G., Danelljan, M., Van Gool, L., Timofte, R., 2019. Learning discriminative model prediction for tracking. In: Proceedings of the IEEE International Conference on Computer Vision, vol. 2019-Octob, pp. 6181–6190.

Bhat, G., Lawin, F.J., Danelljan, M., Robinson, A., Felsberg, M., Van Gool, L., Timofte, R., 2020. Learning what to learn for video object segmentation. In: Vedaldi, A., Bischof, H., Brox, T., Frahm, J.-M. (Eds.), Computer Vision – ECCV 2020. Springer International Publishing, Cham, pp. 777–794.

Bolme, D.S., Beveridge, J.R., Draper, B.A., Lui, Y.M., 2010. Visual object tracking using adaptive correlation filters. In: Proceedings of the IEEE Computer Society Conference on Computer Vision and Pattern Recognition, pp. 2544–2550.

Bracewell, R.N., 1995. Two-Dimensional Imaging. Prentice Hall Signal Processing Series. Prentice Hall, Englewood Cliffs.

Chandaria, J., Thomas, G., Bartczak, B., Koeser, K., Koch, R., Becker, M., Bleser, G., Strieker, D., Wohlleber, C., Felsberg, M., Gustafsson, F., Hol, J.D., Schön, T.B., Skoglund, J., Slycke, P.J., Smeitz, S., 2007. Realtime camera tracking in the MATRIS project. SMPTE Motion Imaging Journal 116 (7–8), 266–271.

Chatfield, K., Simonyan, K., Vedaldi, A., Zisserman, A., 2014. Return of the devil in the details: delving deep into convolutional nets. In: British Machine Vision Conference.

Chen, Q.-S., Defrise, M., Deconinck, F., 1994. Symmetric phase-only matched filtering of Fourier-Mellin transforms for image registration and recognition. IEEE Transactions on Pattern Analysis and Machine Intelligence 16, 1156–1168.

Chéron, G., Laptev, I., Schmid, C., 2015. P-CNN: pose-based CNN features for action recognition. In: 2015 IEEE International Conference on Computer Vision (ICCV), pp. 3218–3226.

Dalal, N., Triggs, B., 2005. Histograms of oriented gradients for human detection. In: Computer Vision and Pattern Recognition, 2005. CVPR 2005. IEEE Computer Society Conference on, vol. 1, pp. 886–893.

Danelljan, M., Häger, G., Khan, F.S., Felsberg, M., 2014a. Accurate scale estimation for robust visual tracking. In: BMVC 2014 – Proceedings of the British Machine Vision Conference 2014.

Danelljan, M., Khan, F.S., Felsberg, M., Van De Weijer, J., 2014b. Adaptive color attributes for real-time visual tracking. In: Proceedings of the IEEE Computer Society Conference on Computer Vision and Pattern Recognition, pp. 1090–1097.

Danelljan, M., Hager, G., Khan, F.S., Felsberg, M., 2015. Learning spatially regularized correlation filters for visual tracking. In: Proceedings of the IEEE International Conference on Computer Vision, Vol. 2015 Inter, pp. 4310–4318.

Danelljan, M., Häger, G., Khan, F.S., Felsberg, M., 2016a. Adaptive decontamination of the training set: a unified formulation for discriminative visual tracking. Proceedings - IEEE Computer Society Conference on Computer Vision and Pattern Recognition 2016-Decem, 1430–1438.

Danelljan, M., Hager, G., Khan, F.S., Felsberg, M., 2016b. Convolutional features for correlation filter based visual tracking. In: Proceedings of the IEEE International Conference on Computer Vision, vol. 2016-Febru, pp. 621–629.

Danelljan, M., Robinson, A., Khan, F., Felsberg, M., 2016. Beyond correlation filters: Learning continuous convolution operators for visual tracking. LNCS, vol. 9909.

Danelljan, M., Hager, G., Khan, F.S., Felsberg, M., 2017. Discriminative scale space tracking. IEEE Transactions on Pattern Analysis and Machine Intelligence 39 (8), 1561–1575.

Danelljan, M., Bhat, G., Gladh, S., Khan, F.S., Felsberg, M., 2019a. Deep motion and appearance cues for visual tracking. Pattern Recognition Letters 124, 74–81.

Danelljan, M., Bhat, G., Khan, F.S., Felsberg, M., 2019b. Atom: accurate tracking by overlap maximization. Proceedings - IEEE Computer Society Conference on Computer Vision and Pattern Recognition 2019-June, 4655–4664.

Dendorfer, P., Osep, A., Milan, A., Schindler, K., Cremers, D., Reid, I., Roth, S., Leal-Taixé, L., 2020. MOTChallenge: a benchmark for single-camera multiple target tracking. International Journal of Computer Vision.

Everingham, M., Van Gool, L., Williams, C.K., Winn, J., Zisserman, A., 2010. The Pascal visual object classes (VOC) challenge. International Journal of Computer Vision 88 (2), 303–338.

Felsberg, M., 1998. Signal Processing Using Frequency Domain Methods in {C}lifford Algebra. Diploma thesis Institute of Computer Science and Applied Mathematics. Christian-Albrechts-University of Kiel.

Felsberg, M., 2009. Spatio-featural scale-space. In: Lecture Notes in Computer Science (Including Subseries Lecture Notes in Artificial Intelligence and Lecture Notes in Bioinformatics). In: LNCS, vol. 5567, pp. 808–819.

Felsberg, M., 2013. Enhanced distribution field tracking using channel representations. In: Proceedings of the IEEE International Conference on Computer Vision, pp. 121–128.

Felsberg, M., 2018. Probabilistic and Biologically Inspired Feature Representations. Morgan & Claypool Publishers.

Felsberg, M., Duits, R., Florack, L., 2005. The monogenic scale space on a rectangular domain and its features. International Journal of Computer Vision 64 (2–3), 187–201.

Galoogahi, H.K., Sim, T., Lucey, S., 2015. Correlation filters with limited boundaries. In: 2015 IEEE Conference on Computer Vision and Pattern Recognition (CVPR), pp. 4630–4638.

Garon, M., Lalonde, J.-F., 2017. Deep 6-DOF tracking. IEEE Transactions on Visualization and Computer Graphics 23, 2410–2418.

Girshick, R., Donahue, J., Darrell, T., Malik, J., 2016. Region-based convolutional networks for accurate object detection and segmentation. IEEE Transactions on Pattern Analysis and Machine Intelligence 38, 142–158.

Grelsson, B., Robinson, A., Felsberg, M., Khan, F.S., 2020. GPS-level accurate camera localization with HorizonNet. Journal of Field Robotics 37 (6), 951–971.

Häger, G., Bhat, G., Danelljan, M., Khan, F.S., Felsberg, M., Rudl, P., Doherty, P., 2016. Combining visual tracking and person detection for long term tracking on a UAV. In: Lecture Notes in Computer Science (Including Subseries Lecture Notes in Artificial Intelligence and Lecture Notes in Bioinformatics). In: LNCS, vol. 10072, pp. 557–568.

Häger, G., Felsberg, M., Khan, F., 2018. Countering bias in tracking evaluations. In: VISIGRAPP 2018 – Proceedings of the 13th International Joint Conference on Computer Vision, Imaging and Computer Graphics Theory and Applications, vol. 5, pp. 581–587.

He, K., Zhang, X., Ren, S., Sun, J., 2016. Deep residual learning for image recognition. In: CVPR.

Henriques, J.F., Caseiro, R., Martins, P., Batista, J., 2012. Exploiting the circulant structure of tracking-by-detection with kernels. In: Lecture Notes in Computer Science (Including Subseries Lecture Notes in Artificial Intelligence and Lecture Notes in Bioinformatics). In: LNCS, vol. 7575, pp. 702–715.

Horner, J.L., Gianino, P.D., 1984. Phase-only matched filtering. Applied Optics 23 (6), 812–816.

Jaccard, P., 1912. The distribution of the flora in the Alpine zone. New Phytologist 11 (2), 37–50.

Johnander, J., Danelljan, M., Brissman, E., Khan, F.S., Felsberg, M., 2019. A generative appearance model for end-to-end video object segmentation. Proceedings - IEEE Computer Society Conference on Computer Vision and Pattern Recognition 2019-June, 8945–8954.

Källhammer, J.E., Eriksson, D., Granlund, G., Felsberg, M., Moe, A., Johansson, B., Wiklund, J., Forssén, P.E., 2007. Near zone pedestrian detection using a low-resolution FIR sensor. In: IEEE Intelligent Vehicles Symposium, Proceedings, pp. 339–345.

Khan, F.S., Anwer, R.M., van de Weijer, J., Bagdanov, A., Vanrell, M., Lopez, A.M., 2012. Color attributes for object detection. In: IEEE Conference on Computer Vision and Pattern Recognition.

Koenderink, J.J., 1984. The structure of images. Biological Cybernetics 50, 363–370.

Kristan, M., Pflugfelder, R., Leonardis, A., Matas, J., Porikli, F., Čehovin, L., Nebehay, G., Fernandez, G., Vojíř, T., Gatt, A., Khajenezhad, A., Salahledin, A., Soltani-Farani, A., Zarezade, A., Petrosino, A., Milton, A., Bozorgtabar, B., Li, B., Chan, C.S., Heng, C., Ward, D., Kearney, D., Monekosso, D., Karaimer, H.C., Rabiee, H.R., Zhu, J., Gao, J., Xiao, J., Zhang, J., Xing, J., Huang, K., Lebeda, K., Cao, L., Maresca, M.E., Lim, M.K., ELHelw, M., Felsberg, M., Remagnino, P., Bowden, R., Goecke, R., Stolkin, R., Lim, S.Y.Y., Maher, S., Poullot, S., Wong, S., Satoh, S., Chen, W., Hu, W., Zhang, X., Li, Y., Niu, Z., 2013. The visual object tracking VOT2013 challenge results. In: Proceedings of the IEEE International Conference on Computer Vision, pp. 98–111.

Kristan, M., Pflugfelder, R., Leonardis, A., Matas, J., Čehovin, L., Nebehay, G., Vojíř, T., Fernández, G., Lukežič, A., Dimitriev, A., Petrosino, A., Saffari, A., Li, B., Han, B., Heng, C.K., Garcia, C., Pangeršič, D., Häger, G., Khan, F.S., Oven, F., Possegger, H., Bischof, H., Nam, H., Zhu, J., Li, J.J., Choi, J.Y., Choi, J.W., Henriques, J.F., van de Weijer, J., Batista, J., Lebeda, K., Öfjäll, K., Yi, K.M., Qin, L., Wen, L., Maresca, M.E., Danelljan, M., Felsberg, M., Cheng, M.M., Torr, P., Huang, Q., Bowden, R., Hare, S., Lim, S.Y.Y., Hong, S., Liao, S., Hadfield, S., Li, S.Z., Duffner, S., Golodetz, S., Mauthner, T., Vineet, V., Lin, W., Li, Y., Qi, Y., Lei, Z., Niu, Z.H., 2015a. The visual object tracking VOT2014 challenge results, vol. 8926.

Kristan, M., Matas, J., Leonardis, A., Felsberg, M., Čehovin, L., Fernández, G., Vojír̃, T., Häger, G., Nebehay, G., Pflugfelder, R., Gupta, A., Bibi, A., Lukežič, A., Garcia-Martin, A., Saffari, A., Petrosino, A., Montero, A., Varfolomieiev, A., Baskurt, A., Zhao, B., Ghanem, B., Martinez, B., Lee, B., Han, B., Wang, C., Garcia, C., Zhang, C., Schmid, C., Tao, D., Kim, D., Huang, D., Prokhorov, D., Du, D., Yeung, D.-Y., Ribeiro, E., Khan, F., Porikli, F., Bunyak, F., Zhu, G., Seetharaman, G., Kieritz, H., Yau, H., Li, H., Qi, H., Bischof, H., Possegger, H., Lee, H., Nam, H., Bogun, I., Jeong, J.-C., Cho, J.-I., Lee, J.-Y., Zhu, J., Shi, J., Li, J., Jia, J., Feng, J., Gao, J., Choi, J., Kim, J.-W., Lang, J., Martinez, J., Choi, J., Xing, J., Xue, K., Palaniappan, K., Lebeda, K., Alahari, K., Gao, K., Yun, K., Wong, K., Luo, L., Ma, L., Ke, L., Wen, L., Bertinetto, L., Pootschi, M., Maresca, M., Danelljan, M., Wen, M., Zhang, M., Arens, M., Valstar, M., Tang, M., Chang, M.-C., Khan, M., Fan, N., Wang, N., Miksik, O., Torr, P., Wang, Q., Martin-Nieto, R., Pelapur, R., Bowden, R., Laganière, R., Moujtahid, S., Hare, S., Hadfield, S., Lyu, S., Li, S., Zhu, S.-C., Becker, S., Duffner, S., Hicks, S., Golodetz, S., Choi, S., Wu, T., Mauthner, T., Pridmore, T., Hu, W., Hübner, W., Wang, X., Li, X., Shi, X., Zhao, X., Mei, X., Shizeng, Y., Hua, Y., Li, Y., Lu, Y., Li, Y., Chen, Z., Huang, Z., Chen, Z., Zhang, Z., He, Z., Hong, Z., 2015b. The visual object tracking VOT2015 challenge results. In: Proceedings of the IEEE International Conference on Computer Vision, vol. 2015-Febru.

Kristan, M., Leonardis, A., Matas, J., Felsberg, M., Pflugfelder, R., Čehovin, L., Vojír, T., Häger, G., Lukežič, A., Fernández, G., Gupta, A., Petrosino, A., Memarmoghadam, A., Martin, A.G., Montero, A.S., Vedaldi, A., Robinson, A., Ma, A.J., Varfolomieiev, A., Alatan, A., Erdem, A., Ghanem, B., Liu, B., Han, B., Martinez, B., Chang, C.M., Xu, C., Sun, C., Kim, D., Chen, D., Du, D., Mishra, D., Yeung, D.Y., Gundogdu, E., Erdem, E., Khan, F., Porikli, F., Zhao, F., Bunyak, F., Battistone, F., Zhu, G., Roffo, G., Sai Subrahmanyam, G.R., Bastos, G., Seetharaman, G., Medeiros, H., Li, H., Qi, H., Bischof, H., Possegger, H., Lu, H., Lee, H., Nam, H., Chang, H.J., Drummond, I., Valmadre, J., Jeong, J.C., Cho, J.I., Lee, J.Y., Zhu, J., Feng, J., Gao, J., Choi, J.Y., Xiao, J., Kim, J.W., Jeong, J., Henriques, J.F., Lang, J., Choi, J., Martinez, J.M., Xing, J., Gao, J., Palaniappan, K., Lebeda, K., Gao, K., Mikolajczyk, K., Qin, L., Wang, L., Wen, L., Bertinetto, L., Rapuru, M.K., Poostchi, M., Maresca, M., Danelljan, M., Mueller, M., Zhang, M., Arens, M., Valstar, M., Tang, M., Baek, M., Khan, M.H., Wang, N., Fan, N., Al-Shakarji, N., Miksik, O., Akin, O., Moallem, P., Senna, P., Torr, P.H., Yuen, P.C., Huang, Q., Nieto, R.M., Pelapur, R., Bowden, R., Laganière, R., Stolkin, R., Walsh, R., Krah, S.B., Li, S., Zhang, S., Yao, S., Hadfield, S., Melzi, S., Lyu, S., Li, S., Becker, S., Golodetz, S., Kakanuru, S., Choi, S., Hu, T., Mauthner, T., Zhang, T., Pridmore, T., Santopietro, V., Hu, W., Li, W., Hübner, W., Lan, X., Wang, X., Li, X., Li, Y., Demiris, Y., Wang, Y., Qi, Y., Yuan, Z., Cai, Z., Xu, Z., He, Z., Chi, Z., 2016. The visual object tracking VOT2016 challenge results. LNCS, vol. 9914.

Kristan, M., Matas, J., Leonardis, A., Vojir, T., Pflugfelder, R., Fernandez, G., Nebehay, G., Porikli, F., Čehovin, L., 2016b. A novel performance evaluation methodology for single-target trackers. IEEE Transactions on Pattern Analysis and Machine Intelligence.

Kristan, M., Leonardis, A., Matas, J., Felsberg, M., Pflugfelder, R., Zajc, L., Vojír, T., Häger, G., Lukežič, A., Eldesokey, A., Fernández, G., García-Martín, Á., Muhic, A., Petrosino, A., Memarmoghadam, A., Vedaldi, A., Manzanera, A., Tran, A., Alatan, A., Mocanu, B., Chen, B., Huang, C., Xu, C., Sun, C., Du, D., Zhang, D., Du, D., Mishra, D., Gundogdu, E., Velasco-Salido, E., Khan, F., Battistone, F., Subrahmanyam, G., Bhat, G., Huang, G., Bastos, G., Seetharaman, G., Zhang, H., Li, H., Lu, H., Drummond, I., Valmadre, J., Jeong, J.-C., Cho, J.-I., Lee, J.-Y., Noskova, J., Zhu, J., Gao, J., Liu, J., Kim, J.-W., Henriques, J., Martínez, J., Zhuang, J., Xing, J., Gao, J., Chen, K., Palaniappan, K., Lebeda, K., Gao, K., Kitani, K., Zhang, L., Wang, L., Yang, L., Wen, L., Bertinetto, L., Poostchi, M., Danelljan, M., Mueller, M., Zhang, M., Yang, M.-H., Xie, N., Wang, N., Miksik, O., Moallem, P., Pallavi Venugopal, M., Senna, P., Torr, P., Wang, Q., Yu, Q., Huang, Q., Martín-Nieto, R., Bowden, R., Liu, R., Tapu, R., Hadfield, S., Lyu, S., Golodetz, S., Choi, S., Zhang, T., Zaharia, T., Santopietro, V., Zou, W., Hu, W., Tao, W., Li, W., Zhou, W., Yu, X., Bian, X., Li, Y., Xing, Y., Fan, Y., Zhu, Z., Zhang, Z., He, Z., 2017. The visual object tracking VOT2017 challenge results. In: Proceedings – 2017 IEEE International Conference on Computer Vision Workshops, ICCVW 2017, vol. 2018-Janua.

Kristan, M., Leonardis, A., Matas, J., Felsberg, M., Pflugfelder, R., Zajc, L., Vojír, T., Bhat, G., Lukežič, A., Eldesokey, A., Fernández, G., García-Martín, Á., Iglesias-Arias, Á., Alatan, A., González-García, A., Petrosino, A., Memarmoghadam, A., Vedaldi, A., Muhič, A., He, A., Smeulders, A., Perera, A., Li, B., Chen, B., Kim, C., Xu, C., Xiong, C., Tian, C., Luo, C., Sun, C., Hao, C., Kim, D., Mishra, D., Chen, D., Wang, D., Wee, D., Gavves, E., Gundogdu, E., Velasco-Salido, E., Khan, F., Yang, F., Zhao, F., Li, F., Battistone, F., De Ath, G., Subrahmanyam, G., Bastos, G., Ling, H., Galoogahi, H., Lee, H., Li, H., Zhao, H., Fan, H., Zhang, H., Possegger, H., Li, H., Lu, H., Zhi, H., Li, H., Lee, H., Chang, H., Drummond, I., Valmadre, J., Martin, J., Chahl, J., Choi, J., Li, J., Wang, J., Qi, J., Sung, J., Johnander, J., Henriques, J., Choi, J., van de Weijer, J., Herranz, J., Martínez, J., Kittler, J., Zhuang, J., Gao, J., Grm, K., Zhang, L., Wang, L., Yang, L., Rout, L., Si, L., Bertinetto, L., Chu, L., Che, M., Maresca, M., Danelljan, M., Yang, M.-H., Abdelpakey, M., Shehata, M., Kang, M., Lee, N., Wang, N., Miksik, O., Moallem, P., Vicente-Moñivar, P., Senna, P., Li, P., Torr, P., Raju, P., Ruihe, Q., Wang, Q., Zhou, Q., Guo, Q., Martín-Nieto, R., Gorthi, R., Tao, R., Bowden, R., Everson, R., Wang, R., Yun, S., Choi, S., Vivas, S., Bai, S., Huang, S., Wu, S., Hadfield, S., Wang, S., Golodetz, S., Ming, T., Xu, T., Zhang, T., Fischer, T., Santopietro, V., Štruc, V., Wei, W., Zuo, W., Feng, W., Wu, W., Zou, W., Hu, W., Zhou, W., Zeng, W., Zhang, X., Wu, X., Wu, X.-J., Tian, X., Li, Y., Lu, Y., Law, Y., Wu, Y., Demiris, Y., Yang, Y., Jiao, Y., Li, Y., Zhang, Y., Sun, Y., Zhang, Z., Zhu, Z., Feng, Z.-H., Wang, Z., He, Z., 2019. The sixth visual object tracking VOT2018 challenge results. LNCS, vol. 11129.

Kristan, M., Matas, J., Leonardis, A., Felsberg, M., Pflugfelder, R., Kämäräinen, J.-K., Zajc, L., Drbohlav, O., Lukezic, A., Berg, A., Eldesokey, A., Kapyla, J., Fernández, G., Gonzalez-Garcia, A., Memarmoghadam, A., Lu, A., He, A., Varfolomieiev, A., Chan, A., Tripathi, A., Smeulders, A., Pedasingu, B., Chen, B., Zhang, B., Baoyuanwu, B., Li, B., He, B., Yan, B., Bai, B., Li, B., Li, B., Kim, B., Ma, C., Fang, C., Qian, C., Chen, C., Li, C., Zhang, C., Tsai, C.-Y., Luo, C., Micheloni, C., Zhang, C., Tao, D., Gupta, D., Song, D., Wang, D., Gavves, E., Yi, E., Khan, F., Zhang, F., Wang, F., Zhao, F., De Ath, G., Bhat, G., Chen, G., Wang, G., Li, G., Cevikalp, H., Du, H., Zhao, H., Saribas, H., Jung, H., Bai, H., Yu, H., Peng, H., Lu, H., Li, H., Li, J., Li, J., Fu, J., Chen, J., Gao, J., Zhao, J., Tang, J., Li, J., Wu, J., Liu, J., Wang, J., Qi, J., Zhang, J., Tsotsos, J., Lee, J., Van De Weijer, J., Kittler, J., Ha Lee, J., Zhuang, J., Zhang, K., Wang, K., Dai, K., Chen, L., Liu, L., Guo, L., Zhang, L., Wang, L., Wang, L., Zhang, L., Wang, L., Zhou, L., Zheng, L., Rout, L., Van Gool, L., Bertinetto, L., Danelljan, M., Dunnhofer, M., Ni, M., Kim, M., Tang, M., Yang, M.-H., Paluru, N., Martinel, N., Xu, P., Zhang, P., Zheng, P., Zhang, P., Torr, P., Wang, Q., Guo, Q., Timofte, R., Gorthi, R., Everson, R., Han, R., Zhang, R., You, S., Zhao, S.-C., Zhao, S., Li, S., Li, S., Ge, S., Bai, S., Guan, S., Xing, T., Xu, T., Yang, T., Zhang, T., Vojír, T., Feng, W., Hu, W., Wang, W., Tang, W., Zeng, W., Liu, W., Chen, X., Qiu, X., Bai, X., Wu, X.-J., Yang, X., Chen, X., Li, X., Sun, X., Chen, X., Tian, X., Tang, X., Zhu, X.-F., Huang, Y., Chen, Y., Lian, Y., Gu, Y., Liu, Y., Chen, Y., Zhang, Y., Xu, Y., Wang, Y., Li, Y., Zhou, Y., Dong, Y., Xu, Y., Zhang, Y., Li, Y., Luo, Z., Zhang, Z., Feng, Z.-H., He, Z., Song, Z., Chen, Z., Zhang, Z., Wu, Z., Xiong, Z., Huang, Z., Teng, Z., Ni, Z., 2019b. The seventh visual object tracking VOT2019 challenge results. In: Proceedings – 2019 International Conference on Computer Vision Workshop, ICCVW 2019.

Kristan, M., Leonardis, A., Matas, J., Felsberg, M., Pflugfelder, R., Kamarainen, J.-K., Zajc, L.Č., Danelljan, M., Lukezic, A., Drbohlav, O., He, L., Zhang, Y., Yan, S., Yang, J., Fernandez, G., et al., 2020. The eighth visual object tracking VOT2020 challenge results.

Lowe, D.G., 2004. Distinctive image features from scale-invariant keypoints. International Journal of Computer Vision 60 (2), 91–110.

Lucas, B.D., Kanade, T., 1981. An iterative image registration technique with an application to stereo vision. In: Proceedings of International Joint Conference on Artificial Intelligence.

Lukežič, A., Zajc, L.Č., Kristan, M., 2018. Fast Spatially Regularized Correlation Filter Tracker.

Matthews, L., Ishikawa, T., Baker, S., 2004. The template update problem. IEEE Transactions on Pattern Analysis and Machine Intelligence 26 (6), 810–815.

Neumann, U., Park, J., 1998. Extendible object-centric tracking for augmented reality. In: Proceedings. IEEE 1998 Virtual Reality Annual International Symposium (Cat. No. 98CB36180), pp. 148–155.

Neumann, U., You, S., Cho, Y., Lee, J., Park, J., 1999. Augmented reality tracking in natural environments. In: International Symposium on Mixed Realities.

Perazzi, F., Pont-Tuset, J., McWilliams, B., Van Gool, L., Gross, M., Sorkine-Hornung, A., 2016. A benchmark dataset and evaluation methodology for video object segmentation. In: Computer Vision and Pattern Recognition.

Perazzi, F., Khoreva, A., Benenson, R., Schiele, B., Sorkine-Hornung, A., 2017. Learning video object segmentation from static images. In: Proceedings – 30th IEEE Conference on Computer Vision and Pattern Recognition, CVPR 2017, vol. 2017-Janua, pp. 3491–3500.

Pietikäinen, M., Zhao, G., 2015. Two decades of local binary patterns: a survey. Advances in Independent Component Analysis and Learning Machines abs/1612.0, 175–210.

Reiter, R., 1978. On Closed World Data Bases. Springer US, Boston, MA, pp. 55–76.

Ripley, D.L., Politzer, T., 2010. Vision disturbance after TBI. NeuroRehabilitation 27, 215–216.

Robinson, A., Lawin, F.J., Danelljan, M., Khan, F.S., Felsberg, M., 2020. Learning fast and robust target models for video object segmentation. In: Proceedings of the IEEE Computer Society Conference on Computer Vision and Pattern Recognition, pp. 7404–7413.

Shi, Jianbo, Tomasi, 1994. Good features to track. In: 1994 Proceedings of IEEE Conference on Computer Vision and Pattern Recognition, pp. 593–600.

Simonyan, K., Zisserman, A., 2015. Very deep convolutional networks for large-scale image recognition. In: International Conference on Learning Representations.

Skoglund, J., Felsberg, M., 2006. Evaluation of subpixel tracking algorithms. LNCS, vol. 4292.

Skoglund, J., Felsberg, M., 2007. Covariance estimation for SAD block matching. LNCS, vol. 4522.

Stauffer, C., Grimson, W.E.L., 2000. Learning patterns of activity using real-time tracking. IEEE Transactions on Pattern Analysis and Machine Intelligence 22 (8), 747–757.

Szegedy, C., Liu, W., Jia, Y., Sermanet, P., Reed, S., Anguelov, D., Erhan, D., Vanhoucke, V., Rabinovich, A., 2015. Going deeper with convolutions. In: Proc. Conf. Computer Vision and Pattern Recognition, pp. 1–9.

Van De Weijer, J., Schmid, C., Verbeek, J., Larlus, D., 2009. Learning color names for real-world applications. IEEE Transactions on Image Processing 18 (7), 1512–1523.

Vojíř, T., Matas, J., 2017. Pixel-wise object segmentations for the VOT 2016 dataset. In: Research Reports of CMP, no. 1.

Wang, B., Qi, Z., Chen, S., 2016. Motion-based feature selection and adaptive template update strategy for robust visual tracking. In: Proceedings – 2016 3rd International Conference on Information Science and Control Engineering, ICISCE 2016, vol. 1, pp. 462–467.

Wu, Y., Lim, J., Yang, M.H., 2013. Online object tracking: a benchmark. In: Proceedings of the IEEE Computer Society Conference on Computer Vision and Pattern Recognition, pp. 2411–2418.

Yang, L., Fan, Y., Xu, N., 2019. Video instance segmentation. In: 2019 IEEE/CVF International Conference on Computer Vision (ICCV), pp. 5187–5196.

Yao, R., Lin, G., Xia, S., Zhao, J., Zhou, Y., 2019. Video object segmentation and tracking: a survey. arXiv 1 (1).

Zhu, Z., Wang, Q., Li, B., Wu, W., 2018. Distractor-aware Siamese networks for visual object tracking. In: Eccv 2018, pp. 1–17. arXiv:1808.06048v1 [cs.CV].

Biographies

Michael Felsberg is a full professor in Computer Vision at the Department of Electrical Engineering, Linköping University, Sweden. He is also honory professor at the School of Engineering, University of KwaZulu-Natal in Durban, South Africa. He holds a PhD degree from Kiel University, Germany (2002) and a docent degree from Linköping University (2005). He received various awards, among others the Olympus award (2005) from the DAGM and best paper awards from ICPR (2016) and VISAPP (2021). His Google Scholar h-index is 44. His research interests include, besides visual object tracking, video object and instance segmentation, point cloud processing, and efficient machine learning techniques.

C H A P T E R

10

Long-term deep object tracking

Efstratios Gavves and Deepak Gupta

Informatics Institute, University of Amsterdam, Amsterdam, Netherlands

CHAPTER POINTS

- A formulation of video object tracking from a machine learning and optimization point of view.

- A summary of visual, learning, and engineering challenges in tracking.

- An overview of state-of-the-art short-term visual object trackers, including their limitations when it comes to longer and more complex video spatiotemporal sequences.

- An introduction of long-term visual object tracking and the catastrophic effect of model decay and target appearance and disappearance.

- A description of deep Siamese trackers, which represent the state-of-the-art in visual object tracking, especially for long-term videos.

- An overview of representation invariance and equivariance, discussing how it relates to deep visual object trackers.

I would like to dedicate this work to my wife Katerina, my son Iasonas, my mother Antonia, my brother Manolis, and of course my father, Gavriil, who will always be in our hearts and minds.

10.1 Introduction

The advent of Deep Learning has in the past decade transformed Computer Vision, from object recognition, semantic segmentation and 3D surface reconstruction to video understanding. Being one of the oldest Computer Vision challenges, video object tracking has also enjoyed great progress in the Deep Learning era, albeit at a delayed pace for reasons that

Copyright © 2022 Elsevier Inc. All rights reserved.

we will describe later in this chapter. The impact of Deep Learning on video object tracking has not just been quantitative, that is improving accuracies in standardized benchmarks. Importantly, the improvement has also been qualitative, generalizing video object tracking from few-second long videos, better known as short-term tracking (Section 10.2) to videos spanning several dozens minutes, better known as long-term tracking (Section 10.3).

Visual object tracking, in its most general form, can be described as learning the model f_ϕ parameterized by ϕ, which attempts to predict the future location of a target object given an initial target definition, that is

$$f_\phi : \mathcal{Y}_1 \times \mathcal{X}_1 \times \mathcal{X}_t \to \mathcal{Y}_t. \tag{10.1}$$

$\mathcal{Y}_t$ denotes the space of all possible predictions, be they in the form of a bounding box or a segmentation mask. $\mathcal{X}_t$ denotes the space of all possible input frames at time t. The only ground truth given to the model is the location of the object at the first frame.

Often, tracker models rely on their own intermediate predictions $y_i, i < t$, to update the model parameters ϕ. In that case the model parameters also change through time and are better represented with ϕ_t. As we will see later, however, it is important to remember that even if the intermediate predictions y_i are used as the target variables to update the models with supervised learning algorithms, these variables are not the same as the true location of the object, $y_i \neq y_i^*$.

To optimize any visual object tracker, the most common framework is empirical risk minimization. Specifically, a loss function is defined by the respective model of choice

$$\mathcal{L} = \mathcal{L}\big(\phi_t; \mathbf{x}_{1:t}, y_1^*, y_{1:t-1}\big) \tag{10.2}$$

minimized with (stochastic) gradient descent or variants,

$$\underset{\phi_t}{\arg\min}\, \mathcal{L} \Rightarrow \phi_{t+1} = \phi_t - \epsilon \frac{\partial \mathcal{L}}{\partial \phi}. \tag{10.3}$$

Optionally, the loss function is augmented by a regularization function $\Omega(\phi)$ that penalizes weights according to predetermined design principles, for instance, penalizing overfitting or nonsparse solutions.

While the basic formulation of trackers is always the same, a recent division of trackers is into short-term and long-term visual object trackers. According to the definition of Kristan et al. (2016), short-term tracking does not require tracker methods to perform redetection of the target. That is, it is assumed the target object does not disappear and reappear from the frame. In contrast, in long-term visual object tracking the target object may disappear and reappear at the frame at any moment. Perhaps more importantly, long-term tracking concerns itself with complex and long videos, where the object might undergo severe appearance disturbances. These disturbances may be due to changes caused by the object itself, *e.g.*, the pose of the target object changes significantly and constantly, thus altering the appearance of the object as well, since the tracker has in its availability only the target appearance at the first frame. These disturbances can also be due to changes in the environment, for instance in the way the scene is illuminated or due to occlusions by other objects in the scene. These disturbances appear also in short-term tracking, naturally, and at first sight, addressing them looks like a straightforward extension for longer videos. However, as we shall see next, these disturbances have further *indirect* effects that are unique in long-term tracking simply due to the longer durations, effects that are often gradual, cumulative and may even prove catastrophic to the models at hand.

10.1.1 Challenges in video object tracking

Intuitively, visual object tracking appears to be a rather straightforward problem. After all, humans can easily follow objects persistently, no matter the object, its appearance and the settings it appears in. In fact, tracking is one of the hardest and most elusive problems of computer vision and applied machine learning. This is largely because it is a severely underconstrained problem: there exist almost no constraints on the types of objects or variations throughout the video in the general case.

Next, we identify and describe three types of challenges in visual object tracking: *visual*, *learning*, and *engineering* challenges. It is important to emphasize that these challenges are rarely – if ever – found in isolation.

10.1.1.1 Visual challenges in tracking

We identify two types of visual modeling challenges: intrinsic visual challenges caused by changes of the target object itself, and extrinsic visual challenges caused by changes in the tracking environment.

Starting from the intrinsic visual challenges, extensive studies (Smeulders et al., 2014; Wu et al., 2015) have grouped the object specific variations: *scale variation, shape deformation, in-plane rotation, out-of-plane rotation*. The scale variation occurs either when the object changes size or when the camera moves closer or further away from the target object. Shape deformation is typically observed when the target object is nonrigid and changes shape over time. In-plane rotations are the rotations of either the target object or the camera that are on the plane perpendicular to the axis between the camera and the target object. An example of an in-plane rotation is when pedestrians are recorded from a birds-eye view, *e.g.*, a drone, and the pedestrians change directions. Out-of-plane rotations, in contrast, are all other rotations taking place in the 3d space, for instance when tracking a person dancing.

In the extrinsic visual challenges that are caused by the environment, we find *illumination variation, occlusion, target disappearance and reappearance* and *background variation*. Illumination variation happens when the degree of illumination on the target changes throughout the sequence, leading to possible very different appearance and even color distributions. For instance, a car navigating through roads and then tunnels will change in color because of the decrease in contrast. Partial or complete occlusion is caused by other objects in the scene, which can be moving or static. Interestingly, occlusion is a type of variation unlike others, in the sense that it about the absence of information and cannot easily be learned explicitly. Similar to occlusion, the disappearance and reappearance of the target happen when the target exits the frame and becomes again visible at a random future point, possibly with different exit and entry points, *e.g.*, a car exiting from the left side and reentering from the right side of the frame. Although short-term tracking oftentimes assumes no target disappearance and reappearance (Kristan et al., 2016), in long-term tracking no such assumption can be made. Last, background variation is often a challenge because from a statistical perspective, it offers ample opportunity of discovering random visual patterns that may temporarily be confused with target object patterns. This can be particularly problematic with models that rely on heavy model updating.

10.1.1.2 *Learning challenges in tracking*

In the general case, visual object tracking does not place any constraints on the types or appearances of objects. One video might be about a car cruising the highway, while another video might be about following an octopus that deforms in any way imaginable. If the video is from a biomedical recording, the tracked object might not even have the regular appearance, shape or motion patterns of everyday objects. In principle, the tracker model must be able to follow all these objects throughout the sequences without any further guidance other than the definition of the target by the user in the first frame. This brings us to the following learning challenge: unlike standard machine learning tasks like object classification, detection or segmentation, in visual object tracking the target objects that are being modeled are arbitrary and cannot be defined in advance. That is, in visual object tracking we cannot have training examples from the tracked object, for it would not be tracking anymore but object detection. This is unlike most machine learning problems, where usually there are training and test sets containing different data samples from the same object categories. Seen otherwise, visual object tracking is akin to the recently popular few-shot and one-shot learning paradigms (Bertinetto et al., 2016a), which are to among the hardest settings for machine learning.

Visual object tracking is primarily modeled by supervised learning methodologies. However, due to the very absence of sufficient positive samples to learn robust models with good generalization, typically tracker models rely on data augmentation at the first frame and model updating in future frames. For model updating, the predictions of the model are used as *pseudo-positives*, which gradually lead to adding bias to the tracker models. We will discuss model decay in further detail later in the chapter, as it is perhaps the single most difficult challenge when it comes to long-term object tracking.

Visual object tracking is as much about localizing the target object as is about separating it from the background. However, the scenes in which objects appear can change dramatically over the duration of the video. What makes things harder is when there exist multiple objects with similar or even identical appearance, for instance when tracking an athlete during a soccer match, or a particular person in a marching band. During tracking, the background can be wildly different from one sequence to another and even more variable through time also. This brings us to the second learning challenge: in visual object tracking the model must learn how to model the positive samples, that is the target object, and ignore the negative samples, that is the background (Bhat et al., 2019). The background may be significantly different from what has been observed in previous videos. In fact, the background might constantly change over time within one video and be highly distracting or even confusing. In that sense, visual object tracking is also similar to anomaly detection, where the machine learning model is exposed only, or mostly, to positive examples and must learn to discriminate them from all other possible negative examples.

Last, a critical learning challenge is the tuning of hyperparameters. Unlike other computer vision and machine learning problems, visual object tracking often refers to scenarios where the inputs are nonstationary and not available in advance. As a result, hyperparameters that are optimal for one video might be grossly suboptimal for others. In fact, a frequent happenstance is that by making small changes in tracker models, the performance improves in several videos but drops in others. Model selection and hyperparameter tuning is often critical to make sure the final tracker model is robust and generalizes.

10.1.1.3 *Engineering challenges in tracking*

Besides the challenges associated with the visual appearance of the target and background, and with the machine learning models, in visual object tracking there exist engineering challenges that relate to the way recording the video sequences or tuning the tracker model are carried out.

An engineering challenge that often causes a tracker to fail is *fast motion*, where the speed of the target object is higher than the frame rate. Fast motion is a problem with tracker models that rely on search radii around their previous predictions to constrain their search. Typically, the search radius is taken to be large enough according to what is expected in the available video sequences. However, it can happen that the target object moves so fast that it is located outside the search radius in the following frame. In that case the tracker will simply fail no matter how accurate its model is. Failure due to fast motion can be particularly problematic with models that perform frequent model updates, as they will use *de facto* misclassiffied patches to update the tracker model.

Another challenge associated with the fast motion of the target is *motion blur*. When the target object moves very fast, the camera sensor receives appearances of the target object at multiple time stamps. This results in an averaging effect that blurs the image and, quite importantly, obscures the target appearance. As a result, the tracker model is likely to miss the target because certain high frequency details that are necessary for precisely representing the target object suddenly disappear.

In the recent years visual object trackers have attempted to address the aforementioned challenges with the power of deep neural networks and large training sets. Next, we will examine the main deep learning approaches for doing so.

10.2 Short-term visual object tracking

Before explaining the emergence of long-term tracking in the recent years, we first give a short introduction to short-term tracking that was the predominant paradigm for a long time. Short-term tracking methods primarily focus on visual instances of short episodes, typically ranging from 10–20 seconds. The primary goal in short-term tracking is to locate the target with most precision for as long as possible, until the target is lost. In general, short-term trackers do not integrate a recovery mechanism that can retrace the target once it is lost by the tracker, mostly for reasons of computational efficiency.

Short-term trackers are judged on basically two grounds: *(i)* how well the target can be localized, and *(ii)* how fast can the inference be done. A general consensus (Kristan et al., 2017) is to evaluate short-term trackers with a focus on either solely their accuracy, or the accuracy given real-time performance, in which case inference speed must meet a minimum threshold. Such a constraint depends on the hardware design and can vary across different evaluation protocols. When real-time constraints are not required, several variants of short-term tracking exist, including tracking on RGB images only (Smeulders et al., 2014), using RGB-D images (Zheng et al., 2017), or even augmenting the RGB data with thermal imagery (Li et al., 2019b).

From a methodological point of view, there are several different ways in which the short-term tracking methods can be categorized. In view of the revolution brought by deep learning and the fact that most state-of-the-art trackers nowadays rely on deep neural architectures in one way or the other, we will divide trackers in two families: shallow (early) and deep trackers. For the shallow trackers, categorization is based on the seminal review paper of Smeulders et al. (2014).

10.2.1 Shallow trackers

Shallow trackers comprise the majority of tracking methods relying on standard computer vision methods before the advent of deep learning. The different categories are:

Tracking using matching. This group of models detects the location of the target through matching the template built from the previous frames with different candidate regions of the search image. Several early trackers using traditional computer vision methods fall under this category. Examples include methods based on normalized cross-correlation (NCC) matching (Briechle and Hanebeck, 2001), Lucas-Kanade Tracker (KLT) (Baker and Matthews, 2004), Kalman Appearance Tracker (KAT) (Nguyen and Smeulders, 2004), Mean-shift Tracker (MST) (Comaniciu et al., 2000) and Locally Orderless Tracking (LOT) method (Oron et al., 2015). These methods differ mostly in how the different candidate regions are sampled and how each of them is matched with the template image. For example, NCC uses intensity values in the template for matching, and performs uniform sampling over the search image.

Tracking using matching with extended appearance models. The idea for this class of trackers is to maintain an extended model of the target's appearance over the previous frames. Generally, such models are slow in performance, especially because at every frame, the candidate samples are matched with the template image as well as the frames stored in the appearance model. An example of this approach is the Incremental Visual Tracking (IVT) (Ross et al., 2008), where eigenimages of the target are computed by incremental PCA over the target's intensity-value template. These are stored in a leaking memory where the old images are slowly forgotten. Other examples of this group are Tracking on the Affine Group (TAG) (Kwon et al., 2009) and Tracking by Sampling Trackers (TST) (Kwon and Lee, 2011).

Tracking using matching with constraints. This category of trackers reduces the representation of the target to a sparse representation. Sparse optimization is performed to identify the right candidate sample in the search frames. Such methods primarily focus on scenarios where the appearance changes rapidly over time, and the appearance model needs to quickly adapt. To prevent the model from drifting while the frequent updates are made, these methods use additional constraints on top of the traditional tracking by matching.

Tracking using discriminative classification. The trackers in this group provide a different view to the problem – they build the model on the distinction of the target foreground against the background. Also referred to as tracking-by-detection, these methods build a classifier to distinguish target pixels from the background pixels, and update the classifier based on new samples coming in. The old, yet popular tracker from this category is the Foreground-Background Tracker (Nguyen and Smeulders, 2006) that uses feature vectors to differentiate

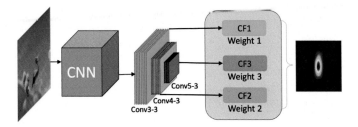

FIGURE 10.1 Schematic representation of a correlation filter learning using a CNN encoding function. Credit: (Fiaz et al., 2019).

the target region against the local background. Other similar methods include Hough-Based Tracking (Godec et al., 2013) and Tracking, Learning and Detection method (Kalal et al., 2010).

Tracking using discriminative classification with constraints. For tracking methods based on discriminative classification, it is important that the samples from the target and local background are properly sampled, otherwise it can adversely affect the performance of the trackers. This class of methods integrates the labeling procedure into the learning process itself. An example of such a strategy includes Structured Output Tracking with Kernels (STR), where new training data is sampled from the surroundings of the target's position in the previous frame. While training, the model enforces the constraint that the confidence associated with the sample at the original position stays the maximum.

10.2.2 Deep trackers

Deep learning-based trackers, referred here as deep trackers, have many advantages over their shallow, mostly handcrafted, counterparts. This is due to their potential to encode multilevel information and exhibit more invariance and equivariance to target appearance variations. There are several ways to categorize deep trackers. Several works tend to separate them as Siamese-based and discriminative trackers. In the following, we use the recent survey paper (Fiaz et al., 2019) to group different trackers into the following two primary groups.

10.2.2.1 Correlation filter-based tracking

Correlation filter-based tracking (CFT) methods perform identification-related computation in the frequency domain to manage the cost associated with tracking. These are based on a tracking-by-detection paradigm, and an example of constructing the correlation filter with deep learning is shown in Fig. 10.1. A target patch is cropped from the template image and used to initialize the correlation filters at the beginning of tracking. For effective representation of the target, feature maps are constructed using appropriate extraction methods. Early methods computed the response map using elementwise multiplication between adaptive learning filter and extracted features, and by using a Discrete Fourier Transform (DFT). Deep learning allows to encode these features such that after cross-correlation, the confidence is directly obtained in the spatial domain. The maximum confidence score estimates the new target position. At the outcome, the target appearance at the newly predicted location is updated by extracting features and updating correlation filters. CFT methods based on various techniques exist in the tracking literature, and these can be described as follows.

Basic correlation filtering (CFT). The basic underlying feature of this class of trackers is that they use kernelized correlation filters in some form in their structure. While the early shallow trackers used conventional features such as the HOG, color names, and so on, the deep trackers use recurrent neural networks or convolutional networks to learn these kernels. Ma et al. (2015) proposed the first such tracker that exploited rich convolutional features using convolutional filters, as demonstrated in Fig. 10.1. It computes independent adaptive correlation filters for each CNN featere, and response maps are computed. Improved variants of this include Hierarchical Correlation Feature-based Tracker (Ma et al., 2015), Hedge Deep Tracking (Qi et al., 2016) and Multitask Correlation Particle Filter (Zhang et al., 2017), among others.

Regularized correlation filtering (R-CFT). Correlation filter trackers face several limitations, such as the need for the same filter and patch sizes, sensitivity of learning to the negative samples, and response maps being less accurate when moving away from the centre of the frame. RCFTs overcome this issue through the use of regularization over the learning process of the filter. For example, Spatially Regularized DCF (SRDCF) (Danelljan et al., 2015) uses spatial regularization, and during the tracking, the regularization component weakens the background information thereby making the tracker less sensitive to surrounding noise. Another example is the STRCF (Li et al., 2018b), which uses an additional temporal regularization in SRDCF to avoid abrupt predictions, and only allows smooth transitions in the tracking trajectory. A more recent example of this class is ECO (Danelljan et al., 2017), which regularizes through the construction of a smaller set of fibers to efficiently capture target representation using matrix factorization.

Siamese-based correlation. A Siamese network joins two inputs and produces a single output. With the twin subnetworks, it can learn the deep representations for the template as well as the search image. This class of trackers combine Siamese networks with CFTs. Siamese Fully Convolutional (SiamFC) networks (Bertinetto et al., 2016b) utilize a convolutional embedding and a correlation layer to integrate the deep features that are obtained from the template as well as those from the candidate image. Other improved variants of such methods include SiamRPN (Li et al., 2018a) and SiamRPN++ (Li et al., 2019a) trackers that improve the precision of tracking by matching through the use of region-proposal networks. A recent improved version is the Discriminative Model Prediction (DiMP) (Bhat et al., 2019) tracker, which continuously adapts the template representation to learn an improved correlation filter from the previous frames.

Part-based correlation filtering. Unlike other CFTs, this category of trackers learns the target representation in parts. Such trackers track individually several parts of the image, and each part can have a different correlation filter. The output response maps are aggregated to construct a final response, from which the new location of the target is estimated using methods such as particle filtering.

Fusion-based correlation filter trackers. The performance of trackers can be improved for certain application problems through fusing information from complementary domains. This can include combining visual, thermal and depth information from images, or even combining low-level and high-level features of a deep network. For example, the deep feature fusion technique by Wang et al. (2017) uses a combination of local detection network (LDN) and global detection network (GDN). LDN employs VGG-16 and fuses information from different parts of the network to generate the response map. If LDN fails to detect, GDN comes into action, parameters of which are rarely updated.

10.2.2.2 Noncorrelation filter-based tracking

Noncorrelation filter-based tracking (NCFT) methods do not employ correlation filters and can be based on concepts of patch learning, sparsity, superpixel, graphs, part-based matching or Siamese matching.

Patch learning. Such trackers extract information independently from different parts of the image. Most such trackers comprise combinations of shared layers and domain-specific layers. While the shared layers exploit generic target representation from all the sequences, the domain specific layer is responsible for identification of the target using binary classification for a sequence. Example deep trackers from this category are Structure Aware Networks (SANet) (Fan and Ling, 2017) and Convolutional Networks without Training (CNT) (Zhang et al., 2016). SANet combines convolutional and recurrent features using a skip concatenation strategy to encode the rich information. CNT employs a hierarchical structure with two feed-forward layers of convolutional network to accurately generate a representation of the target. The lower layer captures local features, and a global representation of target is formed by stacking a simple cell feature map which encodes local as well as geometric layout information.

Siamese-based. Siamese networks have also been used in a noncorrelation filtering context, where the goal is to learn embeddings that can yield similarities between given patches. In general, the template and search regions are fed to a set of convolutional layers that are shared between the two subnetworks of the Siamese model, and the deep features from the two are then fused into a set of sequential fully-connected layers. Examples are SINT (Tao et al., 2016) and GOTURN (Held et al., 2016). For more accurate localization, the output boxes from SINT are refined using four ridge bounding box regressions trained over the bounding box from the initial frame. Siamese trackers including correlation-based ones will be detailed in length later, as they are particularly apt for long-term tracking.

Graph-based. Graph learning methods, in general, are used to predict the labels of unlabeled vertices in a graph. Some recent trackers, such as Tree Structure CNN (TCNN) (Nam et al., 2016), build spatial or temporal graphs to identify accurate tracking trajectories. There exist methods which use both the information and they construct spatiotemporal graphs for tracking. To further model higher-order complex relationships, Structure Aware Tracker (Du et al., 2016) constructs hypergraphs in the temporal dimension. The hypergraph is constructed using candidate parts as nodes, and the hyperedges denote relationships among the parts.

All in all, short-term tracking has fostered very broad and diverse research, as can be seen by the variety of tracker methods. What is common and interesting among most of the short-term trackers is their uninterrupted focus on maximizing localization accuracy in the short duration of the videos, placing longer-term performance as a secondary requirement.

10.3 Long-term visual object tracking

Fueled by the availability of standard datasets (Kristan et al., 2020; Wu et al., 2015; Smeulders et al., 2014), tracking has made great progress over the last few years. These datasets have mainly been designed to tackle challenges encountered in short-term tracking. For example, the average lengths of videos in ALOV (Smeulders et al., 2014) and OTB (Wu et al., 2015) are only about 10 seconds and 20 seconds, respectively. The rationale behind designing these datasets was to select hard moments such as changes in illumination conditions, abrupt motion, clutter, large deformations, sudden occlusions, among others.

However, with longer videos there exist additional challenges and requirements other than the trivial one that the model must perform predictions for longer times. Earlier, we examined various tracking challenges and grouped them into ones about modeling visual appearance, learning and engineering challenges. All challenges that exist for short-term trackers also exist for long-term trackers. Among all the challenges there exist two, that is model decay and target disappearance and reappearance, which have a special place in long-term tracking. While these two challenges are not unique to long-term tracking, their consequences are much more pronounced in a nontrivial manner in trackers that operate on longer sequences.

10.3.1 Long-term model decay

In general, performing well on tracking datasets has often been interpreted as overcoming all major challenges of tracking. In practice, however, additional and different problems occur when the duration of tracking is longer, *e.g.*, half an hour. When considering the practical applications of tracking, long durations are much more frequent than short videos, as in human interactions, sports, ego-documents and TV shows. However, too many model updates can eventually decay the inherent tracker model and the target might get lost. While the decay might not be noticeable in short-term videos, the accumulated effect is very prominent in long-term cases.

We motivate this by a synthetic experiment, illustrated in Fig. 10.2. To demonstrate the severe effects of model decay on long videos, a sample video is chosen randomly from the OTB50 dataset (Wu et al., 2015), and extended artificially in the following periodic manner:

$$x' = [x_1, ..., x_{T-1}, x_T, x_T, x_{T-1}, ..., x_2, x_1, x_1, x_2, ..., x_T, ...]. \tag{10.4}$$

The extended video ensures that any differences in tracking accuracy observed in the later parts of the long sequence will be caused solely due to increased video length and not because of additional visual challenges. In the figure we see the predictions of ECO (Danelljan et al., 2017) – one of the most successful nonSiamese trackers that relies on frequent updates – for 3 different frames observed in 3 different repetitions of the original video. We see that the tracker predictions become less and less accurate over time for increasing repetitions, although the frames are exactly identical. The reason is model decay, which is caused by the gradual but erroneous and heavy updating. The tracker drift caused by model decay has been a long known phenomenon (Smeulders et al., 2014), but, in the context of short-term tracking, this issue has not been very relevant. For long-term tracking, model decay is more often than not catastrophic, even when small mistakes are made at each model update, as shown by Gavves et al. (2020).

To develop circumventing measures against it for long-term tracking, a theoretical understanding is needed that can provide a mathematical definition to the underlying mechanism.

FIGURE 10.2 Predictions from ECO (Danelljan et al., 2017) on an artificially extended video created from OTB50 data (red box: tracker prediction, yellow box: ground truth prediction). *Model decay* is prevalent here although the appearance variation remains intact. Due to the heavy updating involved, model decay is noticeable from the very early stages, even for clearly visible target objects moving slowly. Credit: (Gavves et al., 2020).

Extending on the mathematical definition of trackers, assuming that we update the model parameters at every frame, we have that

$$\phi_{t+1} = \arg\min_{\phi} \mathcal{L}(x_{1:t}, y_{1:t}) \tag{10.5}$$

$$y_{t+1} = f(x_{t+1}; \phi_{t+1}), \tag{10.6}$$

where f is the ϕ-parameterized tracker model that minimizes the tracker loss $\mathcal{L}$ over the dataset $D = [x_{1:t}, y_{1:t}]$ at the timestep $t+1$. The dataset is composed of frames $x_{1:t} = [x_1, ..., x_t]$, and the tracker model f returns as output the bounding box predictions $y_{1:t} = [y_1, ..., y_t]$. To reduce notation clutter, we use $f_{i,t}$ to refer to the output of the tracker model with parameters ϕ_t applied on frame x_i. In the simplest form, the model parameters are updated by taking small steps towards the gradient direction of the loss surface, namely using gradient descent approach (or its variants),

$$\frac{\partial \phi}{\partial t} = -\eta \nabla_\phi \mathcal{L}_t \Rightarrow \tag{10.7}$$

$$\phi_{t+1} = \phi_t - \eta \nabla_\phi \mathcal{L}_t. \tag{10.8}$$

Central to the tracking learning problem, therefore, is the gradient of the tracking loss with respect to the model parameters. Extending on $\nabla_\phi \mathcal{L}$ and using the expectation over t timesteps,

$\mathbb{E}[\cdot] = \frac{1}{t}\sum_{i=1}^{t}[\cdot]$, we have

$$\nabla_\phi \mathcal{L}_t = \nabla_\phi \mathbb{E}\big[(y_i - f_{i,t})^2\big] \tag{10.9}$$

$$= 2\mathbb{E}[f_t \nabla_\phi f_t] - 2\mathbb{E}[y_t \nabla_\phi f_t], . \tag{10.10}$$

To go from Eq. (10.9) to (10.10) we rely on that the bounding box coordinates y_i predicted in previous frames, become input variables with constant values. Thus, they are independent of ϕ and $\mathbb{E}[\nabla_\phi y_i^2] = 0$. This is a strong assumption, given that in practice y_i are determined by the model with parameters ϕ.

By substituting Eq. (10.10) in (10.8), the model parameters update can be described as

$$\phi_{t+1} - \phi_t = -2\eta\big[\mathbb{E}[f_{i,t}\nabla_\phi f_{i,t}] - \mathbb{E}[y_i \nabla_\phi f_{i,t}]\big]. \tag{10.11}$$

An interesting but often overlooked reality is that while tracking is casted as a supervised learning problem, there is only one data sample in the learning dataset that is definitely correct. This one and only correct sample is the pair (x_1, y_1^*) defined by the user in the first frame, where y_1^* represents the coordinates of the user specified bounding box describing the object. While all other bounding boxes $y_i \ \forall \ i > 1$ are used for retraining and fine-tuning the tracker, there is no guarantee that the bounding boxes y_i are indeed correct or even good enough for learning. In fact, had the predictions y_i been good enough for retraining the tracker, the tracker would not need to be retrained.

Based on the argument presented above, it is reasonable to expect that the predictions, which also serve as future training samples for the retraining of the tracker, are noisy measurements of the true bounding box coordinates y_i^*. Assuming Gaussian noise with variance σ_i^2, we can state:

$$y_i = y_i^* + \delta_i, \quad \text{and} \quad \delta_i \sim N\big(0, \sigma_i^2\big), \tag{10.12}$$

Through substituting Eq. (10.12) in (10.11) and some rearrangement of terms, we have

$$\phi_{t+1} - \phi_t = -2\eta\big[\mathbb{E}[f_{i,t}\nabla_\phi f_{i,t}] - \mathbb{E}\big[(y_i^* + \delta_i)\nabla_\phi f_{i,t}\big]\big] \tag{10.13}$$

$$= \underbrace{-2\eta\mathbb{E}\big[(f_{i,t} - y_i^*) \cdot \nabla_\phi f_{i,t}\big]}_{\text{Perfect parameter update}} + \underbrace{2\eta\mathbb{E}[\delta_i \cdot \nabla_\phi f_{i,t}]}_{\text{Parameter bias}}. \tag{10.14}$$

It is easy to recognize the two components in the parameter update of the tracker. The first term in Eq. (10.14) corresponds to the perfect model update component, as it corrects the error made by the model prediction $f_{i,t}$ as compared to the perfect box y_i^*. The second term corresponds to the parameter bias component, as this term depends directly on the error made by past predictions δ_i. If $\delta_i = 0$, then there would be no error, and the parameter updates would also be perfect.

Model dynamics. Having computed the effect of the past errors on the parameter updates of the tracker, we can next examine the effect on the model dynamics $\frac{\partial f}{\partial t}$ over time. Specifically, after updating the parameters, the relation between the past $f_{i,t}$ and the next model $f_{i,t+1}$ is

$$\frac{\partial f}{\partial t} \propto f_{i,t+1} - f_{i,t} = \frac{\partial f}{\partial \phi}\frac{\partial \phi}{\partial t} \Rightarrow \tag{10.15}$$

$$f_{i,t+1} = f_{i,t} + \frac{\partial f}{\partial \phi}\frac{\partial \phi}{\partial t}. \tag{10.16}$$

As $\frac{\partial \phi}{\partial t} \propto \phi_{t+1} - \phi_t$, combining Eq. (10.16) and (10.14) we have that

$$f_{i,t+1} = f_{i,t} \underbrace{-2\eta\mathbb{E}\left[\left(f_{i,t} - y_i^*\right) \cdot \|\nabla_\phi f_{i,t}\|^2\right]}_{\text{Perfect model update}} \tag{10.17}$$

$$\underbrace{+2\eta\mathbb{E}\left[\delta_{i,t} \cdot \|\nabla_\phi f_{i,t}\|^2\right]}_{\text{Model decay}}$$

From Eq. (10.17), we make the following observation. Due to the continuous updates, the tracker model offshoots its predictions by a quantity that is linearly proportional to past errors. We refer to this quantity as model decay.

Long-term tracking & model decay. As the model dynamics in Eq. (10.17) are recursive, it is implied that the bias term accumulates and in fact, worsens over time. For as long as the cumulative model decay is small enough, usually in the early iterations, the model dynamics is sufficiently accurate. The early errors δ_i, are small not only because the tracker is still accurate, but also because the number of summands t is small. This is the reason why model decay is not a problem, and often goes even unnoticed, in short videos. However, in longer videos where t and the number of summands grow, the $\delta_{i,t}$ errors grow as well and the cumulative model decay becomes noticeable.

10.3.2 Target disappearance and reappearance

By assuming that the target is always present short-term trackers can always return their most likely prediction, regardless of whether the likelihood of the prediction is high or low. The learned classifier of such a short-term tracker, therefore, does not require a minimum likelihood score before declaring that a certain location contains the target object. In turn, this means that the learned classifier does not need to calibrate its confidence.

In longer video sequences, however, the target is likely to disappear and reappear from the frame, often multiple times. As a consequence, any long-term tracker must be able to model absence of the target from the frame. While this sounds like a small difference, in reality it can have important theoretical and practical modeling repercussions. For one, the tracker model must balance between expecting that the target object might not be visible due to disappearance or due to severe appearance changes. What is more, in case the model relies on updates to account for appearance changes of the target object, the similarity function of the model will change dynamically with time. This means that one cannot fix a predetermined minimum detection threshold. Instead, the threshold must either be determined dynamically, or the model must calibrate its predictions such that the minimum detection threshold remains valid. Last, in the presence of long-term model decay adapting the threshold dynamically can be hard without directly affecting the accuracy of the tracker and the added bias to the model, especially in the case of online updates.

10.3.3 Long-term trackers

Long-term trackers must not only address the common tracking challenges faced by short-term trackers, but additionally handle long-term decay as well as target disappearance and target reappearance. In theory, model decay is inevitable when online learning is added to short-term trackers. Addressing long-term decay receives the most attention due to the fundamental and catastrophic nature of the challenge. We identify three families of approaches for long-term tracking, organized by way of learning and updating the tracker model given long-term decay: offline learning, online learning and hybrid learning.

10.3.3.1 Offline learning with Siamese trackers

At the core of all tracking algorithms is the model function $f_\phi : \mathcal{Y}_1 \times \mathcal{X}_1 \times \mathcal{X}_t \to \mathcal{Y}_t$ from Eq. (10.1), by which the image of the target is matched to the incoming frames. The tracker model returns predictions $y_{1:t}$, which hopefully should be as close as possible to the true locations $y_{1:t}^*$. The matching function for tracking ideally provides good matching even if the target in the video is occluded, changes its scale, rotates in and out-of-plane or, undergoes uneven illumination, camera motion and other disturbing factors (Smeulders et al., 2014; Wu et al., 2015).

As we discussed earlier, to address these tracking challenges models often rely on online updates using the model predictions as ground truth. This, in turn, leads to model decay that is especially prevalent in the case of long and challenging videos, effectively because we assume that the model predictions are equivalent to the ground-truth predictions during the updates, $y_{1:(t-1)} \equiv y_{1:(t-1)}^*$. By this assumption we acknowledge the following peculiar paradox, which we coin the *tracking paradox*. On one hand, if the model is so accurate so that to assume $\hat{y}_{1:(t-1)} \equiv y_{1:(t-1)}$, then the model requires no further retraining. In other words, by retraining there is no additional benefit as the model predictions are already perfect. On the other hand, if the model is not as accurate as per the assumption, by setting $\hat{y}_{1:(t-1)} \equiv y_{1:(t-1)}$ we only gain temporary improvements in accuracy but gradually make the model worse and worse, and set it to fail eventually.

To break the paradox, we could define a tracker model that relies on *no model updating*. In this case, the tracker model must be able to locate the target simply by the information it has available prior to tracking, that is *only* the bounding box in the first frame. This is clearly challenging, as in complex videos there exist severe challenges that the tracker model must be impervious to, as discussed in Section 10.1.1. In combination with online updates, the traditional way to address these challenges is by modeling each of these distortions explicitly by introducing affine transformations (Lucas and Kanade, 1981), probabilistic matching (Comaniciu et al., 2000), eigenimages (Ross et al., 2008), illumination invariants (Nguyen and Smeulders, 2006), occlusion detection (Pan and Hu, 2007), and so on. Unfortunately, modeling explicitly single challenges – while ignoring all others – yields trackers that may be optimal for that one type of distortion but suboptimal for many others, thus resulting in worse performance.

Rather than explicitly modeling all possible distortions with the tracker model f, another possibility is to cast tracking as an image matching problem, where the query image is always the target at the first frame. Because the query remains fixed and does not alter due to online updates using imperfect predictions for supervision, the model is guaranteed to work consistently and robustly for any length of time. The model definition of the tracker becomes

$$y_{t+1} = \arg\min_{y} f\big(h_\phi(\mathbf{z}), h_\phi(\mathbf{x}_t[y])\big). \tag{10.18}$$

In this definition $\mathbf{z}$ is the patch of the target object in the first frame, $\mathbf{x}_t[y]$) corresponds to the patch in the t-th frame that corresponds to the y bounding box coordinates, h_ϕ are the convolutional neural networks that compute the representations from $\mathbf{z}$ and $\mathbf{x}_t[y]$), and f is the similarity matching function of the tracker. There are no constraints in the types of convolutional neural networks h_ϕ, although in practice some work better than others (Tao et al., 2016; Li et al., 2019a). While for simplicity we use the same parameters ϕ for both neural networks, the parameters do not have to be shared (Li et al., 2019a).

In the case of tracking by target matching the critical component is the matching function, which must be robust against all undesirable tracking challenges. Traditionally, this matching function is learned online given the specific target $\mathbf{z}$. An alternative is to learn the matching function externally in an offline manner, specifically by comparing object appearances recorded at different time stamps, that is (x_i, x_j, y_{ij}^*), where $y_{ij}^* = +1$ if x_i and x_j depict the same object and -1 otherwise. The matching function f can be implemented by a multilayer perceptron (fully connected layer), and trained to minimize the contrastive loss (Tao et al., 2016)

$$\arg\min_{\phi} \frac{1}{2} y_{ij}^* d_{ij}^2 + \frac{1}{2}(1 - y_{ij}^*) \max\big(0, \epsilon - d_{ij}^2\big) \tag{10.19}$$

$$d_{ij} = \big\| h_\phi(x_i) - h_\phi(x_j) \big\|^2 \tag{10.20}$$

or by the cosine similarity to minimize the logistic loss (Bertinetto et al., 2016b)

$$\arg\min_{\phi} \log\big(1 + \exp\big(-y_{ij}^* h_\phi(x_i)^T h_\phi(x_j)\big)\big). \tag{10.21}$$

At the heart of it, for both losses the distances, either the euclidean norm or the cosine similarity, are inner products. Considering that in the minimization of the search in Eq. (10.18) we iterate over bounding box locations that are densely packed on top of the feature maps, the iterative inner products can be more compactly written as convolutions, that is

$$y_{t+1} = \arg\min_{y} h_\phi(\mathbf{z}) * h_\phi(\mathbf{x}_t[y]). \tag{10.22}$$

Eq. (10.22) allows for using fully convolutional Siamese networks (Bertinetto et al., 2016b) with great efficiency.

Visual object trackers that track by a target matching paradigm using Siamese deep neural networks are called *Siamese trackers* – see Fig. 10.3 for a graphical pipeline. Siamese trackers have two branches of convolutional neural networks. The first branch is the template branch, which convolves the target patch to as determined in the first frame. The output is a template with which every future frame can be further convolved to localize the target. The second branch is the candidate branch, processsing the new frames in the video sequences and trying to match the appearance of the target.

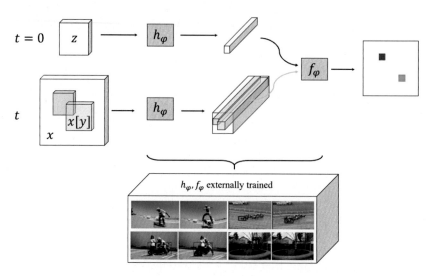

FIGURE 10.3 The Siamese tracker comprises two branches, each modeled by a convolutional neural network. The first branch always contains the target object patch at $t = 0$. The second branch analyzes any other frame in the video. The representations by the two branches are then compared with a similarity function. The similarity function is learned externally and offline, using tracked objects from different datasets. Although the similarity function has not seen the future targets, it can still match their appearance accurately for performing tracking by matching. Figure inspired by Bertinetto et al. (2016b); Tao et al. (2016).

Search region. As with short-term trackers, instead of searching for the target in the whole frame one can restrict the model to search for the next locations of the target within a predefined radius ρ. Setting a search region helps with improving the computational efficiency of the tracker algorithm, as the tracker must only analyze a much smaller image area. Limiting search into a certain region can also reduce the chances of false positives that could accidentally fire when examining the whole image.

The radius r_0 is a hyperparameter that is depends on the maximum velocity of the target. As in most cases we cannot know the maximum velocity of the target. Therefore, if we choose to set a search region, we must do so cautiously. If the target moves faster than $r_0/\Delta t$, where Δt is the interval between any two frames (inverse of the recording speed in frames per second), the target will be outside the search region and the tracker will definitely miss the target. If the radius r_0 is set such that $r_0 = \max(H, W)$, where H and W are the height and the width of the frame, then the tracker will process the whole frame.

In determining the size of the search region, one must balance between improving precision and worsening recall. By setting a small search region the precision may be improved by avoiding excess false positives in the background. However, a small search region may also hurt recall as all locations outside the search region are automatically labeled as negatives; these will be false negatives when the target lines outside the search region. A similar argument can be made for large search regions.[1]

---

[1] Precision is computed as the ratio of true positives over true positives and false positives. Recall is computed as the ratio of true positives over the sum of true positives and false negatives.

Advantages of Siamese trackers. Siamese networks present themselves with certain advantages. By and large, the most important advantage is avoiding any model decay due to the absence of model updates throughout the video. The reason is that although prediction errors δ_i occur, the model is never updated, so $\nabla_\phi f_{i,t} = 0$, see Section 10.3.1. This means that the modeler does not need to worry about the model collapsing and not recovering again. This makes Siamese trackers ideal for long-term tracking.

Comparing Siamese trackers on convolutional neural networks (Tao et al., 2016) with fully convolutional Siamese trackers (Bertinetto et al., 2016b), the former tend to have considerably higher accuracies and better robustness at the cost of higher computational costs. The reason is that convolutional neural networks with fully connected layers can be aggressively pretrained on image databases like ImageNet (Russakovsky et al., 2015) and, thus, complement the Siamese model with strong object priors. In contrast, due to the absence of fully connected layers, fully convolutional neural networks have the benefit of considerably higher efficiency and real-time performance but lower discriminative power (Valmadre et al., 2018).

By solely relying on a pretrained similarity function trained in an end-to-end manner, Siamese trackers require only offline and no online learning at all. This means we can design our offline learning to learn any type of invariants we want our tracker to have. We can do this by simply providing during training to the similarity function examples of variations that we want to match together. What is more, since the training is offline we can leverage very large datasets and improve accuracy without increasing the computational or memory cost.

Another advantage of Siamese trackers is that the similarity function can be trained to combine feature maps from different layers optimally. The feature maps from different layers capture different types of cues, with earlier layers capturing low-level geometric patterns like edges or corners, and later layers capturing higher-level semantic patterns that look like faces or types of objects. As a consequence, Siamese trackers inherently use both low-level geometric and high-level semantic information when determining a match, thus being able to perform fine-grained search on the new frames. This property is particularly important in case of confusing backgrounds when objects may look visually similar but have different semantics; daunting for models that can only use data from the same sequence to disambiguate objects.

A peculiar property of Siamese trackers is that they ignore time during tracking, if no search region is defined and the search is done in the whole frame. The reason is that in the algorithm there is no component that depends on the time variable: the target object appearance is defined in the first, frame, the similarity matching model is learned offline and used throughout the whole process without alterations during intermediate frames and predictions, and for each new frame the search region is also independent of the previous target location. While this property may seem counter-intuitive, it does not impede tracking if the similarity matching function can match appearances near-perfectly between different views of the same object. What is more, by setting the whole frame as a search region, even if the target is missed in one frame, the model still has the opportunity to relocalize the target in the follow-up frames.

Failure cases. Siamese trackers are becoming more and more popular for their favorable properties of good accuracy and remarkable robustness against model drift. However, their very design inherits them with some vices that challenges them in certain scenarios.

The most obvious disadvantage of bare Siamese trackers is confusion in the presence of multiple similar objects. This confusion is particularly noticeable when the objects are not just similar but identical, for instance when attempting to track a single person in a marching band. While tracking one particular person in a marching band is perhaps hard even for a human, the Siamese tracker is inherently unable to distinguish between objects that are identical or nearly identical. The reason is that bare Siamese trackers do not incorporate a motion modeling component.

Another challenge for Siamese trackers is the complete absence of model updates. Even if we assume that the similarity function is perfect, there are cases where the target object changes appearance either by itself (*e.g.*, a pedestrian changing their clothes) or due to the environment (*e.g.*, significant illumination changes that would alter the appearance of the target). In this case, the similarity function puts more emphasis on the shape of the candidate object that should mostly be similar to the shape of the original target, as well as on the semantics of the object, which do not change. This is not a bad property, given that convolutional neural networks have shown great generalization power in object recognition. Indeed, due to this generalization Siamese trackers like (Tao et al., 2016) have shown robustness against the common tracking challenges (Valmadre et al., 2018). That said, this generalization increases the chances of confusion in the presence of multiple similar objects.

Last, as Siamese trackers rely on convolutional neural networks, they face the same challenges that convolutional neural networks have. Specifically, standard convolutional neural networks are not strictly invariant or equivariant to scale and rotations, be it in-plane or out-of-plane rotations. To account for scale and rotation variations, standard Siamese networks rely on heavy pretraining, data augmentation, or postprocessing strategies such as repeating search over multiple scales or rotations. These approaches are costly and work only for as long as the variations are sufficiently represented in the data and its augmentations thereof.

10.3.4 Representation invariance and equivariance

The input data are always affected by noise and, therefore, no two data points are ever precisely alike. For instance, no two images are ever precisely identical as the object might have changed location, the object might have changed its appearance or pose, the illumination in the scene might be different, and so on. Some of these variations are irrelevant to the predictive task at hand, that is object identification and localization in the case of visual object tracking. For instance, how the object is illuminated and whether it lies under a shadow is irrelevant; the tracker must be able to localize the target object regardless. Some of these variations, however, can be relevant. For instance, the change in the scale of an object in the frame is irrelevant to the identification of the object but might be relevant in inferring that the object is moving towards the camera or far away from it. And certainly, acknowledging different variations of the same object are, in fact, critical to ensure generalization.

Representations are invariant to certain variations when they are not influenced by the presence of these variations. Being invariant to illumination, for instance, means that the representation will be largely the same under changes in illumination. Representations are equivariant to certain variations when the representations change proportionally to the amount of variation. Being invariant to rotation, for instance, means that a $\pi/4$ rotation will have twice the effect to the representation over a $\pi/8$ rotation, loosely speaking. Next, we describe how modern visual object trackers include invariances and equivariances in their representations.

10.3.4.1 Invariance in tracking

In Siamese trackers, the similarity matching function is of paramount importance. The reason is that the tracker relies solely on the similarity function to compare two arbitrary patches and declare whether they depict the same target object or not, irrespective of any possible changes in appearance. It is expected, however, that in the duration of the video the target may suffer from severe disturbances and alterations in its appearance. These disturbances can either be caused by the object itself, *e.g.*, a person changing their clothes or a car rotating at a different angle than what was observed in the first frame; or, they can be caused by the environment, *e.g.*, an occlusion by another object, change in the lighting or a shadow from a nearby object. The similarity function must be able to recognize that the object is the same despite all such disturbances. That is, the similarity function must learn to be invariant to all such disturbances commonly found in tracking, as we described them in Section 10.1.1.

Building invariances can either be accomplished by designing them in the method or by having them learned from data. In the former case, the method is hard coded to ignore certain transformations on the visual inputs, for instance, how rotation (Gupta et al., 2021) or scale (Sosnovik et al., 2021) changes affect the visual appearance. In the latter case, the method observes the different variations on big data sets and learns to ignore those variations that are irrelevant to the matching. The advantage of hard-coding invariances is that one can learn the invariances with much smaller datasets. In contrast, learning invariances require much larger datasets. The reason is that the model must pick up all possible appearance variations that relate to these invariances and generalize, thereafter.

Hard-coding invariances can result in smaller models, depending on the mathematical model of the invariance. What is more, if the mathematical model of the invariance makes no or few assumptions, the invariances are more precisely recognized in the data. On the other hand, learning invariances require no explicit mathematical modeling of said invariances. This is much better suited when different variations are entangled with each other, which is usually the case with tracking data. Also, because no explicit mathematical model is required, implementing models that learn invariances are typically simpler and one needs only to focus on collecting sufficiently large and sufficiently diverse data.

Bare Siamese trackers (Tao et al., 2016; Bertinetto et al., 2016b) opt for the latter option and effectively learn from data to ignore invariances that are irrelevant to the identification of the target object. For instance, for the case of illumination changes or deformations, Siamese trackers learn to predict whether any two patches depict the same object or not after training on many examples of objects that underwent changes to their illumination or deformations. This means that Siamese trackers must rely on large-scale datasets to learn to match.

Using large datasets online is not possible for tracking. To learn such invariances, therefore, Siamese trackers must learn the similarity function offline and externally. Offline training means that the training does not take place at the same time as the tracking. External training means that the training does not use the target data that the user would define in the first frame of the video, since the tracker algorithm cannot have access to such data. The similarity function, therefore, must be learned from external data and before the tracker is deployed.

Offline and external learning comes in stark contrast with the design philosophy of most traditional trackers, whose similarity function is typically learned online. Online learning for a tracker means that the tracker uses only the appearance of the target at the first frame (single positive sample), the predictions of the tracker throughout the video (*pseudo-positive* samples), as well as the appearance of the background (negative samples) and potentially other appearances from different videos (negative samples).

To learn invariances from data, a large set of videos of moving objects from different categories is required. These tracked objects must, furthermore, be annotated with their locations throughout the video. To create the dataset, thereafter, one must simply collect all possible combinations of patches (x_i, x_j) at time steps i and j. The pairs (x_i, x_j) must undergo the variations that one wants to model to learn to recognize or learn to ignore. In other words, the videos should be diverse and not contain similar objects and scenarios, otherwise, the model will not be able to generalize.

An important point is that during training there must be absolutely no overlap between the videos used for the external training and any of the tracking videos. Namely, one *must not* involve in the external learning any of the actual tracking targets, since in that case the model would essentially learn an object detector. Instead, in the external learning the model should focus on learning the generic set of disturbances and variations, and not object-specific patterns. Once the matching function has been learned on the external data, it is not adapted anymore and is applied on previously unseen targets and videos as is.

Siamese tracking is reminiscent of the instance search (Tao et al., 2014, 2015; Philbin et al., 2007; Tolias et al., 2015) paradigm, where a given patch in the query image is searched for in a pile of images. Introducing matching learning (Tao et al., 2015) allows for accurate instance search of generic objects even when the search object is visible under a completely different view than the target image. In tracking, the learning of the matching function is performed entirely from tracking examples. After learning, the matching function is capable of comparing patches from different viewpoints for new objects, or, even for new object types that the function has not seen before.

10.3.4.2 *Equivariance in tracking*

First, we give a brief introduction to equivariant models. Then, we describe popular types of equivariance that are built-in Siamese trackers. For a more general overview on equivariance, we refer the interested reader to (Weiler et al., 2018).

The property of equivariance requires functions to commute with the actions of a symmetry group acting on its domain and codomain. For any given transformation group G, a mapping function $h : X \rightarrow Y$ is equivariant if it satisfies

$$h\left(\rho_g^X(x)\right) = \rho_g^Y\left(h(x)\right) \quad g \in G, x \in X, \tag{10.23}$$

where $\rho_g^{(\cdot)}$ denotes a group action in the respective space. For invariance, ρ_g^Y will be an identity mapping.

For clarity, we give translation equivariance as an example in Fig. 10.4. In this example, h stands for the convolutional neural network function and ψ_g denotes the translation group. Example actions from this group include for example, moving one pixel left, or one towards right, or an action comprising shift of several pixels. In this manner, an infinite number of

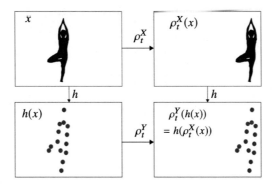

FIGURE 10.4 Schematic representation of patch-wise translation equivariance in CNNs arising from translational weight tying, so that a translation ρ_t^X of the input image x leads to a corresponding translation $\rho_t^Y(f(x))$ of the feature maps $h(x)$. Here $h(\cdot)$ and $\rho_t(\cdot)$ denote the neural network encoding function and translation function respectively. Adapted from Worrall et al. (2017).

actions can be defined within the translation group. Making the network equivariant to translations leads to reduced sample complexity and facilitates generalization of the model against translational variations.

It is important to note that there are several other transformations beyond translation that can be built in the model to improve robustness, if the effects of these transformations are present in the data and the task. Examples include rotations, reflections and scale change. For generalization over any of these transformations, equivariance needs to be enforced on the respective transformation group.

10.3.4.3 *Translation equivariance*

In image recognition in general, and visual object tracking in specific, translation variations of the object are irrelevant to the category of the object but relevant to the location of the object. A tracking model must be translation equivariant (Li et al., 2019a), so that it can accurately return the location of the target object in any future frame.

Fortunately, translation equivariance is almost seamlessly integrated into all tracker models that rely on convolutional neural networks, due to the nature of the convolution that is translation equivariant to the shift operator (Bronstein et al., 2017). In practice, however, and especially in the context of visual object tracking, translation equivariance may not be perfectly built in due to limiting assumptions. Convolutional neural networks are precisely translation equivariant when there is no image padding at the image borders, as noted by Li et al. (2019a). If there is image padding at the borders, however, the output of the convolution is not precisely equal under shift. This becomes particularly relevant when introducing bigger depth to the convolutional neural network used to implement the Siamese matching function. Due to the increasing effective receptive field (Simonyan and Zisserman, 2014), features in deeper layers will effectively depend on the pixel inputs near the image borders, and thus the model stops being truly translation equivariant. As a result, the model for the similarity function is spatially biased to return as most confident predictions patches that are near the center of the image. For a demonstration see Fig. 10.5, in which Li et al. (2019a) place the

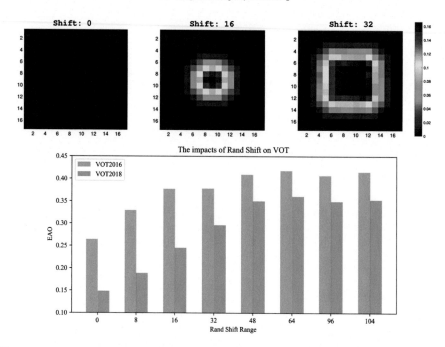

FIGURE 10.5 Above: when not adding any random shift (shift: 0) in the location of the bounding box during learning, the model tends to predict center locations, implying that, strictly speaking, translation equivariance is absent. Adding noise to the ground bounding box (shift: 16 or 32) helps with spreading predictions more uniformly. Below: Adding some noise to the ground truth predictions brings back strict translation equivariance, benefitting visibly tracking accuracy. Credit: (Li et al., 2019a).

targets at random locations sampled in the range of $(0, m)$, where 0 is the center of the patch (that is, the center of the target object during training of the Siamese similarity function) and $m = 0, 16, 32$ is the maximum displacement of the target in the patch. For $m = 0$ the target is always centered in the patch, showing that the tracker will have a tendency to predict center locations regardless of the actual location of the target.

Introducing a spatial bias to the tracker model is not desirable, as the target might be anywhere on the image frame. Perhaps more importantly, the biggest consequence of this spatial bias is that conventional Siamese trackers cannot make use of very deep convolutional neural networks. However, very deep convolutional neural networks, such as ResNet neural networks (He et al., 2016), have time and again been shown to be critical in improving object recognition accuracy.

As translation equivariance is not precisely implemented in deep neural networks due to imperfect assumptions, Li et al. (2019a) propose a practical solution of *spatially aware sampling*. Specifically, in the creation of the training set for training the Siamese similarity function, they propose to add noise to the location of the target object so that it is not centered anymore. By doing so, the spatial bias is reduced, see Fig. 10.5, and deeper neural networks like a ResNet-50, can be used. The use of deeper networks in turn leads to remarkable accuracy

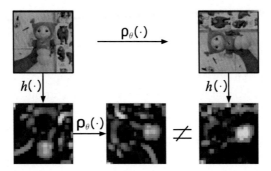

FIGURE 10.6 Example demonstrating rotation nonequivariance in regular CNN models used in object tracking, $\rho_\theta(h(\cdot)) \neq h(\rho_\theta(\cdot))$. Here $h(\cdot)$ and $\rho_\theta(\cdot)$ denote the neural network encoding function and rotation transform, respectively. Credit: (Gupta et al., 2021).

improvements. For further details regarding the precise architecture, we refer the interested reader to the original publication.

As a closing note, it is important to illuminate why deeper networks were successful in object classification but not visual object tracking. In image and object classification strict translation equivariance is not necessarily needed, because the purpose is not to predict the precise location of the object. Therefore, spatial bias is not as hurtful for as long as the model learns to recognize the object category successfully. In contrast, in visual object tracking the very task is to predict the location of the object accurately, for which spatial bias is hurtful.

10.3.4.4 Rotation equivariance

Rotation is among the most prevailing, yet still unresolved, hard challenges encountered in visual object tracking. It can commonly occur in real-life scenarios, especially when the camera records from the top, as in drones, where either the object or the camera itself are rotating. Egocentric videos are another example, where objects can undergo severe in-plane rotations frequently.

Deep learning-based tracking algorithms rely on deep convolutional neural networks that are – in theory – translation equivariant, but not designed to tackle in-plane rotations. The implication is that the model may perform well on object orientations that are represented in the training set but fail on other previously unseen orientations, as in Fig. 10.6.

Similar to Section 10.3.4.3, a straightforward approach to enforce learning of rotated variants is to use training datasets where in-plane rotations occur naturally or through data augmentation. However, as highlighted by Laptev et al. (2016), there are several limitations of data augmentation strategies. First, such procedures would require learning separate representations for different rotated variants of the data. Second, the more variations are considered, the more flexible the tracker model needs to be to capture them all. This means a significant increase in training data and computational budget. Further, such an approach would make the model invariant to rotations but not equivariant to them. Thus, predictions will be unreliable when the target is surrounded by similar objects undergoing different rotations, *e.g.*, tracking a fish in a school of fishes.

An alternative is to build in rotation equivariance to the neural network with the use of group-equivariant CNNs (Cohen and Welling, 2016), and use steerable filters (Weiler et al., 2018) to make the Siamese trackers equivariant to rotations. This way of incorporating rotation equivariance induces a built-in sharing of weights among the different groups of rotations and adds an internal notion of rotation in the model.

Steerable filters. Steerable filters allow for efficiently computing responses for an arbitrary number of discrete filter rotations Λ. What is more, they also exhibit strong expressive power as well. A filter Ψ is rotationally steerable if its rotation by an arbitrary angle θ can be expressed in terms of a fixed set of atomic functions (Freeman et al., 1991; Weiler et al., 2018). Gupta et al. (2021) define the basis of steerable functions using circular harmonics ψ_{jk}

$$\psi_{jk}(r, \omega) = \tau_j(r)e^{ik\omega}, \tag{10.24}$$

where $\omega \in (-\pi, \pi]$ and $j = 1, 2, \ldots, J$ allows to control the radial part of the basis functions. Further, the term (r, ω) refers to transformed version of (x_1, x_2) in polar coordinates and $k \in \mathbb{Z}$ denotes the angular frequency. The benefit of circular harmonics is that one can simply express rotations on ψ_{jk} as a multiplication with a complex exponential,

$$\rho_\theta \psi_{jk}(x) = e^{-ik\theta} \psi_{jk}(x), \tag{10.25}$$

where $\theta \in (-\pi, \pi]$. Note that for clarity purpose, we express $\psi_{jk}(\cdot)$ as $\psi_{jk}(x)$. Each learnt filter is then constructed as a linear combination of the elementary filters,

$$\Psi(x) = \sum_{j=1}^{J} \sum_{k=0}^{K} w_{jk} \psi_{jk}(x), \tag{10.26}$$

with weights $w_{jk} \in \mathbb{C}$. For rotation by θ, the composed filter can be steered through phase manipulation of the elementary filters,

$$\rho_\theta \Psi(x) = \sum_{j=1}^{J} \sum_{k=0}^{K} w_{jk} e^{-ik\theta} \psi_{jk}(x). \tag{10.27}$$

A single orientation of the filter can be obtained by taking real part of Ψ, denoted as Re $\Psi(x)$.

Siamese trackers have two branches of convolutional neural networks. The two branches in standard Siamese trackers receive a single input, that is either the target patch at the first frame or every new frame in the sequence. To have a rotationally equivariant Siamese tracker, the convolutional neural networks in both branches use rotationally equivariant steerable filters.

Rotation equivariant input. Focusing on the template branch, Gupta et al. (2021) modify the template head to contain multiple rotated variants of the first frame defined by the set Z, where $Z = \{z_1, z_2, \ldots, z_\Lambda\}$. Each rotated input I, comprises C channels, where each channel is represented by I_c and $c \in \{1, 2, \ldots, C\}$. This input is convolved with $\hat{C}$ rotated filters $\rho_\theta \Psi_{\hat{c}c}^{(1)}$, where $\hat{c} \in \{1, 2, \ldots, \hat{C}\}$. Based on Eq. (10.27), the resultant features obtained before applying nonlinear activation will be

$$y_{\hat{c}}^{(1)}(x,\theta) = \mathrm{Re} \sum_{c=1}^{C} \sum_{j=1}^{J} \sum_{k=0}^{K_j} w_{\hat{c}cjk} e^{-ik\theta} (I_c * \psi_{jk})(x), \qquad (10.28)$$

where the filters are rotated variants at equidistant orientations θ represented by the set $\Theta = \{0, \Lambda, \ldots, 2\pi \frac{\Lambda-1}{\Lambda}\}$. The bias terms $\beta_{\hat{c}}^{(1)}$ and nonlinearity σ are then applied to obtain the feature map at the first layer $\zeta_{\hat{c}}^{(1)}$.

Note that it is only needed to rotate the inputs on one of the two branches, as the rotations in the convolutions in Eq. (10.28) are relative. Rotating the input in the template branch makes sense, as this can be done once at the first frame.

Note also that in theory, instead of taking all possible rotation versions Z of the template target, it is also possible to first compute the feature $h(z)$ of the original target, then rotate $h(z)$. In practice, however, the spatial resolution of $h(z)$ is very low, typically 6×6 or 7×7 pixels. As a result, there will be artifacts at the corners and edges because of the crudeness of the transformation. Rather, it is better to first rotate the whole frame (not just the target) centering about the target, then crop and compute $h(z)$. Since this is only performed on the target branch only, this step can be precomputed and only contribute a negligible additional cost.

Rotation equivariant convolutions. Feature maps resulting from Eq. (10.28) are processed further using group convolutions, generalizing spatial convolutions over a wider set of transformation groups. Similar to the first layer, steerable filters are defined on the group as

$$y_{\hat{c}}^{(l)}(x,\theta) = \mathrm{Re} \sum_{c=1}^{C} \sum_{\omega \in \Omega} \sum_{j,k} w_{\hat{c}cjk,\theta-\phi} e^{-ik\theta} \left(h_c^{(l-1)}(\cdot,\omega) * \psi_{jk} \right)(x), \qquad (10.29)$$

where $h_c^{(l-1)}$ denotes the cth channel of the feature map at the $(l-1)^{\text{th}}$ layer, now replaced with h. The additional index $\theta - \omega$ introduced in Eq. (10.29) for the weight tensor facilitates the group convolution operation along the rotation dimension. It involves transforming the functions on the group by rotating them spatially.

Rotation equivariant pooling. The output of the last group convolutional layer is further processed through pooling over the rotation dimension. Unlike the conventional classification tasks, pooling is not performed along the spatial dimension. The reason is that we want to preserve the rotation equivariance.

Rotation equivariant Siamese trackers. From the two branches we obtain two sets of feature maps, $\{h(z)\}$ and $h(x)$, where $\{h(z)\}$ is the set containing feature maps at Λ orientations. Next, $\{h(z)\}$ and $h(x)$ are convolved to obtain $\{f(z,x)\}$, a set of Λ heatmaps, where $f_i(z,x) = h(z_i) * h(x)$ for all $z_i \in Z$. Next, $\{f(z,x)\}$ is processed with a global max pooling operation to obtain the final output heatmap $f(Z,x)$, where Z is a set of z values for the multiple orientations of the template. The global max pooling operation identifies the maximum value in $\{f(z,x)\}$ and selects the feature map that contains it.

The aforementioned modules are all the necessary building blocks for the rotation equivariant Siamese tracker. All convolutional and pooling layers must be replaced by their rotation equivariant counterparts. One must first identify the precision of the tracker in terms of discriminating between different orientations of the rotational degree of freedom. With Λ rotation groups, the tracker would be perfectly equivariant to angles defined by the set $\Omega = \{(i-1) \cdot 360/\Lambda\}_{i=1}^{\Lambda}$. To generate $h(z, x)$, Λ group convolutions are performed to generate Λ different heatmaps. Last, global max-pooling is performed over the feature maps to generate $h(Z, x)$, which is then processed to localize the target.

The final tracker is equivariant to *in-plane rotations* only, as out-of-plane rotations require knowledge of the 3D scene. An additional side-benefit of rotation equivariant Siamese trackers, compared to rotation invariant models, is that they can be used to compute the relative orientation changes to the target in an unsupervised way. What is more, they can be used to place additional regularizations and constraints to maintain rotational motion consistency. For further details regarding the construction of rotation equivariant Siamese trackers, we refer the interested reader to Gupta et al. (2021).

10.3.4.5 Scale equivariance

Another type of variations that is often found in video object tracking is scale variations. Measuring scale precisely is crucial when the camera zooms its lens or when the target moves into depth. As noted by Sosnovik et al. (2021), scale is also important in distinguishing among different objects in tracking, especially when many objects in the video have a similar appearance like videos of sports games or crowds. In such circumstances, space-scale equivariance provides a more discriminative representation, which is essential to differentiate among several similar candidates. Importantly, scale-space equivariance helps with maintaining better stability in the size of the predicted bounding boxes, even if the scale changes in the video are modest.

The common way to implement scale into a tracker is to train the network on a large dataset where scale variations occur naturally. As noted by Laptev et al. (2016) and motivated in the previous subsection, however, such training procedures may lead to learning groups of rescaled duplicates of almost the same filters, which renders interscale similarity estimation unreliable. Scale-equivariant models have an internal notion of scale and built-in weight sharing among different filter scales. Thus, scale equivariance aims to produce the same representation for all sizes.

Scale equivariant convolutions. Given a function $\rho_s : \mathbb{R} \to \mathbb{R}$, a scale transformation is defined as follows:

$$L_s[f](t) = f(\rho_s^{-1} t), \quad \forall s \geq 0 \tag{10.30}$$

where cases with $\rho_s > 1$ are referred to as upscaling and with $\rho_s < 1$ as downscaling. Standard convolutional layers and convolutional networks are translation equivariant but not scale-equivariant (Sosnovik et al., 2020).

To build scale-equivariant convolutional networks, Sosnovik et al. (2020) start with choosing a complete basis of functions defined on multiple scales of the following form, choosing the center of the function to be the point $(0, 0)$ in coordinates (u, v):

$$\psi_{\sigma nm}(u, v) = A \frac{1}{\sigma^2} H_n\left(\frac{u}{\sigma}\right) H_m\left(\frac{v}{\sigma}\right) e^{-\frac{u^2 + v^2}{2\sigma^2}} \tag{10.31}$$

Different from steerable filters, here H_n is a Hermite polynomial of the n-th order, and A is a constant used for normalization. A basis of N functions is possible by iterating over increasing pairs of n and m. For a complete and fixed basis the number of functions N is set equal to the number of pixels in the original filter with a chosen set of equidistant scales σ:

$$\Psi_\sigma = \{\psi_{\sigma 00}, \ \psi_{\sigma 01}, \ \psi_{\sigma 10}, \ \psi_{\sigma 11} \ldots\}. \tag{10.32}$$

In the end, the convolutions are learned as weighted combinations of $\psi_{\sigma i}$ using trainable weights w:

$$\kappa_\sigma = \sum_i \Psi_{\sigma i} w_i. \tag{10.33}$$

As a result the final kernels are defined on multiple scales and no image interpolation is needed. Given a function of scale and translation $f(s,t)$, the scale convolution with a kernel $\kappa_\sigma(s,t)$ is equal to

$$[f \star_H \kappa_\sigma](s,t) = \sum_{s'} \left[f(s', \cdot) \star \kappa_{s \cdot \sigma}(s^{-1}s', \cdot) \right](t), \tag{10.34}$$

where $\star_H$ denotes scale equivariant convolution. The output of this convolution is a stack of features each of which corresponds to a different scale.

Scale equivariant pooling and padding. To capture correlations between different scales and to transform a three-dimensional signal into a two-dimensional one, global max pooling must be applied along the scale axis. This operation does not eliminate the scale-equivariant properties of the network. In practice, Sosnovik et al. (2021) recommend to additionally incorporate scale equivariant pooling in places where conventional CNNs have spatial max pooling or strides.

Padding is shown to degrade the localization prowess of convolutional trackers (Li et al., 2019a; Zhang and Peng, 2019), as discussed also in Section 10.3.4.3. Scale equivariant convolutions, however, rely on large receptive fields that consequently yield smaller feature maps. Thus, for scale equivariant trackers with very deep models padding is necessary. To ensure that strict translation equivariance is not affected while still obtaining favorable results, Sosnovik et al. (2021) use circular padding during training and zero padding at inference time.

Scale equivariant Siamese trackers. The convolutional operation that results in the heatmap of a tracker is nonparametric. Both the input and the kernel come from neural networks. Thus, the approach described in Eq. (10.34) is not suitable for this case. Given two functions f_1, f_2 of scale and translation the nonparametric scale convolution is defined as follows:

$$[f_1 \star_H f_2](s,t) = L_{s^{-1}} \left[L_s[f_1] \star f_2 \right](t) \tag{10.35}$$

Here L_s is rescaling implemented as bicubic interpolation. Although it is a relatively slow operation, it is used only once in the tracker and it does not heavily affect the inference time.

The aforementioned modules are all the necessary building blocks for the scale equivariant Siamese tracker. Specifically, one must first identify to what degree objects change in size in this domain, and then accordingly select a set of scales $\sigma_1, \sigma_2, \ldots \sigma_N$. This is a domain-specific hyperparameter. For the networks represented by $h(x)$ and $h(z)$, all convolutional layers need to be replaced with scale-convolutional layers. The basis for these layers is based on the chosen scales $\sigma_1, \sigma_2, \ldots \sigma_N$. Optionally, scale-pooling can be included to additionally capture interscale correlations between all scales. The parameteric scale equivariant convolution needs to be replaced by the nonparametric one from Eq. (10.35) to obtain the final prediction map.

The obtained tracker produces a heatmap $f(z, x)$ defined on scale and translation Each position is assigned a vector of features that encodes both the measure of similarity and the scale relation between the candidate and the template. If additional scale-pooling is included, then all the scale information is aggregated in the similarity score. As the overall structure of the tracker does not change, and the training and inference procedures also do not change, the additional computational cost by introducing scale equivariance is small. For further details regarding the construction of rotation equivariant Siamese trackers, we refer the interested reader to Sosnovik et al. (2021).

10.3.4.6 Efficiency of Siamese trackers

The first Siamese tracker by Tao et al. (2016) relied on the fast RCNN (Girshick, 2015) architecture and the region-of-interest pooling to perform the local template matching. Despite the speed improvements that the region-of-interest pooling provides, searching over boxes is still a computationally expensive procedure. Bertinetto et al. (2016b) note that search over convolutional feature maps may as well be described as a convolution itself, relying on the idea of fully convolutional neural networks (Long et al., 2015). This simple change made Siamese tracking much more efficient and comparable with previous alternatives.

As searching over all possible locations in the image is expensive, Li et al. (2018a) propose to introduce the region proposal network from Ren et al. (2015) to their Siamese tracker. The region proposal network learns to regress on the location and scale box coordinates that are likely to contain the target. Therefore, the Siamese tracker can focus on the relevant regions only for search. To further reduce computations, Li et al. (2019a) propose depth-wise convolutions rather than regular convolutions, obtaining tenfold fewer parameters and considerable computational improvements.

10.3.4.7 Hybrid learning with Siamese trackers

Siamese trackers have shown great benefits in predicting the target locations accurately, especially in long sequences, due to not suffering from model decay. This is possible by relying solely on offline training and not including any online learning component. Ignoring all future appearances of the target and having no online learning, however, is counter-intuitive. Online learning is needed to address scenarios where the appearance of the target changes to a significant degree, such that relying solely on the first frame is not adequate. For instance, imagine the case of a pedestrian taking off their jacket. An online component is also critical in the case multiple similar objects are present in the frame and the model must specialize, a scenario that is challenging for bare Siamese trackers. In that case, online learning can help guide the tracker model to discriminate the target object from all other similar objects in the scene.

To make sure that the online learning does not have a detrimental effect to the tracker model due to model decay, model updates must be cautiously performed. There are two aspects to consider when designing model update mechanisms. The first aspect is when and how frequently a model update should be performed. The more frequent the model updates are, the more likely it is that the tracker model will become more and more biased with time, as shown by Gavves et al. (2020). The second aspect is which part of the model should be updated to ensure the incurred bias is either small or has a minimal effect in the long term to the template matching. The updates can be either directly on the similarity function of the Siamese tracker (Bhat et al., 2019) or on auxiliary networks that complement the similarity function (Tao et al., 2017).

To introduce an online learning component, many approaches propose a metalearning strategy. With a metalearning strategy, the tracker model is initially trained in an offline manner to optimize the parameter weights on seen data. Then, at inference time the metalearner model predicts new parameter weights that are optimal for the new data point.

Bhat et al. (2019) propose a metalearning component that predicts new parameter weights that discriminate between the target and the background. The proposed model follows the standard Siamese architecture for the similarity function. Unlike standard Siamese trackers though, the last convolutional layer in the template branch is followed by a *model initializer* and a *model optimizer* module. The model initializer provides an initial estimate of the model weights, using only the target appearance. These weights are then further processed by the model optimizer module, which uses both the target and the background to derive the final weight parameters. To avoid any potential overfitting, the model optimizer contains very few trainable parameters. In the end, the model is trained both offline and online.

In the offline training phase, the model is trained using pairs of sets $(M_{\text{train}}, M_{\text{test}})$. Each set contains N frames, $M = \{I_j, b_j\}_j$, where b_j are the target bounding boxes available during training. To construct the M_{train} and M_{test}, a random segment is sampled per training sequence and then divided in two. The model predictor then receives as inputs the convolutional feature maps computed from M_{train} to predict the parameter weights. Then, to ensure good generalization, these parameter weights are used to make tracking predictions only on M_{test}. In the online training phase, the initial target appearance is perturbed using data augmentation techniques to generate novel training samples for the training set. This set is complemented with new patches of the target whenever the target is predicted with sufficient confidence. The online learning is effectively similar in nature to the offline one, but done at regular intervals every 20 frames.

Continuing on the same line of work, Danelljan et al. (2020) propose to incorporate probabilistic learning to Bhat et al. (2019). To achieve this, the model learns to minimize the KL divergence between its output $p_\theta(y|x)$ with the ground truth distribution $p(y|y_i)$. The probabilistic output of the model is defined as an energy-based distribution, $p_\theta(y|x) = \frac{1}{Z_\theta} \exp(s_\theta(y, x))$, where $Z_\theta = \int_y \exp(s_\theta(y, x))$. The ground truth distribution $p(y|y_i)$ is estimated empirically as a Gaussian $p(y|y_i) \, \mathcal{N}(y_i, \sigma^2)$, where σ^2 is the empirical variance estimated by a small data sample.

Differently, (Tao et al., 2017; Gavves et al., 2020) approach the problem focusing on when the tracker should perform an update. The idea is that although Siamese trackers can benefit from model updates, they have to be cautious so as not to bias permanently the similarity function. As seen in (10.14), an accurate estimation of model bias would require computing the term $\mathbb{E}[\delta_i \nabla_\phi f_{i,t}]$. However, the state parameters ϕ_{t+1} are not known beforehand, and re-detecting the target in the earlier frames for every model update step (to obtain $f_{i,t}$) would be computationally prohibitive. To this end, a cascaded neural network is proposed to determine whether to perform a model update. First, a Siamese tracker is used to estimate the similarity map after convolving the template feature map with the feature map from every new frame.

The Siamese similarity function is followed by a *decay recognition network*, implemented by an LSTM-based binary classifier. The LSTM classifier receives as inputs the previous K similarity maps returned by the Siamese tracker. To make sure that the LSTM itself is not biased, it is trained also offline. When the LSTM classifier returns a positive prediction, the model weights are updated. Additionally, instead of relying on a search region to speed up search, the model performs global search every T frames, such that the Siamese tracker does not lose the target in case it lies out of the search region. It is important to emphasize that in that model the global search is done at regular intervals, to disassociate model updates from the model predictions. If the updates would depend on the model predictions, that would create a dependency of updates on the model behavior, leading to self-fulfilling updates and eventually model decay.

Similarly, Dai et al. (2020) propose an offline-trained *metaupdater*, which aims to solve the problem of whether the tracker model should be updated or not. The metaupdater relies on the sequence of *(i)* bounding box geometries, *(ii)* confidence scores, *(iii)* and appearance changes over time. During training, the tracker relies on the metaupdater to decide whether its weight parameters should be updated or not. The parameters of the metaupdater are then optimized so as to make correct tracking predictions in the future sequences.

Designing hybrid tracker models that learn both offline and online, while *fundamentally* avoiding – or minimizing – model decay is still an open research question.

10.3.4.8 Online learning beyond Siamese trackers

Before Siamese trackers, there were also a few important attempts in modeling trackers in long video sequences. In their seminal work, Kalal et al. (2012) propose to decompose long-term tracking into a tracking, learning and detection modeling task. The tracker is responsible for estimating the motion of the object from one frame to the other. The assumption of the tracker component is that frame-to-frame motion is limited and the object is visible. The detector treats each frame independently so as to recover any mistakes performed by the tracker. The false positives and false negatives returned by the tracker and the predictor are then monitored and corrected by the learning component. The learning relies on a pair of experts. The P-expert identifies only false negatives and the N-experts identifies only false positives. While both experts can make errors themselves, they are kept independent to ensure that their individual errors are mutually compensated. The proposed tracker is particularly apt in long-term tracking, because of the self-correction mechanism deployed by the independent detector and learning components.

Pernici and Del Bimbo (2013) also propose a tracker that is well suited for long-term tracking. To achieve this, the tracker relies on oversampling of local invariant SIFT (Lowe, 2004) representations, which are used as the training samples passed to a pair of nearest neighbor discriminative classifiers. The first, target object classifier attempts to model the appearance of the target object, by comparing appearance and shape with other neighboring patches. The second, context classifier attempts to model the appearance of the spatiotemporal background. The tracker further relies on random search when the object is absent for a consecutive amount of frames. An important component of the tracker is geometric matching based on RANSAC-like voting. When the number of matches is smaller than expected, the object is presumably occluded. The model is updated in every successful detection, unless there is occlusion.

Compared to Siamese trackers, these earlier methods did not learn an offline similarity function and had to rely on sophisticated mechanisms to make sure model drift was contained. That said, the type of cues used of these approaches bears certain similarities to Siamese trackers. By relying on convolutional neural architectures, similar to Tao et al. (2016); Bertinetto et al. (2016b) take into account both appearance and weak geometric cues in determining the similarity of the target object with candidate locations in new frames, like Kalal et al. (2012); Pernici and Del Bimbo (2013). What is more, both Tao et al. (2017); Kalal et al. (2012) rely on full image search to minimize false negatives.

10.3.5 Datasets and benchmarks

Tracking benchmarks (Smeulders et al., 2014; Wu et al., 2015; Kristan et al., 2016; Liang et al., 2015; Li et al., 2016; Mueller et al., 2016; Valmadre et al., 2018) have played a huge role in the advancement of the field, enabling the objective comparison of different techniques and driving impressive progress in recent years. These benchmarks have focused on the problem of short-term tracking according to the definition of Kristan et al. (2016), which does not require methods to perform redetection. This implies that the object is always present in the video frame. However, existing benchmarks are also *short-term* in the literal sense that the average video length does not exceed 20–30 seconds. The short length of video durations has several consequences in the evaluation of long-term tracking algorithms.

One consequence is that with short videos for tracking, the adverse effects of model decay are not easily noticeable simply because not enough model updates are performed. Even with moderately long sequences, short-term trackers more often than not tend to become so biased so that they miss the target entirely and, perhaps more importantly, without the ability to recover (Gavves et al., 2020). Moreover, as the target objects rarely exit the frame, it is not possible to evaluate whether algorithms handle well cases of object disappearance and reappearance. Last, with longer sequences the metrics for what comprises a successful tracker may not be the same as for shorter sequences. For instance, having pixel-perfect localization of the target object in short videos may yield very high scor es for some of the metrics. However, this pixel-perfect localization becomes less important if the tracker misses the target completely soon after.

For this reason, new datasets, benchmarks and types of evaluation are required specifically for long-term tracking (Fan et al., 2019; Valmadre et al., 2018; Mueller et al., 2016; Huang et al., 2019; Mueller et al., 2018), where the videos are typically much longer, in the range of several minutes up to half an hour, and the target objects frequently appear and disappear.

10.4 Discussion

Visual object tracking is one of the oldest tasks in computer vision. In long sequences model decay and target disappearance and reappearance pose significant challenges to short-term tracker models. The success of deep learning has influenced visual object tracking, especially in the context of long-term sequences. The reason is that with the deep Siamese tracker design one can relay all appearance comparisons to an offline learning of a similarity function and avoid model decay. Although Siamese trackers are impervious to model decay, this comes at the cost of possibly missing the target object in cases where appearance of the target changes significantly compared to the first frame. To address this problem, Siamese trackers with built-in invariances and equivariances, as well as hybrid Siamese trackers that rely on both offline and online learning, are proposed. These Siamese tracker variants can account for large variations in the appearance of the target object, while still exhibiting little model decay. All considering, Siamese trackers perform well and are recommended in long-term visual object tracking.

References

Baker, S., Matthews, I., 2004. Lucas-Kanade 20 years on: a unifying framework. International Journal of Computer Vision.

Bertinetto, L., Henriques, J.a.F., Valmadre, J., Torr, P.H.S., Vedaldi, A., 2016a. Learning feed-forward one-shot learners. In: Advances in Neural Information Processing Systems.

Bertinetto, L., Valmadre, J., Henriques, J.F., Vedaldi, A., Torr, P.H.S., 2016b. Fully-convolutional Siamese networks for object tracking. In: European Conference on Computer Vision Workshops.

Bhat, G., Danelljan, M., Gool, L.V., Timofte, R., 2019. Learning discriminative model prediction for tracking. In: IEEE International Conference on Computer Vision, pp. 6182–6191.

Briechle, K., Hanebeck, U.D., 2001. Template matching using fast normalized cross correlation. In: Optical Pattern Recognition XII, International Society for Optics and Photonics, pp. 95–102.

Bronstein, M.M., Bruna, J., LeCun, Y., Szlam, A., Vandergheynst, P., 2017. Geometric deep learning: going beyond Euclidean data. IEEE Signal Processing Magazine.

Cohen, T., Welling, M., 2016. Group equivariant convolutional networks. In: International Conference on Machine Learning, pp. 2990–2999.

Comaniciu, D., Ramesh, V., Meer, P., 2000. Real-time tracking of non-rigid objects using mean shift. In: IEEE Conference on Computer Vision and Pattern Recognition.

Dai, K., Zhang, Y., Wang, D., Li, J., Lu, H., Yang, X., 2020. High-performance long-term tracking with meta-updater. In: IEEE Conference on Computer Vision and Pattern Recognition.

Danelljan, M., Bhat, G., Shahbaz Khan, F., Felsberg, M., 2017. ECO: efficient convolution operators for tracking. In: IEEE Conference on Computer Vision and Pattern Recognition.

Danelljan, M., Hager, G., Shahbaz Khan, F., Felsberg, M., 2015. Learning spatially regularized correlation filters for visual tracking. In: IEEE International Conference on Computer Vision, pp. 4310–4318.

Danelljan, M., Van Gool, L., Timofte, R., 2020. Probabilistic regression for visual tracking. In: IEEE Conference on Computer Vision and Pattern Recognition.

Du, D., Qi, H., Li, W., Wen, L., Huang, Q., Lyu, S., 2016. Online deformable object tracking based on structure-aware hyper-graph. IEEE Transactions on Image Processing 25, 3572–3584.

Fan, H., Lin, L., Yang, F., Chu, P., Deng, G., Yu, S., Bai, H., Xu, Y., Liao, C., Ling, H., 2019. Lasot: a high-quality benchmark for large-scale single object tracking. In: IEEE Conference on Computer Vision and Pattern Recognition, pp. 5374–5383.

Fan, H., Ling, H., 2017. Sanet: structure-aware network for visual tracking. In: IEEE Conference on Computer Vision and Pattern Recognition, pp. 42–49.

Fiaz, M., Mahmood, A., Javed, S., Jung, S.K., 2019. Handcrafted and deep trackers: recent visual object tracking approaches and trends. ACM Computing Surveys (CSUR) 52, 1–44.

Freeman, W.T., Adelson, E.H., et al., 1991. The design and use of steerable filters. IEEE Transactions on Pattern Analysis and Machine Intelligence 13, 891–906.

Gavves, E., Gupta, D., Tao, R., Smeulders, A., 2020. Model decay in long-term tracking. In: IEEE International Conference on Pattern Recognition.

Girshick, R., 2015. Fast R-CNN. In: IEEE International Conference on Computer Vision.

Godec, M., Roth, P.M., Bischof, H., 2013. Hough-based tracking of non-rigid objects. Computer Vision and Image Understanding 117, 1245–1256.

Gupta, D., Arya, D., Gavves, E., 2021. Rotation equivariant Siamese networks for tracking. In: IEEE Conference on Computer Vision and Pattern Recognition.

He, K., Zhang, X., Ren, S., Sun, J., 2016. Deep residual learning for image recognition. In: IEEE Conference on Computer Vision and Pattern Recognition.

Held, D., Thrun, S., Savarese, S., 2016. Learning to track at 100 fps with deep regression networks. In: European Conference on Computer Vision. Springer, pp. 749–765.

Huang, L., Zhao, X., Huang, K., 2019. Got-10k: a large high-diversity benchmark for generic object tracking in the wild.

Kalal, Z., Matas, J., Mikolajczyk, K., 2010. Pn learning: bootstrapping binary classifiers by structural constraints. In: IEEE Conference on Computer Vision and Pattern Recognition. IEEE, pp. 49–56.

Kalal, Z., Mikolajczyk, K., Matas, J., 2012. Tracking-learning-detection. IEEE Transactions on Pattern Analysis and Machine Intelligence.

Kristan, M., Leonardis, A., Matas, J., Felsberg, M., Pflugfelder, R., Čehovin, L., Vojíř, T., Hager, G., Lukezic, A., Eldesokey, A., Fernandez, G., 2017. The visual object tracking VOT2017 challenge results. In: IEEE International Conference on Computer Vision Workshops.

Kristan, M., Leonardis, A., Matas, J., Felsberg, M., Pflugfelder, R., Kamarainen, J.K., Čehovin Zajc, L., Danelljan, M., Lukezic, A., Drbohlav, O., He, L., Zhang, Y., Yan, S., Yang, J., Fernandez, G., et al., 2020. The eighth visual object tracking vot2020 challenge results. In: ECCV workshops.

Kristan, M., Matas, J., Leonardis, A., Vojíř, T., Pflugfelder, R., Fernandez, G., Nebehay, G., Porikli, F., Čehovin, L., 2016. A novel performance evaluation methodology for single-target trackers. IEEE Transactions on Pattern Analysis and Machine Intelligence.

Kwon, J., Lee, K.M., 2011. Tracking by sampling trackers. In: IEEE International Conference on Computer Vision. IEEE, pp. 1195–1202.

Kwon, J., Lee, K.M., Park, F.C., 2009. Visual tracking via geometric particle filtering on the affine group with optimal importance functions. In: IEEE Conference on Computer Vision and Pattern Recognition. IEEE, pp. 991–998.

Laptev, D., Savinov, N., Buhmann, J.M., Pollefeys, M., 2016. Ti-pooling: transformation-invariant pooling for feature learning in convolutional neural networks. In: IEEE Conference on Computer Vision and Pattern Recognition, pp. 289–297.

Li, A., Lin, M., Wu, Y., Yang, M.H., Yan, S., 2016. NUS-PRO: a new visual tracking challenge. IEEE Transactions on Pattern Analysis and Machine Intelligence.

Li, B., Yan, J., Wu, W., Zhu, Z., Hu, X., 2018a. High performance visual tracking with Siamese region proposal network. In: IEEE Conference on Computer Vision and Pattern Recognition.

Li, F., Tian, C., Zuo, W., Zhang, L., Yang, M.H., 2018b. Learning spatial-temporal regularized correlation filters for visual tracking. In: IEEE Conference on Computer Vision and Pattern Recognition, pp. 4904–4913.

Li, B., Wu, W., Wang, Q., Zhang, F., Xing, J., Yan, J., 2019a. Siamrpn++: evolution of Siamese visual tracking with very deep networks. In: IEEE Conference on Computer Vision and Pattern Recognition, pp. 4282–4291.

Li, C., Liang, X., Lu, Y., Zhao, N., Tang, J., 2019b. Rgb-t object tracking: benchmark and baseline. Pattern Recognition 96, 106977.

Liang, P., Blasch, E., Ling, H., 2015. Encoding color information for visual tracking: algorithms and benchmark. IEEE Transactions on Image Processing.

Long, J., Shelhamer, E., Darrell, T., 2015. Fully convolutional networks for semantic segmentation. In: IEEE Conference on Computer Vision and Pattern Recognition.

Lowe, D.G., 2004. Distinctive image features from scale-invariant keypoints. International Journal of Computer Vision.

Lucas, B.D., Kanade, T., 1981. An iterative image registration technique with an application to stereo vision. In: International Joint Conferences on Artificial Intelligence.

Ma, C., Huang, J.B., Yang, X., Yang, M.H., 2015. Hierarchical convolutional features for visual tracking. In: IEEE International Conference on Computer Vision, pp. 3074–3082.

Mueller, M., Bibi, A., Giancola, S., Alsubaihi, S., Ghanem, B., 2018. Trackingnet: a large-scale dataset and benchmark for object tracking in the wild. In: European Conference on Computer Vision, pp. 300–317.

Mueller, M., Smith, N., Ghanem, B., 2016. A benchmark and simulator for uav tracking. In: European Conference on Computer Vision.

Nam, H., Baek, M., Han, B., 2016. Modeling and propagating cnns in a tree structure for visual tracking. arXiv preprint. arXiv:1608.07242.

Nguyen, H.T., Smeulders, A.W., 2004. Fast occluded object tracking by a robust appearance filter. IEEE Transactions on Pattern Analysis and Machine Intelligence.

Nguyen, H.T., Smeulders, A.W., 2006. Robust tracking using foreground-background texture discrimination. International Journal of Computer Vision 69, 277–293.

Oron, S., Bar-Hillel, A., Levi, D., Avidan, S., 2015. Locally orderless tracking. International Journal of Computer Vision 111, 213–228.

Pan, J., Hu, B., 2007. Robust occlusion handling in object tracking. In: IEEE Conference on Computer Vision and Pattern Recognition.

Pernici, F., Del Bimbo, A., 2013. Object tracking by oversampling local features, pp. 2538–2551.

Philbin, J., Chum, O., Isard, M., Sivic, J., Zisserman, A., 2007. Object retrieval with large vocabularies and fast spatial matching. In: IEEE Conference on Computer Vision and Pattern Recognition.

Qi, Y., Zhang, S., Qin, L., Yao, H., Huang, Q., Lim, J., Yang, M.H., 2016. Hedged deep tracking. In: IEEE Conference on Computer Vision and Pattern Recognition, pp. 4303–4311.

Ren, S., He, K., Girshick, R.B., Sun, J., 2015. Faster r-cnn: towards real-time object detection with region proposal networks. In: Advances in Neural Information Processing Systems.

Ross, D.A., Lim, J., Lin, R.S., Yang, M.H., 2008. Incremental learning for robust visual tracking. International Journal of Computer Vision.

Russakovsky, O., Deng, J., Su, H., Krause, J., Satheesh, S., Ma, S., Huang, Z., Karpathy, A., Khosla, A., Bernstein, M., Berg, A.C., Fei-Fei, L., 2015. ImageNet large scale visual recognition challenge. International Journal of Computer Vision.

Simonyan, K., Zisserman, A., 2014. Very deep convolutional networks for large-scale image recognition. arXiv preprint. arXiv:1409.1556.

Smeulders, A.W.M., Chu, D.M., Cucchiara, R., Calderara, S., Dehghan, A., Shah, M., 2014. Visual tracking: an experimental survey. IEEE Transactions on Pattern Analysis and Machine Intelligence.

Sosnovik, I., Moskalev, A., Smeulders, A., 2021. Scale equivariance improves Siamese tracking.

Sosnovik, I., Szmajam, M., Smeulders, A., 2020. Scale-equivariant steerable networks.

Tao, R., Gavves, E., Smeulders, A.W.M., 2016. Siamese instance search for tracking. In: IEEE Conference on Computer Vision and Pattern Recognition.

Tao, R., Gavves, E., Smeulders, A.W.M., 2017. Tracking for half an hour. arXiv preprint. arXiv:1711.10217.

Tao, R., Gavves, E., Snoek, C., Smeulders, A., 2014. Locality in generic instance search from one example. In: IEEE Conference on Computer Vision and Pattern Recognition.

Tao, R., Smeulders, A.W.M., Chang, S.F., 2015. Attributes and categories for generic instance search from one example. In: IEEE Conference on Computer Vision and Pattern Recognition.

Tolias, G., Avrithis, Y., Jégou, H., 2015. Image search with selective match kernels: aggregation across single and multiple images. International Journal of Computer Vision.

Valmadre, J., Bertinetto, L., Henriques, J., Tao, R., Vedaldi, A., Smeulders, A., Torr, P., Gavves, E., 2018. Long-term tracking in the wild: A benchmark. In: European Conference on Computer Vision.

Wang, G., Wang, J., Tang, W., Yu, N., 2017. Robust visual tracking with deep feature fusion. In: 2017 IEEE International Conference on Acoustics, Speech and Signal Processing. IEEE, pp. 1917–1921.

Weiler, M., Hamprecht, F.A., Storath, M., 2018. Learning steerable filters for rotation equivariant cnns. In: IEEE Conference on Computer Vision and Pattern Recognition, pp. 849–858.

Worrall, D.E., Garbin, S.J., Turmukhambetov, D., Brostow, G.J., 2017. Harmonic networks: deep translation and rotation equivariance. In: IEEE Conference on Computer Vision and Pattern Recognition, pp. 5028–5037.

Wu, Y., Lim, J., Yang, M.H., 2015. Object tracking benchmark. IEEE Transactions on Pattern Analysis and Machine Intelligence.

Zhang, K., Liu, Q., Wu, Y., Yang, M.H., 2016. Robust visual tracking via convolutional networks without training. IEEE Transactions on Image Processing 25, 1779–1792.

Zhang, T., Xu, C., Yang, M.H., 2017. Multi-task correlation particle filter for robust object tracking. In: IEEE Conference on Computer Vision and Pattern Recognition, pp. 4335–4343.

Zhang, Z., Peng, H., 2019. Deeper and wider Siamese networks for real-time visual tracking. In: IEEE Conference on Computer Vision and Pattern Recognition, pp. 4591–4600.

Zheng, W.L., Shen, S.C., Lu, B.L., 2017. Online depth image-based object tracking with sparse representation and object detection. Neural Processing Letters 45, 745–758.

Biographies

Dr. Efstratios Gavves is an Associate Professor with the University of Amsterdam in the Netherlands, Scientific Director of the QUVA Deep Vision Lab, Scientific Director of the POP-AART Lab on using AI for adaptive radiotherapy, and an ELLIS Scholar. He is a recipient of the ERC Career Starting Grant 2020 and NWO VIDI grant 2020 to research the Machine Learning of Temporality for spatiotemporal sequences. He is a cofounder of Ellogon.AI, a University spinoff and in collaboration with the Dutch Cancer Institute (NKI), with the mission of using AI for pathology and genomics. Efstratios has authored several papers in the top computer vision and machine learning conferences and journals. He is also the author of several patents. His research focus is on Temporal Machine Learning and Dynamics, Efficient Computer Vision, and Machine Learning for Oncology.

Deepak K. Gupta currently works as a Research Scientist at Transmute AI Research in Netherlands, where his focus is on fundamental research problems from the field of computer vision and deep learning. Earlier, he worked for two years as a postdoctoral researcher at QUVA Lab and Informatics Institute at the University of Amsterdam where he primarily worked on improving visual object tracking methods. Deepak completed his PhD in Computational Science in 2017 and BS and MS in Geophysics in 2013. He worked for more than a year at Royal Dutch Shell as a Research Scientist (2017–2019) solving problems of geophysics using AI. He is particularly interested in developing efficient algorithms for object tracking and segmentation in videos. In addition, he is also involved in research projects at the intersection of physics, mathematics and machine learning focused towards applications in medical imaging and geophysics.

Learning for action-based scene understanding

Cornelia Fermüller[a] *and Michael Maynord*[b]

[a]University of Maryland, Institute for Advanced Computer Studies, Iribe Center for Computer Science and Engineering, College Park, MD, United States [b]University of Maryland, Computer Science Department, Iribe Center for Computer Science and Engineering, College Park, MD, United States

CHAPTER POINTS

- An action-centric framework for scene and activity interpretation.

- Studies on object affordances and functionalities and their use in the context of action recognition and robot learning.

- Studies on activity recognition as an interplay between cognition and perception.

- The merging of vision and language through embedding spaces.

- Discussion on the future of action and activity understanding through the lens of the action-centric framework.

11.1 Introduction

The purpose of Computer Vision (CV) is to produce interpretations of images and video which are of use to humans. Action is *important* to model because it is a primary means through which others and ourselves interact with our environment, and it is largely through interaction that the environment becomes meaningful. Because of the centrality of action in what humans find meaningful in their environment, humans structure their environments around action. So to fully understand human environments requires an understanding of their relation to actual and possible action. Most contemporary CV methods are not action

Copyright © 2022 Elsevier Inc. All rights reserved.

based in their approach – in this chapter we present methods and frameworks which in modeling the observed scene employ an *action based* or *functional* interpretation.

The centering of action in perception aligns with embodied cognition theories (Varela et al., 1993; Barsalou, 2008), which argue that many aspects of cognition take their origin in motor behavior and action. In a computational approach we can leverage action based representation at multiple time scales for a hierarchical approach to scene understanding. At the early hierarchical levels are static components, the objects, humans, and simple movements of the limbs. These are then combined into increasingly more complex notions that involve interactions between scene components. Temporally *actions* chain together in structured ways to constitute *activities*.

The use of action based representations in computational perception approaches is *challenging*. The classic approach to CV is to recognize scene constituents based purely on their appearance. However, the aspects of the scene related to action are often semantic and relational rather than appearance based in nature. To better model interactions, more complicated architectures are required that not only model visual appearance but which leverage a more cognitive understanding of the intermediate semantic and relational structure of action in the input.

Classic end-to-end visual learning becomes intractable with larger input state spaces, as are found with video and action of increasing duration. This is because the variability in visual appearance increases, presenting challenges both in data and in modeling. In order to scale, a more cognitive approach which models not the appearance but the action structure of the activity is necessary.

A primary advantage of the action based approach to scene interpretation is generalization, i.e., the ability to recognize scene quantities beyond those visually observed in the training set. For example, if we can recognize what makes an object usable for cutting, this will allow us to recognize new kinds of cutting tools, such as an Alaskan ulu, although this object has not been in our training set. Similarly, if we can interpret an observed human activity by understanding the interaction of constituents and by understanding the underlying goal, we can be more robust. Individual constituents may be difficult to recognize because of occlusions, size, unfavorable viewing angles or variability in visual appearance and movements, but reasoning about the cognitive plausibility of the activity can allow the recovering from classification errors. Furthermore, action modeling provides the potential to predict far into the future.

This chapter presents CV learning-based approaches and concepts centered on action. We now outline in brief the contents of the rest of this chapter.

Section 11.2 covers *affordances* – a variety of action-based object description. Affordances have been of great interest to Robot Vision. But also classic, nonembodied CV can benefit from the use of affordances. They reflect how the different objects in a scene can be used, and they are an essential component for action understanding. They carry information on the possible cooccurrences of observed objects, humans, and other scene constituents. Section 11.2 covers the best known CV works on the topic, which include early studies that reason about affordances via geometric measurements, studies that learn affordance maps using algorithms for object detection and semantic segmentation on depth and geometric feature maps, and studies that combine affordances with other constituents for action recognition.

Section 11.3 is devoted primarily to our own work on understanding manipulation activities. We argue that activity interpretation should be implemented as a continuous interplay between reasoning and perception processes. Activities are modeled hierarchically. At the lower level are modules for objects, actions, spatial relations, etc, which are merged at the higher level via a grammatical formulation. The grammar and selected modules supporting an action-driven understanding are described.

Section 11.4 focuses on methods that can achieve a tighter integration of appearance and semantic and relational constraints. We consider the integration within the context of the task of Zero-Shot Learning. We cover first simple methods involving engineered attributes, and proceed through more sophisticated approaches involving merging language and vision through shared embedding spaces, capturing semantic and relational information.

This is followed by a discussion on how these concepts could be applied to action and activity understanding in Section 11.5, and Conclusions in Section 11.6.

11.2 Affordances of objects

Psychologist James Gibson coined the term "affordance" (Gibson, 1977), referring to the action possibilities that an object presents based on humans' (or animals') physical capabilities. For instance, a knife affords "cutting," "stabbing," "poking," "slicing," "throwing," etc. (to a human). The notion of affordance has recently received great interest in the cognitive science and neuroscience literature, strengthened by brain imaging evidence that showed that observing tools activate motor areas of the brain (for a review see Martin, 2007). The concept has been studied in different areas, including developmental psychology, industrial design, sport science, and human computer interaction, and there have been many interpretations and discussions on its meaning. Most distinguish between "affordance" and "function," with the former meaning properties of objects and the latter referring to the role that an object plays in satisfying some purpose. For example the handle of a cup affords "grasping," and its interior "containing," while an electricity plug supports the function to "powering kitchen appliances," or "charging devices," and a water faucet supports the function of "getting drinking water." However, a formal definition does not exist.

In this section, we first motivate the use of affordances in CV (Sec. 11.2.1). Then different works from the literature are discussed: Sec. 11.2.2 is about the earlier approaches, which selected geometric features computed from 3D data to classify affordances of chairs or everyday objects. Sec. 11.2.3 describes works on learning affordances of objects and their parts using CV recognition algorithms applied to depth data or geometric feature maps. Sec. 11.2.4 describes approaches using affordances together with other detectors for scene and action recognition, and approaches that learn affordances for embodied agents. Sec. 11.2.5 concludes with suggestions for future work.

11.2.1 Why would computer vision be interested in affordances?

Looking at objects and scene surfaces from the viewpoint of affordances provides information for visual scene interpretation that is complimentary to the classic cues and aids in robustness and generalizability of learned representations. This information is about the "actionability" that the scene presents at multiple spatial and temporal scales relating to objects, groups of objects, and the complete spatio-temporal scene. Therefore affordances provide information and constraints for scene understanding both in the present and in projecting into the future – thus aiding recognition in addition to prediction, as detailed next.

Models of affordances learned over some objects are transferable to novel object categories. I.e., if our recognition modules can recognize an affordance, they can detect it in objects never seen before, even in a stone that has the right properties. This is because how an object is used depends on physical properties such as its shape, size, material, and weight (Hermans et al., 2011), and we can design processes that pick up these physical properties from images, depth maps, and other modalities, independent of previously encountered object categories. In contrast, classifying objects in images in a conventional end-to-end fashion does not give insight into how visual features such as affordances relate to the object.

Affordances provide valuable information to visual object understanding, such as in understanding the "valid functionality" of objects (Hassanin et al., 2018) – e.g., an inverted cup cannot be used to pour into, or similarly a broken chair cannot be used for sitting (Grabner et al., 2011). Another example is the subcategorization of the classical visual object categories, such as differentiating between chairs for different uses (Stark and Bowyer, 1991).

Since affordances represent the possible actions that can be performed with an object, they carry valuable information for predicting future actions (Koppula and Saxena, 2015; Qi et al., 2018) – because actions relate to each other over time. For example, a bread knife as a whole presents affordances ("graspable", "cut with") allowing the action of "slicing bread," and slicing bread is part of the activity "preparing the bread basket" – an activity consisting of multiple actions extended through time with temporal dependencies. Knowledge of the possibility of "slicing bread" informs possible subsequent actions such as putting the basket on the table. To summarize, affordances and functionalities at the object level also contain information about possible object interactions, spatio-temporal relations, and activities at longer temporal scales. Modeling these relations to get explicit or implicit relations at multiple time scales and semantic levels of abstraction has value for the task of activity understanding.

The concept of affordances has been central to Robot Vision and to research along the Active Vision Paradigm (Bajcsy, 1988). The latter advocates that the vision of systems should not be considered a passive process. Biological systems "move their eyes to select what they see" in an active process. Similarly, artificial embodied systems should be able to change the viewpoint of their cameras in order to select what information to gather from their environment, as different viewpoints present different information. Going further, the paradigm also suggests that embodied systems should avoid employing heavy general-purpose vision processes for all purposes, and only process the information necessary to solve the task at hand (Fermüller and Aloimonos, 1995). Therefore, when a robot or artificial system interacts with objects, often it is more effective to compute what an object can be used for – i.e., compute its affordance and how it can be used – rather than to classify the object according to our language representations. Thus, while the advantages of affordances discussed in this section apply to the classic passive CV formulation, where there is no agent interacting with the environment, a great portion of the research on affordances focuses on Robot Vision.

11.2.2 Early affordance work

Affordances relate to actions. As a consequence they are also grounded in action related physical quantities. For example, an object to sit on or an object to pour into have certain physical quantities, e.g., a certain shape, size, or certain material, etc. All of the earlier approaches utilized such explicit physically meaningful representations in affordance recognition modules.

The first studies used shape and geometry. Stark and Bowyer (1991) proposed the first affordance-based approach to object recognition using 3D CAD models as input. A knowledge-graph, similar to a decision tree, was created to classify chairs and subcategories of chairs (e.g., conventional chair, balance chair, high chair, lounge chair), where the leaves of this graph were procedures for classification of geometric features. These features included relative orientation between surfaces, object dimension, stability, and proximity of surfaces.

Grabner et al. (2011) detected surfaces that afford "sitting," by checking the geometry of a 3D human skeleton model in a sitting pose against the object's geometry. Their features include distance and the intersection of the human's mesh with the object's mesh. The detector was evaluated on Google Warehouse models as well as real 3D data collected with a time of flight camera. For best performance the method was combined with an image based classifier. Similarly, Gupta et al. (2011) modeled affordances in 3D indoor scenes by detecting the regions of the space which allow a human to use it for one of three functions: "laying down," "sitting upright," and "sitting reclined." Like Gupta et al. (2011), they also used constraints based on the occupied 3D space and the contact with a human skeleton. However, their method can take as input images, from which it first derives 3D geometry via learning-based regression methods such as Hedau et al. (2009); Lee et al. (2010).

Hermans et al. (2011) learned the affordances of everyday objects via intermediate representations that encode visual and physical characteristics. Visual characteristics included color, discrete shape, and texture, and physical characteristics included weight and size. Standard classifiers were used in the pipeline, and the approach was demonstrated on seven affordance classes in the robotics domain.

11.2.3 Affordance detection, classification, and segmentation

The problem of recognizing affordances associated with objects and scene surfaces is conceptually similar to the problem of object recognition. A number of recent approaches have used tools from object detection, classification, segmentation, and semantic labeling for affordance localization and recognition. However, these techniques usually were not applied to images, but instead either to RGBD data or to feature maps computed from depth data. This section discusses a few such approaches.

11.2.3.1 Affordance detection from geometric features

This section describes the work of Myers et al. (2015), the first approach applying modern machine learning tools on geometric features. The section details the approach to affordance detection and discusses computational implications.

The focus of the study were tools used in everyday workspaces, and specifically the detection of tool parts associated with different affordances. A dataset (the RGB-D Part Affordance

Dataset) of 105 kitchen, workshop and garden tools was collected. Objects were put on a revolving turntable and recorded with a Kinect camera from a full 360° field of view, about 300 frames for each object, out of which 10,000 RGB-D images were annotated at the pixel level. Fig. 11.1 shows example objects for five of the seven affordances, along with the annotation for one of the objects. It should be noted that affordance is associated with surfaces, for example the inner surface of a cup is "contain" while the outer surface is "wrap-grasp."

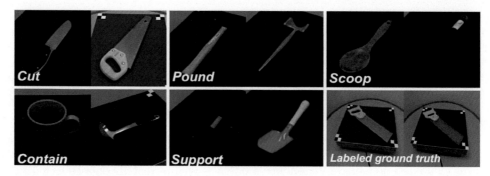

FIGURE 11.1 Sample objects from the RGB-D Part Affordance Dataset, and an example of a full frame image with hand-labeled ground truth (at the lower right). The ground truth labels include rankings for multiple affordances (from Myers et al., 2015).

From the raw depth data, shape features were computed patchwise, specifically, the surface normal, principal curvature, shape index, and the HoG-Depth descriptor (histogram of depth gradients). Using these features as input, two classification approaches were proposed: first, a Structured Random Forest (SRF), which creates point-wise classification; and second the S-HMP (Superpixel Hierarchical Matching Pursuit) algorithm (Bo et al., 2013). The latter works by first oversegmenting the RGB-D image into superpixels. Then, using a dictionary learning technique, the shape features are sparsely encoded at multiple scales per superpixel. Finally, the features are max-pooled over the superpixels and classified via an SVM. Example results are shown in Fig. 11.2 for both the S-HMP and the SRF method, where the gray value encodes the probability for the affordance assignment.

There are two computational aspects to the approach discussed above, that deserve special attention. First, sometimes overlooked, the assignment of affordances to object surfaces in general may not be unique. The same object part may be used for multiple purposes. Assigning affordances is thus a multiclass labeling problem. In Myers et al. (2015) this issue was addressed by having multiple annotators rank how close other affordances were with respect to the essential affordance, from which an ordinal scale for affordance assignment at testing was derived.

Second, a main advantage of the approach is its good generalization to new objects and surfaces. Referring to Fig. 11.2 (Bottom), one can see that the bottom of a cup is classified with the affordance "Pounding" and the edge of a spatula with the affordance "Cutting." This is because the shape of these objects indicates these properties. However, shape by itself would not be sufficient for classification in a practical system. One would have to add additional properties, the most obvious is material. This would allow to decide that a paper cup cannot be used for pounding, or an object with a soft edge cannot be used for cutting.

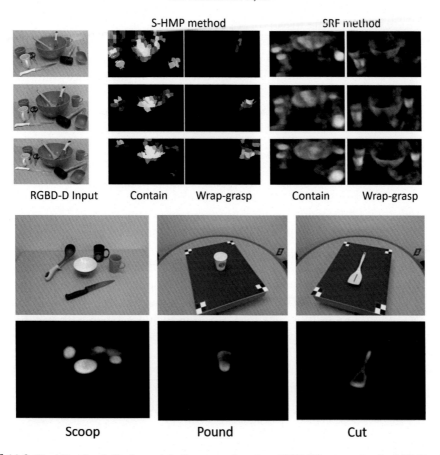

FIGURE 11.2 (Top) Results of affordance detection across three input RGB-D frames using the S-HMP and the SRF method over a cluttered sequence for the target affordances "contain" and "wrap-grasp". Brighter means higher probability of the target affordance. (from Myers et al., 2015) (Bottom) Demonstration of the generalization of the method for new objects for the SRF-method: the bottom of the cup was detected with high probability for "Pounding" and the edge of the spatula with high probability for "Cutting."

11.2.3.2 *Semantic segmentation, and classification from images*

Many of the works that followed Myers et al. (2015) employed neural network approaches using geometric feature maps as input. Affordance detection at pixel level, thus became a semantic labeling problem. However, different from Myers et al. (2015), these approaches often used 2D images as input. In a preprocessing step, depth maps or feature maps were regressed via neural networks. Furthermore, some considered natural images with multiple objects, and employed object detection algorithms to localize the objects, before assigning affordances.

For example, Nguyen et al. (2017) created a dataset with ten object categories and nine affordance categories from the household and workshop domain. It consists of both RGB-D scans and natural images (a subset of ImageNet (Russakovsky et al., 2015)) – for the latter depth maps were created using the CNN approach of Liu et al. (2015). The images were annotated with bounding boxes and affordances at the pixel level. The paper's method first applied an object detector, then within each region computed affordances using a modified VGG-16 network trained for semantic labeling, and finally the affordance values were post-processed with a CRF.

Srikantha and Gall (2016) used the dataset of Koppula and Saxena (2014), which features rich contextual information in terms of human-object interactions, and curated it with pixel-level affordance annotations. The work explored different levels of supervision for semantic segmentation, using a deep convolutional neural network within an expectation maximization framework to take advantage of weakly labeled data like image level annotations or keypoint annotations, as well as human pose as context.

Roy and Todorovic (2016) worked with the indoor scenes from the NYU dataset (Silberman et al., 2012). Their approach first infers the depth map, surface normals, and coarse-level semantic segmentation using a multiscale CNN as mid-level cues, which are then jointly fed as inputs to another multiscale CNN for prediction of the affordance maps.

Ye et al. (2017) designed a method for localizing and recognizing functional areas in indoor scenes. An ontology, as shown in Fig. 11.3 (Left), was defined to categorize image regions according to their affordance or functionality. Categories include: "open with spherical grasp" (such as a door knob), "open with wrap grasp or drag to open" (such as an oven door), "turn on electricity" (such as light switch), etc., as shown in the second last column of the figure. The dataset has 500 images featuring kitchens from the SUN dataset (Xiao et al., 2010), which were curated. The method first runs a CCN-based detector trained to detect the region and then a classifier based on a VGG architecture. Fig. 11.3 (Right) shows example results.

	Functional Area	Main Function	End Category	Symbol	Color Code
Functional area ontology	Small part on a furniture/ appliance/ wall	Open	Spherical grasp to open		
			Wrap grasp to open		
		Turn on/off	Turn on/off electricity		
			Turn on/off water		
			Turn on/off fire		
	objects (vessels and tools)	Move	Two hands raise and move		
			Cylindrical grasp to move		
			Hook grasp to move		
			Pinch grasp to move		
		Manipulate	Manipulate elongated tools		
	Furniture	Use of furniture	To sit, to place and etc.		

FIGURE 11.3 (Left) Functionality Ontology (Right) Sample detection results (from Ye et al., 2017).

11.2.4 Affordance in the context of action recognition and robot learning

In this section we highlight a few approaches that used affordances in conjunction with other quantities for scene and action understanding. We then discuss approaches that addressed the learning of affordances by robots.

11.2.4.1 Action recognition

Affordances encode features of the possible interactions a human can have with the environment. Thus, naturally they provide a glue between different quantities in the scene in space-time, such as between different objects, or between objects and actions. A number of studies have built on this idea, and used affordance relationships as a context for activity and action recognition and prediction. These methods employed various models to encode relations between the different quantities involved, including CRFs, MRFs, And-Or-Graphs, and probabilistic state automata.

Kjellström et al. (2011) investigated the problem of learning action-object interactions from demonstration, which they define as affordances. Hand actions were classified in the context of the manipulated objects using a CRF that gets as input object and hand features. Objects were modeled using hand-crafted features, and actions were modeled by the hand's global velocity, orientation, and joint angles, which were computed from the output of a 3D hand reconstruction and tracking method.

Koppula et al. (2013) considered the problem of learning sequences of subactivties performed by humans and their interactions with objects. They jointly modeled the human activities and object affordances in a Markov Random Field where the nodes represent objects and subactivities, and the edges represent the relationships between object affordances, their relations with subactivities, and their evolution over time. Affordance-subactivity relations were computed from relative geometric features between the object and the human's skeletal joints, and affordance relations between objects from spatial relations. The description was demonstrated for a PR2 robot in performing assistive tasks. In Koppula and Saxena (2015), Koppula and Saxena added to the Markov Model also possible future states in order to predict the next action.

Qi et al. (2017) used a Spatial-Temporal And-Or Graph (ST-AOG) to represent the structure of activities and predict future actions in RGBD video input. Their model is hierarchical: subactivities are modeled by the human action, the objects, and their affordances in spatial graphs, and a stochastic grammar defined over the subactivities encodes the activity. Dutta and Zielinska (2017) also considered the problem of predicting the next action based on object affordances and human interaction. They employed spatio-temporal based probabilistic state automata to model the interactions. In addition to the action class, they also computed the possible action trajectory. Depending on where an object is relative to the human it has different affordances, and depending on the affordance, its orientation and distance to the possible action trajectories were encoded as heatmaps.

11.2.4.2 Affordance learning in robot vision

Affordance has been a central concept in the field of neurorobotics, which aims to understand the cognition of a system whose body is embedded in the environment. In this research, robots acquire increasingly more complex skills using perception and interaction with the environment. Through interaction robots learn affordances, and build upon these hierarchically an understanding of actions, activities and the environment. This research in developmental robotics was enabled by the development of robotic platforms, best known among them, the humanoid robot, iCub (Metta et al., 2008).

In Fitzpatrick et al. (2003) the authors discussed three broad stages in the development of a robot: first learning a body image, second learning the interactions with external objects, and third learning to interpret object-object interactions. Affordances are central to the latter two stages. The humanoid robot through pushing and pulling actions in different directions learned to interact, and by observing affected objects' movements learned affordances, such as whether a spherical object is rollable and a cuboid is slide-able. Finally, the robot also learned to mimic an observed action.

Similarly, the authors of Montesano et al. (2008) defined the three main stages in the architecture of a developing humanoid robot as sensory–motor coordination, world interaction, and imitation. Affordances play a central role for world interactions. In this approach, the system started with basic vision and motor skills from which more complex vision and motor skill were acquired using clustering algorithms. Then, during interaction, effects were observed using perception, such as the changes in object position, velocity, and tactile sensing. A Bayesian network was used to learn affordances, which in this case were encoded as probabilistic relations between actions and percepts (object features and effects). The system was demonstrated to imitate the actions of humans by performing movements with similar effect.

Ugur et al. (2011) also demonstrated a robot learning object affordances through interaction and self-observation. In a first step the robot discovered commonalities in its action-effect experience by discovering effect categories. Building upon these, in a second step, affordance predictors for different behaviors were obtained by learning the mapping from the object features to the effect categories. Ugur and Piater (2016) went a step further and studied mechanisms that produce hierarchical structuring of affordance learning tasks. Guided by intrinsic motivation, the robot started with easy tasks, and building on its knowledge of interactions progressively learned more complex tasks by selecting to explore the object and action most different from previously explored ones. For the experiments the robot could compute the visual features of object dimension, surface patch shape, and surface normals, and its actions were poking from three different directions and stacking. In earlier stages it explored the poking actions to observe their effects on single objects. Building upon these, it then explored in a second stage the stacking of two objects and resulting effects.

11.2.5 Discussion on affordance learning

This section discussed approaches to affordance learning, many of which fall into the domain of Robot Vision and have been conducted with few examples and limited amounts of data. So far, deep learning approaches have not been much used for affordance understanding. The major reason is the lack of large annotated datasets in this domain, necessary for deep learning.

However, we expect that as research shifts away from supervised to unsupervised and self-supervised approaches, we will see learning approaches building on the concept of affordances and observed interactions between humans and objects. This will be facilitated by datasets, such as the EPIC Kitchens dataset, which features a variety of manipulation actions in natural scenes (Damen et al., 2018).

Affordances and functionalities at the object level also encode information about possible object interactions, spatio-temporal relations and possible activities at longer temporal scales. We have discussed in Section 11.2.4 approaches using affordances for action modeling. However, in future work, we could model these relations to get explicit or implicit relations at longer time scales for the problem of activity understanding, the topic of Section 11.3.

Finally, when creating mappings from perception to action for robot learning, we may ground them in affordances. Humans can learn manipulation actions using their perception only. When we see somebody performing actions with a tool unfamiliar to us we can understand the tool's affordance and perform the same action. Similarly, we could approach robot motor learning using perception and action in a tight loop, grounding them in affordances, something that has not yet been done. The robot would learn the task by observing the action and affordances and issuing commands (based on its existing skill set constrained by affordances) to generate the action approximating the observed one, and then adapt gradually to improve performance. The suggested research tasks then amount to developing self-supervised learning and reinforcement-learning approaches grounded in affordance-based representations.

11.3 Functional parsing of manipulation actions

This section describes work – largely from our group – on the interpretation of manipulation activities. Inspired by the embodied cognition paradigm (Varela et al., 1993), this work considers the understanding of human activities a process that involves perception, cognition, and the motor system. The major components are a formalism to combine the different modalities, and CV modules for obtaining semantically meaningful descriptors of action.

11.3.1 The active interplay between cognition and perception

Understanding human actions and activities is the most challenging task currently studied in CV. It is not a task of vision only. Humans can understand what others are doing, because they have models of actions and activities. They understand the goals of actions, and this allows them to interpret their observations despite the large variations in which actions can be executed and variations in visual conditions. Knowledge of some form comes into the interpretation process quite early.

We observe that human behavior is active and exploratory. We continuously shift our gaze to different locations in the scene. We recognize objects and actions and this in turn leads us to fixate at new locations. In this process, perception continuously interacts with cognition at different levels of abstraction: to guide attention, to make predictions, to constrain the search space for recognition, and to reason over what is being perceived. We call this interaction between perception and higher level processes the *Cognitive Dialogue* (Aloimonos and Fermüller, 2015), as it amounts to an iteration of questions and answers, with the cognitive or linguistic processes asking questions about the what and where of quantities in the scene, and the visual processes performing localization, detection, recognition, and reconstruction. A possible simple way to selecting the next question would be by using information-theoretic criteria (Yu et al., 2011).

The reasoning can be implemented through knowledge-based engineering (Aditya et al., 2018) or the use of language. There has been much interest in CV to introduce additional higher-level knowledge about image relationships into the interpretation process. While many studies get this additional information from captions or accompanying text, others (as discussed in Section 11.4) use advanced language processing to obtain additional high level information. In current research, most commonly, the Word2vec space (Mikolov et al., 2013) (see Sec. 11.4.3) is used as language representation, which encodes similarity about linguistic concepts. Alternatively, one could use older, hand-crafted resources encoding lexical semantics, for example the Word-Net database (Miller et al., 1990), which relates words through synonymy (words having the same meaning, like "argue" and "contend") and hypernymy ("is–a" relationships, as between "car" and "vehicle"), among many others. Verbnet (Schuler, 2005), which organizes verb classes, is particularly interesting for action understanding.

11.3.2 Grammars of action

Various mechanisms have been used to encode relationships between the different semantic concepts, that is between actors, objects, verbs, spatio-temporal relations, and attributes. Section 11.2, discussed the use of and-or-graphs and Markov models. Others include Markov Logic Networks (Tran and Davis, 2008), and planning tools (Guha et al., 2013). In this section we describe work on grammars, which can capture the composition of observed activities as sequences of scene constituents, and their recursive structure.

The main motivation for the use of grammars originates from the idea that actions observed in a video have syntactic structure. By considering the goal of actions, the video can be broken into meaningful segments, and these segments together can be organized in the form of a simple grammar. Thus, interpreting the action that is taking place in a video is like understanding a sentence that we read or hear. To parse the video into the primitive actions that constitute complex tasks, the segments of the video are mapped to particular symbols involving objects, tools, movement, and spatial relations. Importantly, the action grammar temporally segments a video at contact, that is, when the hand touches or releases an object, or objects merge or separate. At these points in time a new subaction starts. In applying the grammar for the analysis of a video, a parse tree is produced, which we call the *activity tree*. Fig. 11.4 illustrates the concept. From a video recording of a person performing the activity "cutting a plank," a graph is created, in which nodes of hands, objects, and tools merge into a common node, whenever they touch, or nodes split when the objects and hands separate. Referring to the figure, to compute the quantities involved, different processes (shown in the four subareas of the video frames) extract the human's body, the hands, objects, and geometric relations.

We now discuss several grammar approaches in Section 11.3.2.1, and then discuss in Section 11.3.2.2 whether such grammar representations are sufficiently expressive to capture action and activity structure, and sufficiently parsimonious to be preferred over other representations.

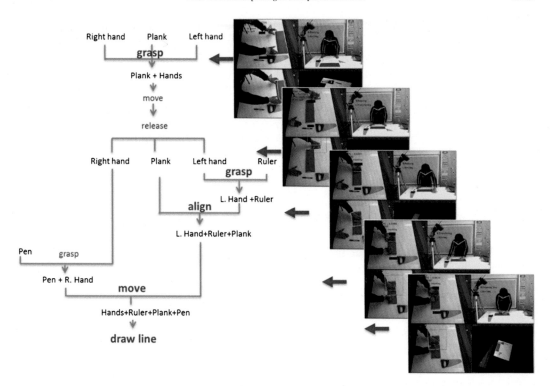

FIGURE 11.4 Illustration of activity description. The camera monitored a person cutting a plank. Four parallel processes computed essential components: (left up) detection of the hand and classification of the grasp type, (right up) gross motion via skeleton fitting, (left down) segmentation of the object, (right down) 3D shape description of the scene.

11.3.2.1 Different implementations of the grammar

The descriptions are based on context-free grammars, originally introduced in Pastra and Aloimonos (2012). Summers-Stay et al. (2012) implemented the idea for parsing assembly actions from RGBD video using only one symbol for all actions, and Yang et al. (2014) enhanced the description introducing grasp into the description and differentiating hand-to-object contact and object-to-object contact. The grammar describes actions at a level of abstraction that is useful for both video interpretation and robot execution. In Yang et al. (2015b) this was demonstrated with some examples. By automatically parsing videos that feature cooking instructions from the Youcook dataset (Das et al., 2013), actions were parsed, which then were performed by a Baxter robot equipped with the necessary motion capabilities.

Abstract descriptions of actions/verbs are necessary to achieve generalization, and perform what is called in the current terminology, few-shot learning or zero-shot learning (see Section 11.4). The basic action grammar reduces the description of actions to only the sequence of "touching relations," that is when the hand touches an object, two objects touch, the hand releases an object, or two objects or pieces of a single object separate (Dessalene et al., 2021).

Wörgötter et al. explore this concept to formulate an ontology (Wörgötter et al., 2013) considering one-handed actions. At the first level, actions are categorized according to the sequence of relations that the hand and one or two objects can have, into six classes, which are: rearrange, destroy, break, take-down, hide, construct. From there they iterate possible actions, and they come up with about 30 fundamental manipulations. Yang et al. (2013) proposed a related concept. They metaclassified actions according to the consequence an action has on an object, that is, what happens geometrically or topologically to an object. They proposed six categories – divide an object, merge two parts, transfer an object, deform an object, object appears, object disappears from the scene – and they also provided algorithms that combine tracking with segmentation to detect topological changes to detect the essential events in video.

11.3.2.2 Are grammars expressive and parsimonious descriptions?

An important question is whether the grammatical representations actually are sufficiently rich to allow for classification of many activities. The authors of Wörgötter et al. (2020) performed psychophysical and computational experiments to answer this question. They described, as above, actions by the sequence of contacts, using five quantities: the hand, the ground, and three objects. In addition they considered ten spatial relations, i.e., above, below, between, etc., to differentiate between altogether 35 different configurations or actions. A subset of ten of these actions (put, shake, stir, take, uncover, chop, cut, hide, lay, and push) was performed in a virtual environment, but instead of the actual objects, cubes were used. Experiments found that humans can recognize these actions, and so can their algorithms. Even more, the description was found to be very powerful for prediction; subjects on average only required 56% of the action duration to recognize the action. Thus, it appears that a description relying only on contact and spatial relations is very powerful for visual recognition.

11.3.3 Modules for action understanding

Individual vision processes are required to recognize discrete components, which then can be combined in higher level reasoning processes – such as the grammars from Section 11.3.2 – to achieve activity recognition and prediction. The descriptors which we discuss in this section differ from those heavily covered in the literature (for an evaluation of successful concepts in current approaches see Sigurdsson et al. (2017)). Specifically, in Section 11.3.3.1 we discuss representations of grasp, and in Section 11.3.3.2 we discuss explicit representation of geometry.

11.3.3.1 Grasping: an essential feature for action understanding

The grasp type provides crucial information about actions. As a motivational example, consider the two scenes in Fig. 11.5 from the VOC challenge (Everingham et al., 2010). Standard CV systems have object and human detectors to recognize the bicycle and the cyclist and pose detectors to confirm that these two cyclists are riding a bike. But humans can tell that the cyclist on the left side is not racing (since his hands are in a "Rest or Extension" grasp), whereas the one on the right is intent on racing (since the hands firmly hold the handlebar in a "Power Cylindrical" grasp).

FIGURE 11.5 (Left) Rest or Extension on the handlebar vs. (Right) Firm power cylindrical grasp of the handlebar (from Yang et al., 2015a).

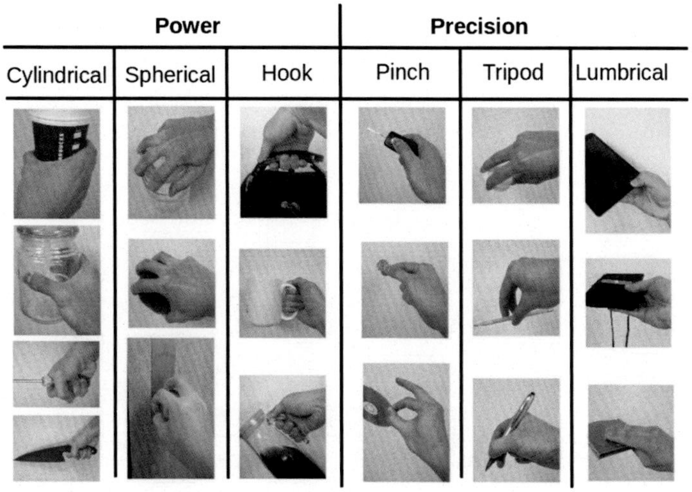

Power			Precision		
Cylindrical	Spherical	Hook	Pinch	Tripod	Lumbrical

FIGURE 11.6 Basic classification (Cutkosky, 1989) of active grasps with examples. At the highest level, grasps are categorized into power and precision grips. Power grips are used when an object is held with force, and can be classified as cylindrical, spherical, and hook. Precision grips provide fine movement and accuracy, and are subdivided into pinch, tripod, and lumbrical.

We cover here two papers, the first employs a basic ontology of grasps types in action understanding tasks, the second studies subtle changes in grasp for differentiating between similar manipulation actions, and develops learning approaches for online action prediction and regression of associated finger forces.

The recognition of grasp type provides essential information for a more detailed analysis of action. (See Fig. 11.6.) Researchers in several areas, including robotics, developmental medicine, and biomechanics, have developed grasping taxonomies that represent a hierarchy of the most common hand postures used for object grasping, with each taxonomy based on the needs of the tasks in the field. In Yang et al. (2015a) a basic classification of the main functional grasps (Cutkosky, 1989) in manipulation tasks was used and then demonstrated as a useful feature in two tasks: for segmentation of activities involving fine motor actions, and for characterizing action intention, i.e., whether the task is casual, or requires skills or forces.

Cognitive studies showed that an actor's intention shapes his/her movement kinematics during movement execution (Ansuini et al., 2015). For example, when subjects grasped a bottle for pouring, the middle and the ring fingers were more extended than when they grasped the bottle with the intent of displacing, throwing, or passing it. Inspired by these findings, in Fermüller et al. (2018), the authors developed a recurrent neural network architecture that monitors hands for predicting actions. Specifically, they considered sets of actions with the same object, such as "squeezing", "flipping", "washing", "wiping" and "scratching" with a sponge (see Fig. 11.7). They analyzed the system, which predicted in real-time the ongoing action to determine at what point in time the classification became accurate, and they also performed a psychophysical experiment, evaluating human performance on the same task. At 10 frames after the contact of the hand with the object, the system and the humans started understanding the action (75% classification accuracy for the sponge actions), and at 25 frames the judgment was very good (95% accuracy for the sponge actions). The visual architecture was an RNN using as input tracked image patches around the hand from which VGG-16 features (Simonyan and Zisserman, 2014) were computed.

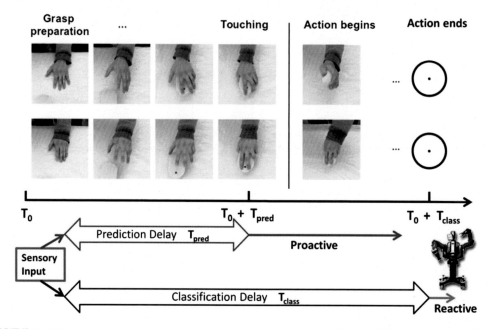

FIGURE 11.7 Examples demonstrating that early movements are strong indicators of the intended manipulation actions. Early prediction of action significantly reduces the delay in real-time interaction, which is fundamentally important for a proactive system.

In addition the paper also demonstrated association of vision with forces. Data was recorded from subjects that performed the same action with both hands. Sensors on the fingers of one hand recorded the forces, and the other hand was recorded visually. A recurrent neural network was trained to regress from vision to forces. It was then shown that using as input video only, when the visual classifier was combined with regressed forces, improved performance could be achieved. The concept appears promising. As shown, learning the mapping from vision to forces creates a bi-modal space that can aid visual recognition. Furthermore, there are immediate implications for robotics. Currently robots rely on haptic devices or force and torque sensors to learn tasks. If we can predict the forces exerted by the human demonstrator visually, it would allow us to teach robots much more efficiently.

11.3.3.2 *Geometry to robustify*

The use of geometry is important because it provides robust information for scene description and additional information for recognition. Geometry is computed using reconstruction processes, which are low-level (requiring only the image features and knowledge of camera positions), but no machine learning or training data is required. With the advent of cheap RGB-D sensors more than a decade ago, reconstructing scene geometry has become much easier and more accurate, and thus these sensors have become the standard vision sensors in robotics. Their use facilitates computing accurately and fast the distances to control the robot's movement, as well as computing the geometry and shape of objects to aid scene interpretation. This section discusses three geometric methods: accurate tracking of nonrigid object transformations and detection of topological changes; computation of pairwise spatial relations of objects over time; and computation of object symmetry and its use for better foreground-background segmentation.

Building on an efficient point cloud library (Zampogiannis et al., 2018), Zampogiannis et al. (2019) developed a technique for accurately tracking nonrigid object transformations and detecting the topological changes, that is, contacts and separations of body parts and objects needed for the grammatical description (see Section 11.3.2). The gist of the method lies in a warp field estimation that considers forward and backward warps between consecutive frames to detect regions of the deformed geometry that undergo topological changes.

Activity descriptions can also profit from descriptors of spatial relations between objects. In Zampogiannis et al. (2015), the authors introduced a representation for manipulation actions based on the evolution of the spatial relations between objects in the scene. The method was implemented by tracking objects in RGBD video, and reasoning over the spatial relations of observed object pairs. The resulting descriptor amounts to a sequence of spatial relation predicates (e.g., in, left, right, front, behind, below, above, touch), and this descriptor was shown to be sufficiently expressive for distinguishing between four different actions.

Another concept is to exploit general knowledge of object shape properties. For example, symmetry detection can help with segmentation both in 2D (Teo et al., 2015) and 3D (Ecins et al., 2016). Imagine looking at a cluttered scene. Since most objects we work with are symmetric, either bilateral or rotational, we can "fill in the back of the object" that is not visible, and this aids the segmentation and recognition.

11.3.4 Discussion on activity understanding

Activity Understanding is a very challenging problem. End-to-end solutions do not scale well because of the large variations in appearance at high levels of abstraction and temporal extension.

We discussed hierarchical approaches, and we detailed one higher level description – action grammars. Action grammars can segment activities at times of contact and capture the recursive structure of action sequences, similar to that found in language. We described experiments demonstrating the expressive power of action grammars. We also described processes at the lower level, which have not received much attention in CV, but which are essential for supporting an action-based approach. These include affordances, grasp-type and the use of geometry to aid temporal and spatial segmentation and descriptions of spatial relations.

In this section we emphasized the necessity to utilize meaningful action-based representations at different levels of the hierarchy. To elaborate further, it is very important that these representations need to be robust, because of the many challenges involved in activity understanding. We can achieve robustness in part through the use of geometry – geometry does not require memorization, and can be estimated from low-level measurements. Thus, we should introduce geometry into the pipeline whenever possible before starting with recognition. Beyond geometry, any concept that provides universally true information is meaningful for activity understanding. We may model physical laws. We may include model ontologies to aid generalization, for example by grouping verbs according to the effect they have on objects (Yang et al., 2013). We also may include processes that model causality; actions constrain each other causally, some combinations are not possible physically. These representations capture more knowledge and better constraints for activity interpretation.

The integration of vision and cognition or language is hard. This is because of the semantic gap, i.e., the disparity between the symbolic or linguistic representations and the visual representations based on signals. We want an integration of the two which is not brittle. Thus, we need to avoid setting thresholds or converting to purely symbolic representation too early within the pipeline. This is because if vision fails to return the right quantities, then imprecisions compound through further abstraction, resulting in failed reasoning. The next research challenge is to study learning approaches that relate perception with higher-level reasoning for a deeper integration. Section 11.4 covers deep learning approaches useful for such integration. Currently such approaches are primarily constrained to object recognition, and to an extent to action recognition. Activity understanding would benefit from these methods.

11.4 Functional scene understanding through deep learning with language and vision

Here we consider the merging of vision and language and associated representations – the merging of "signal" and "symbol". The merging of multiple representations is important due to the suitability of different representations for capturing different characteristics of the world. Here we seek to allow information from lower level representations of appearance and information from higher level representations of relations and semantics to complement each other.

Systems involving symbolic and continuous representations often have hard boundaries, below which the system is continuous, and above which the system is symbolic. Precisely where this boundary is set varies, but generally does not fall below the level of abstraction reflected in human language as language is a primary source of symbolically represented world knowledge.

Symbolic representation is more important for action and activity understanding than for other CV tasks such as object detection as the nature of this task is more abstract and less appearance based. Action has temporal structure at multiple scales, and is structured around satisfaction of conditions – the defining characteristics of action are semantic and relational.

Many computer vision tasks can benefit from the integration of vision and language. However, one task which is ideal for the study of the integration of the two is Zero Shot Learning (ZSL) – this is because unlike other tasks it cannot be solved without the introduction of nonvisual knowledge such as is reflected in language.

ZSL is a task with two sets: a training set, and a test set. The categories of these sets break into "seen" and "unseen" categories. The training set consists of only "seen" categories, while the test set contains "unseen" categories as well as, optionally, "seen" categories. To illustrate, there could be a ZSL task which includes "run" and "stand" in seen categories, and "walk" in unseen categories. The task would then be to learn to visually recognize and properly categorize walking, when walking has never previously been visually encountered, but running and standing have been visually encountered.

There are multiple approaches to ZSL. Early work on ZSL focuses on *attributes* – visually recognizable characteristics with differential class (e.g., object classes or action classes) associations. Attributes are generalizable in that attribute detectors trained only on the seen set are able to detect attributes in samples from both the seen and unseen set.

More recent work on ZSL tends to focus on *semantic embedding spaces*. These are Euclidean vector spaces where semantic categories, such as reflected in language, are associated with vectors – or points in space. These vector representations of words are of significantly lower dimension than naive 1-hot encodings (vectors where each dimension corresponds to a class, and all but 1 dimension have 0 values), and have the quality that words which are similar in semantics are associated with similar vectors nearby in the embedding space.

With semantic embedding spaces, symbolic representations are vectorized in such a way that the semantic relations of the symbol categories are preserved. This allows for integration of symbolic semantics into deep architectures, whose internal representations consist of vectors. This integration amounts to properly aligning the visual vector representations with the vectorized symbolic representations.

Different ZSL methods use different shared embeddings – some embed visual features into the semantic space, some embed the semantic space into the visual feature space, and some embed both into a third shared space. Once both visual and semantic representations lie in the same space, categorizing visual input is a matter of finding the nearest semantic label in this space.

State-of-the-art contemporary ZSL methods often rely on CNNs for visual features and produce shared embedding spaces with pretrained semantic embedding spaces such as word2vec (Mandal et al., 2019; Xian et al., 2018). Some shared embedding based ZSL methods structure and train their models in an end-to-end fashion, which is more challenging, but provides dividends in performance (Zhang et al., 2017).

The remainder of this section is structured as follows: In Section 11.4.1 we detail a simple use of attributes for ZSL; in Section 11.4.1 we detail a more nuanced *relative attribute* formulation; in Section 11.4.2 we cover the use of shared semantic spaces in ZSL; in Section 11.4.3 we cover basic approaches to semantic vector space construction; in Section 11.4.4 we cover the incorporation of knowledge in the form of graphs for ZSL action classification.

11.4.1 Attributes in zero-shot learning

The use of attributes – including action centric attributes such as affordances covered in Section 11.2 – in recognition allows the construction of classifiers which are humanly interpretable and specifiable. Attributes mitigate the issue of opacity through use of an explicit predefined mid-level representation below the level of class categories.

Use of attributes allows an easy mechanism through which to learn visual representations from the available training data and to transfer those representations onto the classes for which no training data is available. Attributes are general in that they have presence across multiple class categories, and through taking multiple attributes with different distributions across classes allow representation of select classes.

The use of attributes in ZSL is as follows: Attribute detectors are trained over the seen set and associated attribute labels. These detectors are generalizable across both the seen and unseen set. Due to the different class coverage (e.g., object or action class) of different attribute categories, different combinations of attributes can represent different classes – class detectors can then be instantiated on top of attribute detectors. This instantiation follows a specification of which attributes are associated with which classes. When given a specification for unseen categories, detectors can be built even though no visual samples of the unseen categories have been encountered.

Conventional use of attributes in computer vision is binary: an attribute is represented as either present, or absent. This limits the representational power of attribute representations. However, binary representations can be generalized to scalar representations where each attribute is associated with a scalar degree rather than a binary category. This is both more flexible representationally, and allows the inclusion of attributes which do not so cleanly fall into a binary categorization. For example, while an attribute of "indoors / outdoors" often is clearly binary, an attribute of "moving fast / moving slow" has a more even distribution over gradations in visual input.

Parikh and Grauman (2011) presents one approach to generalizing binary attributes to scalar attributes. A challenge in generalizing from binary to scalar attributes is inconsistency in annotations, as different annotators may have different understandings of what different attributes' degrees correspond to in the scalar representation. They resolve this challenge by requesting that annotators not assign scalar values to attributes, but rank images in terms of attribute degree. After images are ranked, scalar attribute values can be derived from their annotated relative attribute degrees.

For binary attributes, attribute detectors can be trained using conventional classifiers, but producing scalar attributes requires other methods. Parikh and Grauman (2011) train a ranking function over images for attributes over the seen set, and use that ranking function in the production of a scalar value.

Relative attribute representations allow for greater flexibility in class specifications. With consideration to the attribute of "moving fast / moving slow" one can specify that an unseen category "running" is faster than the seen category of "walking", or that the unseen category of "standing" is slower than the seen category of "walking". This is done without a need to define either a binary specification or intuit a scalar value with which to describe the unseen categories.

11.4.2 Shared embedding spaces

The generalization of binary attributes to scalar attributes increases their representational power. However, the increased representational power came at the cost of increased difficulty in annotating attributes and specifying classes. Relative Attributes (Parikh and Grauman, 2011) introduced one solution to these challenges.

Attributes can also be abstracted away. It need not be the case that visual representations move through humanly understandable attributes. Abstracting attributes away has a couple advantages:

- Avoid the imprecision introduced by passing through engineered, rather than learned, representations.
- Avoid the overhead and imprecision of annotating attributes.

The question is then how to construct a system such that training for classification of seen classes results in a classifier which works not only for seen classes, but for unseen classes as well, without leveraging an engineered mid-level representation which allows the specification of unseen classes in terms of that mid-level representation.

One approach that can be used is to define unseen classes in terms of their similarity relations to seen classes, where these relations are learned from text corpora. Extensive work exists in Natural Language Processing (NLP) producing *semantic vector spaces* which represent similarity relations among words – word2vec (Mikolov et al., 2013) is one popular example. The intuition is that terms in these spaces are located in proximity to other terms with which they share semantic similarity. Trained semantic spaces are publicly available – these can be taken and used without a need to produce them from scratch. See Section 11.4.3 for details on construction of such spaces.

Terms with semantic similarity often share similarity in the visual space as well – e.g., "jogging" is both semantically and visually part way between "running" and "walking". And so, it is often the case that if the visual similarity to known categories can be determined, then the semantic similarity relations can be established as well. Then, from these semantic relations we can infer semantic categories of visual inputs.

To illustrate: Consider a set of seen classes including "running" and "walking", and a set of unseen classes including "jogging", a semantic vector space capturing semantic proximity between these categories, and a computer vision architecture, such as a CNN, which produces visual feature representations of input. Then, consider an input sample of unknown class. Say that the visual representation of this input as produced by the CNN is part way between the visual representation for "running" and for "walking". We then go to the semantic language space and see that the label that is located part way between "running" and "walking" is jogging and assign this label to the input.

Most often, comparing similarities between samples of unknown class and known classes is not done in the visual space. It is more common to map the input into the semantic space and perform comparisons to known classes there. This requires embedding one space into another.

The mechanism of embedding one space into another can be as simple as a linear transformation applied over one space, which is then trained over a similarity loss between two spaces. DeViSe (Frome et al., 2013) is a good example of an architecture using this method. This involves two pretrained representations – visual features taken from, for instance, a CNN trained over classification, and word vectors in an embedding space, which may be produced through means discussed in Section 11.4.3.

One method of embedding visual features into the semantic space of the word vectors is then illustrated in Fig. 11.8. A layer of nodes is appended to the top of the pretrained CNN features, and then trained. The loss that this layer is trained over is the similarity (e.g., cosine similarity) between the output of this layer, and the vectors in the semantic embedding space corresponding to the labels of the visual input. This learns a simple linear mapping from visual feature vectors onto text derived semantic features.

More sophisticated mappings than linear translations are often used. Using multiple layers of neurons, in conjunction with nonlinear activations, produce nonlinear mappings (e.g., Kato et al., 2018 use such a method). Kodirov et al. (2017) use an autoencoder with semantic constraints to produce the embedding, and find that the constraint of visual reconstruction leads to better generalization to unseen classes in ZSL.

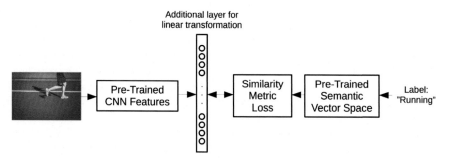

FIGURE 11.8 A simple approach to embedding visual representations into a semantic space, as is used by methods such as DeViSe (Frome et al., 2013). Two pretrained models are used: 1) pretrained visual features, such as produced by a CNN trained over a visual task, and 2) a semantic vector space as produced through methods discussed in Section 11.4.3. A simple linear transformation is produced through appending a layer of nodes on top of the visual features, and training them w.r.t. their similarity to the semantic word vector of the labels corresponding to visual input. Once this linear transformation is learned, it constitutes a mapping from the space of visual features into the semantic vector space.

Once both visual input and semantic representations are placed in the same space, then determining the class of novel visual input is as simple as representing visual input in that space and then finding the nearest semantic label in that same space.

11.4.3 Construction of semantic vector spaces

11.4.3.1 word2vec

Here we wish to construct a vector space into which words are embedded in such a way that their semantics are captured spatially. The end results are a space where, for example, vectors for "jog", "run", and "walk" are located in proximity, with "jog" being placed in the middle of the three.

How can a word's semantics be defined? One answer is through the interactions that word has with other words in text corpora – semantics of words can be defined in relation to other words. How do we model words' relations to other words? One simple approach is cooccurrence – if two words occur in proximity, they are taken to be related. The more frequently that they cooccur together, the more strongly they are related. This is the principle on which methods embedding words into vector spaces such as word2vec are based.

We start from the simplest vector representation for words – a 1-hot encoding, where each vector position corresponds to one word in a vocabulary V. This is a high-dimensional, inefficient representation, where there are no meaningful spatial relations among words. We can embed words into a lower dimensional space with said desirable properties through the solving of one of two related tasks:

1. Predicting a target word based on context
2. Predicting context based on a target word

Here "context" C is defined as the set of terms "nearby" the target term. These are defined as the other terms present in an n-gram associated with the target term, without consideration for word order.

Each of these tasks can be solved using simple architectures, and the solving of these tasks produce in the process lower dimensional representations which can then be taken and used for other tasks.

A method – *Continuous Bag Of Words* (CBOW) – for solving Task 1 (Mikolov et al., 2013) is shown in Fig. 11.9(a). Input consists of multiple words of context, each represented as a 1-hot vector of dimension $|V|$. These vectors are summed to produce a vector of size $|V|$. This is fed through a single layer of N neurons, where N is the dimension of the embedding space. On top of this we have one more layer, of size $|V|$, whose job it is to predict, in 1-hot representation, the word associated with the context consisting of the terms fed as input to the first layer.

A method – *Skip Gram* – for solving Task 2 (Mikolov et al., 2013) is shown in Fig. 11.9(b). Input consists of a single term, represented as a 1-hot vector of dimension $|V|$. This is fed into a single layer of dimension N, where N is the dimension of the embedding space. On top of this layer we have an output layer, consisting of $|C|$ sets of $|V|$ nodes, each associated with one term of context C in the n-gram associated with the input term.

Both architectures are trained through serially feeding in n-grams extracted from large text corpora, and applying a soft-max loss to the final layer to enforce alignment with expected terms.

After these architectures are trained, the hidden layer then constitutes a mapping of V from a 1-hot representation of size $|V|$ to an embedded representation of size N where terms are spatially located in proximity to terms with similar semantics.

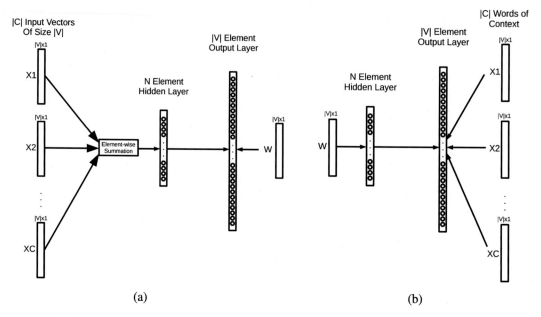

(a)   (b)

FIGURE 11.9   (a) The Continuous Bag Of Words method to solving the task of predicting a target word based on that words context (Mikolov et al., 2013). $|C|$ input vectors of context are fed through two layers, the first a hidden layer of size N, the second an output layer of size V. Loss is computed w.r.t. the word W. After training the output layer can be discarded, and the hidden layer used as a translation from 1-hot word encodings into an embedded space of size N. (b) Skip Gram method to solving the task of predicting the context of a term (Mikolov et al., 2013). Word W is fed through two network layers, the first a hidden layer of size N, the second an output layer of size V. Loss is computed w.r.t. $|C|$ words of context. After training, the output layer can be discarded, and the hidden layer used as a translation from 1-hot word encodings into an embedded space of size N.

11.4.4 Shared embedding spaces and graphical models

ZSL of action can benefit from additional structure beyond what simple mapping into a shared embedding space can provide. Additional structure can be represented in the form of graphs. Multiple works (Ghosh et al., 2020a,b; Kato et al., 2018; Yan et al., 2018) employ graphs for the purpose of recognition of actions, processing these through use of a Graph Convolutional Network (GCN) to produce vector representations of action categories suitable for ZSL.

Ghosh et al. (2020a) evaluate 3 different graphical representations, the last of which is applicable to Few-Shot Learning rather than ZSL due to its use of visual features from a small number of samples. Kato et al. (2018) construct a graph based on Subject Verb Object triplets derived from knowledge corpora. All of these are processed through a GCN.

Throughout (Ghosh et al., 2020a) sentence2vec (Pagliardini et al., 2017) is used rather than word2vec, as authors find that action categories are better represented through phrases than through single words which can be confounded by multiple meanings. The first graph is composed of nodes taking values of sentence2vec embeddings of action category phrases. These nodes are linked together based on cosine similarity between vector representations – the top N closest neighbors per node are given edges. The second graph associates verbs and nouns – derived from Part-Of-Speech tagging of phrases describing actions – with each action class, incorporating nouns as a strong connection between seen and unseen categories. The third graph incorporates visual representations derived from a small number of samples of the unseen categories – this graph is thus applicable to Few-Shot Learning, rather than ZSL. The reason for the incorporation of visual representations is that categories which are similar in the semantic space may nevertheless have distinct visual appearances – authors give the example of "pommel horse" and "horse walking", which have similar word embedding representations but dissimilar visual appearances.

Kato et al. (2018) construct a graph consisting of three types of nodes: noun nodes, object nodes, and action nodes. Action nodes are linked to the verbs and nouns which the associated action involves.

Values of graph nodes are generally initialized with values derived from semantic vector spaces – this is important as it establishes the initial relations which the GCN iterates over. This iteration incorporates relations defined by the edges in the graph, and allows information to transfer from node-to-node along edges. E.g., in Kato et al. (2018) action nodes, which are initially set to zero vectors, acquire a representation determined by the vectors of neighboring noun and verb nodes, which have been initialized with semantic vectors.

Like previously detailed methods, these methods learn a mapping from visual features (taken from a CNN pretrained on a separate task) into a shared embedding space – though here that space is shared with representations produced by the GCN. In Kato et al. (2018) that mapping is produced through two layers of neurons with sigmoid activations, resulting in a nonlinear mapping from the visual features into the shared semantic space. Similar to previous work, these layers are trained through applying a loss measuring the similarity between the visual features after embedding, and the vectors associated with the labels of the input as produced by the GCN.

The mapping from pretrained visual features onto the GCN produced action vectors can then be trained over the seen training set. To obtain action predictions when applying the network to novel actions during testing, nearest neighbor can be performed between the visual features and the action vectors.

11.5 Future directions

Here we discuss implications and future work implied by the action based framework outlined in the rest of this chapter. Action has implications for *tasks and datasets* – it enables conceptual modeling conductive to generalization and longer term temporal prediction. Action benefits from modeling of *concepts* beyond those from conventional CV. As CV progresses the scalability of fully supervised methods becomes an increasing issue, and action based methods help mitigate this issue while benefiting from *semi- and unsupervised* paradigms. Finally action helps enable an integration of *cognition* and symbolic modeling into *perception*, including across multiple perceptual modalities.

Tasks and datasets: Activities span long time spans. Thus when seeking solutions to visual activity understanding, we face problems much more challenging than those we encounter in current action recognition tasks. Objects, actions, affordances, and other scene constituents relate to each other semantically over multiple time scales, and we need to find ways to model these relations. We think that this capability is not well demonstrated with the task of recognizing actions. Instead, we should pick tasks that demonstrate generalization and a conceptual understanding of action (as opposed to a purely appearance based understanding). Such tasks include zero-shot learning and the prediction of future actions, and translation from one viewpoint to another (e.g., from first person to third person). Today's CV research is largely driven by the collection of new datasets and definition of new challenges – existing datasets do not sufficiently cover long term and conceptual modeling of activities. The datasets ideally should have recordings from multiple views, because this opens possibilities for interesting research, for example, to solve the challenge of transferring knowledge between the first person and the third person view. Lastly, most datasets feature indoor scenes. It will be interesting to collect outdoor scenes and analyze them, as discussed above, by looking at the relations between affordances, interactions and long-term relations. We could also attempt an action-based analysis on data from the autonomous driving domain.

Concepts for long-term activity understanding: Activity understanding requires action-based concepts at multiple time scales. We discussed in this chapter such quantities at the single image and short-term time scales, including affordances, hand grasps, geometric relations, and we emphasized the use of geometric reconstruction processes because of their robustness (see discussion in Section 11.3.4). The next step will be to add further constraints and include robust constraints for modeling temporal-relations at longer scales. We could make use of ontologies in categorizing objects and actions. On verbs we can impose classifications based on action effects (Yang et al., 2013), ergonomic principles, or force and location related constraints. Longer-term relations include causal relations, and constraints on possible and impossible action sequences. We can also model physics constraints, and use physics engines – but in order to integrate these into deep architectures we need to include these constraints into vector spaces that relate perception to cognition (Section 11.4).

Reducing supervision: Early approaches on integrating language with vision (see Sections 11.4.1 and 11.4.4) have relied heavily on supervision. For example, visual attribute recognition or object recognition has been implemented via supervised learning. Graph-based models using shared embeddings for ZSL need to incorporate in advance the categories of the "unseen" set which they may encounter during test time. Naturally, evolving approaches will find their way into the action-based framework, including unsupervised and self-supervised learning, transfer learning, metalearning, and eventually never-ending learning (Mitchell et al., 2015). For example, in constructing visual ontologies, we don't want to rely fully on supervised visual learning of metaverb classes. One solution to such modeling is dictionary learning (Zheng et al., 2016), but so far these approaches have been limited to simple actions. We will need methods that scale to more complex human manipulation actions. Generative Adversarial Networks (GANs) and Variational Autoencoders (VAEs) have been shown successful in modeling fine-grained action images and videos. We could, for example, use VAEs to learn the underlying distribution of data into a discretized latent space that encodes the metacategories.

Involving the whole brain: Humans' understanding of the world is grounded in our motoric cognition and all our senses. Similarly, our models should include other sensory modalities like auditory, tactile, or proprioception signals in addition to modeling of vision and cognition representations. Different modalities provide complementary information, and because of this allow for different ways of organizing concepts. In addition to questions of learning with different modalities, we also need methods for accessing stored concepts when given perception from any modality. Studying the integration of diverse modalities will also benefit from studying memory. A framework known as Vector Symbolic Architectures (VSA) (Plate, 1995; Eliasmith, 2013), which includes Hyperdimensional (HD) Computing (methods making use of very high dimensional vector spaces) (Pentti, 2009), has been proposed as a theoretical model for artificial intelligence. HD Computing combines advantages of neural-based AI approaches with systematic compositionality and rule-like behavior from classical symbolic AI (Levy and Gayler, 2008). In this framework concepts are encoded into vector spaces, and algebraic operations are defined on these vector spaces. These operations include the addition of related concepts and the binding of vectors of different origin – for example this could be sound and vision, or vision and motor (Mitrokhin et al., 2019). These operations maintain the separability of one modality from the other. We may build on this framework and integrate it with neural network approaches. The goal will be to retain an explicit memory encoding different perception modalities, while maintaining the capability of recalling information from any modality.

11.6 Conclusions

The purpose of Computer Vision is to produce interpretations which are of use to humans. Action is central to our understanding of the world, yet is underutilized in contemporary CV. This chapter covered scene and activity understanding that has the concept of action and interaction at its center. We covered action based approaches to scene understanding involving modeling at multiple temporal scales, starting from object interpretation in terms of affordances at the instantaneous level, up to basic actions, then up to full activities at the longer temporal scale. We described the well developed area of affordance learning, and described works on activity understanding combining cognitive and linguistic approaches with humanly interpretable modules essential in characterizing activities and segmenting video temporally. We discussed methods for the integration of visual representations with knowledge, both engineered and derived from text corpora. We covered the integration of vision, centered on action, with graphical constraints, and discussed future directions including creating new challenges and datasets, adding concepts for encoding long-term relations, adapting semi- and unsupervised learning approaches, and incorporating memory as a central component to the action-based framework.

Acknowledgment

The support of the National Science Foundation under grants BCS 1824198 and OISE 2020624 is gratefully acknowledged.

References

Aditya, Somak, Yang, Yezhou, Baral, Chitta, Aloimonos, Yiannis, Fermüller, Cornelia, 2018. Image understanding using vision and reasoning through scene description graph. Computer Vision and Image Understanding 173, 33–45.

Aloimonos, Yiannis, Fermüller, Cornelia, 2015. The cognitive dialogue: a new model for vision implementing common sense reasoning. Image and Vision Computing 34, 42–44.

Ansuini, Caterina, Cavallo, Andrea, Bertone, Cesare, Becchio, Cristina, 2015. Intentions in the brain: the unveiling of mister Hyde. The Neuroscientist 21 (2), 126–135.

Bajcsy, Ruzena, 1988. Active perception. Proceedings of the IEEE 76 (8), 966–1005.

Barsalou, Lawrence W., 2008. Grounded cognition. Annual Review of Psychology 59, 617–645.

Bo, Liefeng, Ren, Xiaofeng, Fox, Dieter, 2013. Unsupervised feature learning for rgb-d based object recognition. In: Experimental Robotics. Springer, pp. 387–402.

Cutkosky, Mark R., 1989. On grasp choice, grasp models, and the design of hands for manufacturing tasks. IEEE Transactions on Robotics and Automation 5 (3), 269–279.

Damen, Dima, Doughty, Hazel, Maria Farinella, Giovanni, Fidler, Sanja, Furnari, Antonino, Kazakos, Evangelos, Moltisanti, Davide, Munro, Jonathan, Perrett, Toby, Price, Will, et al., 2018. Scaling egocentric vision: the EPIC-kitchens dataset. In: Proceedings of the European Conference on Computer Vision (ECCV), pp. 720–736.

Das, Pradipto, Xu, Chenliang, Doell, Richard F., Corso, Jason J., 2013. A thousand frames in just a few words: lingual description of videos through latent topics and sparse object stitching. In: Proceedings of the IEEE Conference on Computer Vision and Pattern Recognition, pp. 2634–2641.

Dessalene, Eadom, Devaraj, Chinmaya, Maynord, Michael, Fermüller, Cornelia, Aloimonos, Yiannis, 2021. Forecasting action through contact representations from first person video. IEEE Transactions on Pattern Analysis and Machine Intelligence.

Dutta, V., Zielinska, T., 2017. Action prediction based on physically grounded object affordances in human-object interactions. In: Proceedings of the 11th International Workshop on Robot Motion and Control.

Ecins, Aleksandrs, Fermüller, Cornelia, Aloimonos, Yiannis, 2016. Cluttered scene segmentation using the symmetry constraint. In: IEEE International Conference on Robotics and Automation (ICRA), pp. 2271–2278.

Eliasmith, Chris, 2013. How to Build a Brain: A Neural Architecture for Biological Cognition. Oxford University Press.

Everingham, M., Van Gool, L., Williams, C.K.I., Winn, J., Zisserman, A., 2010. The Pascal visual object classes (VOC) challenge. International Journal of Computer Vision 88 (2), 303–338.

Fermüller, Cornelia, Aloimonos, Yiannis, 1995. Vision and action. Image and Vision Computing 13 (10), 725–744.

Fermüller, Cornelia, Wang, Fang, Yang, Yezhou, Zampogiannis, Konstantinos, Zhang, Yi, Barranco, Francisco, Pfeiffer, Michael, 2018. Prediction of manipulation actions. International Journal of Computer Vision 126 (2), 358–374.

Fitzpatrick, Paul, Metta, Giorgio, Natale, Lorenzo, Rao, Sajit, Sandini, Giulio, 2003. Learning about objects through action-initial steps towards artificial cognition. In: IEEE International Conference on Robotics and Automation, vol. 3, pp. 3140–3145.

Frome, Andrea, Corrado, Greg, Shlens, Jonathon, Bengio, Samy, Dean, Jeffrey, Ranzato, Marc'Aurelio, Devise, Tomas Mikolov, 2013. A deep visual-semantic embedding model. Advances in Neural Information Processing Systems 26.

Ghosh, Pallabi, Saini, Nirat, Davis, Larry S., Shrivastava, Abhinav, 2020a. All about knowledge graphs for actions. arXiv preprint. arXiv:2008.12432.

Ghosh, Pallabi, Yao, Yi, Davis, Larry, Divakaran, Ajay, 2020b. Stacked spatio-temporal graph convolutional networks for action segmentation. In: Proceedings of the IEEE/CVF Winter Conference on Applications of Computer Vision, pp. 576–585.

Gibson, James J., 1977. The theory of affordances. In: Bransford, John, Shaw, Robert E. (Eds.), Perceiving, Acting, and Knowing: Toward an Ecological Psychology. Lawrence Erlbaum Associates, Hillsdale, NJ, pp. 67–82.

Grabner, Helmut, Gall, Jürgen, Van Gool, Luc, 2011. What makes a chair a chair? In: IEEE Conference on Computer Vision and Pattern Recognition, pp. 1529–1536.

Guha, Anupam, Yang, Yezhou, Fermüller, Cornelia, Aloimonos, Yiannis, 2013. Minimalist plans for interpreting manipulation actions. In: IEEE/RSJ International Conference on Intelligent Robots and Systems, pp. 5908–5914.

Gupta, Abhinav, Satkin, Scott, Efros, Alexei A., Hebert, Martial, 2011. From 3d scene geometry to human workspace. In: IEEE Conference on Computer Vision and Pattern Recognition, pp. 1961–1968.

Hassanin, Mohammed, Khan, Salman, Tahtali, Murat, 2018. Visual affordance and function understanding: a survey. arXiv preprint. arXiv:1807.06775.

Hedau, Varsha, Hoiem, Derek, Forsyth, David, 2009. Recovering the spatial layout of cluttered rooms. In: IEEE International Conference on Computer Vision, pp. 1849–1856.

Hermans, Tucker, Rehg, James M., Bobick, Aaron, 2011. Affordance prediction via learned object attributes. In: IEEE International Conference on Robotics and Automation (ICRA): Workshop on Semantic Perception, Mapping, and Exploration.

Pentti, Kanerva, 2009. Hyperdimensional computing: an introduction to computing in distributed representation with high-dimensional random vectors. Cognitive Computation 1, 139–159.

Kato, Keizo, Li, Yin, Gupta, Abhinav, 2018. Compositional learning for human object interaction. In: Proceedings of the European Conference on Computer Vision (ECCV), pp. 234–251.

Kjellström, Hedvig, Romero, Javier, Kragić, Danica, 2011. Visual object-action recognition: inferring object affordances from human demonstration. Computer Vision and Image Understanding 115 (1), 81–90.

Kodirov, Elyor, Xiang, Tao, Gong, Shaogang, 2017. Semantic autoencoder for zero-shot learning. In: Proceedings of the IEEE Conference on Computer Vision and Pattern Recognition, pp. 3174–3183.

Koppula, Hema S., Saxena, Ashutosh, 2014. Physically grounded spatio-temporal object affordances. In: European Conference on Computer Vision. Springer, pp. 831–847.

Koppula, Hema S., Saxena, Ashutosh, 2015. Anticipating human activities using object affordances for reactive robotic response. IEEE Transactions on Pattern Analysis and Machine Intelligence 38 (1), 14–29.

Koppula, Hema Swetha, Gupta, Rudhir, Saxena, Ashutosh, 2013. Learning human activities and object affordances from rgb-d videos. The International Journal of Robotics Research 32 (8), 951–970.

Lee, David C., Gupta, Abhinav, Hebert, Martial, Kanade, Takeo, 2010. Estimating spatial layout of rooms using volumetric reasoning about objects and surfaces. In: Advances in Neural Information Processing Systems, pp. 1288–1296.

Levy, S.D., Gayler, R., 2008. Vector symbolic architectures: a new building material for artificial general intelligence. In: Proceedings of the First Conference on Artificial General Intelligence (AGI-08). IOS Press.

Liu, Fayao, Shen, Chunhua, Lin, Guosheng, 2015. Deep convolutional neural fields for depth estimation from a single image. In: Proceedings of the IEEE Conference on Computer Vision and Pattern Recognition, pp. 5162–5170.

Mandal, Devraj, Narayan, Sanath, Kumar Dwivedi, Sai, Gupta, Vikram, Ahmed, Shuaib, Shahbaz Khan, Fahad, Shao, Ling, 2019. Out-of-distribution detection for generalized zero-shot action recognition. In: Proceedings of the IEEE/CVF Conference on Computer Vision and Pattern Recognition, pp. 9985–9993.

Martin, A., 2007. The representation of object concepts in the brain. Annual Review of Psychology 58, 25–45.

Metta, Giorgio, Sandini, Giulio, Vernon, David, Natale, Lorenzo, Nori, Francesco, 2008. The iCub humanoid robot: an open platform for research in embodied cognition. In: Proceedings of the 8th Workshop on Performance Metrics for Intelligent Systems, pp. 50–56.

Mikolov, Tomas, Chen, Kai, Corrado, Greg, Dean, Jeffrey, 2013. Efficient estimation of word representations in vector space. arXiv preprint. arXiv:1301.3781.

Miller, George A., Beckwith, Richard, Fellbaum, Christiane, Gross, Derek, Miller, Katherine J., 1990. Introduction to wordnet: an on-line lexical database. International Journal of Lexicography 3 (4), 235–244.

Mitchell, T., Cohen, W., Hruschka, E., Talukdar, P., Betteridge, J., Carlson, A., Dalvi, B., Gardner, M., Kisiel, B., Krishnamurthy, J., Lao, N., Mazaitis, K., Mohamed, T., Nakashole, N., Platanios, E., Ritter, A., Samadi, M., Settles, B., Wang, R., Wijaya, D., Gupta, A., Chen, X., Saparov, A., Greaves, M., Welling, J., 2015. Never-ending learning. In: Proceedings of the Twenty-Ninth AAAI Conference on Artificial Intelligence (AAAI-15).

Mitrokhin, A., Sutor, P., Fermüller, C., Aloimonos, Y., 2019. Learning sensorimotor control with neuromorphic sensors: toward hyperdimensional active perception. Science Robotics 4 (30), eaaw6736.

Montesano, Luis, Lopes, Manuel, Bernardino, Alexandre, Santos-Victor, José, 2008. Learning object affordances: from sensory–motor coordination to imitation. IEEE Transactions on Robotics 24 (1), 15–26.

Myers, Austin, Teo, Ching L., Fermüller, Cornelia, Aloimonos, Yiannis, 2015. Affordance detection of tool parts from geometric features. In: 2015 IEEE International Conference on Robotics and Automation (ICRA), pp. 1374–1381.

Nguyen, Anh, Kanoulas, Dimitrios, Caldwell, Darwin G., Tsagarakis, Nikos G., 2017. Object-based affordances detection with convolutional neural networks and dense conditional random fields. In: IEEE/RSJ International Conference on Intelligent Robots and Systems (IROS), pp. 5908–5915.

Pagliardini, Matteo, Gupta, Prakhar, Jaggi, Martin, 2017. Unsupervised learning of sentence embeddings using compositional n-gram features. arXiv preprint. arXiv:1703.02507.

Parikh, Devi, Grauman, Kristen, 2011. Relative attributes. In: International Conference on Computer Vision, pp. 503–510.

Pastra, Katerina, Aloimonos, Yiannis, 2012. The minimalist grammar of action. Philosophical Transactions of the Royal Society B: Biological Sciences 367 (1585), 103–117.

Plate, Tony A., 1995. Holographic reduced representations. IEEE Transactions on Neural Networks 6 (3), 623–641.

Qi, Siyuan, Huang, Siyuan, Wei, Ping, Zhu, Song-Chun, 2017. Predicting human activities using stochastic grammar. In: Proceedings of the IEEE International Conference on Computer Vision, pp. 1164–1172.

Qi, Siyuan, Wang, Wenguan, Jia, Baoxiong, Shen, Jianbing, Zhu, Song-Chun, 2018. Learning human-object interactions by graph parsing neural networks. In: Proceedings of the European Conference on Computer Vision (ECCV), pp. 401–417.

Roy, Anirban, Todorovic, Sinisa, 2016. A multi-scale cnn for affordance segmentation in rgb images. In: European Conference on Computer Vision. Springer, pp. 186–201.

Russakovsky, Olga, Deng, Jia, Su, Hao, Krause, Jonathan, Satheesh, Sanjeev, Ma, Sean, Huang, Zhiheng, Karpathy, Andrej, Khosla, Aditya, Bernstein, Michael, et al., 2015. Imagenet large scale visual recognition challenge. International Journal of Computer Vision 115 (3), 211–252.

Schuler, Karin Kipper, 2005. VerbNet: a broad-coverage, comprehensive verb lexicon. PhD thesis. Computer and Information Science Department, University of Pennsylvania, Philadelphia, PA.

Sigurdsson, Gunnar A., Russakovsky, Olga, Gupta, Abhinav, 2017. What actions are needed for understanding human actions in videos? In: Proceedings of the IEEE International Conference on Computer Vision, pp. 2137–2146.

Silberman, Nathan, Hoiem, Derek, Kohli, Pushmeet, Fergus, Rob, 2012. Indoor segmentation and support inference from rgbd images. In: European Conference on Computer Vision. Springer, pp. 746–760.

Simonyan, Karen, Zisserman, Andrew, 2014. Very deep convolutional networks for large-scale image recognition. arXiv preprint. arXiv:1409.1556.

Srikantha, Abhilash, Gall, Jürgen, 2016. Weakly supervised learning of affordances. arXiv preprint. arXiv:1605.02964.

Stark, Louise, Bowyer, Kevin, 1991. Achieving generalized object recognition through reasoning about association of function to structure. IEEE Transactions on Pattern Analysis and Machine Intelligence 13 (10), 1097–1104.

Summers-Stay, Douglas, Teo, Ching L., Yang, Yezhou, Fermüller, Cornelia, Aloimonos, Yiannis, 2012. Using a minimal action grammar for activity understanding in the real world. In: IEEE/RSJ International Conference on Intelligent Robots and Systems, pp. 4104–4111.

Teo, Ching Lik, Fermüller, Cornelia, Aloimonos, Yiannis, 2015. Detection and segmentation of 2d curved reflection symmetric structures. In: Proceedings of the IEEE International Conference on Computer Vision, pp. 1644–1652.

Tran, Son D., Davis, Larry S., 2008. Event modeling and recognition using Markov logic networks. In: European Conference on Computer Vision. Springer, pp. 610–623.

Ugur, Emre, Erhan, Oztop, Erol, Sahin, 2011. Goal emulation and planning in perceptual space using learned affordances. Robotics and Autonomous Systems 59 (7–8), 580–595.

Ugur, Emre, Piater, Justus, 2016. Emergent structuring of interdependent affordance learning tasks using intrinsic motivation and empirical feature selection. IEEE Transactions on Cognitive and Developmental Systems 9 (4), 328–340.

Varela, Francisco J., Rosch, Eleanor, Thompson, Evan, 1993. The Embodied Mind: Cognitive Science and Human Experience. MIT Press.

Wörgötter, Florentin, Erdal Aksoy, Eren, Krüger, Norbert, Piater, Justus, Ude, Ales, Tamosiunaite, Minija, 2013. A simple ontology of manipulation actions based on hand-object relations. IEEE Transactions on Autonomous Mental Development 5 (2), 117–134.

Wörgötter, Florentin, Ziaeetabar, F., Pfeiffer, S., Kaya, O., Kulvicius, T., Tamosiunaite, M., 2020. Humans predict action using grammar-like structures. Scientific Reports 10 (1), 1–11.

Xian, Yongqin, Lorenz, Tobias, Schiele, Bernt, Akata, Zeynep, 2018. Feature generating networks for zero-shot learning. In: Proceedings of the IEEE Conference on Computer Vision and Pattern Recognition, pp. 5542–5551.

Xiao, Jianxiong, Hays, James, Ehinger, Krista A., Oliva, A., Torralba, A., 2010. Sun database: large-scale scene recognition from abbey to zoo. IEEE Computer Society Conference on Computer Vision and Pattern Recognition, 3485–3492.

Yan, Sijie, Xiong, Yuanjun, Lin, Dahua, 2018. Spatial temporal graph convolutional networks for skeleton-based action recognition. In: Proceedings of the AAAI Conference on Artificial Intelligence, vol. 32.

Yang, Yezhou, Fermüller, Cornelia, Aloimonos, Yiannis, 2013. Detection of manipulation action consequences (MAC). In: Proceedings of the IEEE Conference on Computer Vision and Pattern Recognition, pp. 2563–2570.

Yang, Yezhou, Guha, Anupam, Fermüller, Cornelia, Aloimonos, Yiannis, 2014. A cognitive system for understanding human manipulation actions. Advances in Cognitive Systems 3, 67–86.

Yang, Yezhou, Fermüller, Cornelia, Li, Yi, Aloimonos, Yiannis, 2015a. Grasp type revisited: a modern perspective on a classical feature for vision. In: Proceedings of the IEEE Conference on Computer Vision and Pattern Recognition, pp. 400–408.

Yang, Yezhou, Li, Yi, Fermüller, Cornelia, Aloimonos, Yiannis, 2015b. Robot learning manipulation action plans by "watching" unconstrained videos from the world wide web. In: Proceedings of the AAAI Conference on Artificial Intelligence, vol. 29.

Ye, Chengxi, Yang, Yezhou, Mao, Ren, Fermüller, Cornelia, Aloimonos, Yiannis, 2017. What can I do around here? Deep functional scene understanding for cognitive robots. In: IEEE International Conference on Robotics and Automation (ICRA), pp. 4604–4611.

Yu, Xiaodong, Fermüller, Cornelia, Teo, Ching Lik, Yang, Yezhou, Aloimonos, Yiannis, 2011. Active scene recognition with vision and language. In: 2011 International Conference on Computer Vision, pp. 810–817.

Zampogiannis, Konstantinos, Fermüller, Cornelia, Cilantro, Yiannis Aloimonos, 2018. A lean, versatile, and efficient library for point cloud data processing. In: Proceedings of the 26th ACM International Conference on Multimedia, pp. 1364–1367.

Zampogiannis, Konstantinos, Fermüller, Cornelia, Aloimonos, Yiannis, 2019. Topology-aware non-rigid point cloud registration. IEEE Transactions on Pattern Analysis and Machine Intelligence.

Zampogiannis, Konstantinos, Yang, Yezhou, Fermüller, Cornelia, Aloimonos, Yiannis, 2015. Learning the spatial semantics of manipulation actions through preposition grounding. In: IEEE International Conference on Robotics and Automation (ICRA), pp. 1389–1396.

Zhang, Li, Xiang, Tao, Gong, Shaogang, 2017. Learning a deep embedding model for zero-shot learning. In: Proceedings of the IEEE Conference on Computer Vision and Pattern Recognition, pp. 2021–2030.

Zheng, J., Jiang, Z., Chellappa, R., 2016. Cross-view action recognition via transferable dictionary learning. IEEE Transactions on Image Processing 25 (6), 2542–2556.

Biographies

Cornelia Fermüller is a Research Scientist at the University of Maryland Institute for Advanced Computer Studies. She holds a Ph.D. from the Vienna University of Technology, Austria and an M.S. from the Graz University of Technology, both in Applied Mathematics. Her research interest has been to understand principles of active vision systems and develop biological-inspired methods, especially in the area of motion. Her recent work has focused on human action interpretation and the development of 3D motion algorithms for extreme conditions using event-based sensors.

Michael Maynord is a PhD candidate in the department of Computer Science at the University of Maryland College Park, advised by Yiannis Aloimonos and Cornelia Fermüller. His background encompasses symbolic Artificial Intelligence, including cognitive architectures, Computer Vision, including action understanding, and methods integrating AI and CV.

Self-supervised temporal event segmentation inspired by cognitive theories

Ramy Mounir[a], *Sathyanarayanan Aakur*[b], *and Sudeep Sarkar*[a]

[a]Computer Science and Engineering, University of South Florida, Tampa, FL, United States
[b]Computer Science, Oklahoma State University, Stillwater, OK, United States

CHAPTERPOINTS

- We can use cognitive science theories in event segmentation to design highly effective computer vision algorithms for Spatio-temporal segmentation of events in videos that do not require any labeled data.

- We discuss three perceptual prediction models from EST in three progressive versions: temporal segmentation using perceptual prediction framework, temporal segmentation along with event working models based on attention maps, and finally spatial and temporal localization of events.

- The approaches can learn robust event representations from only a single-pass through an unlabeled streaming video.

- They show state-of-the-art performance in unsupervised temporal segmentation and spatial-temporal action localization while offering competitive performance with fully supervised baselines that require extensive amounts of annotation.

Advanced Methods and Deep Learning in Computer Vision
https://doi.org/10.1016/B978-0-12-822109-9.00021-7

Copyright © 2022 Elsevier Inc. All rights reserved.

12.1 Introduction

How do we detect and segment events? How do we represent events? How do we perceive events? And, more importantly, what is an event? In computer vision research, the terms "actions", "activities", and "events" are often conflated. Most works lump these terms to represent "some activity happening" in the scene involving objects and actors labeled with a phrase, i.e., jumping, chopping an onion, changing tires, cooking a meal, and so on. There is no clear distinction among activity recognition, action recognition, or event recognition in existing computer vision literature. There is also not much clarity into the nature of events. On the contrary, event perception is a mature field of cognitive science research (Radvansky and Zacks, 2014; Shipley and Zacks, 2008; Richmond and Zacks, 2017). We will start by summarizing some of the cognitive science findings that we use as inspiration to build the solutions outlined in this chapter. Ideally, one should read the references cited and not just rely on our summary, which is just a brief introduction to a much richer body of work. We will highlight some ideas that we use to construct computer vision solutions to perform self-supervised temporal event segmentation.

Events are the central units of our experiences. The brain receives a continuous stream of sensory input, both from the outside world and inside the body, and segments them into discrete units or packet representations called events. Each event is indicative of key moments in the input sensory stream. Hence, an event is defined as *"a segment of time at a given location that is perceived by an observer to have a beginning and an end"* (Zacks and Tversky, 2001). Note that this definition of events makes it distinct from activities. For example, cooking is an activity, but "cooking a meal" is an event. Just like a story, the latter has a beginning, a middle, and an end. Events can be of different types and duration, depending on the agent and acted upon environments. Some events are short, e.g., making a bed. Some events are extended, e.g., a cricket match. Events involving animate agents, such as humans and animals, are often goal-directed. While the goal of these events may not always be immediately visible to the observer, it exists as the purpose that guides the event. There are also events that do not involve animate agents nor have goals, such as natural events.

There is substantial support for the notion that the human cognition uses *"structured representations of events, called event models, to capture information about the Spatio-temporal framework, entities and objects, and other salient features of a situation"* (Richmond and Zacks, 2017). These event models are a compositional representation of the event and its constituent elements and have partonomies, i.e., hierarchies formed by the "part of" relation. Fig. 12.1 shows an example of a "make sandwich" event and its hierarchical structure. At the lowest level of the hierarchy are elementary actions, such as "carry knife" and "open lid". These actions are part of longer activity units, such as "cut bun", which are in turn part of the "make sandwich" event. This partonomy is analogous to that for objects, which can also be described as a composition of individual parts.

Just like for objects parts, that have boundaries, events have segmentation boundaries at multiple timescales. Event segmentation and the grouping of the segments into a hierarchy is a continuous activity that happens simultaneously at multiple timescales. Evidence from neuroscience experiments (Zacks et al., 2001a) indicates that the posterior temporal, parietal cortex, and lateral frontal cortex become active during event boundaries. Some experiments (Kurby and Zacks, 2008) show that event segmentation plays a significant role in the

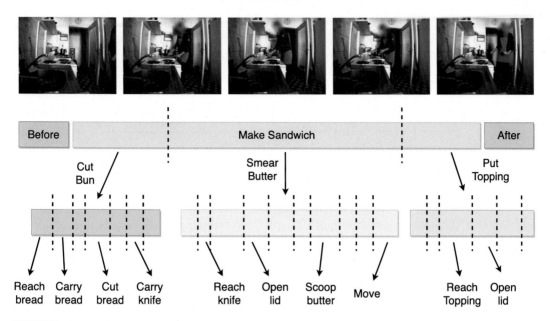

FIGURE 12.1 The events hierarchy consists of multiple levels of segmentation. The higher-level event "Make Sandwich" can be segmented into lower-level events "Cut Bun", "Smear Butter", and "Put Topping". Each lower-level event can be further segmented into its constituent events at a lower level. The event "Make Sandwich" can be part of an even higher level event "Have Breakfast". This compositional relationship between events forms "the events hierarchy". Images are from the Breakfast Actions Dataset (Kuehne et al., 2014).

core functions of cognitive control and memory encoding. However, the process of event segmentation does not require conscious attention. Event segmentation can be processed purely with bottom-up, signal-based features extracted from motion and appearance. This conclusion was based on a movie experiment by Zacks et al. (2001a), Zacks and Magliano (2011). A set of participants were shown movie clips of events and were asked to mark the boundaries of events using a clicker as they watched. A different set of participants were asked to do the same but were shown the movie in reverse! The boundaries marked by both groups of participants were remarkably similar. Reversing movie direction prevents viewers from relying on familiar schemas to interpret events, making high-level information – such as goal attainments – difficult to identify (Hard et al., 2006). This is evidence for the claim that high-level information about the event is not necessary to segment events. It also challenges us to design computer vision systems that do not need prior training labels for event segmentation, i.e., to design unsupervised event segmentation algorithms in a continual online manner.

In this chapter, we consider the event segmentation problem, i.e., how do we mark the temporal boundaries (when one event ends, and another begins Zacks et al., 2001b) of the event *and* spatially locate them in the image. Building the event model and the partonomic hierarchy is a separate process that we do not consider here. First, we will introduce the Event Segmentation Theory (EST) model based on the perceptual prediction model to compute the event boundaries proposed by Zacks et al. (2014). Then we present our computer vision solution based on the perceptual prediction model from EST in three progressive versions:

temporal segmentation using a perceptual prediction framework, temporal segmentation along with event working models based on attention maps, and finally spatial and temporal localization of events. We end with a discussion of other solutions for event segmentation in the literature and how they relate to ours.

12.2 The event segmentation theory from cognitive science

Event Segmentation Theory (EST), developed by Zacks et al. (2007) based on cognitive neuroscience experiments, posits that humans maintain a stable representation of "what is happening now" that is updated based on transient increases in the *perceptual prediction error*. This representation of the current event is called the working *event model* and is used to anticipate the next incoming sensory input. This process of making predictions to anticipate the next input is the key element of this theory; predictions play a central role in building event representations.

Our brains are constantly making predictions of what features our senses expect to see next. When we observe someone cooking, we continuously make predictions. These predictions are made at different temporal granularities. We predict the trajectory of motion at fine temporal scales to anticipate hand location in the next frame. And, at coarse scales to anticipate what utensils could be used next. Indeed, there is a definition of artificial intelligence that puts the ability to make predictions as the central feature (Hawkins and Blakeslee, 2004). The extent to which an agent can be considered intelligent is determined by the temporal window and spatial extent to which it can predict with precision. Thus, a small insect can make predictions of its immediate surrounding and its near future. Humans, armed with high-level scientific reasoning and logic, can predict events over much larger space and longer temporal windows. The content of the prediction depends on the task at hand. In our current case, it is about making predictions about aspects of a visual event.

The error in the prediction drives the segmentation process to group events into discrete time intervals. As depicted in Fig. 12.2, EST is a continual perceptual process that segments a continuous stream of multimodal sensory input into a discrete and coherent set of events, which can be further segmented to form a hierarchy of events at multiple timescales. The process of event segmentation does not require conscious attention. Instead, it emerges as a side effect of the ongoing perceptual prediction process.

As sensory inputs are perceived, a perceptual processing unit receives, filters, and encodes the incoming features to useful higher-level representations. Computationally, the perceptual processing unit could be represented by a deep learning stack to process and extract features. A key component of the perceptual processing in EST is that the encoded features are conditioned upon on the representations from the working event model. This conditioning makes the feature representations robust to small variations from one instant to another. The working event model guides the perceptual processing to extract relevant features for the event being observed. Working event models are very specific and limited in capacity; they get updated at event boundaries.

The perceptual processing unit sends the extracted features (conditioned on the working event model) to a prediction unit to anticipate future perceptual features. Any mismatch in

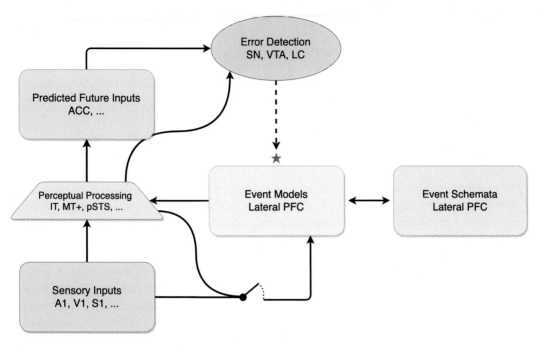

FIGURE 12.2 Information flow, according to the Event Segmentation Theory (EST), adapted from Zacks et al. (2007). The letter acronyms refer to the different brain regions where these activities are happening, as discovered by cognitive neuroscience experiments. The "Perceptual Processing" unit uses "Sensory Inputs" to extract higher-level features relevant to the predictive task. The extracted features are used to predict future perceptual features and compared to the actual future perceptual features. The "Error Detection" module computes the prediction error, which generates a reset signal (denoted by ⋆) at high prediction errors. The reset signal updates the "Event Model" by accepting input from the "Sensory Inputs" and "Perceptual Processing" blocks.

the predicted and actual features for the next time instance is monitored continuously and is termed the *prediction error*. The prediction error is a measure of the quality of predictions and hence an indicator of the suitability of the working event model. A transient increase in the prediction error is an indicator of an event boundary, i.e., the current event could be changing. A new event model is needed to make future predictions. Thus, the prediction error works as a gating mechanism to update the working model with an updated representation of the new event. This update of the working event model involves integrating sensory input and prior expectations from long-term memory about the next event. This long-term event memory is referred to as a *event schemata*, which encodes a more stable representation of the event compared to the working event model. Event schemata store sequential information in terms of distinctive physical features about objects and actors, likely next events, and actors' goals, all of which are learned from past occurrences of the event. The changes in the event schemata happen at a slower learning rate than the working event memory.

The prediction quality depends on the accuracy of the working event model in representing what is being perceived. Typically, the working model is a good fit, and the prediction error is low. However, from time to time, the current observations can become less predictable, resulting in an increase in the prediction error and the need for an update in the working event model. In the EST framework, this update is mediated via a gating mechanism, a function of the error signal. When the error signal increases, the working event model is updated based on the sensory signal and the event schemata information until the error reduces. Thus, the overall system operates mostly at a stable state with low prediction errors and transient periods of high errors signaling the event boundaries.

12.3 Version 1: single-pass temporal segmentation using prediction

The event segmentation theory offers us a mechanism for temporal segmentation of events, i.e., breaking a video into parts, without the need for training labels and in a continual, online manner. In this section, we demonstrate the power of EST using a simple incarnation of the EST that outperforms more complex state-of-the-art supervised approaches with only a single pass through the video.

Computationally, we implement the EST framework using well-known deep learning components. The perceptual prediction block is implemented as an encoder module, based on convolutional neural networks (CNNs) (LeCun et al., 1995). A Long Short-Term Memory (LSTM) cell (Hochreiter and Schmidhuber, 1997) is used to aggregate the features temporally and predict future perceptual features. The LSTM's inner cell structure offers an ideal implementation alternative for perceptual feature prediction for two reasons. First, the LSTM's recurrent nature allows us to integrate over past frames to predict the future perceptual frames. Second, the hidden state of the LSTM acts as an internal event model that can be used to condition the extracted features on a stable event representation built from previous frames. An adaptive learning mechanism provides a simple and practical method for implementing the gating mechanism used to update the event model based on the perceptual prediction error.

In Aakur and Sarkar (2019), we focused on building representations for the working event model and did not implement the event schemata block. We begin with a discussion on frame encoding and feature extraction, in Section 12.3.1, followed by an explanation of how we utilize a recurrent cell to compute a stable representation of previous frames (Section 12.3.2). We also briefly discuss the role of the reconstruction layer in converting the predicted representation to future perceptual features in Section 12.3.3. We introduce the gating mechanism and adaptive learning functions for boundary detection in Sections 12.3.5 and 12.3.6. Algorithm 1 shows the pseudocode for the perceptual prediction framework.

At the core of the approach, illustrated in Fig. 12.3, is a predictive processing platform that encodes a visual input I_t into a higher-level abstraction I'_t using an encoder network. The abstracted feature is used as a prior for predicting the feature I'_{t+1} at time $t + 1$. The future prediction module, LSTM, combines the extracted features with a stable representation (hidden state) of the event based on previous frames to predict a future perceptual representation. The reconstruction or decoder network transforms the predicted representation into actual features (same dimensions as I_{t+1}), which are used to detect the event boundaries between successive activities in streaming, input video.

Algorithm 1 Temporal Event Segmentation Model. The input is an untrimmed/streaming video $\mathbb{I}$, which is a set of frames $\{I_1, ..., I_t, I_{t+1}, ...I_T\}$. The output is a set of event boundaries $\{b_1, b_2, ...b_{T-1}\}$.

Input: Video frames $\{I_1, ..., I_t, I_{t+1}, ...I_T\} \in \mathbb{R}^{TxCxWxH}$
Output: Event boundary values $\mathbb{B} = \{b_1, b_2, ...b_{T-1}\}$

 1: **procedure** LSTM(h_{t-1}, I'_t)
 2:     $i_t \leftarrow \sigma(W_i I'_t + W_{hi} h_{t-1} + b_i)$ $\triangleright$ Input gate
 3:     $f_t \leftarrow \sigma(W_f I'_t + W_{hf} h_{t-1} + b_f)$ $\triangleright$ Forget gate
 4:     $o_t \leftarrow \sigma(W_o I'_t + W_{ho} h_{t-1} + b_o)$ $\triangleright$ Output gate
 5:     $g_t \leftarrow \phi(W_g I'_t + W_{hg} h_{t-1} + b_g)$
 6:     $m_t \leftarrow f_t \cdot m_{t-1} + i_t \cdot g_t$
 7:     $h_t \leftarrow o_t \cdot \phi(m_t)$
 8:     **return** h_t
 9: **end procedure**

10: **procedure** GATE$(E_P(t))$
11:     $P_q(t) = P_q(t-1) + \frac{1}{n}(E_P(t) - P_q(t-1))$ $\triangleright$ Running average
12:     **if** $\frac{E_P(t)}{P_q(t-1)} > \psi_e$ **then**
13:         $P_q(t-1) \leftarrow P_q(t)$
14:         **return** $True$
15:     **else**
16:         $P_q(t-1) \leftarrow P_q(t)$
17:         **return** $False$
18:     **end if**
19: **end procedure**

20: **procedure** SEGMENT(I_t, I_{t+1}, h_{t-1}) $\triangleright$ Main segmentation layer
21:     $I'_t \leftarrow$ ENCODER(I'_t) $\triangleright$ Basic CNN encoder
22:     $I'_{t+1} \leftarrow$ ENCODER(I'_{t+1}) $\triangleright$ Basic CNN encoder
23:     $h_t \leftarrow$ LSTM(h_{t-1}, I'_t)
24:     $y_{t+1} \leftarrow$ DECODER(h_t) $\triangleright$ Single dense layer
25:     $E_P(t) \leftarrow \sum_{i=1}^{n} ||I'_{t+1} - y'_{t+1}||^2_{\ell_1}$
26:     $b_t \leftarrow$ GATE$(E_P(t))$
27:     **return** h_t, b_t
28: **end procedure**

29: $h_t \leftarrow 0$
30: **for** $\{I_t, I_{t+1}\} \in \{I_1, I_2\}, \{I_2, I_3\}, ..., \{I_{T-1}, I_T\}$ **do**
31:     $h_t, b_t \leftarrow$ SEGMENT(I_t, I_{t+1}, h_t)
32:     $\mathbb{B}.\text{append}(b_t)$
33: **end for**

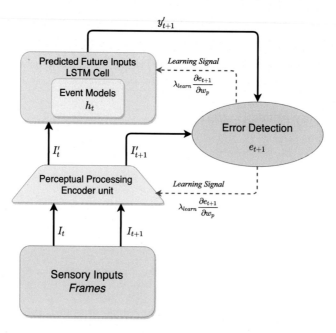

FIGURE 12.3 **Version 1 – Single-pass Temporal Segmentation using Perceptual Prediction:** The model architecture is a stripped down version of the blocks proposed for the EST. It consists of four essential components: an encoder network, a predicting unit, a decoding network, an error detection and boundary decision unit.

12.3.1 Feature extraction and encoding

We encode the input frame at each time step into abstracted, higher-level visual features and use the features as a basis for perceptual processing rather than the raw input at the pixel level (for reduced network complexity) or higher-level semantics (which require training data in the form of labels). An optimized encoder only extracts features relevant to the task being solved, in this case, prediction. The encoding process entails learning a function $g(I(t), \omega_e)$ that transforms an input frame I_t from pixel space into a higher dimensional feature space I'_t.

The transformation function $g(I(t), \omega_e)$ extracts relevant spatial features into a condensed vector representation using a set of learnable parameters ω_e. In practice, a pretrained backbone architecture can be used here for encoding raw pixels to useful features. In this work, we use a VGG16 (Simonyan and Zisserman, 2014) CNN model pretrained on ImageNet (Russakovsky et al., 2015) as the transformation function $g(\cdot)$. The pretrained weights are used to provide good initialization parameters but are further finetuned using the prediction error.

12.3.2 Recurrent prediction for feature forecasting

Given the perceptual features at time t (I'_t), the next step is the prediction of the perceptual features at time $t + 1$. The predicted features are a function of the extracted features I'_t, and an internal, working model of the current event. The internal model modulates the input sensory signal at every frame, similar to the conditioning of the perceptual processing unit on the event model in Section 12.2. Formally, this can be defined by a generative model $P(I'_{t+1}|\omega_p, I_t)$, where ω_p is the set of hidden parameters characterizing the internal state of the current observed event.

To capture the temporal dependencies among *intra*-event frames and *inter*-event frames, we use a Long Short Term Memory Network (LSTM) (Hochreiter and Schmidhuber, 1997). The LSTM model is mathematically expressed in Algorithm 1, where σ is a nonlinear activation function, the dot-operator ($\cdot$) represents Hadamard (element-wise) multiplication, ϕ is the hyperbolic tangent function (*tanh*) and W_x and b_x represent the trained weights and biases for each of the gates. Collectively, $\{W_{hi}, W_{hf}, W_{ho}, W_{hg}\}$ and their respective biases constitute the learnable parameters ω_p.

The LSTM cell, formally defined in Algorithm 1, consists of three main gates; the input gate i_t, forget gate f_t and output gate o_t. The forget and input (combined with g_t memory layer) gates work together to update the internal event model as specified by their learnable parameters. The output gate processes the input signal by conditioning frame-wise perceptual input on the current event's internal memory. It is worth noting that other recurrent models, such as a Gated Recurrent Unit (GRU), could also be used.

The event state h_t is a representation of the event observed at time instant t and hence is more sensitive to the observed input $I'(t)$ than the event layer, which is more persistent across events. The event layer is a gated layer, which receives input from the encoder and the recurrent event model. However, the inputs to the event layer are modulated by a self-supervised gating signal (Section 12.3.5), which is indicative of the quality of predictions made by the recurrent model. The gating allows for updating the weights quickly but also maintains a coherent state within the event.

12.3.3 Feature reconstruction

The goal of the perceptual processing unit (or rather the reconstruction network) is to reconstruct the predicted feature y'_{t+1} given a source prediction h_t, which maximizes the probability

$$p(y'_{t+1}|h_t) \propto p(h_t|y'_{t+1})\, p(y'_{t+1}) \tag{12.1}$$

where the first term is the likelihood and the second is the feature prior model. However, we model $\log p(y'_{t+1}|h_t)$ as a log-linear model $f(\cdot)$ conditioned upon the weights of the recurrent model ω_p and the observed feature I'_t and characterized by

$$\log p(y'_{t+1}|h_t) = \sum_{n=1}^{t} f(\omega_p, I'_t) + \log Z(h_t) \tag{12.2}$$

where $Z(h_t)$ is a normalization constant that does not depend on the weights ω_p and is used to provide a more concrete estimate of the probability to account for uncertainty. In practice, it is ignored as the predictive learning provides necessary regularization. The reconstruction model completes the generative process for forecasting the feature at time $t+1$ and helps construct the self-supervised learning setting to identify event boundaries.

12.3.4 Self-supervised loss function

One distinctive property of features within the same event is that they are predictable given previous perceptual inputs and a stable event model. We rely on this property to detect event boundaries. Therefore, we can define a prediction error function that receives, as input, the model's prediction y'_t, and the actual future sensory features I'_t as targets. The resulting error, termed as the perceptual prediction error, is calculated as shown in Eq. (12.3).

$$E_P(t) = \sum_{i=1}^{n} \|I'_t - y'_t\|^2_{\ell_1} \tag{12.3}$$

The perceptual prediction error is reasonably indicative of the relevance of the recurrent model's internal state to the actual event being observed. When the event changes by crossing an event boundary, the internal model becomes unusable, leading to incorrect predictions. The prediction quality increases after updating the internal model with a newer representation capable of predicting future features. Fig. 12.4 illustrates an instance of this effect. The minimization of the perceptual prediction error serves as the objective function for training the network.

12.3.5 Error gating

According to the Event Segmentation Theory, the perceptual features can become highly unpredictable at event boundaries since the working event model will need to be updated to process the new event. For example, in Fig. 12.4, we can see that the visual representation of the features learned by the encoder network for the activities *take bowl* and *crack eggs* are closer together than the features between the activities *take bowl* and *spoon flour*. At event boundaries, the diverging feature space becomes inexplicable by the recurrent internal model, causing a transient increase in the perceptual prediction error. We see that the perceptual prediction error (second plot from the bottom) matches well with the ground truth segmentation (second from the top) for the video "Make Pancake". As illustrated, the error rates are higher at the event boundaries and lower within the "in-event" frames.

To model the prediction error and identify event boundaries, we can use a low pass filter to maintain a running average of the perceptual prediction error made over the last n input time steps. An n value of 5 is used here based on the average human perception response time (≈ 200 ms) (Thorpe et al., 1996). We maintain a running average of the prediction error, called the prediction quality and given by

$$P_q(t) = P_q(t-1) + \frac{1}{n}\big(E_P(t) - P_q(t-1)\big) \tag{12.4}$$

where P_q is the prediction quality and $E_P(t)$ is the prediction error from Eq. (12.3). A gating signal, $G(t)$, is triggered when the current prediction error exceeds the average prediction quality metric by at least 50%. Formally, we can define the gating function as

$$G(t) = \begin{cases} 1, & \frac{E_P(t)}{P_q(t-1)} > \psi_e \\ 0, & \text{otherwise} \end{cases} \tag{12.5}$$

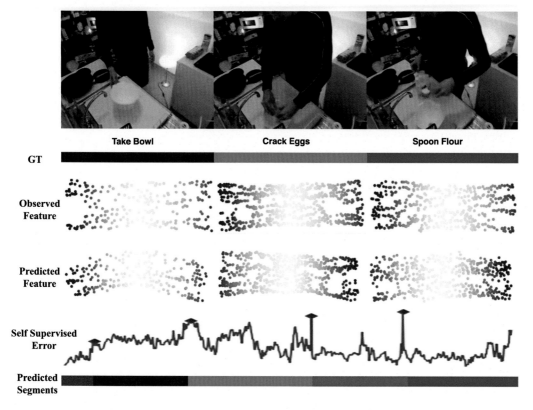

FIGURE 12.4 A visualization of the predictive learning process in the Event Segmentation Theory is shown here. The current sensory inputs are abstracted into features of lower variability that are conducive to prediction. A working event model is constructed and is used to continuously predict the features observed at the next time step. Predictions are compared continuously to observed features, and the resulting prediction error serves as an indicator of the suitability of the event model. A gating mechanism, a function of the prediction error, modulates the learning process, and provides cues for event segmentation. ♦ refers to predicted event boundaries. The features were visualized using T-SNE (van der Maaten and Hinton, 2008) for presentation. Reproduced with permission from Aakur and Sarkar (2019).

where $E_P(t)$ is the perceptual prediction error at time t, $G(t)$ is the value of the gating signal at time t, $P_q(t-1)$ is the average prediction quality metric at time t and Ψ_e is the prediction error threshold for boundary detection. For optimal prediction, the perceptual prediction error would be very high at the event boundary frames and very low at all within-event frames. Ψ_e is a hyperparameter which can be used to set the timescale at which we will be detecting event boundaries.

12.3.6 Adaptive learning for plasticity

Learning a robust event representation is at the core of event segmentation approaches using predictive learning. An event representation is considered to be robust when the perceptual prediction error is low for *intra*-event frames and high at *inter*-event frames. If the learned event representations *overfit* to intra-event observations, then minor perturbations in the raw pixel space can increase the prediction error. This would negate the underlying assumption that transient errors indicate a change in the observed event. Additionally, the event representation must be stable across events with varying temporal duration to avoid catastrophic forgetting, i.e., the condition where the predictions do not capture intra-event variations in long event sequences. Hence, it is necessary to ensure that the model does not overfit to short-term perceptual features while maintaining a robust event representation representing the *entire* event.

To allow for some plasticity and avoid catastrophic forgetting in the network, we use adaptive learning. Adaptive learning is similar to the learning rate schedule, a commonly used technique for training deep neural networks. However, instead of using predetermined intervals for changing the learning rates, we use the prediction error to modulate the learning rate. The learning rate can be tuned to control the propagation of the error back to the learnable parameters.

For example, when the perceptual prediction rate is lower than the average prediction rate, the predictor model is considered a good, stable representation of the current event. Propagating the prediction error when there is a good representation of the event can lead to overfitting of the predictor model to that particular event and does not help generalize. Hence, lower learning rates are used for time steps when there are negligible prediction error and a relatively higher (by a magnitude of 100) for higher prediction error.

Intuitively, this adaptive learning rate allows the model to adapt much quicker to new events (at event boundaries where there are likely to be higher errors) and learn to maintain the internal representation for within-event frames. Formally, the learning rate is defined as the result of the adaptive learning rule described as a function of the perceptual prediction error defined in Section 12.3.4 and is defined as

$$\lambda_{learn} = \begin{cases} \Delta_t^- \lambda_{init}, & E_P(t) > \mu_e \\ \Delta_t^+ \lambda_{init}, & E_P(t) < \mu_e \\ \lambda_{init}, & \text{otherwise} \end{cases} \quad (12.6)$$

where Δ_t^-, Δ_t^+ and λ_{init} refer to the scaling of the learning rate in the negative direction, positive direction and the initial learning rate respectively and $\mu_e = \frac{1}{t_2 - t_1} \int_{t_1}^{t_2} E_P \, dE_P$. The learning rate is adjusted based on the quality of the predictions characterized by the perceptual prediction error between a temporal sequence between times t_1 and t_2, typically defined by the gating signal.

12.3.7 Results

12.3.7.1 Datasets

We evaluate and analyze the performance of the perceptual prediction framework on three large, publicly available datasets – Breakfast Actions (Kuehne et al., 2014), INRIA Instructional Videos dataset (Alayrac et al., 2016) and the 50 Salads dataset (Stein and McKenna, 2013). Each dataset offers a different challenge to the approach allowing us to evaluate its performance on various challenging conditions.

Breakfast actions dataset is a large collection of 1712 videos of 10 breakfast activities performed by 52 actors. Each activity consists of multiple subactivities that possess visual and temporal variations according to the subject's preferences and style. Varying qualities of visual data and complexities such as occlusions and viewpoints increase the complexity of the temporal segmentation task.

INRIA instructional videos dataset contains 150 videos of 5 different activities collected from YouTube. Each of the videos is, on average, 2 minutes long and has around 47 subactivities. A "background" class denotes a sequence where there does not exist a clear subactivity that is visually discriminable. This offers a considerable challenge for approaches that are not explicitly trained for such visual features.

50 Salads dataset is a multimodal dataset collected in the cooking domain. The dataset contains over four (4) hours of annotated data of 25 people preparing two mixed salads each. It provides data in different modalities such as RGB frames, depth maps, and accelerometer data for devices attached to different items such as knives, spoons, and bottles, to name a few. The annotations of activities are provided at different levels of granularities – high, low, and eval. We use the "eval" granularity following evaluation protocols in prior works (Lea et al., 2016, 2017).

12.3.7.2 Evaluation metrics

We use two commonly used evaluation metrics for analyzing the performance of this approach. We use the same evaluation protocol and code as in Alayrac et al. (2016); Sener and Yao (2018). We utilize the Hungarian matching algorithm to obtain the one-to-one mappings between the predicted segments and the ground truth to evaluate the performance due to the unsupervised nature of the predictive approach. We use the mean over frames (MoF) to evaluate the network's ability to temporally localize the subactivities. We evaluate the divergence of the predicted segments from the ground truth segmentation using the Jaccard index (Intersection over Union or IoU). We also use the F1 score to evaluate the quality of the temporal segmentation. The evaluation protocol for the recognition task in Section 12.3.7.4.1 is the unit level accuracy for the 48 classes as seen in Table 3 from Kuehne et al. (2014) and compared in Kuehne et al. (2014); Aakur et al. (2019); de Souza et al. (2016); Huang et al. (2016).

12.3.7.3 Ablative studies

We evaluate different variations of our framework to compare the effectiveness of each component. We varied the prediction history n, and the prediction error threshold Ψ. Increasing frame window tends to merge frames and smaller clusters near the event boundaries to the prior activity class due to a transient increase in error. This results in a higher IoU and a lower MoF. Low error threshold results in oversegmentation as boundary detection becomes sensitive to small changes. The number of predicted clusters decreases as the window size and threshold increases. We also trained four models (Fig. 12.5), with different predictor units. We trained two recurrent neural networks (RNN) as the predictor units with and without adaptive learning (AL) described in Section 12.3.6 indicated as *RNN + No AL* and *RNN + AL*, respectively. We also trained LSTM without adaptive learning (*LSTM + No AL*) to compare against our main model (*LSTM + AL*). We use RNNs as a possible alternative due to the short-term future predictions (1 frame ahead) required.

FIGURE 12.5 Ablative Studies: Illustrative comparison of variations in the model architecture, using RNNs and LSTMs with and without adaptive learning on the INRIA Instructional Videos Dataset on a video with ground truth *Change Tire*. It can be seen that complex visual scenes with activities of shorter duration pose a significant challenge to the framework and cause fragmentation and oversegmentation. However, the use of adaptive learning helps alleviate this to some extent. Note: Temporal segmentation timelines are shown without the background class for better visualization.

12.3.7.4 *Quantitative evaluation*

Breakfast actions dataset We evaluate the performance of our full model $LSTM + AL$ on the breakfast actions dataset and compare against fully supervised, weakly supervised, and unsupervised approaches. We show the performance of the SVM (Kuehne et al., 2014) approach to highlight the importance of temporal modeling. As shown in Table 12.1, the perceptual prediction approach outperformed all other unsupervised and all weakly supervised approaches and some fully supervised approaches.

TABLE 12.1 Segmentation Results on the Breakfast Action dataset. MoF refers to the Mean over Frames metric and IoU is the Intersection over Union metric.

Supervision	Approach	MoF	IoU
Full	SVM (Kuehne et al., 2014)	15.8	–
	HTK(64) (Kuehne et al., 2016)	56.3	–
	ED-TCN (Lea et al., 2017)	43.3	42.0
	TCFPN (Ding and Xu, 2018)	52.0	**54.9**
	GRU (Richard et al., 2017)	**60.6**	–
Weak	OCDC (Bojanowski et al., 2014)	8.9	23.4
	ECTC (Huang et al., 2016)	27.7	–
	Fine2Coarse (Richard and Gall, 2016)	33.3	**47.3**
	TCFPN + ISBA (Ding and Xu, 2018)	**38.4**	40.6
None	KNN+GMM (Sener and Yao, 2018)	34.6	**47.1**
	Ours (LSTM + AL)	**42.9**	46.9

It should be noted that the other unsupervised approach (Sener and Yao, 2018) requires the number of clusters (from ground truth) to achieve the reported performance. In contrast, our approach does not require such knowledge and is done in a streaming fashion. Additionally, the weakly supervised methods (Huang et al., 2016; Richard and Gall, 2016; Ding and Xu, 2018) require both the number of actions and an ordered list of sub-activities as input. ECTC (Huang et al., 2016) is based on discriminative clustering, while OCDC (Bojanowski et al., 2014) and Fine2Coarse (Richard and Gall, 2016) are also RNN-based methods.

50 Salads dataset We also evaluate our approach on the 50 Salads dataset, using only the visual features as input. We report the Mean of Frames (MoF) metric for a fair comparison. As can be seen from Table 12.2, the predictive approach significantly outperforms the other unsupervised approach, improving by 6.6% on the MoF metric. We also show the performance of the frame-based classification approaches VGG and IDT (Lea et al., 2016) to demonstrate the effectiveness of temporal modeling.

TABLE 12.2 Segmentation Results on the 50 Salads dataset, at granularity 'Eval'. **Models were intentionally reported without temporal constraints for ablative studies.

Supervision	Approach	MoF
Full	VGG** (Lea et al., 2016)	7.6%
	IDT** (Lea et al., 2016)	54.3%
	S-CNN + LSTM (Lea et al., 2016)	66.6%
	TDRN (Lei and Todorovic, 2018)	68.1%
	ST-CNN + Seg (Lea et al., 2016)	72.0%
	TCN (Lea et al., 2017)	**73.4%**
None	LSTM + KNN (Bhatnagar et al., 2017)	54.0%
	Ours (LSTM + AL)	**60.6%**

It should be noted that the fully supervised approaches require significantly more training data – both in the form of labels as well as training epochs. Additionally, the TCN approach (Lea et al., 2017) uses accelerometer data to achieve the state-of-the-art performance of 73.4%

INRIA instructional videos dataset: Finally, we evaluate our approach on the INRIA Instructional Videos dataset, which posed a significant challenge in the form of high amounts of background (noise) data. We report the F1 score for a fair comparison to the other state-of-the-art approaches. As can be seen from Table 12.3, the predictive model outperforms the other unsupervised approach (Sener and Yao, 2018) by 7.5%, the weakly supervised approach (Bojanowski et al., 2014) by 7.9% and has a competitive performance to the fully supervised approaches (Malmaud et al., 2015; Alayrac et al., 2016; Sener and Yao, 2018).

We also evaluate the performance of the models with and without adaptive learning. The effectiveness of LSTMs in capturing long-term temporal dependencies is significant, as can be seen in Table 12.3, primarily due to the long duration of activities in the dataset. Additionally, adaptive learning has a significant improvement in the segmentation framework, improving the performance by 9% and 11% for the RNN-based model and the LSTM-based model, respectively, indicating reduced overfitting of the model to the visual data.

TABLE 12.3 Segmentation Results on the INRIA Instructional Videos dataset. We report F1 score for fair comparison.

Supervision	Approach	F1
Full	HMM + Text (Malmaud et al., 2015)	22.9%
	Discriminative Clustering (Alayrac et al., 2016)	41.4%
	KNN+GMM (Sener and Yao, 2018) + GT	**69.2%**
Weak	OCDC + Text Features (Bojanowski et al., 2014)	28.9%
	OCDC (Bojanowski et al., 2014)	**31.8%**
None	KNN+GMM (Sener and Yao, 2018)	32.2%
	Ours (RNN + No AL)	25.9%
	Ours (RNN + AL)	29.4%
	Ours (LSTM + No AL)	36.4%
	Ours (LSTM + AL)	**39.7%**

TABLE 12.4 Activity recognition results on Breakfast Actions dataset. HCF and AL refer to handcrafted features and Adaptive Learning, respectively.

Approach	Precision
HCF + HMM (Kuehne et al., 2014)	14.90%
HCF + CFG + HMM (Kuehne et al., 2014)	31.8%
RNN + ECTC (Huang et al., 2016)	35.6%
RNN + ECTC (Cosine) (Huang et al., 2016)	36.7%
HCF + Pattern Theory (de Souza et al., 2016)	38.6%
HCF + Pattern Theory + ConceptNet (Aakur et al., 2019)	42.9%
VGG16 + LSTM	33.54%
VGG16 + LSTM + Predictive Features(AL)	37.87%

12.3.7.4.1 Improved features for action recognition

To evaluate the network's ability to learn highly discriminative features for recognition, we evaluated the performance of the predictive approach in a recognition task. We pretrain the model for temporal segmentation using the Breakfast Actions dataset and use the hidden layer of the LSTM as input to a fully connected layer to minimize a cross-entropy loss for training. We also trained another network with the same structure – VGG16 + LSTM – without pretraining to show the usefulness of the learned features using self-supervision.

As can be seen from Table 12.4, using self-supervision to pretrain the network before the recognition task improves the recognition performance of the network and has yielded comparable performance to the other state-of-the-art approaches. It improves the recognition accuracy by 4.3% over the network without predictive pretraining.

12.3.7.5 Qualitative evaluation

Through the predictive, self-supervised framework, we can learn the sequence of visual features in streaming video. We visualize the model's segmentation performance on the Breakfast Actions Dataset in Fig. 12.6. It can be seen that the predictive framework has high temporal coherence and does not suffer from oversegmentation, especially when the segments are long. Long activity sequences allow the model to learn from observation by providing more "*intra*-event" samples. Additionally, weakly supervised approaches like OCDC (Bojanowski et al., 2014) and ECTC (Huang et al., 2016) suffer from oversegmentation and intra-class fragmentation. This could arguably be attributed to the fact that they tend to enforce semantics in the form of a weak ordering of activities in the video regardless of the changes in visual features. Fully supervised approaches – such as HTK (Kuehne et al., 2016) – perform better, especially due to their ability to assign semantics to visual features. However, they are also affected by unbalanced data and dataset shift, as can be seen in Fig. 12.6 where the background class was segmented into other classes.

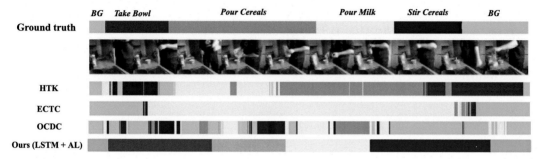

FIGURE 12.6 Illustration of the segmentation performance of main model on the Breakfast Actions Dataset on a video with ground truth *Make Cereals*. The predictive approach does not show the tendency to oversegment and provides coherent segmentation. The approach, however, shows a tendency to take longer to detect boundaries for visually similar activities.

We also qualitatively evaluate the impact of adaptive learning and long term temporal memory in Fig. 12.5, where the performance of the alternative methods described in Section 12.2. It can be seen that the use of adaptive learning during training prevents the model from overfitting to any single class's intra-event frames and helps generalize to other classes regardless of the amount of training data. It is not to say that the problem of unbalanced data is alleviated, but adaptive learning *does* help to some extent.

12.4 Version 2: segmentation using attention-based event models

The simple perceptual prediction framework, introduced in Section 12.3, uses a global representation of input video frames for perceptual prediction. It does not *attend* or rather focus on relevant spatial areas for both perception and prediction and as such may not capture *fine-grained* event representations. This is essentially strengthening the pathway proposed in the EST (Fig. 12.2) for the event models to influence the perceptual processing of the current sensory input. In our earlier model, this pathway was weakly implemented via the memory structure in the LSTM. In this section, we show that the framework can be enhanced with the idea of *spatial attention* to help focus on relevant spatial regions for a more fine-grained segmentation approach that can handle very long video sequences, *without significant training needs*. The full architecture is shown in detail in Fig. 12.7 and formally expressed in Algorithm 2. The spatial attention mechanism helps the network to learn an attention function between the internal current event model in the recurrent memory and the input at each time step. The resulting attention map can be used to localize the event within each processed frame spatially.

Algorithm 2 Temporal Event Segmentation Model with Attention-based Spatial Event Localization. The input is an untrimmed/streaming video $\mathbb{I}$, which is a set of frames $\{I_1, ..., I_t, I_{t+1}, ...I_T\}$. The output is a set of event boundaries $\{b_1, b_2, ...b_{T-1}\}$.

Input: Video frames $\{I_1, ..., I_t, I_{t+1}, ...I_T\} \in \mathbb{R}^{TxCxWxH}$
Output: Event boundary values $\mathbb{B} = \{b_1, b_2, ...b_{T-1}\}$

  1: **procedure** ATTENTION(I_t', h_{t-1})   ▷ Attention block
  2:      $a_t \leftarrow linear(tanh(linear(h_{t-1}) + linear(I_t')))$
  3:      $A_t \leftarrow softmax(a_t)$
  4:      $I_t'' \leftarrow A_t \odot I_t'$
  5:      **return** I_t''
  6: **end procedure**

  7: **procedure** SEGMENT(I_t, I_{t+1}, h_{t-1}, y_{t-1})   ▷ Main segmentation layer
  8:      $I_t' \leftarrow$ ENCODER(I_t)   ▷ Basic CNN encoder
  9:      $I_{t+1}' \leftarrow$ ENCODER(I_{t+1})   ▷ Basic CNN encoder
10:      $I_t'' \leftarrow$ ATTENTION(I_t', h_{t-1})
11:      $h_t \leftarrow$ LSTM(h_{t-1}, $linear(concat(I_t'', y_{t-1}))$)
12:      $y_t \leftarrow$ DECODER(h_t)   ▷ Single dense layer
13:      $e_t \leftarrow ||(I_{t+1}' - y_t)^{\odot 2} \odot (I_{t+1}' - I_t')^{\odot 2}||^2$ ▷ Motion weighted loss
14:      $b_t \leftarrow$ GATE(e_t)
15:      **return** h_t, b_t, y_t
16: **end procedure**

17: $h_t \leftarrow 0$
18: $y_{t-1} \leftarrow 0$
19: **for** $\{I_t, I_{t+1}\} \in \{I_1, I_2\}, \{I_2, I_3\}, ..., \{I_{T-1}, I_T\}$ **do**
20:      $h_t, b_t, y_t \leftarrow$ SEGMENT(I_t, I_{t+1}, h_t, y_{t-1})
21:      $\mathbb{B}.append(b_t)$
22: **end for**

12.4.1 Feature extraction

The feature extraction process in this architecture differs from the previous perceptual framework. We need to maintain the spatial arrangement of feature vectors for the attention mechanism to work. In this architecture, the encoder outputs a grid of feature vectors with the same spatial resolution as the resulting attention map. In other words, the attention mechanism dictates the value of attention assigned to each of the encoded feature vectors. We only use convolution operations in the encoder since kernels (weights) maintain feature maps' spatial configuration.

The raw input images are transformed from pixel space into a higher-level feature space using an encoder (CNN) model. This encoded feature representation allows the network to extract features of higher importance to the task being learned. The encoder network transforms an input image with dimensions $W \times H \times D$ to output features with dimensions $N \times N \times M$, where $N \times N$ are the spatial dimensions and M is the feature vector length.

12.4.2 Attention unit

Attention units have successfully been applied in supervised tasks such as image captioning (Xu et al., 2015) and for various natural language processing tasks such as translation and language modeling (Vaswani et al., 2017; Bahdanau et al., 2014; Luong et al., 2015; Devlin et al., 2018; Yang et al., 2019). Attention, in auto-regressive language models, is used to expose different temporal – or spatial – segments of the input to the decoding recurrent cell at every time step using fully supervised architectures. We use attention in a slightly different form, where the LSTM is decoded only once (per input frame) to predict future features and uses attention-weighted input to do so. Unlike (Xu et al., 2015; Vaswani et al., 2017; Bahdanau et al., 2014; Luong et al., 2015; Devlin et al., 2018; Yang et al., 2019), the attention weights and biases are trained using unsupervised loss functions.

In this framework, we utilize Bahdanau attention (Bahdanau et al., 2014) to spatially localize the event in each processed frame. The attention unit receives as an input the encoded features and outputs a set of attention weights (A_t) with dimensions $N \times N \times 1$. The hidden feature vectors (h_{t-1}) from the prediction layer of the previous time step are used to calculate the output attention weights (expressed visually in Fig. 12.7), as

$$A_t = \gamma \left(FC \left(tanh \left(FC(h_{t-1}) + FC(I_t') \right) \right) \right) \tag{12.7}$$

where FC represents a single fully connected neural network layer and γ represents a softmax function. The weights (A_t) are then multiplied by the encoded input feature vectors (I_t') to generate the masked feature vectors (I_t''). Some visualization of the resulting spatial attention masks is shown in Fig. 12.8. The attention mask is extracted from A_t, linearly scaled and resized, then overlayed on the raw input image (I_t).

12.4.3 Motion weighted loss function

The prediction loss we discussed in Section 12.3.4 applies an L2 loss function over the prediction of the whole frame. In this section, we introduce a motion weighted loss function which extracts motion related features from the feature vectors. The motion weighted loss acts upon the encoded frames in feature space rather than raw pixel space and is calculated using a continuous mask applied to the prediction loss. This modification aims to increase the prediction loss at moving features while reducing the loss of static/background features. Fig. 12.8 shows a comparison between prediction and motion weighted loss for an event with the bird moving in and out of a nest. The motion weighted loss is defined formally as

$$e_t = ||(I_{t+1}' - y_t)^{\odot 2} \odot (I_{t+1}' - I_t')^{\odot 2}||^2 \tag{12.8}$$

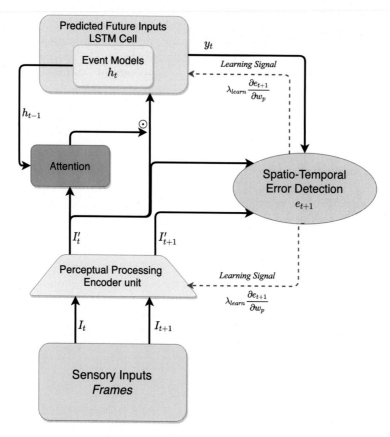

FIGURE 12.7 **Version 2: Temporal Segmentation using Attention-based Event Models:** The self-learning perceptual prediction algorithm's architecture enhanced with an attention mechanism that offers a stronger way for the event model to influence the current sensory input. Input frames from each time instant are encoded into high-level features using a deep-learning stack, followed by an attention overlay based on inputs from previous time instants, which is input to an LSTM. The training loss is composed based on the predicted and computed features from the current and next frames.

where $\odot$ denotes Hadamard (element-wise) operation. Note that the motion-weighted vector (the second term) is computed at the *feature level* and not at the pixel level and hence is more robust to minor changes due to sensor noise.

12.4.4 Results

In this section, we will look into the results of processing videos spanning over several days to flag spatial locations and temporal boundaries of events. Unlike other datasets, events in such extended videos only take place over a few frames, while the rest of the video contains no significant events or motion (background activities). We modify the perceptual prediction framework to include an attention mechanism for spatial localization, and we also mask the prediction loss function to extract motion related features. We begin by defining the wildlife

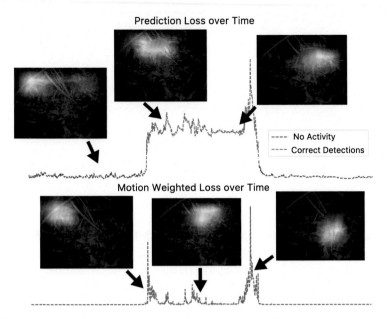

FIGURE 12.8 Plots of the prediction and motion-weighted losses before, during, and after an activity: (top) feature prediction loss over the frames, (bottom) motion weighted feature prediction loss over the frames. Errors for some selected frames are shown for both plots, overlaid with the corresponding attention map.

extended video dataset used for testing the modified approach, followed by an explanation of the evaluation metrics used to quantify performance. We discuss the model variations evaluated and conclude by presenting quantitative and qualitative results.

12.4.4.1 Dataset

We analyze the performance of our model on a wildlife monitoring dataset. The dataset consists of 10 days (254 hours) of continuous monitoring of a nest of the Kagu, a flightless bird of New Caledonia. The labels include four unique bird activities, {feeding the chick, incubation/brooding, nest building while sitting on the nest, nest building around the nest}. Start and end times for each instance of these activities are provided with the annotations. We modified the annotations to include the "walk-in" and "walk-out" events representing the transitioning events from an empty nest to incubation and vice versa. Our approach can flag the nest building (on and around the nest), feeding the chick, walk in and out events. Other events based on climate, time of day, lighting conditions are ignored by our segmentation network. Fig. 12.9 shows a selection of images from the dataset.

12.4.4.2 Evaluation metrics

We provide quantitative results, in the form of receiver operating characteristic (ROC) charts, for both frame level (Fig. 12.10) and activity level (Figs. 12.11 and 12.12) event segmentation. Frame window size (ϕ) is defined as the maximum joining window size between events; a high ϕ value can cause separate detected events to merge, which decreases the overall performance.

FIGURE 12.9 Samples of images from the Kagu bird wildlife monitoring dataset.

12.4.4.2.1 Frame level

The recall value in frame-level ROC is calculated as the ratio of true positive frames (event present) to the number of positive frames in the annotations dataset, while the false positive rate is expressed as the ratio of the false positive frames to the total number of negative frames (event not present) in the annotation dataset. The threshold value (ψ) is varied to obtain a single ROC line while varying the frame window size (ϕ) results in a different ROC line.

12.4.4.2.2 Activity level

The Hungarian matching (Munkres assignment) algorithm is utilized to achieve a one-to-one mapping between the ground truth labeled events and the detected events. Recall is defined as the ratio of the number of correctly detected events (overlapping frames) to the total number of ground-truth events. For the activity level ROC chart, the recall values are plotted against the false positive rate per minute, defined as the ratio of the total number of false-positive detected events to the total duration of the dataset in minutes. The false-positive rate per minute evaluation metric is also used in the ActEV TRECVID challenge (ActEV, 2019) to quantify activity detection systems' performance. Frame window size value (ϕ) is varied to obtain a single ROC line while varying the threshold value (ψ) results in a different ROC line.

12.4.4.3 Ablative studies

Different variations of our framework have been evaluated to quantify the effect of individual components on the overall performance. In our experiments, we tested the base model, which trains the perceptual prediction framework – including an attention unit – using the prediction loss function for backpropagation of the error signal. We refer to the base model as *LSTM+ATTN*. We also experimented with the effect of removing the attention unit from the model architecture, on the overall segmentation performance; results of this variation are reported under the model name *LSTM*. Further testing includes using the motion weighted loss for backpropagation of the error signal. We refer to the motion weighted model as *LSTM+ATTN+MW*. Each of the models has been tested extensively; results are reported in Sections 12.4.4.4 and 12.4.4.5, as well as visually expressed in Figs. 12.10, 12.11, 12.12, 12.13 and 12.14.

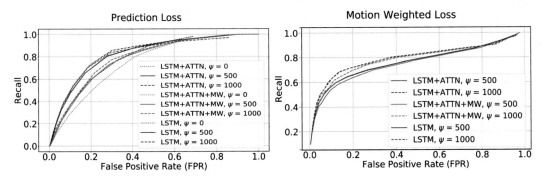

FIGURE 12.10 Frame-level event segmentation ROCs when activities are detected based on simple thresholding of the prediction and motion weighted loss signals. Plots are shown for different ablation studies.

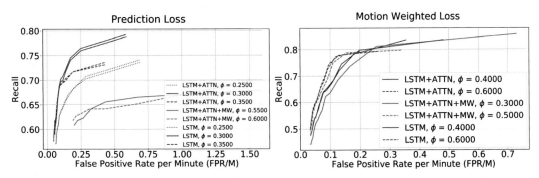

FIGURE 12.11 Activity-level event segmentation ROCs when activities are detected based on simple thresholding of the prediction and motion weighted loss signals. Plots are shown for different ablation studies.

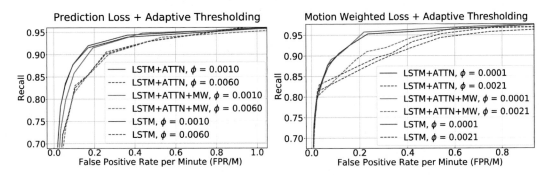

FIGURE 12.12 Activity-level event segmentation ROCs when activities are detected based on adaptive thresholding of the prediction and motion weighted loss signals. Plots are shown for different ablation studies.

12.4.4.4 *Quantitative evaluation*

We trained three different models, *LSTM*, *LSTM+ATTN*, and *LSTM+ATTN+MW*, for frame level and activity level event segmentation. Simple and adaptive gating functions were applied to prediction and motion weighted loss signals (Section 12.4.3) for frame level and activity level experiments. We present ROC curves for each model in Figs. 12.10, 12.11 and 12.12 by varying parameters such as the threshold value ψ and the frame window size ϕ.

It is to be noted that thresholding a loss signal does not necessarily imply that the model was trained to minimize this particular signal. In other words, the loss functions used for backpropagating the error to the models' learnable parameters are identified only in the model name (Section 12.4.4.3); however, thresholding experiments have been conducted on different types of loss signals, regardless of the backpropagating loss function used for training.

The best performing model, for frame level segmentation, ($LSTM+ATTN$, $\psi = 1000$) is capable of achieving {40%, 60%, 80%} frame recalls value at {5%, 10%, 20%} frame false positive rate respectively. Activity level segmentation can recall {80%, 90%, 95%} of the activities at {0.02, 0.1, 0.2} activity false positive rate per minute, respectively, for the model ($LSTM+ATTN$, $\phi = 0.0021$) as presented in Fig. 12.12. A 0.02 false positive activity rate per minute can also be interpreted as one false activity detection every 50 minutes of training (for a recall of 80% of the groundtruth activities).

Comparing the results shown in Figs. 12.11 and 12.12 indicates a significant increase in the overall performance when using an adaptive threshold for loss signal gating. The efficacy of adaptive thresholding is evident when applied to activity level event segmentation. Results have also shown that the model can effectively generate attention maps (Section 12.4.4.5) without impacting the segmentation performance.

12.4.4.5 Qualitative evaluation

Samples of the qualitative attention results are presented in Figs. 12.13 and 12.14. The attention mask, extracted from the model, has been trained to track the event, in all processed frames, without supervision. Our results show that the events are tracked and localized in various lighting (shadows, day/night) and occlusion conditions. Attention has also learned to indefinitely focus on the bird regardless of its motion state (stationary/Nonstationary), which indicates that the model has acquired a high-level temporal understanding of the events in the scene and learned the underlying structure of the bird in an unsupervised manner. Supplementary results[1] display a timelapse of attention weighted frames during illumination changes and moving shadows. Fig. 12.8 summarizes the visualization of the prediction loss signal, motion weighted loss signal, and attention mask during a walk in and out event.

12.5 Version 3: spatio-temporal localization using prediction loss map

Finally, we show that the framework from the previous section can be further enhanced to localize the event in the images, i.e., mark using bounding boxes where the event is happening. The attention-maps results in Section 12.4 only generate a heat map over the grid size defined by the encoder's output spatial resolution. In other words, the attention map acts as

[1] Available at https://ramyamounir.github.io/projects/EventSegmentation.

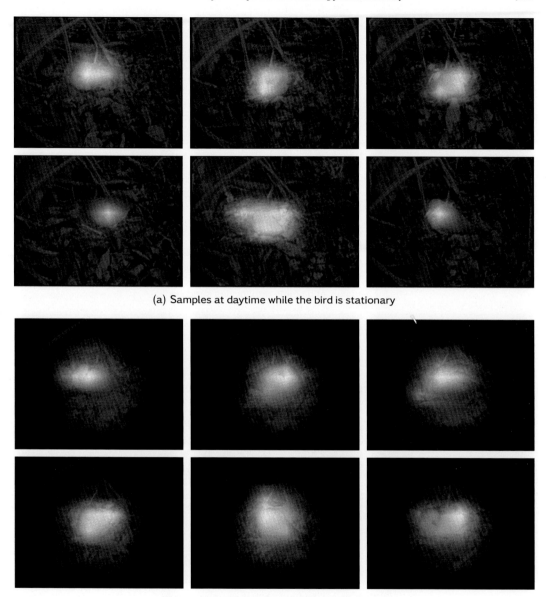

(a) Samples at daytime while the bird is stationary

(b) Samples at night while the bird is stationary

FIGURE 12.13 Samples of Bahdanau attention weights visualized on input images.

a pointer as to where the action is happening; however, it does not generate an accurate mask nor a bounding box. To generate a bounding box, we use a region proposal network and prediction loss map to filter these proposals using a spatio-temporal energy minimization function. The following description is from Aakur and Sarkar (2020).

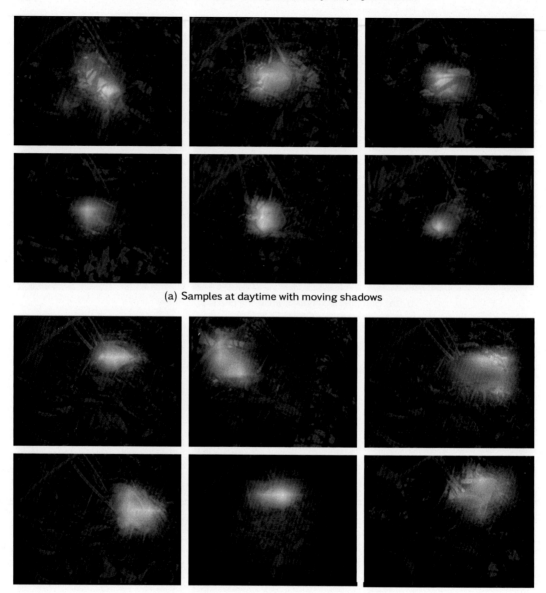

(a) Samples at daytime with moving shadows

(b) Samples at daytime while the bird is moving

FIGURE 12.14 Samples of Bahdanau attention weights visualized on input images.

We start by extracting relevant features (Section 12.5.1) from the raw frames in pixel space. The extracted features will be used as an input to the region proposal network and the prediction modules (Section 12.5.2). An energy function combines the prediction loss and the proposed regions (bounding boxes) to extract action tubes (Section 12.5.4) that are consistent

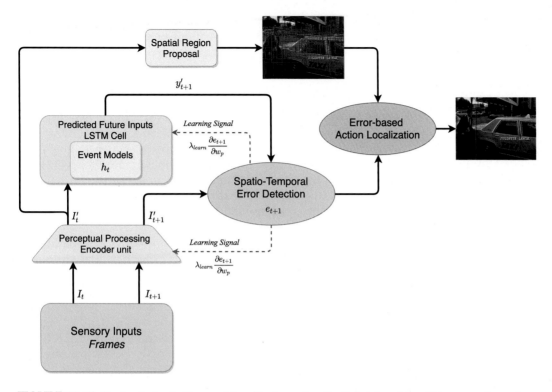

FIGURE 12.15 Version 3: Spatio-Temporal Localization using Prediction Loss Map: This approach envelopes Version 2 with a component that localizes the dominant action based on prediction error. It has four core components: (i) feature extraction and spatial region proposal, (ii) a future prediction framework, (iii) a spatial-temporal error detection module, and (iv) the error-based action localization process.

spatially and temporally. Fig. 12.15 shows the four components of the architecture: (i) feature extraction and spatial region proposal, (ii) a self-supervised future prediction framework, (iii) a spatial-temporal error detection module, and (iv) the error-based action localization process.

12.5.1 Feature extraction

Similar to previous approaches, a CNN encoder is used to extract relevant features. However, the extracted features in this approach are used as an input for the spatial region proposal network and the prediction stack. The region proposal is essentially a set of bounding boxes identifying possible areas of action and associated objects for each frame. A pretrained CNN is used for extracting the features and generating proposals.

We use class-agnostic proposals (i.e., the object category is ignored, and only feature-based localizations are taken into account) for two primary reasons. First, we do not want to make any assumptions on the actor's characteristics, such as label, role, and affordance. Second, despite significant progress in object detection, there can be many missed detections, especially when the object (or actor) performs actions that can transform their physical appearance. It is to be noted that these considerations can result in a large number of region proposals that require careful and robust selection but can yield higher chances of correct localization.

12.5.2 Hierarchical prediction stack

Similar to the approaches mentioned above, this framework learns a predictive function on the high-level features extracted by the encoder. A Long Short-Term Memory (LSTM) network is used to predict the next set of feature vectors, in a temporal sequence, given the encoded input and an internal model of the current event. The internal model effectively captures the spatial-temporal dynamics of the observed event. Similar to the previous attention model, the LSTM processes a set of feature vectors each time step, not a single feature vector. The spatial resolution of the feature maps is determined by the last convolution layer of the encoder.

The choice of an LSTM network is not arbitrary; while other approaches such as convolutional decoders (Jia et al., 2016) and mixture-of-network models (Vondrick et al., 2016) are viable alternatives for future prediction, we propose the use of recurrent networks for the following reasons. First, we want to model the temporal dynamics across *all* frames of the observed action (or event). Second, LSTMs can allow for multiple possible futures and hence will not average the outcomes of these possible futures, as can be the case with other prediction models. Imagine a sequence of frames $I_a = (I_a^1, I_a^2, \ldots I_a^n)$ corresponding to the activity a. Given the complex nature of videos such as those in instructional or sports domains, the next set of frames can be followed by frames of activity b or c with equal probability, given by $I_b = (I_b^1, I_b^2, \ldots I_b^m)$ and $I_c = (I_c^1, I_c^2, \ldots I_c^k)$ respectively. Using a fully connected or convolutional prediction unit is likely to result in the prediction of features that tend to be the average of the two activities a and b, i.e., $I_{avg}^k = \frac{1}{2}(I_b^k + I_c^k)$ for the time k. This is not a desirable outcome because the predicted features can either be an unlikely outcome or, more probably, be outside the plausible manifold of representations. The use of recurrent networks such as RNNs and LSTMs allows for multiple futures that can be possible at time $t + 1$, conditioned upon the observation of frames until time t. Third, since we work with error-based localization, using LSTMs can ensure that the learning process aggregates the spatial-temporal error across time and can yield progressively better predictions, especially for actions of longer duration.

In contrast to the previous approaches, the prediction unit here consists of a stack of LSTMs. The output of one LSTM is used as the input to another LSTM. Each LSTM in the stack has its own parameters, defining a different internal model of the event based on its position in the hierarchy. This modification allows for modeling both spatial and temporal dependencies since each higher-level LSTM acts as a progressive decoder framework that captures the temporal dependencies captured by the lower-level LSTMs. The first LSTM captures the spatial dependency that is propagated up the prediction stack.

The updated hidden state of the first (bottom) LSTM layer (h_t^1) depends on the current observation f_t^S, the previous hidden state (h_{t-1}^1) and memory state (m_{t-1}^1). Each of the higher-level LSTMs at level l takes the output of the bottom LSTM's output h_t^{l-1} and memory state m_t^{l-1} and can be defined as $(h_t^l, m_t^l) = LSTM(h_{t-1}^l, h_t^{l-1}, m_t^{l-1})$. Note this is different from a typical hierarchical LSTM model (Song et al., 2017) in that the higher LSTMs are impacted by the output of the lower level LSTMs at *current time step*, as opposed to that from the previous time step. Collectively, the event model W_e is described by the learnable parameters and their respective biases from the hierarchical LSTM stack.

Hence, the top layer of the prediction stack acts as the decoder whose goal is to predict the next feature f_{t+1}^S given all previous predictions $\hat{f}_1^S, \hat{f}_2^S, \dots \hat{f}_t^S$, an event model W_e and the current observation f_t^S. We model this prediction function as a log-linear model characterized by

$$\log p\big(\hat{f}_{t+1}^s | h_t^l\big) = \sum_{n=1}^{t} f\big(W_e, f_t^S\big) + \log Z(h_t) \tag{12.9}$$

where h_t^l is the hidden state of the *l*th level LSTM at time t and $Z(h_t)$ is a normalization constant. The LSTM prediction stack acts as a generative process for anticipating future features.

12.5.3 Prediction loss

The attention map for this framework is extracted directly from the actual prediction loss (See Fig. 12.7) at the top of the prediction stack. The prediction loss is a factor of the quality of the predictions made and the relative spatial alignment of the prediction errors. The motion weighted loss from Eq. (12.8) is used to compute a weight α_{ij} associated with each spatial location (i, j) in the predicted feature $\hat{f}_{t+1}^S$ as

$$\alpha_{ij} = \frac{\exp(e_{ij})}{\sum_{m=1}^{w_k} \sum_{n=1}^{h_k} \exp(e_{mn})} \tag{12.10}$$

where e_{ij} represents the weighted prediction error at location (i, j) (Eq. (12.8)). It can be considered to be a function $a(f_t^S, h_{t-1}^l)$ of the state of the top-most LSTM and the input feature f_t^S at time t. The resulting matrix is an error-based attention map that allows us to localize the prediction error at a specific spatial location. And the average spatial error over time, $E(t)$, is used for temporal localization.

12.5.4 Action tubes extraction

The actions are extracted as tubes using an energy objective function that is minimized to ensure spatial and temporal consistency. The action localization module receives, as an input, a stream of bounding box proposals (several proposals per frame) and a stream of prediction loss maps (one per frame). An energy function is defined to extract coherent action tubes by filtering the proposals using the prediction loss map and returning the collection of proposals with a higher probability of action localization. This is achieved by assigning an energy term to each of the bounding box proposals ($\mathcal{B}_{it}$) at time t and choosing the top k bounding boxes with the least energy as our final proposals. The energy of a bounding box $\mathcal{B}_{it}$ is defined as

$$E(\mathcal{B}_{it}) = w_\alpha \, \phi(\alpha_{ij}, \mathcal{B}_{it}) + w_t \delta\big(\mathcal{B}_{it}, \{\mathcal{B}_{j,t-1}\}\big) \tag{12.11}$$

where $\phi(\cdot)$ is a function that returns a value characteristic of the distance between the bounding box center and location of the maximum error, $\delta(\cdot)$ is a function that returns the minimum spatial distance between the current bounding box and the closest bounding box from the previous time step. The constants w_α and w_t are scaling factors. Fig. 12.18 shows an example of extracted action tubes for a single activity in a stream of frames.

12.5.5 Results

12.5.5.1 Data

We use three publicly available datasets to evaluate our approach on the action localization task.

UCF Sports (Rodriguez et al., 2008) is an action localization dataset consisting of 10 classes of sports actions, such as skating and lifting, collected from sports broadcasts. It is an interesting dataset since it has a high concentration of distinct scenes and motions that make it challenging for localization and recognition. We use the splits (103 training and 47 testing videos) as defined in Lan et al. (2011) for evaluation.

JHMDB (Jhuang et al., 2013) is composed of 21 action classes and 928 trimmed videos. All videos are annotated with human-joints for every frame. The ground truth bounding box for the action localization task is chosen such that the box encompasses all the joints. This dataset offers several challenges, such as increasing amounts of background clutter, high interclass similarity, complex motion (including camera motion), and occluded objects of interest. We present all results as the average across all three splits.

THUMOS'13 (Jiang et al., 2014) is a subset of the UCF-101 (Soomro et al., 2012) dataset, consisting of 24 classes and 3, 207 videos. Ground truth bounding boxes are provided for each of the classes for the action localization task. It is also known as the **UCF-101-24** dataset. Following prior works (Li et al., 2018; Soomro and Shah, 2017), we perform the experiments and report results on the first split.

We also analyze the approach's generalization ability on egocentric videos by evaluating it on the *unsupervised gaze prediction task*. There is ample evidence from cognitive psychology for a strong correlation between gaze points and action localization (Tipper et al., 1992). Hence, the gaze prediction task would be a reasonable measure of the generalization to action localization in egocentric videos. We evaluate the performance on the **GTEA Gaze** (Fathi et al., 2012) dataset, which consists of 17 sequences of tasks performed by 14 subjects, with each sequence lasting about 4 minutes. We use the official splits for the GTEA datasets as defined in prior works (Fathi et al., 2012).

12.5.5.2 Metrics and baselines

For the **action localization** task, we follow prior works (Li et al., 2018; Soomro and Shah, 2017) and report the mean average precision (mAP) at various overlap thresholds, obtained by computing the Intersection Over Union (IoU) of the predicted and ground truth bounding boxes. We also evaluate the quality of bounding box proposals by measuring the average, perframe IoU, and the bounding box *recall* at varying overlap ratios.

Since ours is an unsupervised approach, we obtain class labels by clustering the learned representations using the *k-means* algorithm. While more complicated clustering may yield better recognition results (Soomro and Shah, 2017), the k-means approach allows us to evaluate the robustness of learned features. We evaluate our approach in two settings K_{gt} and K_{opt}, where the number of clusters is set to the number of ground truth action classes, and an optimal number obtained through the elbow method (Kodinariya and Makwana, 2013), respectively. From our experiments, we observe that K_{opt} is three times the number of ground truth classes, which is not unreasonable and has been a working assumption in other deep learning-based clustering approaches (Hershey et al., 2016). Clusters are mapped to the ground truth clusters for evaluation using the Hungarian method, as done in prior unsupervised approaches (Ji et al., 2019; Xie et al., 2016). We also compare against other LSTM and attention-based approaches (Section 12.5.5.3.3) to the action localization problem for evaluating the effectiveness of our training protocol.

For the **gaze prediction** task, we evaluate the approaches using **Area Under the Curve** (AUC), which measures the area under the curve on saliency maps for true positive versus false-positive rates under various threshold values. We also report the **Average Angular Error** (AAE), which measures the angular distance between the predicted and ground truth gaze positions. Since our model's output is a saliency map, AUC is a more appropriate metric than the average angular error (AAE), which requires specific locations.

12.5.5.3 Quantitative evaluation

In this section, we present the quantitative evaluation of our approach on two different tasks, namely action localization and egocentric gaze prediction. For the action localization task, we evaluate our approach on two aspects – the quality of proposals and spatial-temporal localization.

12.5.5.3.1 Quality of localization proposals

We first evaluate the quality of our localization proposals by assuming perfect class prediction. This allows us to independently assess the quality of localization performed in a self-supervised manner. We present the evaluation results in Table 12.5 and compare them against fully supervised, weakly supervised, and unsupervised baselines. As can be seen, the predictive approach outperforms many supervised and weakly supervised baselines. APT (Van Gemert et al., 2015) achieves a higher localization score. However, it produces, on average, 1500 proposals per video, whereas our approach returns approximately 10 proposals. A large number of localization proposals per video can lead to higher recall and IoU but makes the localization task more challenging, i.e., action labeling per video harder, and can affect the ability to generalize across domains.

Also, it should be noted that our approach produces proposals in *streaming* fashion, as opposed to many of the other approaches, which produce action tubes based on motion computed across the entire video. This can make real-time action localization in streaming videos harder.

12.5.5.3.2 Spatial-temporal action localization

We also evaluate our approach on the spatial-temporal localization task. This evaluation allows us to analyze the robustness of the self-supervised features learned through prediction. We generate video-level class labels through clustering and use the standard evaluation metrics (Section 12.5.5.2) to quantify the performance. The AUC curves with respect to vary-

TABLE 12.5 Comparison with fully supervised and weakly supervised baselines on class-agnostic action localization on UCF Sports dataset. We report the average localization accuracy of each approach i.e. average IoU.

Supervision	Approach	Approach
Full	STPD (Tran and Yuan, 2011)	44.6
	Max Path Search (Tran and Yuan, 2012)	**54.3**
Weak	Ma et al. (Ma et al., 2013)	44.6
	GBVS (Grundmann et al., 2010)	42.1
	Soomro et al. (Soomro and Shah, 2017)	**47.7**
None	IME Tublets (Jain et al., 2014)	51.5
	APT (Van Gemert et al., 2015)	**63.7**
	Predictive Approach	55.7

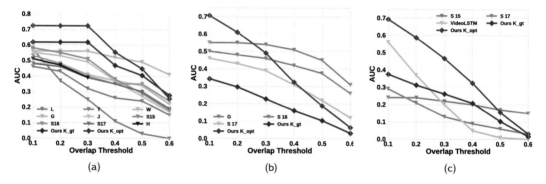

(a)   (b)   (c)

FIGURE 12.16 AUC for the action localization tasks are shown for (a) UCF Sports, (b) JHMDB and (c) THUMOS13 datasets. We compare against baselines with varying levels of supervision such as Lan et al. (2011), Tian et al. (2013), Wang et al. (2014b), Gkioxari and Malik (2015), Jain et al. (2014), Soomro et al. (2015, 2016); Soomro and Shah (2017), Hou et al. (2017), and VideoLSTM (Li et al., 2018).

ing overlap thresholds are presented in Fig. 12.16. We compare against a mix of supervised, weakly-supervised, and unsupervised baselines on all three datasets.

On the **UCF Sports** dataset (Fig. 12.16(a)), we outperform all baselines including several supervised baselines, except for Gkioxari and Malik Gkioxari and Malik (2015) at higher overlap thresholds ($\sigma > 0.4$) when we set number of clusters k to the number of ground truth classes. When we allow for some oversegmentation and use the *optimal* number of clusters, we outperform all baselines till $\sigma > 0.5$.

On the **JHMDB** dataset (Fig. 12.16(b)), we find that our approach, while having high recall even at more stringent thresholds (77.8% @ $\sigma = 0.5$), the large camera motion and intra-class variations have a significant impact on the classification accuracy. Hence, the mAP suffers when we set k to be the number of ground truth classes. When we set the number of clusters to the optimal number of clusters, we outperform other baselines at lower thresholds (mAP @ $\sigma < 0.5$). It should be noted that the other unsupervised baseline (Soomro *et al.* (2017)) uses object detection proposals from a Faster R-CNN backbone to score the *"humanness"* of a

TABLE 12.6 Comparison with other LSTM-based and attention-based approaches on the THUMOS'13 dataset. We report average recall at various overlap thresholds, mAP at 0.2 overlap threshold and the average number of proposals per frame.

| Approach | Annotations | | # Proposals | Average Recall | | | | | mAP |
	Labels	Boxes		0.1	0.2	0.3	0.4	0.5	@0.2
ALSTM (Sharma et al., 2015)	✓	✗	1	0.46	0.28	0.05	0.02	–	0.06
VideoLSTM (Li et al., 2018)	✓	✗	1	0.71	0.52	0.32	0.11	–	0.37
Actor Supervision (Escorcia et al., 2020)	✓	✗	∼ 1000	**0.89**	–	–	–	**0.44**	0.46
Our Approach	✗	✗	∼ 10	0.84	**0.72**	**0.58**	**0.47**	0.33	**0.59**

proposal. This assumption tends to make the approach biased towards human-centered action localization and affects its ability to generalize towards actions with nonhuman actors. On the other hand, we do not make any assumptions on the characteristics of the actor, scene, or motion dynamics.

On the **THUMOS'13** dataset (Fig. 12.16(c)), we achieve consistent improvements over unsupervised and weakly supervised baselines, at $k = k_{gt}$ and achieve state-of-the-art mAP scores when $k = k_{opt}$. It is interesting to note that we perform competitively (when $k = k_{gt}$) with the weakly-supervised attention-based VideoLSTM (Li et al., 2018), which uses a convLSTM for temporal modeling along with a CNN-based spatial attention mechanism. It should be noted that we have a higher recall rate (0.47 @ $\sigma = 0.4$ and 0.33 @ $\sigma = 0.5$) at higher thresholds than other state-of-the-art approaches on THUMOS'13 and shows the robustness of the error-based localization approach to intra-class variation and occlusion.

Clustering quality. Since there is a significant difference in the mAP score when we set a different number of clusters in k-means, we measured the homogeneity (or purity) of the clustering. The homogeneity score measures the "quality" of the cluster by measuring how well a cluster models a given ground-truth class. Since we allow the oversegmentation of clusters when we set k to the optimal number of clusters, this is an essential measure of feature robustness. Higher homogeneity indicates that intra-class variations are captured since all data points in a given cluster belong to the same ground truth class. We observe an average homogeneity score of 74.56% when k is set to the number of ground truth classes and 78.97% when we use the optimal number of clusters. As can be seen, although we oversegment, each of the clusters typically models a single action class to a high degree of integrity.

12.5.5.3.3 Comparison with other LSTM-based approaches

We also compare our approach with other LSTM-based and attention-based models to highlight the importance of the self-supervised learning paradigm. Since LSTM-based frameworks can have highly similar architectures, we consider different requirements and characteristics, such as the level of annotation required for training and the number of localization proposals returned per video. We compare with three approaches similar in spirit to our approach – ALSTM (Sharma et al., 2015), VideoLSTM (Li et al., 2018) and Actor Supervision (Escorcia et al., 2020) and summarize the results in Table 12.6. It can be seen that we significantly outperform VideoLSTM and ALSTM on the THUMOS'13 dataset in both recall and $mAP @ \sigma = 0.2$.

Actor Supervision (Escorcia et al., 2020) outperforms our approach on recall, but it is to be noted that the region proposals are dependent on two factors – (i) object detection-based actor proposals and (ii) a filtering mechanism that limits proposals based on ground truth action classes, which can increase the training requirements and limit generalizability. Also, note that returning a higher number of localization proposals can increase recall at the cost of generalization.

12.5.5.3.4 Ablative studies

This predictive framework has three major components that affect its performance the most – (i) the region proposal module, (ii) the future prediction module, and (iii) the error-based action localization module. We consider and evaluate several alternatives to all three modules. We choose selective search (Uijlings et al., 2013) and EdgeBox (Zhu et al., 2015) as alternative region proposal methods to SSD.

We use an attention-based localization method for action localization as an approximation of the ALSTM (Sharma et al., 2015) to evaluate the effectiveness of using the error-based localization. We also evaluate a 1-layer LSTM predictor with a fully connected decoder network to approximate (Aakur and Sarkar, 2019) on the localization task. We evaluate the effect of attention-based prediction by introducing a Bahdanau (Bahdanau et al., 2014) attention layer before prediction as an alternative to the error-based action localization module.

These ablative studies are conducted on the UCF Sports dataset. The results are plotted in Fig. 12.17(a). It can be seen that the use of the prediction error-based localization has a significant improvement over a trained attention-based localization approach. We can also see that the choice of region proposal methods does have some effect on the performance of the approach, with selective search and EdgeBox proposals doing slightly better at higher thresholds ($\sigma \in (0.4, 0.5)$) at the cost of inference time and additional bounding box proposals (50 compared to the 10 from SSD-based region proposal). Using SSD for generating proposals allows us to share weights across the frame encoder and region proposal tasks and reduce the memory and computational footprint of the approach. We also find that using attention as part of the prediction module significantly impacts the architecture's performance. It could, arguably, be attributed to the objective function, which aims to minimize the prediction error. Note that we use the *error-based attention* to localize events in this approach as opposed to a learned attention vector from Section 12.4. We find that using Bahdanau attention to encode the features could impact the prediction function for this approach.

12.5.5.3.5 Unsupervised egocentric gaze prediction

Finally, we evaluate the ability to generalize to egocentric videos by quantifying the model's performance on the unsupervised gaze prediction task. Given that we do not need any annotations or other auxiliary data, we employ the same architecture and training strategy for this task. We evaluate on the GTEA gaze dataset and compare it with other unsupervised models in Table 12.7. As can be seen, we obtain competitive results on the gaze prediction task, outperforming all baselines on both the AUC and AAE scores. It is to be noted that we outperform the center bias method on the AUC metric. Center bias exploits the spatial bias in egocentric images and always predicts the center of the video frame as the

TABLE 12.7 Comparison with state-of-the-art on the unsupervised egocentric gaze prediction task on the GTEA dataset.

	Itti and Koch (2000)	GBVS (Harel et al., 2007)	AWS-D Leboran et al. (2016)	Center Bias	OBDL (Sayed et al., 2015)	Ours
AUC	0.747	0.769	0.770	0.789	0.801	**0.861**
AAE	18.4	15.3	18.2	**10.2**	15.6	13.6

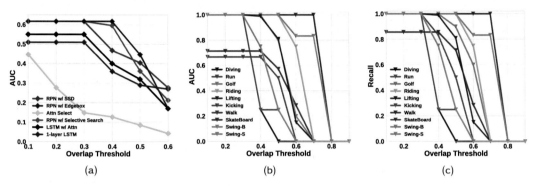

FIGURE 12.17 Qualitative analysis of our approach on UCF Sports dataset (a) ablative variations on AUC. (a) class-wise AUC, and (c) class-wise bounding box recall at different overlap thresholds.

predicted gaze position. The AUC metric's significant improvement indicates that our approach predicts gaze fixations that are more closely aligned with the ground truth than the center bias approach. Given that the model was not designed explicitly for this task, it is a remarkable performance, especially given the performance of fully supervised baselines such as DFG (Zhang et al., 2017), which achieves 10.6 and 88.3 for AUC and AAE.

12.5.5.4 Qualitative evaluation

We find that our approach has a consistently high recall for the localization task across datasets and domains. We consider that an action is correctly localized if the average IoU across all frames is higher than 0.5, which indicates that most, if not all, frames in a video are correctly localized. We illustrate the recall scores and subsequent AUC scores for each class in the UCF sports dataset in Figs. 12.17(b) and (c). For many classes (7/10 to be specific), we have more than 80% recall at an overlap threshold of 0.5. Through visual inspection, we find that the spatial-temporal error is often correlated with the actor but is usually not at the center of the region of interest and thus reduces the quality of the chosen proposals. We illustrate this effect in Fig. 12.18. The first row shows the input frame, the second shows the error-based attention, and the last row shows the final localization proposals. If more proposals are returned (as is the case with selective search and EdgeBox), we can obtain a higher recall (Fig. 12.17(b)) and higher mAP.

Successful Localization

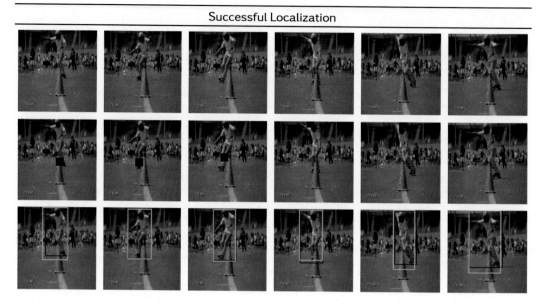

FIGURE 12.18 **Qualitative Examples**: Error-based attention location and the final prediction. Green BB: Prediction, Blue BB: Ground truth.

12.6 Other event segmentation approaches in computer vision

Typically, there are three different classes of approaches for temporal event segmentation: fully supervised, weakly supervised, and fully unsupervised. While fully supervised approaches have more precise localization and better labeling accuracy, the required number of annotations is rather large. The labeled training data demands do not scale well with an increase in label classes. While not requiring frame-level annotations, weakly supervised approaches have the underlying assumption that there exists a large, annotated training set that allows for effective detection of all possible actors (both human and nonhuman) in the set of action classes. Unsupervised approaches, such as ours, do not make any such assumptions but can result in poorer localization performance. We alleviate this to an extent by leveraging advances in region proposal mechanisms and robust self-learning representations. One particular class of unsupervised approach, called self-supervision, uses the data itself, without annotations, for supervision.

12.6.1 Supervised approaches

Fully supervised methods have been the dominant approach to temporal event segmentation. They consider the problem to be fully supervised and use the groundtruth annotations to take a *segment by classification* approach i.e., they assign the labels to semantically coherent "*chunks*", with contiguous frames sharing the same label, to segment the video into its constituent segments. The common pipeline in these approaches has been to extract features (either hand-crafted or automated using deep learning) to perform frame-based labeling support vector machines (Kuehne et al., 2014) or model temporal dynamics using Hidden Markov Models (Kuehne et al., 2014), temporal convolutional neural networks (TCN) (Lea et al., 2017), spatiotemporal convolutional neural networks (CNN) (Lea et al., 2016) and recurrent networks (Richard et al., 2017) to name a few. While fully supervised approaches are appealing due to their strong quantitative performance, obtaining large-scale annotated datasets, especially with frame-level annotations, can become quite expensive and may not always be available. This problem becomes more pronounced as the granularity of the events becomes finer.

Similarly, some approaches tackle action localization through the simultaneous generation of bounding box proposals and labeling each bounding box with the predicted action class. Both bounding box generation and labeling are fully supervised, i.e., they require ground truth annotations of both bounding boxes and labels. Typical approaches leverage advances in object detection to include temporal information (Gkioxari and Malik, 2015; Hou et al., 2017; Jain et al., 2014; Soomro et al., 2015, 2016; Tian et al., 2013; Tran and Yuan, 2012; Wang et al., 2014b) for proposal generation. The final step typically involves the use of the Viterbi algorithm (Gkioxari and Malik, 2015) to link the generated bounding boxes across time.

12.6.2 Weakly-supervised approaches

The underlying concept behind weak supervision is to alleviate the need for direct labeling by leveraging accompanying text scripts or instructions as indirect supervision for learning highly discriminant features. There have been two common approaches to weakly supervised learning for temporal segmentation of videos – (1) using script or instructions for weak annotation (Bojanowski et al., 2014; Ding and Xu, 2018; Alayrac et al., 2016; Malmaud et al., 2015), and (2) following an incomplete temporal localization of actions for learning and inference (Huang et al., 2016; Richard et al., 2017). While such approaches model the temporal transitions using RNNs, they still rely on enforcing semantics for segmenting actions and hence require some supervision for learning and inference.

These approaches have been explored for action localization to reduce the need for extensive annotations (Escorcia et al., 2020; Lan et al., 2011; Li et al., 2018; Sharma et al., 2015). They typically only require video-level labels and rely on object detection-based approaches to generate bounding box proposals. It is to be noted that weakly supervised approaches also use object-level labels and characteristics to guide the bounding box selection process. Some approaches (Escorcia et al., 2020) use a similarity-based tracker to connect bounding boxes across time to incorporate temporal consistency.

12.6.3 Unsupervised approaches

Unsupervised approaches do not need external supervision in terms of annotated training data that can be used to determine when the output is right and wrong. These approaches have not been explored to the same extent as supervised and weakly-supervised approaches. The primary approach is to use clustering as the unsupervised approach using discriminant features (Bhatnagar et al., 2017; Sener and Yao, 2018). The models incorporate a temporal consistency into the segmentation approach by using either LSTMs (Bhatnagar et al., 2017) or generalized mallows model (Sener and Yao, 2018). Garcia et al. (2018) explore the use of a generative LSTM network to segment sequences like we do. However, they handle only coarse temporal resolution in life-log images sampled as far apart as 30 seconds. Consecutive images when events change have more variability making for easier discrimination. Besides, they require an iterative training process, which we do not.

Some approaches do not require any supervision for either labels or bounding boxes. The two more common approaches are to generate action proposals using (i) supervoxels (Jain et al., 2014; Soomro and Shah, 2017) and (ii) clustering motion trajectories (Van Gemert et al., 2015). It should be noted that Soomro and Shah (2017) also uses object characteristics to evaluate the "humanness" of each supervoxel to select bounding box proposals.

Our approach falls into the class of unsupervised action localization approaches. The most closely related approaches (with respect to architecture and theme) to ours are VideoLSTM (Li et al., 2018) and Actor Supervision (Escorcia et al., 2020), which use attention in the selection process for generating bounding box proposals, but require video-level labels. We, on the other hand, do not require any labels or bounding box annotations for training.

12.6.4 Self-supervised approaches

One class of unsupervised approach that has gained attention is self-supervised approaches that use training data, but unannotated. There are two main types of self-supervised approaches: (1) hiding a subset of the data and trying to predict it, i.e., predict the occluded from the visible or predict color from gray-level (Zhang et al., 2016; Vondrick et al., 2018); (2) altering the input and predicting the function (parameters) used for altering it (Gidaris et al., 2018; Doersch et al., 2015; Misra et al., 2016; Fernando et al., 2017). The prediction of part of the input, or the unaltered version of the input, forces the network to learn good semantic features from the dataset. Contrastive learning approaches (Chen et al., 2020; Grill et al., 2020; Caron et al., 2020) fall under the second category where the input is altered and the network is forced to learn a representation that maximizes the similarity between the original and the altered input. These approaches have been found to learn fairly robust representations that can then be coupled with different tasks, i.e., labeling, motion estimation, and so on.

In the context of videos, self-supervision has been mostly prediction of the next couple of image frames (Srivastava et al., 2015; Mathieu et al., 2015; Neverova et al., 2017; Lotter et al., 2016; Finn et al., 2016). Feature-level predictions have been proposed to discover cooccurring visual feature spaces (Wang et al., 2014a) and learn representations for better recognition (Liu et al., 2018; Li et al., 2017). Such approaches propose the discovery of context through cooccurrences of features across temporal sequences to learn either multifeature correlation such as motion and appearance modeling for future frame generation (Li et al., 2017) or for unsupervised learning of cooccurring concepts like the presence of noses and eyes in faces (Wang et al., 2014a).

There is also an increasing body of work on the prediction of future activities (Sun et al., 2019; Luc et al., 2017; Ma et al., 2016; Liang et al., 2019; Kitani et al., 2012; Fragkiadaki et al., 2015; Walker et al., 2014; Alahi et al., 2016; Santoro et al., 2017; Hamilton et al., 2017). However, these approaches all adopt the standard annotated training-based deep-learning framework.

Our approach is similar in spirit to the predictive learning framework from videos espoused by Vondrick et al. (2016), but for higher-level concepts, instead of just action labels.

There are strong indications from brain and cognitive studies, apart from Zack's research group, that predictions play a significant role in how the brain circuits learn new concepts. Based on studies of many neuroscience experiments, Hawkins et al. (2016) have also suggested a repeated architecture of layers of feed-forward, predictive memory, and temporal pooling, which they demonstrated for anomaly detection in 1D signals. Heeger (2017) has also has shown the effectiveness of a layered architecture with bottom-up and top-down connections, but driven by prior predictions for temporal signal processing. Lotter et al. (2016) have implemented a predictive coding stack, but for video frame-level prediction.

12.7 Conclusions

This chapter drew from cognitive science research to define the problem of event segmentation and to design highly effective computer vision algorithms for spatio-temporal segmentation of events in videos. These approaches do not require any annotated data, nor do they require multiple passes through the data. They can process *streaming* video data while learning robust representations for segmenting events.

The main idea is to use predictive learning, as posited in the Event Segmentation Theory in cognitive science, to detect event boundaries. As in the EST, a perceptual processing unit (a CNN stack) sends the extracted features from the current frame (conditioned on the working event model) to a prediction unit (LSTM, a stack of LSTMs, bounding box prediction head), which predicts future perceptual features. A mismatch in the predicted features and the actual features computed for the next time instance generates a prediction error signal. A high error signals an event boundary. A new event model is now needed to make future predictions. The prediction error triggers a gating mechanism to update the working event model (hidden states in the LSTM) with a new model.

Extensive experiments on a variety of domains demonstrate that the predictive learning approach can learn robust event representations from unlabeled *streaming* video sequences, *with only one epoch of training (single pass through video)*. The event representations were able to obtain state-of-the-art results in unsupervised temporal segmentation (Section 12.3) and spatial-temporal action localization (Section 12.5) while offering competitive performance with fully supervised baselines that require extensive amounts of annotated training data. Additionally, we show that the predictive learning framework can process and segment streaming video data of extremely long duration (Section 12.4), at close to real-time processing. Predictive learning can help break the increasing dependency on training data and move towards open-world visual understanding. We hope that these results encourage research in this promising direction and unlatch computer vision research from the increasing dependence on more and more annotated data.

Acknowledgments

This research was supported in part by the US National Science Foundation grants CNS 1513126, IIS 1956050, and IIS 1955230.

References

Aakur, Sathyanarayanan N., Sarkar, Sudeep, 2019. A perceptual prediction framework for self supervised event segmentation. In: The IEEE Conference on Computer Vision and Pattern Recognition (CVPR).

Aakur, Sathyanarayanan N., Sarkar, Sudeep, 2020. Action localization through continual predictive learning. arXiv preprint. arXiv:2003.12185.

Aakur, Sathyanarayanan, de Souza, Fillipe D.M., Sarkar, Sudeep, 2019. Going deeper with semantics: exploiting semantic contextualization for interpretation of human activity in videos. In: IEEE Winter Conference on Applications of Computer Vision (WACV). IEEE.

ActEV: Activities in Extended Video, 2019. https://actev.nist.gov/.

Alahi, Alexandre, Goel, Kratarth, Ramanathan, Vignesh, Robicquet, Alexandre, Fei-Fei, Li, Savarese, Silvio, 2016. Social lstm: human trajectory prediction in crowded spaces. In: Proceedings of the IEEE Conference on Computer Vision and Pattern Recognition, pp. 961–971.

Alayrac, Jean-Baptiste, Bojanowski, Piotr, Agrawal, Nishant, Sivic, Josef, Laptev, Ivan, Lacoste-Julien, Simon, 2016. Unsupervised learning from narrated instruction videos. In: IEEE Conference on Computer Vision and Pattern Recognition (CVPR), pp. 4575–4583.

Bahdanau, Dzmitry, Cho, Kyunghyun, Bengio, Yoshua, 2014. Neural machine translation by jointly learning to align and translate. arXiv preprint. arXiv:1409.0473.

Bhatnagar, Bharat Lal, Singh, Suriya, Arora, Chetan, Jawahar, C.V., CVIT, K.C.I.S., 2017. Unsupervised learning of deep feature representation for clustering egocentric actions. In: International Joint Conference on Artificial Intelligence (IJCAI). AAAI Press, pp. 1447–1453.

Bojanowski, Piotr, Lajugie, R´emi, Bach, Francis, Laptev, Ivan, Ponce, Jean, Schmid, Cordelia, Sivic, Josef, 2014. Weakly supervised action labeling in videos under ordering constraints. In: European Conference on Computer Vision (ECCV). Springer, pp. 628–643.

Caron, Mathilde, Misra, Ishan, Mairal, Julien, Goyal, Priya, Bojanowski, Piotr, Joulin, Armand, 2020. Unsupervised learning of visual features by contrasting cluster assignments. In: Advances in Neural Information Processing Systems, vol. 33.

Chen, Ting, Kornblith, Simon, Norouzi, Mohammad, Hinton, Geoffrey, 2020. A simple framework for contrastive learning of visual representations. arXiv preprint. arXiv:2002.05709.

Devlin, Jacob, Chang, Ming-Wei, Lee, Kenton, Toutanova, Kristina, 2018. Bert: pre-training of deep bidirectional transformers for language understanding. arXiv preprint. arXiv:1810.04805.

Ding, Li, Xu, Chenliang, 2018. Weakly-supervised action segmentation with iterative soft boundary assignment. In: IEEE Conference on Computer Vision and Pattern Recognition (CVPR).

Doersch, Carl, Gupta, Abhinav, Efros, Alexei A., 2015. Unsupervised visual representation learning by context prediction. In: Proceedings of the IEEE International Conference on Computer Vision, pp. 1422–1430.

Escorcia, Victor, Dao, Cuong D., Jain, Mihir, Ghanem, Bernard, Snoek, Cees, 2020. Guess Where? Actor-Supervision for Spatiotemporal Action Localization. Computer Vision and Image Understanding, vol. 192, p. 102886.

Fathi, Alireza, Li, Yin, Rehg, James M., 2012. Learning to recognize daily actions using gaze. In: European Conference on Computer Vision. Springer, pp. 314–327.

Fernando, Basura, Bilen, Hakan, Gavves, Efstratios, Gould, Stephen, 2017. Self-supervised video representation learning with odd-one-out networks. In: Proceedings of the IEEE Conference on Computer Vision and Pattern Recognition, pp. 3636–3645.

Finn, Chelsea, Goodfellow, Ian J., Levine, Sergey, 2016. Unsupervised learning for physical interaction through video prediction. CoRR. arXiv:1605.07157.

Fragkiadaki, Katerina, Levine, Sergey, Felsen, Panna, Malik, Jitendra, 2015. Recurrent network models for human dynamics. In: Proceedings of the IEEE International Conference on Computer Vision, pp. 4346–4354.

Garcia del Molino, Ana, Lim, Joo-Hwee, Tan, Ah-Hwee, 2018. Predicting visual context for unsupervised event segmentation in continuous photo-streams. In: ACM Conference on Multimedia (ACM MM). ACM, pp. 10–17.

Gidaris, Spyros, Singh, Praveer, Komodakis, Nikos, 2018. Unsupervised representation learning by predicting image rotations. arXiv:1803.07728 [cs.CV].

Gkioxari, Georgia, Malik, Jitendra, 2015. Finding action tubes. In: Proceedings of the IEEE Conference on Computer Vision and Pattern Recognition, pp. 759–768.

Grill, Jean-Bastien, et al., 2020. Bootstrap your own latent-a new approach to self-supervised learning. In: Advances in Neural Information Processing Systems, vol. 33.

Grundmann, Matthias, Kwatra, Vivek, Han, Mei, Essa, Irfan, 2010. Efficient hierarchical graph-based video segmentation. In: 2010 IEEE Computer Society Conference on Computer Vision and Pattern Recognition. IEEE, pp. 2141–2148.

Hamilton, William L., Ying, Rex, Leskovec, Jure, 2017. Representation learning on graphs: methods and applications. arXiv preprint. arXiv:1709.05584.

Hard, Bridgette M, Tversky, Barbara, Lang, David S., 2006. Making sense of abstract events: building event schemas. Memory & Cognition 34 (6), 1221–1235.

Harel, Jonathan, Koch, Christof, Perona, Pietro, 2007. Graph-based visual saliency. In: Advances in Neural Information Processing Systems, pp. 545–552.

Hawkins, Jeff, Ahmad, Subutai, 2016. Why neurons have thousands of synapses, a theory of sequence memory in neocortex. Frontiers in Neural Circuits, vol. 10, p. 23.

Hawkins, Jeff, Blakeslee, Sandra, 2004. On Intelligence. Macmillan.

Heeger, David J., 2017. Theory of cortical function. Proceedings of the National Academy of Sciences 114 (8), 1773–1782.

Hershey, John R., Chen, Zhuo, Le Roux, Jonathan, Watanabe, Shinji, 2016. Deep clustering: discriminative embeddings for segmentation and separation. In: 2016 IEEE International Conference on Acoustics, Speech and Signal Processing (ICASSP). IEEE, pp. 31–35.

Hochreiter, Sepp, Schmidhuber, Jürgen, 1997. Long short-term memory. Neural Computation 9 (8), 1735–1780.

Sayed, Hossein Khatoonabadi, Vasconcelos, Nuno, Bajic, Ivan V., Shan, Yufeng, 2015. How many bits does it take for a stimulus to be salient? In: Proceedings of the IEEE Conference on Computer Vision and Pattern Recognition, pp. 5501–5510.

Hou, Rui, Chen, Chen, Shah, Mubarak, 2017. Tube convolutional neural network (T-CNN) for action detection in videos. In: Proceedings of the IEEE International Conference on Computer Vision (ICCV), pp. 5822–5831.

Huang, De-An, Fei-Fei, Li, Niebles, Juan Carlos, 2016. Connectionist temporal modeling for weakly supervised action labeling. In: European Conference on Computer Vision (ECCV). Springer, pp. 137–153.

Itti, Laurent, Koch, Christof, 2000. A saliency-based search mechanism for overt and covert shifts of visual attention. Vision Research 40 (10–12), 1489–1506.

Jain, Mihir, Van Gemert, Jan, J´egou, Herv´e, Bouthemy, Patrick, Snoek, Cees GM, 2014. Action localization with tubelets from motion. In: Proceedings of the IEEE Conference on Computer Vision and Pattern Recognition, pp. 740–747.

Jhuang, Hueihan, Gall, Juergen, Zuffi, Silvia, Schmid, Cordelia, Black, Michael J., 2013. Towards understanding action recognition. In: Proceedings of the IEEE International Conference on Computer Vision, pp. 3192–3199.

Ji, Xu, Henriques, João F., Vedaldi, Andrea, 2019. Invariant information clustering for unsupervised image classification and segmentation. In: Proceedings of the IEEE International Conference on Computer Vision, pp. 9865–9874.

Jia, Xu, De Brabandere, Bert, Tuytelaars, Tinne, Gool, Luc V., 2016. Dynamic filter networks. In: Neural Information Processing Systems, pp. 667–675.

Jiang, Yu-Gang, Liu, Jingen, Roshan Zamir, A., Toderici, George, Laptev, Ivan, 2014. Mubarak Shah, and Rahul Sukthankar. THUMOS challenge: Action recognition with a large number of classes.

Kitani, Kris M., Ziebart, Brian D., Bagnell, James Andrew, Hebert, Martial, 2012. Activity forecasting. In: European Conference on Computer Vision. Springer, pp. 201–214.

Kodinariya, Trupti M., Makwana, Prashant R., 2013. Review on determining number of cluster in k-means clustering. International Journal 1 (6), 90–95.

Kuehne, Hilde, Arslan, Ali, Serre, Thomas, 2014. The language of actions: recovering the syntax and semantics of goal-directed human activities. In: IEEE Conference on Computer Vision and Pattern Recognition (CVPR), pp. 780–787.

Kuehne, Hilde, Gall, Juergen, Serre, Thomas, 2016. An end-to-end generative framework for video segmentation and recognition. In: IEEE Winter Conference on Applications of Computer Vision (WACV). IEEE, pp. 1–8.

Kurby, Christopher A., Zacks, Jeffrey M., 2008. Segmentation in the perception and memory of events. Trends in Cognitive Sciences 12 (2), 72–79.

Lan, Tian, Wang, Yang, Mori, Greg, 2011. Discriminative figure-centric models for joint action localization and recognition. In: 2011International Conference on Computer Vision. IEEE, pp. 2003–2010.

Lea, Colin, Reiter, Austin, Vidal, Ren´e, Hager, Gregory D., 2016. Segmental spatiotemporal cnns for fine-grained action segmentation. In: European Conference on Computer Vision (ECCV). Springer, pp. 36–52.

Leboran, Victor, Garcia-Diaz, Anton, Fdez-Vidal, Xose R., Pardo, Xose M., 2016. Dynamic whitening saliency. IEEE Transactions on Pattern Analysis and Machine Intelligence 39 (5), 893–907.

LeCun, Yann, Bengio, Yoshua, et al., 1995. Convolutional networks for images, speech, and time series.

Lei, Peng, Todorovic, Sinisa, 2018. Temporal deformable residual networks for action segmentation in videos. In: IEEE Conference on Computer Vision and Pattern Recognition (CVPR), pp. 6742–6751.

Li, Ruiyu, Tapaswi, Makarand, Liao, Renjie, Jia, Jiaya, Urtasun, Raquel, Fidler, Sanja, 2017. Situation recognition with graph neural networks. In: Proceedings of the IEEE International Conference on Computer Vision, pp. 4173–4182.

Li, Zhenyang, Gavrilyuk, Kirill, Gavves, Efstratios, Jain, Mihir, Snoek, Cees GM, 2018. Videolstm convolves, attends and flows for action recognition. Computer Vision and Image Understanding 166, 41–50.

Liang, Junwei, Jiang, Lu, Carlos Niebles, Juan, Hauptmann, Alexander G., Fei-Fei, Li, 2019. Peeking into the future: predicting future person activities and locations in videos. In: Proceedings of the IEEE Conference on Computer Vision and Pattern Recognition, pp. 5725–5734.

Liu, Wenqian, Sharma, Abhishek, Camps, Octavia, Sznaier, Mario, 2018. DYAN: a dynamical atoms-based network for video prediction. In: Proceedings of the European Conference on Computer Vision (ECCV), pp. 170–185.

Lotter, William, Kreiman, Gabriel, Cox, David, 2016. Deep predictive coding networks for video prediction and unsupervised learning. arXiv preprint. arXiv:1605.08104.

Luc, Pauline, Neverova, Natalia, Couprie, Camille, Verbeek, Jakob, LeCun, Yann, 2017. Pre-dicting deeper into the future of semantic segmentation. In: Proceedings of the IEEE International Conference on Computer Vision, pp. 648–657.

Luong, Minh-Thang, Pham, Hieu, Manning, Christopher D., 2015. Effective approaches to attention-based neural machine translation. arXiv preprint. arXiv:1508.04025.

Ma, Shugao, Sigal, Leonid, Sclaroff, Stan, 2016. Learning activity progression in lstms for activity detection and early detection. In: Proceedings of the IEEE Conference on Computer Vision and Pattern Recognition, pp. 1942–1950.

Ma, Shugao, Zhang, Jianming, Ikizler-Cinbis, Nazli, Sclaroff, Stan, 2013. Action recognition and localization by hierarchical space-time segments. In: Proceedings of the IEEE International Conference on Computer Vision, pp. 2744–2751.

van der Maaten, Laurens, Hinton, Geoffrey, 2008. Visualizing data using t-SNE. Journal of Machine Learning Research 9, 2579–2605.

Malmaud, Jonathan, Huang, Jonathan, Rathod, Vivek, Johnston, Nick, Rabinovich, Andrew, Murphy, Kevin, 2015. What's cookin'? Interpreting cooking videos using text, speech and vision. arXiv preprint. arXiv:1503.01558.

Mathieu, Michaël, Couprie, Camille, LeCun, Yann, 2015. Deep multi-scale video prediction beyond mean square error. CoRR. arXiv:1511.05440 [abs].

Misra, Ishan, Zitnick, C. Lawrence, Hebert, Martial, 2016. Shuffle and learn: unsupervised learning using temporal order verification. In: European Conference on Computer Vision. Springer, pp. 527–544.

Neverova, Natalia, Luc, Pauline, Couprie, Camille, Verbeek, Jakob J., LeCun, Yann, 2017. Predicting deeper into the future of semantic segmentation. CoRR abs. arXiv:1703.07684.

Radvansky, Gabriel A., Zacks, Jeffrey M., 2014. Event Cognition. Oxford University Press.

Lea, Colin, Flynn, Michael D., Ren´e, Vidal, Austin, Reiter, Hager, Gregory D., 2017. Temporal convolutional networks for action segmentation and detection. In: IEEE International Conference on Computer Vision (ICCV).

Richard, Alexander, Gall, Juergen, 2016. Temporal action detection using a statistical language model. In: IEEE Conference on Computer Vision and Pattern Recognition (CVPR), pp. 3131–3140.

Richard, Alexander, Kuehne, Hilde, Gall, Juergen, 2017. Weakly Supervised Action Learning with RNN Based Fine-to-Coarse Modeling. IEEE Conference on Computer Vision and Pattern Recognition (CVPR), vol. 1.2, p. 3.

Richmond, Lauren L., Zacks, Jeffrey M., 2017. Constructing experience: event models from perception to action. Trends in Cognitive Sciences 21 (12), 962–980.

Rodriguez, Mikel D., Ahmed, Javed, Shah, Mubarak, 2008. Action Mach a spatio-temporal maximum average correlation height filter for action recognition. In: 2008 IEEE Conference on Computer Vision and Pattern Recognition. IEEE, pp. 1–8.

Russakovsky, Olga, et al., 2015. Imagenet large scale visual recognition challenge. International Journal of Computer Vision (IJCV) 115 (3), 211–252.

Santoro, Adam, Raposo, David, Barrett, David G., Malinowski, Mateusz, Pascanu, Razvan, Battaglia, Peter, Lillicrap, Timothy, 2017. A simple neural network module for relational reasoning. In: Advances in Neural Information Processing Systems, pp. 4967–4976.

Sener, Fadime, Yao, Angela, 2018. Unsupervised learning and segmentation of complex activities from video. In: IEEE Conference on Computer Vision and Pattern Recognition (CVPR).

Sharma, Shikhar, Kiros, Ryan, Salakhutdinov, Ruslan, 2015. Action Recognition Using Visual Attention. Neural Information Processing Systems: Time Series Workshop.

Shipley, Thomas F., Zacks, Jeffrey M., 2008. Understanding Events: From Perception to Action, vol. 4. Oxford University Press.

Simonyan, Karen, Zisserman, Andrew, 2014. Very deep convolutional networks for large-scale image recognition. arXiv preprint. arXiv:1409.1556.

Song, Jingkuan, Gao, Lianli, Guo, Liu, Wu, Zhao, Zhang, Dongxiang, Tao Shen, Heng, 2017. Hierarchical LSTM with adjusted temporal attention for video captioning. In: Proceedings of the 26th International Joint Conference on Artificial Intelligence. AAAI Press, pp. 2737–2743.

Soomro, Khurram, Idrees, Haroon, Shah, Mubarak, 2015. Action localization in videos through context walk. In: Proceedings of the IEEE International Conference on Computer Vision, pp. 3280–3288.

Soomro, Khurram, Idrees, Haroon, Shah, Mubarak, 2016. Predicting the where and what of actors and actions through online action localization. In: Proceedings of the IEEE Conference on Computer Vision and Pattern Recognition, pp. 2648–2657.

Soomro, Khurram, Shah, Mubarak, 2017. Unsupervised action discovery and localization in videos. In: Proceedings of the IEEE International Conference on Computer Vision, pp. 696–705.

Soomro, Khurram, Zamir, Amir Roshan, Shah, Mubarak, 2012. UCF101: a dataset of 101 human actions classes from videos in the wild. arXiv preprint. arXiv:1212.0402.

de Souza, Fillipe D.M., Sarkar, Sudeep, Srivastava, Anuj, Su, Jingyong, 2016. Spatially coherent interpretations of videos using pattern theory. International Journal on Computer Vision (IJCV), 1–21.

Srivastava, Nitish, Mansimov, Elman, Salakhutdinov, Ruslan, 2015. Unsupervised learning of video representations using LSTMs. CoRR. arXiv:1502.04681 [abs].

Stein, Sebastian, McKenna, Stephen J., 2013. Combining embedded accelerometers with computer vision for recognizing food preparation activities. In: ACM International Joint Conference on Pervasive and Ubiquitous Computing. ACM, pp. 729–738.

Sun, Chen, Shrivastava, Abhinav, Vondrick, Carl, Sukthankar, Rahul, Murphy, Kevin, Schmid, Cordelia, 2019. Relational action forecasting. In: Proceedings of the IEEE Conference on Computer Vision and Pattern Recognition, pp. 273–283.

Thorpe, Simon, Fize, Denis, Marlot, Catherine, 1996. Speed of processing in the human visual system. Nature 381 (6582), 520.

Tian, Yicong, Sukthankar, Rahul, Shah, Mubarak, 2013. Spatiotemporal deformable part models for action detection. In: Proceedings of the IEEE Conference on Computer Vision and Pattern Recognition, pp. 2642–2649.

Tipper, Steven P., Lortie, Cathy, Baylis, Gordon C., 1992. Selective reaching: Evidence for action-centered attention. Journal of Experimental Psychology. Human Perception and Performance 18 (4), 891.

Tran, Du, Yuan, Junsong, 2012. Max-margin structured output regression for spatio-temporal action localization. In: Advances in Neural Information Processing Systems, pp. 350–358.

Tran, Du, Yuan, Junsong, 2011. Optimal spatio-temporal path discovery for video event detection. In: CVPR 2011. IEEE, pp. 3321–3328.

Uijlings, Jasper R.R., Van De Sande, Koen E.A., Gevers, Theo, Smeulders, Arnold W.M., 2013. Selective search for object recognition. International Journal of Computer Vision (IJCV) 104 (2), 154–171.

Van Gemert, Jan C., Jain, Mihir, Gati, Ella, Snoek, Cees G.M., et al., 2015. APT: action localization proposals from dense trajectories. In: BMVC, vol. 2, p. 4.

Vaswani, Ashish, Shazeer, Noam, Parmar, Niki, Uszkoreit, Jakob, Jones, Llion, Gomez, Aidan N., Kaiser, Łukasz, Polosukhin, Illia, 2017. Attention is all you need. In: Advances in Neural Information Processing Systems, pp. 5998–6008.

Vondrick, Carl, Pirsiavash, Hamed, Torralba, Antonio, 2016. Anticipating visual representations from unlabeled video. In: IEEE Conference on Computer Vision and Pattern Recognition (CVPR), pp. 98–106.

Vondrick, Carl, Shrivastava, Abhinav, Fathi, Alireza, Guadarrama, Sergio, Murphy, Kevin, 2018. Tracking emerges by colorizing videos. In: Proceedings of the European Conference on Computer Vision (ECCV), pp. 391–408.

Walker, Jacob, Gupta, Abhinav, Hebert, Martial, 2014. Patch to the future: unsupervised visual prediction. In: Proceedings of the IEEE Conference on Computer Vision and Pattern Recognition, pp. 3302–3309.

Wang, Hongxing, Yuan, Junsong, Wu, Ying, 2014a. Context-aware discovery of visual co-occurrence patterns. IEEE Transactions on Image Processing 23 (4), 1805–1819.

Wang, Limin, Qiao, Yu, Tang, Xiaoou, 2014b. Video action detection with relational dynamic-poselets. In: European Conference on Computer Vision. Springer, pp. 565–580.

Xie, Junyuan, Girshick, Ross, Farhadi, Ali, 2016. Unsupervised deep embedding for clustering analysis. In: International Conference on Machine Learning (ICML), pp. 478–487.

Xu, Kelvin, Ba, Jimmy, Kiros, Ryan, Cho, Kyunghyun, Courville, Aaron, Salakhudinov, Ruslan, Zemel, Rich, Bengio, Yoshua, 2015. Show, attend and tell: neural image caption generation with visual attention. In: International Conference on Machine Learning, p. 2048. 2057.

Yang, Zhilin, Dai, Zihang, Yang, Yiming, Carbonell, Jaime, Salakhutdinov, Russ R., Le, Quoc V., 2019. Xlnet: generalized autoregressive pretraining for language understanding. In: Advances in Neural Information Processing Systems, pp. 5754–5764.

Zacks, Jeffrey M., Braver, Todd S., Sheridan, Margaret A., Donaldson, David I., Snyder, Abraham Z., Ollinger, John M., Buckner, Randy L., Raichle, Marcus E., 2001a. Human brain activity time-locked to perceptual event boundaries. Nature Neuroscience 4 (6), 651–655.

Zacks, Jeffrey M., Tversky, Barbara, Iyer, Gowri, 2001b. Perceiving, remembering, and communicating structure in events. Journal of Experimental Psychology. General 130 (1), 29.

Zacks, Jeffrey M., Magliano, Joseph P., 2011. Film, narrative, and cognitive neuroscience. Art and the Senses 435, 454.

Zacks, Jeffrey M., Speer, Nicole K., Swallow, Khena M., Braver, Todd S., Reynolds, Jeremy R., 2007. Event perception: a mind-brain perspective. Psychological Bulletin 133 (2), 273.

Zacks, Jeffrey M., Tversky, Barbara, 2001. Event structure in perception and conception. Psychological Bulletin 127 (1), 3.

Zhang, Mengmi, Teck Ma, Keng, Hwee Lim, Joo, Zhao, Qi, Feng, Jiashi, 2017. Deep future gaze: gaze anticipation on egocentric videos using adversarial networks. In: Proceedings of the IEEE Conference on Computer Vision and Pattern Recognition, pp. 4372–4381.

Zhang, Richard, Isola, Phillip, Efros, Alexei A., 2016. Colorful image colorization. arXiv:1603.08511 [cs.CV].

Zhu, Gao, Porikli, Fatih, Li, Hongdong, 2015. Tracking randomly moving objects on edge box proposals. arXiv preprint. arXiv:1507.08085.

Biographies

Ramy Mounir is a Ph.D. candidate in the Computer Science and Engineering department at the University of South Florida (USF) in Tampa. He received his B.Eng and M.S. degrees in Mechanical Engineering from USF in 2015 and 2018, respectively. He graduated Summa Cum Laude and received the Outstanding Graduate Award in 2015. He also received his Robotics Graduate Certificate from USF in 2018. He is the recipient of the Early Innovation Award from Intel Corporation. His research interests include self-supervised learning of hierarchical representations of objects and events, implementing perceptual and cognitive theories using computational deep learning methods and predictive models.

Sathyanarayanan N. Aakur is an Assistant Professor of Computer Science at Oklahoma State University. He received the B.Eng. degree in Electronics and Communication Engineering from Anna University, Chennai in 2013. He received the M.S. degree in Management Information Systems and the Ph.D. degree in Computer Science from the University of South Florida, Tampa, in 2015 and 2019, respectively. His research interests include self-supervised learning, commonsense reasoning for visual understanding, and deep learning applications for genomics.

Sudeep Sarkar is a Professor and Chair of Computer Science and Engineering and the Associate Vice President for Special Programs at the University of South Florida in Tampa. He received his B.Tech. degree from the Indian Institute of Technology, Kanpur, M.S. and Ph.D. degrees in Electrical Engineering from The Ohio State University, Columbus. He has more than 25-year expertise in computer vision and pattern recognition algorithms and systems, holds ten U.S. patents, licensed technologies, and has published high-impact journal and conference papers. He is a Fellow of the AAAS, IEEE, IAPR, AIMBE, and NAI. He has served on many journal boards and is currently the Editor-in-Chief for Pattern Recognition Letters.

Probabilistic anomaly detection methods using learned models from time-series data for multimedia self-aware systems

Carlo Regazzoni, Ali Krayani, Giulia Slavic, and Lucio Marcenaro

DITEN, University of Genoa, Genoa, Italy

CHAPTER POINTS

- This chapter introduces an anomaly detection framework and explains its benefits to self-aware systems to keep learning incrementally.

- We briefly review anomaly detection methods in the state of the art, performing a comparison of them, and highlighting our contribution.

- A self-aware agent dealing with multimedia sensory data can detect hierarchical-level abnormalities and handle both low- and high-dimensional data.

- Generalized States (GSs) are built directly from low dimensional observations. In contrast, a Variational Autoencoder is used to output lower dimension GSs from high dimensional data.

- A Generalized Dynamic Bayesian Network (GDBN) is learned from GSs and used to predict the system's dynamics at multiple levels.

- The predictive and diagnostic messages passing inside the GDBN allow calculating the Generalized Errors to discover new emergent rules.

- We validate the proposed approach using real multi-sensory data from a semi-autonomous self-aware vehicle.

Copyright © 2022 Elsevier Inc. All rights reserved.

13.1 Introduction

The identification of abnormal instances of data represents a relevant issue across several diverse research topics. Through anomaly detection algorithms, video surveillance cameras can recognize when possibly dangerous or violent events are happening, such as break-ins, assaults, armed robberies or traffic incidents; suspicious zones in medical images like mammograms or tomography scans can be identified and can provide a help for doctors to diagnose tumors or nodules; fraudulent credit card transactions can be detected as deviations from the normal usage profile of the clients. In general, anomaly detection methods constitute a necessary component in all applications that require identifying a deviation from a known set of rules or models. It is indeed fundamental to note that anomalies are related to a model and are not absolute: in the video surveillance case, the initial model could describe the normal interactions between customers in a store; in the medical imaging case it could extract the salient features of a healthy organ; in the credit card case it could keep track of user's normal credit card transaction patterns. Abnormal data correspond to a deviation from these learned models and are instead subject to different rules. Therefore, it is evident how anomaly detection, which has many applications, represents a fundamental step also for artificial self-aware systems that can continually learn from new situations, as it allows a system to distinguish the occurrence of a new situation from an already experienced one.

Self-awareness can be defined as the "capacity of becoming the object of one's own attention" (Morin, 2006) and is reached when an agent focuses not only on the external environment, but also on the evolution of its own state. The concept of self-awareness has been typically attributed to biological beings and has been studied by scientists of different backgrounds, such as Haykin and Fuster (2014), Friston et al. (2014) and Damasio (1999). Recent research has also started to transfer this concept to embody artificial agents, such as semiautonomous cars and drones or cognitive radios. For these agents to be at the same time aware of their surroundings and of their internal state, they must be endowed with two different types of sensors: i) exteroceptive sensors to reach situational awareness, such as external cameras; ii) proprioceptive sensors to gain self-awareness, such as a steering wheel sensor in a vehicle. Though their situational and self-awareness capabilities, these agents can recognize the experience they are encountering from a set of learned experiences; this knowledge consequently constitutes the base to conduce a decision making step and perform an action, impressed on the environment through a set of actuators.

As recently proposed by Regazzoni et al. (2020), six basic capabilities should be possessed by an agent for it to be considered self-aware: initialization, inference, anomaly detection, model creation, and interface with control. The self-aware agent is first initialized with a predefined model (i), and is able to memorize the model (ii); it consequently uses the learned model to perform inferences about its future state and the future changes in the environment (iii); it identifies new situations when they appear (iv) and learns a new model to describe them (v); finally, it uses the acquired knowledge to perform decisions and impress changes in the environment (vi). It can be observed how each capability is consecutive to the other and how they form a closed loop, as (v) reconnects itself with (ii). Again, the importance of anomaly detection for self-aware systems appears fundamental, as it triggers the identification of the presence of unknown rules and the creation of a new model. In such a context, it is additionally fundamental to remark how we consider time series data and how the generation of anomalies can be defined through the comparison of the prediction at one time instant for the following one and the observation at the following time instant.

In this chapter, we propose a method for the detection of anomalies on time-series data, to be inserted in the above-described self-awareness cycle. The method leverages Generalized Dynamic Bayesian Networks (GDBNs) and their message-passing capabilities to define a hierarchy of different levels of anomalies. It can be used both for low-dimensional data and for high-dimensional data. In the high-dimensional case (e.g., images), Generative Models such as Variational Autoencoders (VAEs) or Generative Adversarial Networks (GANs) can be used to perform reduction of the high-dimensional data to a probabilistic low-dimensional latent space. In this chapter, we discuss as a case study the use of a VAE for dimensionality reduction.

The rest of the chapter is structured as follows: Section 13.2 presents basic concepts about Generative Models, GDBNs, Variational Autoencoders and state of the art regarding anomaly detection in low-dimensional and high-dimensional data; Section 13.3 describes the proposed method; Section 13.4 shows and discusses results obtained on proposed data; Section 13.5 concludes the chapter and presents some insights about possible future work.

13.2 Base concepts and state of the art

13.2.1 Generative models

Two types of models can be distinguished in the Machine Learning context: *discriminative models* and *generative models*. Discriminative models learn the conditional probability of a class label given some observed input; this is the case of classifiers. On the other hand, generative models are able to learn the underlying hidden distribution of the data they are trained on and can generate samples from the same distribution. Bayesian Networks (BNs) are a type of generative model factorizing the joint distribution of data w.r.t. conditional probabilities depending on hidden model variables; whereas Dynamic Bayesian Networks (DBNs) perform this factorization across time.

When working on high dimensional data such as images, the most commonly used generative models are Variational Autoencoders (VAEs) (Kingma and Welling, 2014) and Generative Adversarial Networks (GANs) (Goodfellow et al., 2014). VAEs learn the probability distribution of data *explicitly*, commonly assuming a Gaussian or Gaussian Mixture Model distribution. VAEs supposing a Gaussian distribution of the data encode each data sample through a mean and a variance. Sampling from the learned distribution, they can generate similar datapoints to the ones they were trained on. Conversely, GANs implicitly learn the distribution of the data without defining any parameters. VAEs and GANs can additionally be used inside a DBN to learn the relation between high-dimensional data (e.g., images) and low-dimensional latent states (i.e., the socalled *observation model*) (Slavic et al., 2021). We can consequently distinguish two cases in our generative model definition problem: in the first case, the observation model is given (e.g., radio-data case); in the second case, the observation model must be learned (e.g., video-data case). In this chapter, we report results on both the partially learned and totally learned case.

13.2.2 Dynamic Bayesian Network (DBN) models

Bayesian Networks (BNs) are directed acyclic graph models where nodes in the graph represent a set of random variables and edges encode a particular factorization of the joint distribution of that set from a single point in time. However, in many real-world applications, most of the events are not detected based on a single point in time but on multiple observations that yield to a certain event (Mihajlovic and Petkovic, 2001).

A Dynamic Bayesian Network (DBN) is an extension of the BNs that can model dynamic processes and describes a system's evolution with time at hierarchical levels. DBNs allow encoding probabilistic dependencies and feedbacks between random variables over different time slices. A DBN is typically represented by two sets of parameters. The first includes the number of nodes in each time slice and the corresponding topology while the second set consists of the conditional probability distributions (CPDs) described by the edges of the network. DBNs generalize linear dynamical systems by representing the hidden and observed states in terms of random state variables providing a graphical structure that specifies the corresponding conditional dependencies and a compact parameterization of the model. DBNs can decompose data with complex and nonlinear dynamics into segments that are explainable by simpler dynamical units. A specific class of DBN models known as Switching linear dynamic systems (SLDS) (Fox et al., 2011) can be used to represent the dynamical units and explain their switching behavior and their dependency on both observations and continuous hidden states. Learning a DBN consists of parameter learning and structure learning. The former is the process of learning the distributions of discrete or continuous hidden variables in the DBN while the latter is the process of using data to learn the links (i.e., the conditional probabilities) among random variables in the DBN. Both parameter and structure learning depend on the state-space model in question. A state-space representation of a time series process defines a prior $P(X_t)$, a state transition function $P(X_t|X_{t-1})$ and an observation function $P(Z_t|X_t)$. Hidden Markov Models (HMMs) and Kalman Filter Models (KFMs) are possible ways to represent state-space models encoded in a simple DBN which includes one hidden variable and one observed node per slice as shown in Fig. 13.1. The hidden discrete variables in HMM are assumed to have a multinomial discrete distribution and evolve according to transition rules parametrized in the model. The discrete variables and the corresponding transition matrix can be learned by using a clustering algorithm to the observed time-series data. In the case of KFMs the state transition evolution becomes a linear-Gaussian such as:

$$X_t = AX_{t-1} + BU_t + w_t \qquad (13.1)$$

where U_t is the control input realizing the parent node of the corresponding X_t. In the Bayesian framework, learning starts with a priori knowledge about the model structure (i.e. links in the DBN) and model parameters. In the case of a simple DBN (Fig. 13.1), the initial knowledge about the hidden state variable X_t in terms of probability distribution $P(X_t)$ can be updated using data Z_t and used to calculate the posterior probability $P(X_t|Z_t)$ using Bayes rule:

$$P(X_t|Z_t) = \frac{P(Z_t|X_t)P(X_t)}{P(Z_t)}. \qquad (13.2)$$

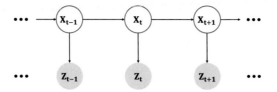

FIGURE 13.1 Simple Dynamic Bayesian Network. The network is an HMM when X is a discrete variable; it is a KFM when X is a continuous variable.

A possible combination of Particle Filter (PF) and Kalman Filter (KF) presented in Baydoun et al. (2018) can be implemented on the DBN as a probabilistic inference algorithm. In the Bayesian framework, there are different types of probabilistic inferences, namely, the top-down (or predictive) inference and the bottom-up (or diagnostic) inference. Each of these inferences relies on decomposing the required computation into local calculations at each node of the network which requires only messages passed (i.e. using the message passing algorithm (Winn and Bishop, 2005)) along the edges connected to that node.

In this chapter, the proposed approach is enriched with advanced capabilities compared to the classical Bayesian inference to exploit the predictive and diagnostic messages incoming into a generic node of the DBN to calculate abnormality measurements at hierarchical levels using appropriate probabilistic distances. Such abnormalities can also be used to learn new models incrementally that describe variations at different abstraction levels.

13.2.3 Variational autoencoder

As observed in Sec. 13.2.1, a VAE is a generative model. In its vanilla version, it is composed of an encoder $Q_\theta(X|Z)$ and a decoder $P_\phi(Z|X)$, which can be estimated through the learning process of the VAE on training data. Through θ and ϕ, we define the parameters of the encoder and decoder, respectively. The encoder $Q_\theta(X|Z)$ allows to represent each sample Z inputted to it through two bottleneck features, i.e., a mean μ and a variance σ^2. On the other hand, the decoder $P_\phi(Z|X)$ synthesizes an observation Z from a latent state X sampled from $\mathcal{N}(\mu, \sigma^2)$. Therefore, the former allows reduction of dimensionality of the observations (i.e., image data) and treatment of anomaly detection at a low-dimensional state level; instead, the latter allows to obtain again high-dimensional data and treat anomaly detection at the observation level. Note how, in the DBN framework described in the previous section, the VAE can be seen as an observation model $P(Z_t|X_t)$.

The learning of the VAE is conducted to optimize the parameters θ and ϕ, maximizing the sum of the lower bound on the marginal likelihood of each observation Z of the dataset D, as described in Kingma and Welling (2014, 2019):

$$\mathcal{L}_{\phi,\theta}(D) = \sum_{Z \in D} \mathcal{L}_{\phi,\theta}(Z), \tag{13.3}$$

where $\mathcal{L}(\theta, \phi; Z)$ is defined as:

$$\mathcal{L}_{\phi,\theta,}(Z) = -D_{KL}\big(Q_\theta(X|Z)||P_\phi(X)\big) + E_{Q_\theta(X|Z)}\big[\log p_\phi(Z|X)\big], \tag{13.4}$$

where the term D_{KL} is the KL divergence. Therefore, the first term measures the difference between the encoder's distribution $Q_\theta(X|Z)$ and the prior $P_\phi(X)$; the prior typically being a standard normal distribution $\mathcal{N}(0, 1)$. The second term is the expected log-likelihood of the observation Z and forces the VAE to reconstruct the input data.

In this work, the ability of the VAE is used to encode the input information in a significant lower-dimensional space that exhibits probabilistic properties. As observed in Section 13.2.1, VAEs can be integrated inside a DBN model, in the case in which the observation model is not given, but must be learned too.

13.2.4 Types of anomalies and anomaly detection methods

As observed in Sec. 13.1 anomalies correspond to any deviation that manifests itself w.r.t. a reference model. The concept of anomaly is vast and dependent on data type, cardinality of relationship between the involved variables, data structure and data distribution (Foorthuis, 2020). Different anomaly types – and consequently anomaly distances and anomaly detection algorithms – can be indentified when working on time-independent or time dependent applications, on low-dimensional data such as trajectories, high dimensional data such as images or structures such as graphs, etc.. To estimate the complexity and vastness of anomaly detection, we reconnect ourselves again to the work of Foorthuis (2020), that through a review of anomaly detection literature recognizes three broad groups of anomaly, 9 basic types and 63 subtypes. Taking as an example video data, Ramachandra et al. (2020) distinguish five main types of anomaly: *i)* appearance anomaly, *ii)* short-term motion anomaly, *iii)* long-term motion anomaly, *iv)* group anomaly, *v)* time-of-day anomaly. To make an example, a car driving along a street could either see an abnormal object as a traffic cone at the center of the street (*i*), detect an abnormal motion as the car in front suddenly braking (*ii*), perform a sequence of abnormal movements, due for example to the driver falling asleep (*iii*), observe two other vehicles interacting in an abnormal way, as in a collision (*iv*); finally, normal motions and speeds during the day could not be legal after sunset (*v*).

The work by Chandola et al. (2009) represents a base point for classification of anomaly detection methods. In this section, we report a brief summary of the defined anomaly classes for point anomalies. The authors of Chandola et al. (2009) distinguish six general methods for point anomaly detection; the first five are nonprobabilistic methods, while the sixth one groups together the probabilistic methods. In the following paragraphs, we give a brief discussion of five out of six of these methods. Additionally, Chandola et al. (2009) notice how additional methods can be distinguished for contextual anomalies such as abnormal sequences in a time series.

Classification based methods (CLA)

In this type of methods, a classifier is learned that can either distinguish normal data from abnormal ones based on a discriminative boundary (one-class classifier) or different classes of normal data from each other (multiclass classifier). In the second case, an instance is detected as abnormal when it can not be associated with enough confidence to any of the normal classes. Any algorithm typically used for classification can be adopted for this type of method, including Neural Networks, Bayesian Networks, Support Vector Machines (SVMs) and rule-based classification methods.

Distance based methods (DB)

These types of method are based on the calculation of the distance of data points from their closest neighbors, leveraging the assumptions that normal data occur in dense neighborhoods, whereas anomalies are far from their closest neighbors. The choice of the distance metric is therefore essential. To account for neighborhoods of various densities, the distance value is often relative to the distances between points in the closest neighborhood.

Clustering based methods (CLU)

These methods rely on assumptions dependent on the chosen clustering techniques: *i)* techniques like DBSCAN, ROCK or SNN assume that anomalies are not assigned to any of the built clusters; *ii)* when using techniques like SOM, one can assume that abnormal data lie far away from the cluster's centroid; *iii)* finally, one could leverage the assumption that normal data belong to big clusters and abnormal ones to small clusters.

Information theoretic methods (IT)

These methods use information theoretic measures such as Kolomogorov Complexity, entropy or relative entropy to analyze the information content of the data and extract the most anomalous subset of data points (i.e., sequence, subgraph, image area) w.r.t. the rest of the data.

Statistic/probabilistic methods (STA)

Probabilistic-based methods aim to model the normal sample features' distributions (inliers) associated with the training data, whereas abnormalities can be detected as data samples known as outliers with low probability (Rivera et al., 2020). The probability that a data sample belongs to a certain distribution can be estimated using probabilistic distances such as Bhattacharyya, Mahalanobis, Hellinger or the Kullback–Leibler divergence. Abnormality detection using probabilistic (statistical) models can be classified into two main groups, parametric as Gaussian Mixture Model (GMM) methods and Regression methods and nonparametric as Histogram or Kernel Density Estimation (KDE) methods (Wang et al., 2019). A significant advantage of the probabilistic approach is that it can be easily generalized to different data types and modalities; whereas its drawback is related to fitting the data to a certain distribution which may not be appropriate in some cases (Aggarwal, 2016).

Additionally, Chandola et al. (2009) identify a further set of anomaly methods, i.e., spectral anomaly detection techniques, based on the assumption that data, especially high dimensional ones, can be brought to a lower dimensional subspace where normal and abnormal data can be more easily distinguished. Examples of such methods could be using PCA on image data or embedding graphs as adjacency matrices. These methods can be used as a preprocessing step to the other methods defined above. As we will talk more in detail on the problem of dimensionality reduction in Sec. 13.2.6, we momentarily ignore this type of method. Similarly, when handling abnormal sequences of data, such as in time-series data, detection can be either performed through an adaptation of the previous methods to contextual anomalies or through methods that learn the structure of the data, for example executing the prediction of the following time instants from the previous ones (*prediction methods (PRE)*).

Table 13.1 summarizes the proposed classification, indicating desired characteristics for each method. Notice how all the proposed methods are applicable to time-series data and are semisupervised or unsupervised, necessitating no training labels. Most methods could also be applied for some variety of incremental learning through the use of anomaly scores, but this problem is rarely tackled in the state of the art. Classification does not give an anomaly score with a confidence range, as in general in this class of method a sample is only described as belonging or not to the class; information theoretic methods also do not have this capability. Distance based methods and information theoretic methods are additionally difficult to extend to high dimensional data. Only prediction based methods and statistical methods can be applied to probability models. Finally, our method has all the defined characteristics and additionally provides an anomaly score on different levels of abstraction.

TABLE 13.1 Comparisons between the different methods. *: unnecessary for one-class version, necessary for multiclass version; **: valid for most methods of this category but not all.

	CLA	DB	CLU	IT	STA	PRE	Ours
Data labeling unnecessary	*	✓	✓	✓	✓	✓	✓
Gives anomaly score with confidence range	✗**	✓	✓	✗	✓	✓	✓
Extendable to HD data	✓	✗	✓	✗	✓	✓	✓
Applicable to probability models	✗	✗	✗	✗	✓	✓	✓
Applicable to time-series data	✓	✓	✓	✓	✓	✓	✓
Allows incremental learning	✓	✓	✓	✗	✓	✓	✓
Gives anomaly on diverse abstraction levels	✗	✗	✗	✗	✗	✗	✓

13.2.5 Anomaly detection in low-dimensional data

In this chapter, we concentrate on semisupervised probabilistic methods based on DBN representations for time-series data. Several works have leveraged BNs and DBNs to perform anomaly detection in various fields (Mascaro et al., 2014; Bronstein et al., 2001; Salotti, 2018). This type of networks, allowing a hierarchical probabilistic representation of the world, is particularly appropriate for autonomous systems that try to reflect human reasoning. BNs additionally allow to build a hierarchy of anomalies.

In particular, Markov Jump Particle Filter (MJPF) has been used as a basic Bayesian filter applied on DBNs to perform anomaly detection on low-dimensional data from different types of sensors of multimodal physical agents like semiautonomous vehicles, as odometry data in Baydoun et al. (2018) and control information in Kanapram et al. (2019).

13.2.6 Anomaly detection in high-dimensional data

As video data is high-dimensional, it cannot be elaborated pixel-wise in a time-efficient way by most anomaly detection algorithms developed for low-dimensional data – including MJPF. Instead, features need first to be extracted from the video data, allowing the problem to be reformulated in a lower dimensionality (Chong and Tay, 2015). The process of feature extraction can either be accomplished through *hand-crafted methods* or through *Deep Learning (DL) based methods*. In the first case, human knowledge about specific problems such as occlusion, variations in orientation, scale, or illumination is exploited to develop feature extractors; one famous example among many is that of the Histogram of Oriented Gradients (HOG) (Dalal and Triggs, 2005). In DL based methods, NNs are trained to extract from data the relevant features for a specific task by minimizing a loss function. As DL based methods have largely proved to outperform hand-crafted ones (Antipov et al., 2015; Alshazly et al., 2019; Nugroho, 2018), in this chapter we only consider the former ones.

Once dimensionality reduction is performed, many anomaly detection methods described in Section 13.2.4 can be performed on the extracted features, as summarized in Table 13.1. Additionally, when employing a feature extractor that also allows to perform reconstruction of the original frames from the low-dimensional features, reconstruction based anomaly methods are typically used too. As defined by Kiran et al. (2018), these types of method learn to reconstruct normal frames or video sequences, and identify badly reconstructed data samples as anomalies; they typically use techniques such as Principal Component Analysis or Autoencoders. Generative models like VAEs and GANs can be similarly used too, additionally displaying the ability to learn the underlying distribution of data at the observation level.

This ability of VAEs and GANs is in line with the spirit of this chapter, as it focuses its attention on the development of a method that is probabilistic and that can consequently be inserted inside a DBN framework for capturing anomalies and exploiting them to perform the continual learning of models in time-series data. By reducing the dimensionality of the observed frames, VAEs and GANs can be used as nonlinear observation models in the DBN framework described in Sec. 13.2.2.

As VAEs are able to explicitly describe the probability distribution of the underlying data, several works have attempted to use them for the creation of a linear switching model for high dimensional data (Watter et al., 2015; Johnson et al., 2016; Fraccaro et al., 2017; Becker-Ehmck et al., 2019). These works however did not consider anomaly detection and continual learning of models as an objective. Additionally, experiments were until now conducted on very simple datasets displaying only basic motions and observations from a static view point.

Another work (Ravanbakhsh et al., 2020) instead coupled GANs and DBNs contemplating at the same time anomaly detection and continual learning. In this work, a hierarchy of coupled cross-modal GANs is developed. When a high anomaly is produced due to the detection of abnormal observations or abnormal dynamics, a new GAN is built for learning the new situation. Instead of a MJPF, a Switching model is developed, consisting of the hierarchy of GANs at the continuous level and of an HMM at the discrete level. However, due to the learning of the data probability without explicit parametrization – as is intrinsic to GANs – this method has limitations and does not contemplate the possibility of using some of the anomaly distances typically used for statistic anomaly detection methods (see Sec. 13.2.4).

The closest works to the one proposed in this chapter are those of Slavic et al. (2021, 2020) and Campo et al. (2020), in which a coupling of VAE and MJPF is introduced, with the objective of performing anomaly detection and continual learning of models.

13.3 Framework for computing anomaly in self-aware systems

13.3.1 General framework description

The proposed framework is described in Fig. 13.2. Low and high dimensional data can be given as input to the framework. After executing preprocessing on low-dimensional data and feature extraction through the VAEs' encoders on high-dimensional data, a Generalized State vector is built including the filtered states as generalized errors obtained from the initial model which is activated during the real-time inference during the initial iteration. In this case, the initial model represents static rules of the surrounding environment (i.e. the expected dynamic changes are quasi null and are affected by only random perturbations). Performing clustering on the Generalized Errors, a Generalized Dynamic Bayesian Network (GDBN) model is learned. When new data is obtained, the learned model is used to perform inference through the GDBN and different levels of anomalies are extracted (i.e., frame anomaly, state anomaly, discrete level anomaly). If a high level of anomaly is detected, the corresponding Generalized States and Generalized Errors are stored, and a new model is built to be used in parallel with the previous ones in the next GDBNs inference.

From Fig. 13.2 we can also deduce the characteristics described in Table 13.1. *i)* First of all, our method does not need any data labeling. Instead, it is semisupervised: a model is learned and anomaly is detected w.r.t. it. *ii)* Both low and high dimensional data can be used as input. In Fig. 13.2, odometry, steering angle, power and rotor velocity data from a vehicle are shown as examples of low-dimensional data; video data from a camera as high-dimensional data; *iii)* Time-series data is considered; *iv)* A probability model in the form of a GDBN is learned; *v)* Anomalies w.r.t. this model is detected at different abstraction levels, specifically at the frame level, continuous level and discrete level; *vi)* The detected anomalies are coupled with a confidence range, as will be evident in Section 13.3.4; *vii)* When an anomaly is detected w.r.t. a learned model, a new model is created, allowing incremental learning and an expansion of previous knowledge.

In the following sections, a more detailed description of the framework is provided. The training of the GDBN model is defined in Section 13.3.2. Sections 13.3.3 and 13.3.4 describe the MJPF testing algorithm and its multilevel anomaly measurements, respectively. Finally, in Section 13.3.5, the use of Generalized Errors obtained through the DBN anomalies is set as a basis for the continual learning of new models. The index of the Section where each concept is explained can also be found in red in Fig. 13.2 near its graphic representation.

13.3.2 Generalized dynamic Bayesian network (GDBN) model

Initially, a Self-Aware agent starts perceiving the surrounding environment using an initial GDBN (i.e., a null force filter with static assumptions about the environmental states) by interpreting the received generalized observations $\tilde{Z}_t = [Z_t \; \dot{Z}_t]$ that comprise the variable and its generalized coordinates of motion coming from the agent's sensors. In this case, the agent detects abnormalities all the time and calculates the Generalized Errors ($GE(\tilde{X}_t)$) expressed as:

$$\tilde{X}_t = [X_t \; \dot{X}_t] \tag{13.5}$$

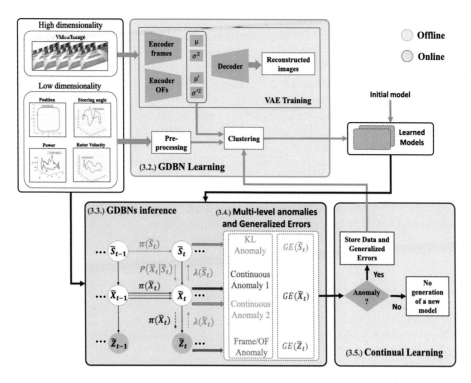

FIGURE 13.2 Block diagram of the proposed method.

where, X_t are the predicted hidden states performed by the null force filter according to the following dynamic model:

$$X_t = X_{t-1} + v_t \tag{13.6}$$

while, $\dot{X}_t$ are errors on derivatives calculated as innovations by the null force filter using current generalized observations as follows:

$$\dot{X}_t = \frac{Z_t - H X_{t-1}}{\Delta t}. \tag{13.7}$$

Consequently, the GEs collected in previous experience are used during the training phase as input to an unsupervised clustering algorithm (e.g. Growing Neural Gas (GNG) (Fritzke, 1994), Self-Organizing Maps (SOMs) (Kohonen, 2001)) which encodes the GEs into discrete components producing a set of discrete variables (**S**) or neurons which we refer to as super-states, such that:

$$\mathbf{S} = \{S_1, S_2, \ldots, S_M\} \tag{13.8}$$

where M is the total number of superstates.

Note how the equations considered above refer to the low-dimensional data case, in which the observation model is known, is linear and is expressed by the matrix H that maps observations to hidden states. In the case of high-dimensional data, instead, we must first bring data to a low-dimensionality through the use of a feature extractor. Therefore, the first part of the GDBN that must be learned is the observation model itself. As observed in Section 13.2.6, this can be performed through the training of networks such as VAEs and GANs. In this chapter, we consider as example case the use of VAEs for such a task. To capture both information of frame content and of motion between subsequent frames, we train a couple of VAEs, one to reconstruct the frame at time t, z_t, and the other to reconstruct the Optical Flow (OF) between frames at time t and $t + 1$, OF_t. The bottleneck is consequently divided in two parts: one, that we call X_t, representing the state/content (from z_t); the other, that we call $\dot{X}_t$, capturing the motion/velocity information (from OF_t). Clustering is performed on the combination of these two variables.

Once the dimensionality reduction process is performed, the rest of the method remains mostly the same, with some adjustments and limitations.

After clustering, the $M \times M$ transition matrix (Π) defined as:

$$\Pi = \begin{bmatrix} \pi(S_t = S_1) \\ \vdots \\ \pi(S_t = S_M) \end{bmatrix} = \begin{bmatrix} \pi_{11} & \cdots & \pi_{1M} \\ \vdots & \ddots & \vdots \\ \pi_{M1} & \cdots & \pi_{MM} \end{bmatrix} \tag{13.9}$$

is learned by estimating the transition probabilities $\pi_{i,j} = P(S_t = i | S_{t-1} = i)$, $i, j \in \mathbf{S}$ over a period of time. Thus, it is possible to create a set of generalized superstates $\tilde{S}_t$ defined as follows:

$$\tilde{S}_t = [S_t \; \dot{S}_t] = \begin{bmatrix} S_t & E(S_t | S_{t-1}) \end{bmatrix} \tag{13.10}$$

which includes the current discrete variable S_t and the event $E(.)$ of transiting to that variable conditioned to be in S_{t-1} in the previous time instant. This set of discrete variables realizes the top or discrete level of the GDBN. Each generalized superstate $\tilde{S}_t$ ($\tilde{S}_t \in \tilde{\mathbf{S}}$) is associated with statistical properties as mean value ($\mu_{\tilde{S}_t}$), covariance matrix ($\Sigma_{\tilde{S}_t}$) and a set of hidden Generalized States (GSs) $\tilde{X}_t$ encoded inside it. The hidden continuous GSs ($\tilde{X}_t$) represent the intermediate or continuous level of the GDBN. The relationship among the hidden states and superstates is characterized by $P(\tilde{X}_t | \tilde{S}_t)$ link in the GDBN. The GDBN represents the hidden variables in generalized coordinates of motion, which allows the SA agent to self-organize by optimizing the joint posterior as new observations arrive and to encode continually new concepts related to an emergent situation after detecting anomalies.

The bottom level in the GDBN stands for the real generalized observations ($\tilde{Z}_t$) measured by the sensors. The path from $\tilde{S}_t$ to $\tilde{Z}_t$ realizes the chain of causality among random variables at hierarchical levels. The probabilistic dependencies among the variables involved in the chain of causality is characterized by the linked edges. Considering in this case the chain rule, the joint probability of $\tilde{S}_t$, $\tilde{X}_t$ and $\tilde{Z}_t$; $P(\tilde{S}_t, \tilde{X}_t, \tilde{Z}_t)$, can be factorized as a product of the conditional probabilities such as:

$$P(\tilde{S}_t, \tilde{X}_t, \tilde{Z}_t) = P(\tilde{S}_t) P(\tilde{X}_t | \tilde{S}_t) P(\tilde{Z}_t | \tilde{X}_t). \tag{13.11}$$

This implies the possibility to use the model to generate new data samples given the direct cause (i.e. the parent node). In addition, links between generalized hidden variables at consecutive time instants represent the corresponding conditional temporal probabilities. Π embeds the dynamic rules at the discrete level that drive the dynamic changes by switching between multiple dynamic models at the continuous level. Such probabilistic causal reasoning in the GDBN, allows to explain events, diagnose causes, and make predictions about future events which improve the decision making/action processes. The transition probability $P(\tilde{S}_t|\tilde{S}_{t-1})$ at the discrete level can be decomposed into:

$$\tilde{S}_t = f(\tilde{S}_{t-1}) + w_t = f(\pi_{ij}) + w_t \tag{13.12}$$

where $f(.)$ is a nonlinear function that determines the superstates temporal evolution based on the learned transition matrix and is subject to a process noise w_t which is assumed to be drawn from a zero multivaraite normal distribution with covariance Σ_t such that $w_t \sim \mathcal{N}(0, \Sigma_t)$.

The dynamic causal models describing the state-space representation of a time-series process – assuming that each observation $\tilde{Z}_t$ is generated from a d-dimensional hidden state $\tilde{X}_t$ which by the way has been generated by a discrete hidden superstate $\tilde{S}_t$ – have the following forms:

$$\tilde{X}_t = g(\tilde{X}_{t-1}, \tilde{S}_{t-1}) + w_t = A\tilde{X}_{t-1} + BU_{\tilde{S}_t} + w_t \tag{13.13}$$

$$\tilde{Z}_t = h(\tilde{X}_t) + v_t = H\tilde{X}_t + v_t. \tag{13.14}$$

The continuous function $g(.)$ in Eq. (13.13) is assumed to be linear and determines the state temporal evolution at the continuous level guided by the discrete level predictions and it is subject to a Gaussian noise w_t. In Eq. (13.13), A and B are the dynamic model matrix and the control model matrix, respectively. A different control vector $U_{\tilde{S}_t}$ is associated to each superstate $\tilde{S}_t$ that depends on the mean derivative of $\tilde{X}_t$ samples encoded in that superstate. So, Π encodes not only the transitions between superstates but also jumps between different linear models at the continuous level. In Eq. (13.14), H is the observation matrix that maps hidden $\tilde{X}_t$ to observation $\tilde{Z}_t$ and v_t is the measurement Gaussian noise with zero mean and covariance R_t such that $v_t \sim \mathcal{N}(0, R_t)$.

Note how, in the case of high-dimensional data, the observation model described in Eq. (13.14) corresponds to the nonlinear transformation applied by the VAEs when performing feature extraction. Therefore, the function h is in this case nonlinear. Additionally, also Eq. (13.13) can not be directly applied to describe the state temporal evolution, due to strong nonlinearities that can be found in the GS. To keep the structure of the high-dimensional case as similar to the low-dimensional one as possible, we choose to learn for each found superstate a Neural Network $N^{(S)}$ describing the temporal evolution of the GS for that superstate. Eq. (13.13) can therefore be substituted with the following expression:

$$\tilde{X}_t = g(\tilde{X}_{t-1}, \tilde{S}_{t-1}) + w_t = N^{\tilde{S}_t}(\tilde{X}_{t-1}) + w_t \tag{13.15}$$

where w_t is the error after convergence of the network and can be approximated as a Gaussian noise.

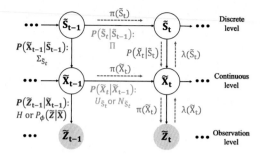

FIGURE 13.3 The proposed Generalized Dynamic Bayesian Network.

Fig. 13.3 summarizes the proposed GDBN. In red, the intraframe connections are linked to their corresponding element: $P(\tilde{Z}_{t-1}|\tilde{X}_{t-1})$ is the observation model and corresponds to matrix H in the low-dimensional case and to the VAE's decoder $P_\phi(\tilde{Z}|\tilde{X})$ in the high-dimensional case; the link $P(\tilde{X}_{t-1}|\tilde{S}_{t-1})$ is correlated with the cluster covariance $\Sigma_{\tilde{S}_t}$. In green, the interframe connections are displayed: the prediction model $P(\tilde{X}_t|\tilde{X}_{t-1})$ is chosen through the local motion $U_{\tilde{S}_t}$ in the low-dimensional case and through the NN N_{S_t} in the high dimensional case; $P(\tilde{S}_t|\tilde{S}_{t-1})$ is encoded by the transition matrix Π.

13.3.3 Real-time inference algorithm

The Markov Jump Particle Filter (MJPF) firstly introduced in Baydoun et al. (2018) is here employed during the real-time process to perform inferences at different hierarchical levels starting from the learned GDBN model. MJPF realizes a switching model that uses a combination of Particle Filter (PF) to predict discrete superstates and a bank of Kalman Filters (KFs) for continuous states' prediction and estimation. In the case of GSs extracted from high-dimensional data, in which we apply a nonlinear model to perform state prediction, a bank of Unscented Kalman Filters (UKF) (Wan and van der Merwe, 2000) can be used instead, as performed in Slavic et al. (2020). Such a switching behavior between the dynamic transitions at discrete/continuous levels and observations permits updating of belief in hidden variables by passing local messages in simultaneous inference modes, namely the predictive or causal inference (top-down) and the diagnostic inference (bottom-up). Temporal predictive messages ($\pi(\tilde{S}_t), \pi(\tilde{X}_t)$); refer to Fig. 13.3) depend on the dynamic rules stored in the model while the hierarchical intraslice top-down messages from $\tilde{S}_t$ to $\tilde{X}_t$ depend on the clustering statistics including mean values and covariance matrices. The bottom-up reasoning is based on likelihood models consisting of messages ($\lambda(\tilde{S}_t), \lambda(\tilde{X}_t)$); refer to Fig. 13.3) passing in a feed-backward manner to adjusting expectations given a sequence of observations. The PF relies on the transition matrix encoded in the dynamic model as a proposal distribution to predict future superstates ($\tilde{S}_t^n$). Initially, it draws N samples from that distribution which are equally weighted and associated to a specific superstate such that:

$$\langle \tilde{S}_t^n, W^n \rangle \sim \langle \pi\left(\tilde{S}_t^n\right), 1/N \rangle. \tag{13.16}$$

Then, for each particle $(.^n)$, a KF is employed to predict the continuous variables $\tilde{X}_t$ which depend on the expected superstate as pointed out in Eq. (13.13) and can be expressed in terms of conditional probability $P(\tilde{X}_t|\tilde{X}_{t-1}, \tilde{S}_t^n)$. The posterior probability associated with the predicted state is defined as:

$$\pi(\tilde{X}_t) = P(\tilde{X}_t, \tilde{S}_t|\tilde{Z}_{t-1}) = \int P(\tilde{X}_t|\tilde{X}_{t-1}, \tilde{S}_t) \overbrace{\lambda(\tilde{X}_{t-1})}^{P(\tilde{Z}_{t-1}|\tilde{X}_{t-1})} d\tilde{X}_{t-1}. \tag{13.17}$$

Accordingly, once a new evidence $\tilde{Z}_t$ is observed, a message back-ward propagated from the bottom-level towards the higher levels can be exploited by the MJPF to estimate the posterior probability $P(\tilde{X}_t, \tilde{S}_t^n|Z_t)$ in the following way:

$$P(\tilde{X}_t, \tilde{S}_t|\tilde{Z}_t) = \pi(\tilde{X}_t)\lambda(\tilde{X}_t). \tag{13.18}$$

Consequently, the weights of the corresponding particle can be updated according to:

$$W_t^n = W_t^n \lambda(\tilde{S}_t) \tag{13.19}$$

and then normalized following the Sequential Importance Resampling (SIR) approach. In Eq. (13.19), $\lambda(\tilde{S}_t)$ is the diagnostic message propagated from bottom towards the top level of hierarchy to update the belief in hidden variables at that level and can be calculated as:

$$\lambda(\tilde{S}_t) = \lambda(\tilde{X}_t)P(\tilde{X}_t|\tilde{S}_t) = P(\tilde{Z}_t|\tilde{X}_t)P(\tilde{X}_t|\tilde{S}_t) \tag{13.20}$$

where $P(\tilde{X}_t|\tilde{S}_t) \sim \mathcal{N}(\mu_{\tilde{S}_k}, \Sigma_{\tilde{S}_k})$ denotes a Gaussian distribution with mean $\mu_{\tilde{S}_k}$ and covariance $\Sigma_{\tilde{S}_k}$. While, $\lambda(\tilde{X}_t) \sim \mathcal{N}(\mu_{\tilde{Z}_t}, R)$ denotes a Gaussian distribution with mean $\mu_{\tilde{Z}_t}$ and covariance R. The multiplication between $\lambda(\tilde{X}_t)$ and $P(\tilde{X}_t|\tilde{S}_t)$ can be obtained by calculating the Bhattacharyya distance (D_B) as follows:

$$D_B\big(\lambda(\tilde{X}_t), P(\tilde{X}_t|\tilde{S}_t = \tilde{S}_k)\big) = -\ln \int \sqrt{\lambda(\tilde{X}_t)P(\tilde{X}_t|\tilde{S}_t = \tilde{S}_k)}d\tilde{X}_t \tag{13.21}$$

where $\tilde{S}_k \in \tilde{S}$. The vector D_λ containing all the D_B values between $\lambda(\tilde{X}_t)$ and all the superstates in the set $\tilde{S}$ is here estimated as:

$$D_\lambda = \big[D_B\big(\lambda(\tilde{X}_t), P(\tilde{X}_t|\tilde{S}_t = \tilde{S}_1)\big), \ldots, D_B\big(\lambda(\tilde{X}_t), P(\tilde{X}_t|\tilde{S}_t = \tilde{S}_L)\big)\big]. \tag{13.22}$$

Therefore, the vector $\lambda(\tilde{S}_t)$ in terms of probability is:

$$\lambda(\tilde{S}_t) = \left[\frac{1/D_\lambda(1)}{1/\sum_{l=1}^{L} D_\lambda(l)}, \ldots, \frac{1/D_\lambda(L)}{1/\sum_{l=1}^{L} D_\lambda(l)}\right]. \tag{13.23}$$

In classical Bayesian filtering, predictive and diagnostic reasoning is used to estimate a joint posterior at different hierarchical levels. However, an indispensable step is missed in this process to evaluate the differences between two messages arriving at a given node based on a probabilistic metric that estimates the surprise caused by observations which can not be explained by the model. The following section shows how to provide a Self-Aware agent with the capability to detect multimodal abnormalities which are the basis to continually update the knowledge about the perceived environment and encode new concepts related to the emergent abnormal behavior.

13.3.4 Multimodal abnormality measurements

In the proposed approach, classical Bayesian inference (i.e., MJPF) is enriched with advanced functionality that exploits a probabilistic distance between top-down and bottom-up messages incoming to a generic node at different levels inside the GDBN to define hierarchical abnormality measurements. Such probabilistic distances quantify the similarity between two probability distributions ($\pi(\mathcal{X})$ and $\lambda(\mathcal{X})$) over the domain $\mathcal{X}$ which can be continuous or discrete distributions. Several important probabilistic distances include the following:

- Bhattacharrya distance (D_B) (Bhattacharyya, 1946): this distance has been proposed to reflect the degree of dissimilarity between two probability distributions. It is based on the Bhattacharyya Coefficient ($\mathcal{BC}$) that approximates the amount of overlap between two distributions. In the case of discrete probability distributions $\mathcal{BC}$ is defined as follows:

$$\mathcal{BC}\big(\pi(\mathcal{X}), \lambda(\mathcal{X})\big) = \sum_{x\in\mathcal{X}} \sqrt{\pi(x)\lambda(x)} \tag{13.24}$$

while, in the case of continuous distributions:

$$\mathcal{BC}\big(\pi(\mathcal{X}), \lambda(\mathcal{X})\big) = \int \sqrt{\pi(\mathcal{X})\lambda(\mathcal{X})}d\mathcal{X}. \tag{13.25}$$

Thus, the Bhattacharrya distance D_B is defined through the $\mathcal{BC}$ in the following way:

$$D_B = -\ln\big[\mathcal{BC}\big(\pi(\mathcal{X}), \lambda(\mathcal{X})\big)\big]. \tag{13.26}$$

In either cases (discrete and continuous distributions), $0 \le \mathcal{BC} \le 1$ and $0 \le D_B \le \infty$.
- Hellinger distance (D_H) (Beran, 1977): this distance (D_H) is also related to the Bhattacharyya Coefficient ($\mathcal{BC}$) and it can be defined as:

$$D_H = \sqrt{1 - \mathcal{BC}\big(\pi(\mathcal{X}), \lambda(\mathcal{X})\big)} \tag{13.27}$$

and satisfies the following property: $0 \le D_H \le \infty$.
- Kullback–Leibler divergence (D_{KL}) (Kullback and Leibler, 1951): the D_{KL} is a way to measure the matching between two probability distributions. If two distributions perfectly match then $D_{KL} = 0$, otherwise it ranges between 0 and ∞. The D_{KL} between two discrete probability distributions can be formulated as:

$$D_{KL}\big(\pi(\mathcal{X})||\lambda(\mathcal{X})\big) = \sum_{x\in\mathcal{X}} \pi(x)\log\frac{\pi(x)}{\lambda(x)} \tag{13.28}$$

while for continuous distributions it is defined in the following way:

$$D_{KL}\big(\pi(\mathcal{X})||\lambda(\mathcal{X})\big) = \int \pi(\mathcal{X}) \log \frac{\pi(\mathcal{X})}{\lambda(\mathcal{X})} d\mathcal{X}. \tag{13.29}$$

- Bregman Divergence (B_D) (Bregman, 1967): B_D measures the distance between two distributions defined in terms of a strictly convex function (ϕ).

$$B_D = \phi\big(\pi(\mathcal{X})\big) - \phi\big(\lambda(\mathcal{X})\big) - \nabla\phi\big(\lambda(\mathcal{X})\big)\big(\lambda(\mathcal{X}) - \pi(\mathcal{X})\big) \tag{13.30}$$

where $\nabla\phi(\lambda(\mathcal{X}))$ is the gradient of ϕ at $\lambda(\mathcal{X})$ and $B_D \geq 0$.

In the following subsections, we use some of the aforementioned probabilistic distances as examples to associate abnormality metrics with the different levels of the GDBN.

13.3.4.1 Discrete level

The abnormality indicator at the discrete level is defined as a distance between the predictive ($\pi(\tilde{S}_t)$) and diagnostic ($\lambda(\tilde{S}_t)$) messages entering to node $\tilde{S}_t$. Such a difference provides an awareness signal indicating of how the real sensory signals received from the surrounding environments behave with respect to the rules encoded in the generalized model. We use the symmetric Kullback-Leibler Divergence as an example to measure the similarity between the two discrete probability distributions ($\pi(\tilde{S}_t)$ and $\lambda(\tilde{S}_t)$). The Kullback-Leibler Divergence Abnormality (KLDA) as defined in Krayani et al. (2020) has the following form:

$$\boldsymbol{KLDA} = D_{KL}\big(\pi(\tilde{S}_t)||\lambda(\tilde{S}_t)\big) + D_{KL}\big(\lambda(\tilde{S}_t)||\pi(\tilde{S}_t)\big). \tag{13.31}$$

At each time instant t the histogram of the predicted particles is extracted and the probability of occurence of each particle is calculated as:

$$p(\tilde{S}_t = i) = \frac{y(\tilde{S}_t = i)}{N} \quad i \in \tilde{S} \tag{13.32}$$

where $y(.)$ is the frequency of occurrence of a specific superstate i and N is the total number of particles propagated by PF. It is worth noting that $\lambda(\tilde{S}_t)$ is unique for all particles at time instant t. Let's define $\mathbb{S}$ as the set of winning particles whose probability of occurrence is greater than zero such that:

$$\mathbb{S} = \big\{i | p(\tilde{S}_t = i) > 0\big\} \quad i \in S. \tag{13.33}$$

D_{KL} is calculated between $\lambda(S_t)$ and specific rows of the transition matrix related to the winning particles in $\mathbb{S}$. Hence, (13.31) becomes:

$$\boldsymbol{KLDA} = \sum_{i \in \mathbb{S}} \left[p(i) \sum_{j=1}^{M} \pi_{ij} \log\left(\frac{\pi_{ij}}{\lambda_j}\right) \right] + \sum_{i \in \mathbb{S}} \left[p(i) \sum_{j=1}^{M} \lambda_j \log\left(\frac{\lambda_j}{\pi_{ij}}\right) \right]. \tag{13.34}$$

13.3.4.2 Continuous level

At this level we will focus on the messages entering to node $\tilde{X}_t$ and calculate the corresponding abnormality measurements defined as statistical metrics based on a probabilistic distance (e.g. D_B, D_H, D_{KL}, etc.). The message $P(\tilde{X}_t | \tilde{S}_t)$ forwarded towards the intermediate level describes the probability of having the prediction $\tilde{X}_t$ in a certain superstate. Thus calculating the difference between the predictive message $\pi(\tilde{X}_t)$ and $P(\tilde{X}_t | \tilde{S}_t)$ allows to evaluate if the predictions performed at the discrete level match the prediction done at the continuous level. For example, such a difference can be obtained through the D_B distance in the following way:

$$Db2 = D_B\left(\pi(\tilde{X}_t), P\left(\tilde{X}_t | \tilde{S}_t^n\right)\right) = -\ln \int \sqrt{P\left(\tilde{X}_t, \tilde{S}_t^n | \tilde{Z}_{t-1}\right) P\left(\tilde{X}_t | \tilde{S}_t^n\right)} d\tilde{X}_t. \tag{13.35}$$

On the other hand, it is important to evaluate how much the real observations support the predictions performed at the continuous level which lead to detect any abnormal behavior occurred in the surrounding environment. The second abnormality at the continuous level can be calculated as the difference between the observations ($\lambda(\tilde{X}_t)$) and the predicted generalized states ($\pi(\tilde{X}_t)$) which is also based on the probabilistic distances defined previously (e.g. D_B, D_H, D_{KL}, etc.). For example using D_B distance, the second abnormality at the continuous level can be formulated as:

$$Db1 = D_B\left(\pi(\tilde{X}_t), \lambda(\tilde{X}_t)\right) = -\ln \int \sqrt{P\left(\tilde{X}_t, \tilde{S}_t^n | \tilde{Z}_{t-1}\right) P(\tilde{Z}_t | \tilde{X}_t)} d\tilde{X}_t. \tag{13.36}$$

13.3.4.3 Observation level

The anomaly at this level is particularly informative in the case of high-dimensional data. In general, two types of observation-level anomalies can be distinguished: *i)* the direct *reconstruction error* due to having learned an observation model that is not suitable for the observed data. This is the case, for example, of anomalies due to previously unseen contents and elements appearing in the scene. These anomalies lead to a high discrepancy between the observed image, X_t, and its reconstruction, $\hat{X}_t$ through the VAE network. *ii)* the distance between the observation, X_t, at t and the predictive message $\pi(\tilde{X}_t)$ forward propagated from the continuous level towards the node Z_t. From the practical point of view, this can be interpreted through the calculation of an frame and OF anomaly such as the Mean Squared Error (MSE) between the predicted image, $\hat{Z}_{t-1}$, at time $t-1$ and the observed image, X_t, at time instant t.

13.3.5 Use of generalized errors for continual learning

In the previous sections, we have seen how the described framework allows to enrich the GDBN architecture and leverage its message-passing concept for the calculation of anomalies and GEs.

The objective of anomaly indicators is to provide the agent with the capability of understanding whether the models that it is using are suitable or not to predict the evolution of the world and of its own state in the situation it is sensing and experiencing. As shown in

Fig. 13.2, in the "Continual Learning" diagram, when high anomalies are detected, the agent is alerted that it should learn a new model. It consequently stores the corresponding GEs.

On the other hand, the purpose of GEs is identifying which actions the filter has to execute in order to correct the predictions it performed. A new model learned from the GEs is one that minimizes the GEs on the same type of sequence as the one on which they had been identified. A clustering process is performed again and a new model is inserted in the vocabulary of learned models.

To summarize, anomalies indicate *that a new model should be created*, whereas GEs instruct the agent on *how* the new model should be created.

It is to note how, in the high-dimensional case, some limitations to the Continual Learning through GEs are present. New observations (e.g., new objects appearing in the scene, new types of environments, etc.) could be detected. In this case, the training of a new VAE is necessary. In this situation, testing requires the use of multiple VAEs (and so, multiple observation models) in parallel.

13.4 Case study results: anomaly detection on multisensory data from a self-aware vehicle

13.4.1 Case study presentation

As a case study on which to present the framework described in Section 13.3, anomaly detection on multisensory data from a moving vehicle is performed. The vehicle, called "iCab" (Marın-Plaza et al., 2016), displayed in Fig. 13.4a, is maneuvered by a human while performing different tasks in the closed environment shown in Fig. 13.4b. Different data types (i.e., video from an onboard front camera, odometry data, control data, etc.) are extracted from the vehicle. We choose to consider odometry data and video data in detail and provide additional examples regarding control data (Steering Angle, Rotor Velocity, Power). As odometry observations have dimension $d = 2$, they constitute the example for a low-dimensional sensor. On the other hand, image data has an initial dimension of $d = 640 \times 480$, providing a study case for the treatment of high-dimensional data. Both these types of sensors are what we defined in the Introduction of the chapter as "exteroceptive sensors".

To perform training on an initial model, as displayed in Fig. 13.2, we take data from a case in which the vehicle moves around the environment without being influenced by external agents, i.e., it performs Perimeter Monitoring (PM) of the courtyard. Two other tasks are proposed, in which the vehicle is hindered in its original PM movement by the presence of pedestrians. Consequently, as the driver must perform new motions to avoid them, anomalies should be detected w.r.t. the first learned model. These two abnormal tasks differentiate themselves based on where the pedestrian is located and, as a consequence, how the driver avoids them: in one case (Fig. 13.5c), the pedestrian, at the center of the courtyard lane, is avoided by moving on its side (PA); in the other case (Fig. 13.5e), as the pedestrian is at a corner of the environment, it is avoided through a U-turn motion (U-turn).

Fig. 13.5 displays the proposed tasks by showing the positions that the vehicle covers across time (left side, i.e., Fig. 13.5a, 13.5c, 13.5e) and some of the images taken from the front camera (right side, i.e., Fig. 13.5b, 13.5d, 13.5f).

(a) 'iCab" vehicle   (b) Environment

FIGURE 13.4 Employed vehicle and environment.

13.4.2 DBN model learning

As a first step of the described framework, an initial model must be learned. In the case of the odometry data, the observation model is known and is given by matrix H, defined as seen in Section 13.3.2. The model's components to be learned are: *i)* the discrete superstates found with a clustering algorithm as the GNG (Fritzke, 1994), with their statistical properties (i.e., mean value μ_{S_k}, covariance matrix Σ_{S_k}); *ii)* the transition matrix Π describing the links between superstates. On the other hand, in the case of video data, the parameters of the encoders and decoder of the VAE must be learned too, which corresponds to estimating the DBN's observation model. Additionally, in this second case, the NNs capturing the prediction model inside each cluster must be trained too.

Through this step, the normality model is learned. In the presented case study, the PM data is used to obtain this initial model.

Fig. 13.6 displays the performed clustering on the odometry and video (13.6b) data. Observing the odometry case (13.6a), the mean values μ_{S_k} for each cluster can be visually recognized as the mean value of the position data of the clusters (shown through the cyan dots) and the mean values of the speed data of the clustered (displayed through the blue arrows). Additionally, the dashed circles are proportional to the standard deviation of the clusters' positions, which can be extracted from the diagonal of Σ_{S_k}. Note that the circumference has been approximated for better visualization purposes through the maximum value along the two directions. Instead, Fig. 13.6a represents clustering in the video case. As opposite sides of the courtyard display similar visual appearance, while the same motions are performed, a symmetry is present. A shadow area is assigned a unique separate cluster (*cluster* 1).

13.4.3 Multilevel anomaly detection

Once all the parameters of the initial model have been learned, it can be used to perform inference when the data of the other two tasks are given. Anomalies can be detected between the predictions performed with the model based on the PM task and the observations from data extracted during the PA and U-turn tasks.

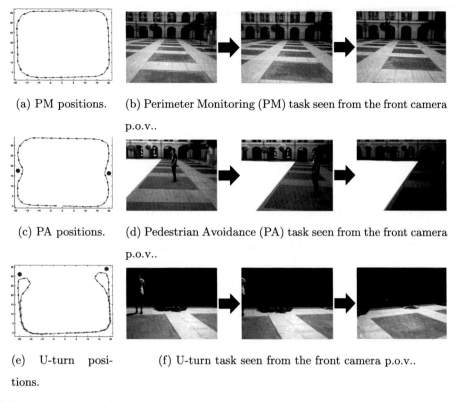

(a) PM positions.   (b) Perimeter Monitoring (PM) task seen from the front camera p.o.v..

(c) PA positions.   (d) Pedestrian Avoidance (PA) task seen from the front camera p.o.v..

(e) U-turn positions.   (f) U-turn task seen from the front camera p.o.v..

FIGURE 13.5 The proposed tasks performed by the self-aware vehicle. The odometry data of the vehicle is displayed in blue on the left, whereas the red dots show where the pedestrian is located. On the right we propose some images from the front camera of the vehicle, while performing the three tasks.

13.4.3.1 *Pedestrian avoidance task*

Figs. 13.7 and 13.8 show the anomalies detected when filtering with the above-described framework on the odometry and on the video data, respectively. In the former case, anomalies at continuous and discrete levels are displayed only, while in the latter, anomalies at the observation level are shown too. Green areas refer to straight motion zones, yellow areas to curving motion zones, and blue areas to zones displaying abnormal actions (i.e., the pedestrian avoidance maneuver).

In Fig. 13.7 we distinguish three plots. The black plot displays the mean value of the GEs, that we call $\tilde{\epsilon}_t$. The red plot shows anomaly $Db2$ (Eq. (13.35)). To notice the similarity between these two plots, as they both represent a comparison between $\pi(X_k)$ and $\lambda(X_k)$. Both of them display high anomaly values corresponding to the Pedestrian avoidance motion and false positives due to slightly irregular curving actions. However, it is to note how $Db1$ gives better results, as it considers the full probability information of $\pi(X_k)$ and $\lambda(X_k)$, while $\tilde{\epsilon}_t$ discards the covariance information Σ_{S_k} and only keeps the mean μ_{S_k}.

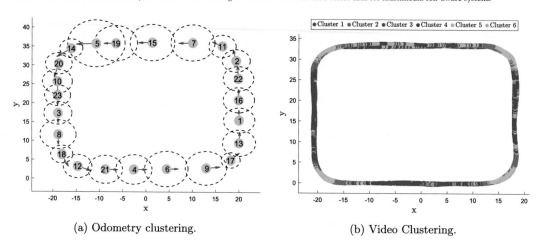

(a) Odometry clustering.   (b) Video Clustering.

FIGURE 13.6 Odometry and video Clustering.

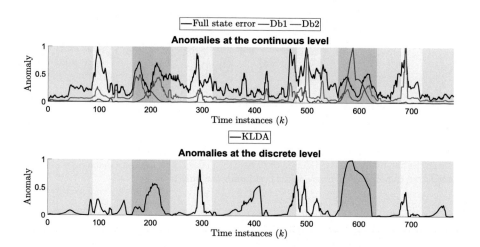

FIGURE 13.7 Odometry multilevel anomaly values for Pedestrian Avoidance task.

Finally, the blue plot in Fig. 13.7 is related to anomaly $Db1$ (Eq. (13.36)). This anomaly becomes particularly relevant at the center of the abnormal motion. For example, we could observe that around time $t = 600$, particles get assigned to clusters relative to the curve on the upper right of Fig. 13.6a, due to the motion mostly resembling this area. However, the likelihood of being in that area is quite low.

The second plot of Fig. 13.7 refers to the $KLDA$ anomaly described in Eq. (13.34). Pedestrian avoidance zones clearly display a very high anomaly value, a consequence of abnormal passages between clusters.

Fig. 13.8 displays these same types of anomalies in the video data case. Additionally, the first plot shows anomalies at the observation level. Black and blue lines refer to the direct VAE

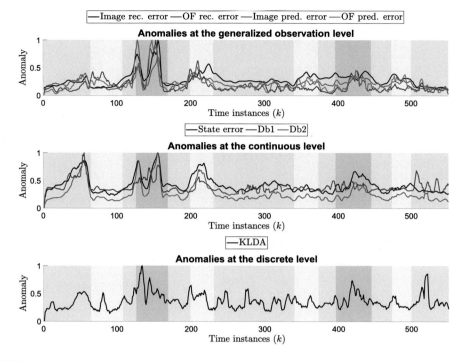

FIGURE 13.8 Video multilevel anomaly values for Pedestrian Avoidance task.

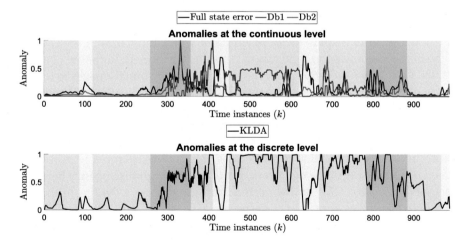

FIGURE 13.9 Odometry multilevel anomaly values for pedestrian avoidance through U-turn task.

reconstruction anomalies, signaling an unsuitable observation model. On the other hand, red and green anomalies are the image-level errors due to discrepancy between observation and prediction.

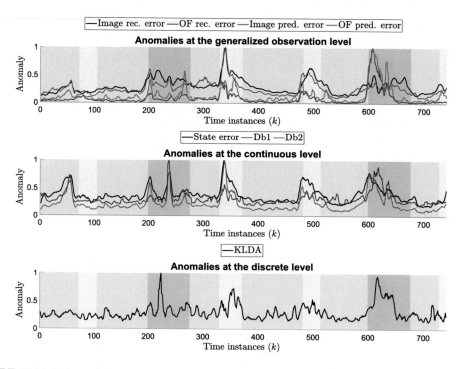

FIGURE 13.10 Video multilevel anomaly values for pedestrian avoidance through U-turn task.

13.4.3.2 U-turn task

Fig. 13.9 and Fig. 13.10 display the odometry and video anomalies in the U-turn task. In this case, the blue area shows the zone where U-turn is performed. In the area between these two zones, the vehicle moves with similar motions observing similar images to those of training, but with opposite direction. It consequently follows from this that: *i)* anomalies $\tilde{\epsilon}_t$ and $Db2$ in the odometry case (Fig. 13.9) display high values only in correspondence to the abnormal U-turn motions and curves on opposite directions to that of training; *ii)* anomaly $Db1$ and $KLDA$ show a high value along all the U-turn motion and while the car is moving with opposite direction to that of training. In both cases, this derives from particles being assigned to clusters on the other side of the courtyard, resulting in a high discrepancy between cluster values and predictions for CLA and abnormal transitions for $KLDA$. *iii)* In the video case (Fig. 13.10), high anomalies appear in correspondence to the U-turn motion and to performing curves in the opposite direction to that of training only.

13.4.3.3 Image-level anomalies

Fig. 13.11 displays examples of image-level behavior. The first two columns and the last two ones report actual frame and OF data at consequent time instants; the third and fourth columns show the direct reconstructions through the VAE; the fifth and sixth columns display the prediction's reconstruction.

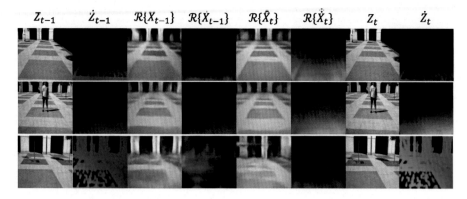

FIGURE 13.11 Examples of image-level anomalies. With the symbol $\mathcal{R}\{X\}$ we express in a concise way the decoding of the latent variable X through the VAE decoder.

Three example cases are shown. *i)* In the first row, an example of normal straight motion taken from the PA dataset is reported. *ii)* In the second row, we display a time instant previous to the start of the avoidance maneuver in the PA dataset. The presence of the pedestrian causes an anomaly that is strictly related to the observation model. A new VAE should be trained to learn the "pedestrian" concept. *iii)* In the third row, we have an anomaly due to performing a curve in the opposite direction w.r.t. training, in the U-turn dataset. To note how, despite not being optimal, the reconstruction is reasonable. On the other hand, the prediction model can capture the anomaly in a more evident way. This is because, when observing the curve-related part of the courtyard, we could expect to be moving left (see the red color of the predicted OF $\mathcal{R}\{\hat{\dot{X}}_t\}$), but instead we perform a turn to the right.

13.4.3.4 Anomaly detection evaluation

To evaluate the anomaly detection method, the anomaly estimation obtained at the different abstraction levels is compared with the Ground Truth (GT). The GT of the PA and U-turn scenarios are built by hand, considering the vehicle's motion for the odometry data. In the odometry U-turn case, the entire sequence from the beginning of the first U-turn to the end of the second one is considered abnormal, as the vehicle moves in the opposite direction w.r.t. training. For the video data, 10 frames before starting the abnormal maneuver are added to the positive GT to account for the pedestrian's presence; additionally, a simple shadow detection method is used to account for abnormal shadow zones.

In Figs. 13.12 and 13.13, the Receiver Operating Characteristic (ROC) curves are shown for the anomaly results on odometry and video data, respectively. A ROC curve plots the values of True Positive Rate (TPR) and False Positive Rate (FPR) at different anomaly thresholds, where $TPR = \frac{TP}{TP+FN}$ and $FPR = \frac{FP}{FP+TN}$. In our case, TPR and FPR consist respectively in the percentage of times in which anomalies are correctly identified and in which they are incorrectly assigned. The AUC (Area Under the Curve) can be used as a measurement of the performance of the method. Another measure of performance is the accuracy (ACC), given by the sum of True Positives (TPs) and True Negatives (TNs). In the field of incremental self-

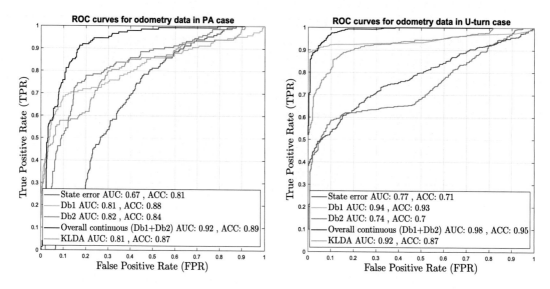

FIGURE 13.12 ROC curves for odometry data.

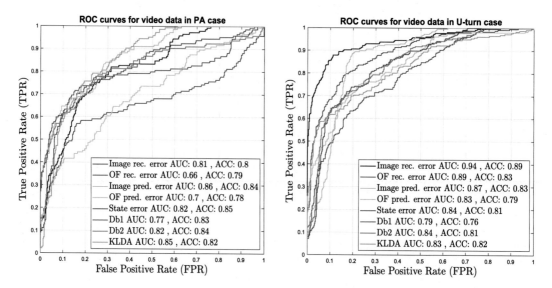

FIGURE 13.13 ROC curves for video data.

aware systems, further evaluation methods could be used, e.g., how good are the anomalies to extract errors allowing to build better models.

In Fig. 13.12, the two continuous anomaly values (i.e., *Db*1 and *Db*2) are normalized and summed together to provide a single anomaly value, displaying much higher performance than its two separate components.

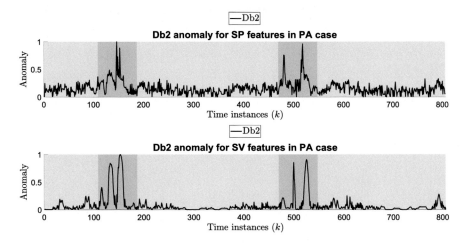

FIGURE 13.14 Control data (SP, SV) anomaly at continuous level for pedestrian avoidance case.

13.4.4 Proprioceptive sensory data anomalies

Some in-detail descriptions of training and testing using odometry and video data were provided in the previous sections. They all constituted examples of exteroceptive data sensors of the vehicle. Besides exteroceptive sensors, in a Self-Awareness framework, proprioceptive sensors of the vehicle should be considered too. These include, for example, Steering Angle (S), Rotor Velocity (V), and Power (P).

Fig. 13.14 and Fig. 13.15 show results of Db2 anomaly on Pedestrian Avoidance and U-turn case, respectively. The three control features – i.e., S, V, and P – can be combined in different ways to form the considered sensory data on which to build the GDBN model. In Fig. 13.14 we report results using the combinations SP (Steering Angle and Power) and SV (Steering Angle and Rotor Velocity) on the PA task; in Fig. 13.14 we show the combinations SP, SV and VP (Rotor Velocity and Power) on the U-turn task. In both cases, the green areas correspond to normal zones and the blue areas to abnormal zones, i.e., Pedestrian Avoidance in the first case and U-turn in the second one. Results related to the control data can be found in Kanapram et al. (2019).

13.4.5 Additional results

The provided study case considered anomalies from a semiautonomous vehicle framework, proposing both low-dimensional, high-dimensional, proprioceptive, and exteroceptive data. A system with a single agent (vehicle) was considered, but systems with multiple agents could be handled too, and the interaction between them modeled at a higher layer, as done by Kanapram et al. (2020).

It is important to note that the described method can be applied to data from other applications too, e.g., self-aware radios. Results in that direction can be examined in the work by Krayani et al. (2020).

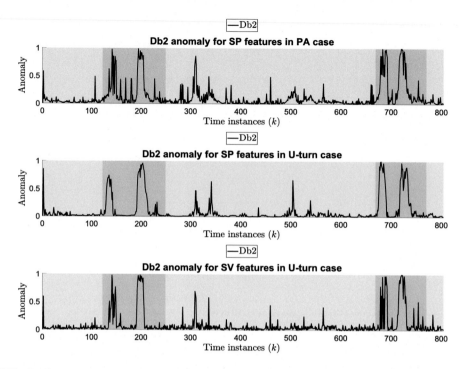

FIGURE 13.15 Control data (SP, SV, VP) anomaly at continuous level for U-turn case.

In all these works, anomalies on a maximum of three levels have been considered. However, higher abstraction levels and, consequently, higher levels of anomalies could be defined. An example in this direction is provided by Zaal et al. (2019): for each task, a graph is obtained connecting adjacent clusters; where a high anomaly on the graph is detected, an emergent concept is found.

13.5 Conclusions

In this chapter, we presented a framework to detect anomalies in time-series data for a Self-Aware system. Under the Self-Awareness framework, an agent dealing with multimedia data, including high and low dimensional data, is capable of detecting anomalies at hierarchical levels, which realize an essential step to continually learn a new dynamic model that encodes the emergent abnormal behavior. For the high dimensional data, a VAE is used to reduce the data-dimensionality. The features are extracted and incorporated in a generalized state vector, whereas in low dimensional data, features are extracted directly from observations and incorporated in a generalized state vector. Training the dynamic model as a probabilistic graphical model structured in a Dynamic Bayesian Network (DBN) forms the bridge between high and low dimensional data. It is shown how the messages passing inside the DBN can be exploited to define a hierarchy of anomalies based on probabilistic distances. A real data-set is used to evaluate the framework, and results shown that multimodal abnormalities can be detected effectively by the Self-Aware agent.

References

Aggarwal, C.C., 2016. Outlier Analysis, 2nd ed. Springer Publishing Company, Incorporated.

Alshazly, H., Linse, C., Barth, E., Martinetz, T., 2019. Handcrafted versus cnn features for ear recognition. Symmetry 11, 1493.

Antipov, G., Berrani, S.A., Ruchaud, N., Dugelay, J.L., 2015. Learned vs. hand-crafted features for pedestrian gender recognition. In: ACM International Conference on Multimedia, pp. 1263–1266.

Baydoun, M., Campo, D., Sanguineti, V., Marcenaro, L., Cavallaro, A., Regazzoni, C., 2018. Learning switching models for abnormality detection for autonomous driving. In: 2018 21st International Conference on Information Fusion (FUSION), pp. 2606–2613.

Becker-Ehmck, P., Peters, J., Smagt, P.V.D., 2019. Switching linear dynamics for variational Bayes filtering. In: International Conference on Machine Learning, pp. 553–562.

Beran, R., 1977. Minimum Hellinger distance estimates for parametric models. The Annals of Statistics 5, 445–463.

Bhattacharyya, A., 1946. On a measure of divergence between two multinomial populations. Sankhyā: The Indian Journal of Statistics (1933–1960) 7, 401–406.

Bregman, L., 1967. The relaxation method of finding the common point of convex sets and its application to the solution of problems in convex programming. U.S.S.R. Computational Mathematics and Mathematical Physics 7, 200–217.

Bronstein, A., Das, J., Duro, M., Friedrich, R., Kleyner, G., Mueller, M., Singhal, S., Cohen, I., 2001. Self-aware services: using Bayesian networks for detecting anomalies in Internet-based services. In: 2001 IEEE/IFIP International Symposium on Integrated Network Management Proceedings. Integrated Network Management VII. Integrated Management Strategies for the New Millennium (Cat. No. 01EX470), pp. 623–638.

Campo, D., Slavic, G., Baydoun, M., Marcenaro, L., Regazzoni, C., 2020. Continual learning of predictive models in video sequences via variational autoencoders. In: IEEE International Conference on Image Processing, pp. 753–757.

Chandola, V., Banerjee, A., Kumar, V., 2009. Anomaly detection: a survey. ACM Computing Surveys 41, 15:1–15:58.

Chong, Y.S., Tay, Y.H., 2015. Modeling representation of videos for anomaly detection using deep learning: a review. arXiv:1505.00523 [abs].

Dalal, N., Triggs, B., 2005. Histograms of oriented gradients for human detection. In: IEEE Conference on Computer Vision and Pattern Recognition, pp. 886–893.

Damasio, A.R., 1999. The Feeling of What Happens: Body and Emotion in the Making of Consciousness. Harcourt Brace.

Foorthuis, R., 2020. On the nature and types of anomalies: a review. arXiv:2007.15634 [abs].

Fox, E., Sudderth, E.B., Jordan, M.I., Willsky, A.S., 2011. Bayesian nonparametric inference of switching dynamic linear models. IEEE Transactions on Signal Processing 59, 1569–1585.

Fraccaro, M., Kamronn, S., Paquet, U., Winther, O., 2017. A disentangled recognition and nonlinear dynamics model for unsupervised learning. In: Conference on Neural Information Processing Systems, pp. 3601–3610.

Friston, K.J., Sengupta, B., Auletta, G., 2014. Cognitive dynamics: from attractors to active inference. Proceedings of the IEEE 102, 427–445.

Fritzke, B., 1994. A growing neural gas network learns topologies. In: Conference on Neural Information Processing Systems, pp. 625–632.

Goodfellow, I.J., Pouget-Abadie, J., Mirza, M., Xu, B., Warde-Farley, D., Ozair, S., Courville, A., Bengio, Y., 2014. Generative adversarial nets. In: Conference on Neural Information Processing Systems, pp. 2672–2680.

Haykin, S., Fuster, J.M., 2014. On cognitive dynamic systems: cognitive neuroscience and engineering learning from each other. Proceedings of the IEEE 102, 608–628.

Johnson, M., Duvenaud, D., Wiltschko, A.B., Adams, R., Datta, S., 2016. Composing graphical models with neural networks for structured representations and fast inference. In: Conference on Neural Information Processing Systems, pp. 2946–2954.

Kanapram, D., Campo, D., Baydoun, M., Marcenaro, L., Bodanese, E., Regazzoni, C., Marchese, M., 2019. Dynamic Bayesian approach for decision-making in ego-things. In: 2019 IEEE 5th World Forum on Internet of Things (WF-IoT), pp. 909–914.

Kanapram, D., Patrone, F., Marín-Plaza, P., Marchese, M., Bodanese, E.L., Marcenaro, L., Gómez, D.M., Regazzoni, C.S., 2020. Collective awareness for abnormality detection in connected autonomous vehicles. IEEE Internet of Things Journal 7, 3774–3789.

Kingma, D.P., Welling, M., 2014. Auto-encoding variational Bayes. In: International Conference on Learning Representations.

Kingma, D.P., Welling, M., 2019. An introduction to variational autoencoders. Foundations and Trends in Machine Learning 12, 307–392.

Kiran, B.R., Thomas, D., Parakkal, R., 2018. An overview of deep learning based methods for unsupervised and semi-supervised anomaly detection in videos. Journal of Imaging 4, 36.

Kohonen, T., 2001. Self-Organizing Maps. Ser. Physics and Astronomy Online Library. Springer, Berlin, Heidelberg.

Krayani, A., Baydoun, M., Marcenaro, L., Alam, A.S., Regazzoni, C., 2020. Self-learning Bayesian generative models for jammer detection in cognitive-uav-radios. In: GLOBECOM 2020–2020 IEEE Global Communications Conference, pp. 1–7.

Kullback, S., Leibler, R.A., 1951. On information and sufficiency. The Annals of Mathematical Statistics 22, 79–86.

Marın-Plaza, P., Beltrán, J., Hussein, A., Musleh, B., Martın, D., de la Escalera, A., Armingol, J.M., 2016. Stereo vision-based local occupancy grid map for autonomous navigation in ros. In: International Joint Conference on Computer Vision, Imaging and Computer Graphics Theory and Applications, pp. 703–708.

Mascaro, S., Nicholson, A., Korb, K., 2014. Anomaly detection in vessel tracks using Bayesian networks. International Journal of Approximate Reasoning 55, 84–98.

Mihajlovic, V., Petkovic, M., 2001. Dynamic Bayesian Networks: a State of the Art. TR-CTIT-34 of CTIT Technical Report Series. University of Twente.

Morin, A., 2006. Levels of consciousness and self-awareness: a comparison and integration of various neurocognitive views. Consciousness Cognition 15, 358–371.

Nugroho, K.A., 2018. A comparison of handcrafted and deep neural network feature extraction for classifying optical coherence tomography (oct) images. In: International Conference on Informatics and Computational Sciences, pp. 1–6.

Ramachandra, B., Jones, M., Vatsavai, R.R., 2020. A survey of single-scene video anomaly detection. ArXiv. arXiv:2004.05993 [abs].

Ravanbakhsh, M., Baydoun, M., Campo, D., Marín, P., Martín, D., Marcenaro, L., Regazzoni, Carlo, 2020. Learning self-awareness for autonomous vehicles: exploring multisensory incremental models. IEEE Transactions on Intelligent Transportation Systems, 1–15.

Regazzoni, C.S., Marcenaro, L., Campo, D., Rinner, B., 2020. Multisensorial generative and descriptive self-awareness models for autonomous systems. Proceedings of the IEEE 108, 987–1010.

Rivera, A.R., Khan, A., Bekkouch, I.E.I., Sheikh, T.S., 2020. Anomaly detection based on zero-shot outlier synthesis and hierarchical feature distillation. IEEE Transactions on Neural Networks and Learning Systems, 1–11.

Salotti, J.M., 2018. Bayesian network for the prediction of situation awareness errors. International Journal of Human Factors Modelling and Simulation 6, 119–126.

Slavic, G., Baydoun, M., Campo, D., Marcenaro, L., Regazzoni, C., 2021. Multilevel anomaly detection through variational autoencoders and Bayesian models for self-aware embodied agents. IEEE Transactions on Multimedia, 1. https://doi.org/10.1109/TMM.2021.3065232.

Slavic, G., Campo, D., Baydoun, M., Marín, P., Martín, D., Marcenaro, L., Regazzoni, C., 2020. Anomaly detection in video data based on probabilistic latent space models. In: IEEE Conference on Evolving and Adaptive Intelligent Systems, pp. 1–8.

Wan, E.A., van der Merwe, R., 2000. The unscented Kalman filter for nonlinear estimation. In: IEEE Adaptive Systems for Signal Processing, Communications, and Control Symposium, pp. 153–158.

Wang, H., Bah, M.J., Hammad, M., 2019. Progress in outlier detection techniques: a survey. IEEE Access 7, 107964–108000.

Watter, M., Springenberg, J.T., Boedecker, J., Riedmiller, M., 2015. Embed to control: a locally linear latent dynamics model for control from raw images. In: Conference on Neural Information Processing Systems, pp. 2746–2754.

Winn, J., Bishop, C., 2005. Variational message passing. Journal of Machine Learning Research 6, 661–694.

Zaal, H., Iqbal, H., Campo, D., Marcenaro, L., Regazzoni, C.S., 2019. Incremental learning of abnormalities in autonomous systems. In: 2019 16th IEEE International Conference on Advanced Video and Signal Based Surveillance (AVSS), pp. 1–8.

Biographies

Carlo Regazzoni: He is full professor of Cognitive Telecommunications Systems at DITEN, University of Genoa, Italy. He has been responsible of several national and EU funded research projects. He is currently the coordinator of international PhD courses on Interactive and Cognitive Environments involving several European universities. He served as general chair in several conferences and associate/guest editor in several international technical journals. He has served in many roles in governance bodies of IEEE SPS and He is serving as Vice President Conferences IEEE Signal Processing Society in 2015–2017.

Ali Krayani: received the bachelor's degree in telecommunication engineering from the Politecnico di Torino, Italy, in 2014 and the master's degree in telecommunication engineering from the University of Florence, Italy, in 2017. He is currently pursuing the Ph.D. degree under the Joint Doctorate in Interactive and Cognitive Environments programme between the University of Genoa and the Queen Mary University of London. His current research interests include cognitive radios, cellular systems, UAV communications, self-awareness, dynamic Bayesian networks, and artificial intelligence.

Giulia Slavic: She obtained her titles of Electronics and Information Technologies Engineer in 2017 and MS of Internet and Multimedia Engineering in 2020 from the University of Genova, Italy. She is currently a PhD student at the University of Genova. Her research interests include the use of deep learning and signal processing algorithms for sequential multisensory data.

Lucio Marcenaro: He has over 20 years of experience in signal processing and image sequence analysis. He is the author of about 160 scientific papers related to signal processing for computer vision and cognitive radio. He graduated in Electronic Engineering in 1999 and obtained a Ph.D. in Computer Science and Electronic Engineering in 2003. He is an Associate Professor in Telecommunications for the Polytechnic School of Engineering of the University of Genoa. His main current research interests are video processing for event recognition, detection, and localization of objects in complex scenes, distributed heterogeneous sensors, and bio-inspired cognitive autonomous systems.

Deep plug-and-play and deep unfolding methods for image restoration

Kai Zhang and Radu Timofte

Computer Vision Lab, ETH Zürich, Zürich, Switzerland

CHAPTER POINTS

- The merits and drawbacks of model-based methods and learning-based methods for image restoration are discussed.

- Deep plug-and-play methods and deep unfolding methods can take advantage of both learning-based methods and model-based methods.

- Experimental results demonstrate the flexibility and effectiveness of deep plug-and-play methods and deep unfolding methods for image restoration.

14.1 Introduction

Image restoration (IR) has been a long-standing problem for its highly practical value in various low-level vision applications (Richardson, 1972; Andrews and Hunt, 1977). This chapter focuses on three representative and fundamental IR problems: denoising, deblurring, and superresoution. In general, the purpose of IR is to recover the latent clean image $\mathbf{x}$ from its degraded observation $\mathbf{y} = (\mathbf{x} \otimes \mathbf{k}) \downarrow_{\mathbf{s}} + \mathbf{n}$, where $\otimes$ represents two-dimensional convolution of $\mathbf{x}$ with blur kernel $\mathbf{k}$, $\downarrow_{\mathbf{s}}$ denotes the standard $\mathbf{s}$-fold downsampler, i.e., keeping the upper-left pixel for each distinct $\mathbf{s} \times \mathbf{s}$ patch and discarding the others, and $\mathbf{n}$ is usually assumed to be additive, white Gaussian noise (AWGN) specified by standard deviation (or noise level) σ. The above degradation model is a general model for single image superreso-

Copyright © 2022 Elsevier Inc. All rights reserved.

lution (SISR). However, if the scale factor **s** is 1, it becomes an image deblurring degradation model; by further fixing **k** as a delta kernel, it turns into a denoising degradation model.

Since IR is an ill-posed inverse problem, the prior, which is also called regularization, needs to be adopted to constrain the solution space (Roth and Black, 2009; Zoran and Weiss, 2011). From a Bayesian perspective, the solution $\hat{\mathbf{x}}$ can be obtained by solving a Maximum A Posteriori (MAP) estimation problem,

$$\hat{\mathbf{x}} = \arg\max_{\mathbf{x}} \log p(\mathbf{y}|\mathbf{x}) + \log p(\mathbf{x}), \tag{14.1}$$

where $\log p(\mathbf{y}|\mathbf{x})$ represents the log-likelihood of observation **y**, $\log p(\mathbf{x})$ delivers the prior of clean image **x** and is independent of degraded image **y**. More formally, (14.1) can be reformulated as

$$\hat{\mathbf{x}} = \arg\min_{\mathbf{x}} \frac{1}{2\sigma^2} \|\mathbf{y} - (\mathbf{x} \otimes \mathbf{k})\downarrow_{\mathbf{s}}\|^2 + \lambda\mathcal{R}(\mathbf{x}), \tag{14.2}$$

where the solution minimizes an energy function composed of a data term $\frac{1}{2\sigma^2}\|\mathbf{y} - (\mathbf{x} \otimes \mathbf{k})\downarrow_{\mathbf{s}}\|^2$ and a regularization or prior term $\lambda\mathcal{R}(\mathbf{x})$ with regularization parameter λ. Specifically, the data term guarantees that the solution accords with the degradation process, while the prior term alleviates the ill-posedness of the problem by enforcing the desired property on the solution.

Generally, the methods to solve Eq. (14.2) can be divided into two main categories, viz, model-based methods and learning-based methods. The former aim to directly solve Eq. (14.2) with some optimization algorithms, while the latter mostly train a predefined parameterized function by optimization of a loss function on a training set containing N degraded-clean image pairs $\{(\mathbf{y}_i, \mathbf{x}_i)\}_{i=1}^{N}$ (Tappen, 2007; Barbu, 2009; Sun and Tappen, 2013; Schmidt and Roth, 2014; Chen and Pock, 2017). In particular, the learning-based methods are usually modeled as the following bi-level optimization problem

$$\begin{cases} \min_{\Theta} \sum_{i=1}^{N} \mathcal{L}(\hat{\mathbf{x}}_i, \mathbf{x}_i) & (14.3a) \\ s.t. \quad \hat{\mathbf{x}}_i = f(\mathbf{y}_i, \Theta), & (14.3b) \end{cases}$$

where Θ denotes the trainable parameters, $\mathcal{L}(\hat{\mathbf{x}}_i, \mathbf{x}_i)$ measures the loss of estimated clean image $\hat{\mathbf{x}}_i$ with respect to ground truth image $\mathbf{x}_i$.

The main difference between model-based methods and learning-based methods is that the former are flexible to handle various IR tasks by simply specifying the degradation operations and can directly optimize on the degraded image **y**, whereas the later require cumbersome training to learn the model before testing and are usually restricted by specialized tasks. Nevertheless, learning-based methods can not only enjoy a fast testing speed but also tend to deliver better performance due to the end-to-end training. In contrast, model-based methods are usually time-consuming with sophisticated priors for the purpose of good performance (Gu et al., 2014). For example, model-based methods such as NCSR (Dong et al., 2013) are flexible to handle denoising, superresolution and deblurring, whereas learning-based methods MLP (Burger et al., 2012), SRCNN (Dong et al., 2016), DCNN (Xu et al., 2014) are designed for these respective tasks. Even for a specific task such as denoising, model-based methods (e.g., BM3D (Dabov et al., 2007) and WNNM (Gu et al., 2014)) can handle different noise levels, whereas the learning-based method of (Jain and Seung, 2009) separately trains a different model for each level. As a result, these two categories of methods have their respective merits and drawbacks, and thus it would be attractive to investigate their integration to leverage their respective merits.

The integration of model-based methods and learning-based methods has resulted in the deep plug-and-play IR method which replaces the denoising subproblem of model-based optimization with the learning-based CNN denoiser prior. The main idea of deep plug-and-play IR is that, with the aid of variable splitting algorithms, such as the alternating direction method of multipliers (ADMM) (Boyd et al., 2011) and half-quadratic splitting (HQS) (Geman and Yang, 1995), it is possible to deal with the data term and prior term separately (Parikh et al., 2014), and particularly, the prior term only corresponds to a denoising subproblem (Danielyan et al., 2010; Heide et al., 2014; Venkatakrishnan et al., 2013) which can be solved via a deep CNN denoiser. Hence, one can treat the deep plug-and-play method as a special case of model-based methods.

Apart from deep plug-and-play methods, deep unfolding methods can also integrate model-based methods and learning-based methods. Their main difference is that the latter optimize the parameters in an end-to-end manner by minimizing the loss function over a large training set, and thus generally produce better results even with fewer iterations. Mathematically, the deep unfolding method can be modeled as

$$
\begin{cases}
\min_{\Theta} \sum_{i=1}^{N} \mathcal{L}(\hat{\mathbf{x}}_i, \mathbf{x}_i) & \text{(14.4a)} \\[2mm]
s.t. \quad \hat{\mathbf{x}}_i = \arg\min_{\mathbf{x}} \dfrac{1}{2\sigma^2} \| \mathbf{y}_i - (\mathbf{x} \otimes \mathbf{k}) \downarrow_{\mathbf{s}} \|^2 + \lambda \mathcal{R}(\mathbf{x}). & \text{(14.4b)}
\end{cases}
$$

In other words, the deep unfolding method aims to replace the predefined function Eq. (14.3b) with the unfolding inference Eq. (14.4b), thus one can treat it as a special case of learning-based methods. The main idea of deep unfolding IR is to first unfold the model-based energy function via a half-quadratic splitting algorithm to obtain an inference which iteratively alternates between solving two subproblems, one related to a data term and the other to a prior term, and then to treat the inference as a deep network, by replacing the solutions to the two subproblems with neural modules. Since the two subproblems correspond respectively to enforcing degradation consistency knowledge and guaranteeing denoiser prior knowledge, the deep unfolding method is well-principled with explicit degradation and prior constraints, which is a distinctive advantage over plain learning-based SISR methods. Fig. 14.1 shows an illustration of the connections between learning-based methods, model-based methods, deep plug-and-play methods, and deep unfolding methods.

The rest of this chapter is organized as follows. Since both deep plug-and-play and deep unfolding methods are highly related to variable splitting algorithms, we thus first introduce one of the most popular ones, i.e., the half quadratic splitting (HQS) algorithm, in Section 14.2. We then present the deep plug-and-play image restoration methods in Section 14.3. Specifically, we propose a highly flexible and effective CNN denoiser prior and then plug the deep

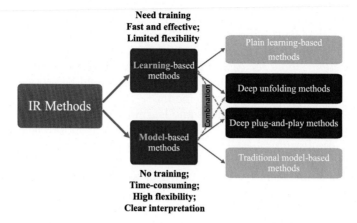

FIGURE 14.1 An illustration of the connections between learning-based methods, model-based methods, deep plug-and-play methods, and deep unfolding methods.

denoiser prior as a modular part into the HQS algorithm to solve various image restoration problems. In Section 14.4, we provide the deep unfolding image restoration methods which can handle deblurring and superresolution with different blur kernels and scale factors via a single model. Section 14.5 shows the quantitative and qualitative results and also provides a thorough analysis of hyper-parameter setting and intermediate results to better understand the working mechanism of deep plug-and-play methods and deep unfolding methods.

14.2 Half quadratic splitting (HQS) algorithm

Although there exist various variable splitting algorithms to solve Eq. (14.2), the half-quadratic splitting (HQS) algorithm owes its popularity to simplicity and fast convergence. We therefore adopt HQS in this chapter. Typically, HQS tackles Eq. (14.2) by introducing an auxiliary variable $\mathbf{z}$, leading to the following approximate equivalence

$$E_\mu(\mathbf{x}, \mathbf{z}) = \frac{1}{2\sigma^2}\|\mathbf{y} - (\mathbf{z} \otimes \mathbf{k})\downarrow_\mathbf{s}\|^2 + \lambda\Phi(\mathbf{x}) + \frac{\mu}{2}\|\mathbf{z} - \mathbf{x}\|^2, \qquad (14.5)$$

where μ is the penalty parameter. Such a problem can be addressed by iteratively solving subproblems for $\mathbf{x}$ and $\mathbf{z}$

$$\begin{cases} \mathbf{z}_k = \arg\min_\mathbf{z} \|\mathbf{y} - (\mathbf{z} \otimes \mathbf{k})\downarrow_\mathbf{s}\|^2 + \mu\sigma^2\|\mathbf{z} - \mathbf{x}_{k-1}\|^2, & (14.6) \\ \mathbf{x}_k = \arg\min_\mathbf{x} \dfrac{\mu}{2}\|\mathbf{z}_k - \mathbf{x}\|^2 + \lambda\Phi(\mathbf{x}). & (14.7) \end{cases}$$

According to Eq. (14.6), μ should be large enough so that $\mathbf{x}$ and $\mathbf{z}$ are approximately equal to the fixed point. However, this would also result in slow convergence. Therefore, a good rule of thumb is to iteratively increase μ. For convenience, in the k-th iteration μ is denoted by μ_k.

It can be observed that the data term and the prior term are decoupled into Eq. (14.6) and Eq. (14.7), respectively. For the solution of Eq. (14.6), the fast Fourier transform (FFT) can be utilized by assuming the convolution is carried out with circular boundary conditions. Notably, it has a closed-form expression (Zhao et al., 2016)

$$\mathbf{z}_k = \mathcal{F}^{-1}\left(\frac{1}{\alpha_k}\left(\mathbf{d} - \overline{\mathcal{F}(\mathbf{k})} \odot_{\mathbf{s}} \frac{(\mathcal{F}(\mathbf{k})\mathbf{d}) \Downarrow_{\mathbf{s}}}{(\overline{\mathcal{F}(\mathbf{k})}\mathcal{F}(\mathbf{k})) \Downarrow_{\mathbf{s}} + \alpha_k}\right)\right), \tag{14.8}$$

where $\mathbf{d}$ is defined as

$$\mathbf{d} = \overline{\mathcal{F}(\mathbf{k})}\mathcal{F}(\mathbf{y} \uparrow_{\mathbf{s}}) + \alpha_k \mathcal{F}(\mathbf{x}_{k-1})$$

with $\alpha_k \triangleq \mu_k \sigma^2$ and where the $\mathcal{F}(\cdot)$ and $\mathcal{F}^{-1}(\cdot)$ denote FFT and inverse FFT, $\overline{\mathcal{F}(\cdot)}$ denotes the complex conjugate of $\mathcal{F}(\cdot)$; $\odot_{\mathbf{s}}$ denotes the distinct block processing operator with element-wise multiplication, i.e., applying element-wise multiplication to the $\mathbf{s} \times \mathbf{s}$ distinct blocks of $\overline{\mathcal{F}(\mathbf{k})}$; $\Downarrow_{\mathbf{s}}$ denotes the distinct block downsampler, i.e., averaging the $\mathbf{s} \times \mathbf{s}$ distinct blocks; $\uparrow_{\mathbf{s}}$ denotes the standard $\mathbf{s}$-fold upsampler, i.e., upsampling the spatial size by filling the new entries with zeros. For the special case of deblurring when $\mathbf{s} = 1$, Eq. (14.8) can be concisely written as

$$\mathbf{z}_k = \mathcal{F}^{-1}\left(\frac{\overline{\mathcal{F}(\mathbf{k})}\mathcal{F}(\mathbf{y}) + \alpha_k \mathcal{F}(\mathbf{x}_{k-1})}{\overline{\mathcal{F}(\mathbf{k})}\mathcal{F}(\mathbf{k}) + \alpha_k}\right). \tag{14.9}$$

In other words, Eq. (14.8) generalizes Eq. (14.9). For the solution of Eq. (14.7), it is known that, from a Bayesian perspective, it actually corresponds to a denoising problem with noise level $\beta_k \triangleq \sqrt{\lambda/\mu_k}$ (Chan et al., 2017).

14.3 Deep plug-and-play image restoration

Plug-and-play IR generally involves two steps. The first step is to decouple the data term and prior term of the objective function via a certain variable splitting algorithm, resulting in an iterative scheme consisting of alternately solving a data subproblem and a prior subproblem. The second step is to solve the prior subproblem with any off-the-shelf denoisers, such as K-SVD (Elad and Aharon, 2006), nonlocal means (Buades et al., 2005), BM3D (Dabov et al., 2007). As a result, unlike traditional model-based methods, which need to specify the explicit and hand-crafted image priors, plug-and-play IR can implicitly define the prior via the denoiser. Such an advantage offers the possibility of leveraging a very deep CNN denoiser to improve effectiveness.

The plug-and-play IR can be traced back to Danielyan et al. (2010); Zoran and Weiss (2011); Venkatakrishnan et al. (2013). Danielyan et al. (2012) used Nash equilibrium to derive an iterative decoupled deblurring BM3D (IDDBM3D) method for image deblurring. In Egiazarian and Katkovnik (2015), a similar method equipped with CBM3D denoiser prior was proposed for single image superresolution (SISR). By iteratively updating a back-projection step and a CBM3D denoising step, the method has an encouraging performance for its PSNR improvement over SRCNN (Dong et al., 2016). In Danielyan et al. (2010), the augmented Lagrangian method was adopted to fuse the BM3D denoiser to solve image deblurring task. With a similar iterative scheme as in Danielyan et al. (2012), the first work that treats the denoiser as "plug-and-play prior" was proposed in Venkatakrishnan et al. (2013). Prior to that, a similar plug-and-play idea is mentioned in Zoran and Weiss (2011) where the HQS algorithm is adopted for image denoising, deblurring and inpainting. Heide et al. (2014) used an alternative to ADMM and HQS, i.e., the primal-dual algorithm (Chambolle and Pock, 2011), to decouple the data term and prior term. Teodoro et al. (2016) plugged a class-specific Gaussian mixture model (GMM) denoiser (Zoran and Weiss, 2011) into ADMM to solve image deblurring and compressive imaging. Metzler et al. (2016) developed a denoising-based approximate message passing (AMP) method to integrate denoisers, such as BLS-GSM (Portilla et al., 2003) and BM3D, for compressed sensing reconstruction. Chan et al. (2017) proposed a plug-and-play ADMM algorithm with BM3D denoiser for single image superresolution and quantized Poisson image recovery for single-photon imaging. Kamilov et al. (2017) proposed a fast iterative shrinkage thresholding algorithm (FISTA) with BM3D and WNNM (Gu et al., 2014) denoisers for nonlinear inverse scattering. Sun et al. (2019b) proposed FISTA by plugging the TV and BM3D denoiser prior for Fourier ptychographic microscopy. Yair and Michaeli (2018) proposed to use a WNNM denoiser as the plug-and-play prior for inpainting and deblurring. Gavaskar and Chaudhury (2020) investigated the convergence of ISTA-based plug-and-play IR with a nonlocal means denoiser.

With the development of deep learning techniques such as network design and gradient-based optimization algorithm, a CNN-based denoiser has shown promising performance in terms of effectiveness and efficiency. Following its success, a flurry of CNN denoiser-based plug-and-play IR works have been proposed. Romano et al. (2017) proposed explicit regularization by TNRD denoiser for image deblurring and SISR. In our previous work (Zhang et al., 2017b), different CNN denoisers are trained to plug into HQS algorithm to solve deblurring and SISR. Tirer and Giryes (2018) proposed iterative denoising and backward projections with IRCNN denoisers for image inpainting and deblurring. Gu et al. (2018) proposed to adopt WNNM and IRCNN denoisers for plug-and-play deblurring and SISR. Tirer and Giryes (2019) proposed use the IRCNN denoisers for plug-and-play SISR. Li and Wu (2019) plugged the IRCNN denoisers into the split Bregman iteration algorithm to solve depth image inpainting. Ryu et al. (2019) provided the theoretical convergence analysis of plug-and-play IR based on a forward-backward splitting algorithm and ADMM algorithm, and proposed spectral normalization to train a DnCNN denoiser. Sun et al. (2019a) proposed a block coordinate regularization-by-denoising (RED) algorithm by leveraging the DnCNN (Zhang et al., 2017a) denoiser as the explicit regularizer.

Although plug-and-play IR can leverage the powerful expressiveness of a CNN denoiser, existing methods generally exploit DnCNN or IRCNN denoisers which do not make full use of CNN. Typically, the denoiser for plug-and-play IR should be nonblind and is required to handle a wide range of noise levels. However, DnCNN needs to separately learn a model for each noise level. To reduce the number of denoisers, some works adopt one denoiser fixed to a small noise level. However, according to Romano et al. (2017), such a strategy tends to require a large number of iterations for a satisfying performance, which would increase the computational burden. While IRCNN denoisers can handle a wide range of noise levels, it consists of 25 separate 7-layer denoisers, among which each denoiser is trained on an interval noise level of 2. Such a denoiser suffers from the following two drawbacks. First, it does not have the flexibility to handle a specific noise level. Second, it is not effective enough due to the shallow layers. Given the above considerations, it is necessary to devise a flexible and powerful denoiser to boost the performance of plug-and-play IR. Inspired by FFDNet (Zhang et al., 2018a), the proposed deep denoiser can handle a wide range of noise levels via a single model by taking the noise level map as input. Moreover, its effectiveness is enhanced by taking advantage of both ResNet (He et al., 2016) and U-Net (Ronneberger et al., 2015). The deep denoiser is further incorporated into HQS-based plug-and-play IR to show the merits of using a powerful deep denoiser. Meanwhile, a novel periodical geometric self-ensemble is proposed to potentially improve the performance without introducing an extra computational burden, and a thorough analysis of parameter setting and intermediate results is provided to better understand the working mechanism of the proposed deep plug-and-play IR.

14.3.1 Learning deep CNN denoiser prior

Although various CNN-based denoising methods have been recently proposed, most of them are not designed for plug-and-play IR. In Lehtinen et al. (2018); Krull et al. (2019); Batson and Royer (2019), a novel training strategy without ground-truth is proposed. In Guo et al. (2019); Brooks et al. (2019); Abdelhamed et al. (2019); Zamir et al. (2020), a real noise synthesis technique is proposed to handle real digital photographs. However, from a Bayesian perspective, the denoiser for plug-and-play IR should be a Gaussian denoiser. Hence, one can add synthetic Gaussian noise to a clean image for supervised training. In Lefkimmiatis (2017); Zhang et al. (2019); Liu et al. (2018); Plötz and Roth (2018), the nonlocal module was incorporated into the network design for better restoration. However, these methods learn a separate model for each noise level. Perhaps the most suitable denoiser for plug-and-play IR is FFDNet (Zhang et al., 2018a) which can handle a wide range of noise levels by taking the noise level map as input. Nevertheless, FFDNet only has a comparable performance to DnCNN and IRCNN, thus lacking effectiveness to boost the performance of plug-and-play IR. For this reason, we propose to improve FFDNet by taking advantage of the widely-used U-Net (Ronneberger et al., 2015) and ResNet (He et al., 2016) for architecture design.

14.3.1.1 Denoising network architecture

It is well-known that U-Net (Ronneberger et al., 2015) is effective and efficient for image-to-image translation, while ResNet (He et al., 2016) is superior in increasing the modeling capacity by stacking multiple residual blocks. Following FFDNet (Zhang et al., 2018a), which takes the noise level map as input, the proposed denoiser, namely DRUNet, further integrates residual blocks into U-Net for effective denoiser prior modeling. Note that this work does not focus on designing a new denoising network architecture. The similar idea of combining U-Net and ResNet can also be found in other works such as Zhang et al. (2018); Venkatesh et al. (2018).

Like FFDNet, DRUNet has the ability to handle various noise levels via a single model. The backbone of DRUNet is U-Net which consists of four scales. Each scale has an identity skip connection between 2×2 strided convolution (SConv) downscaling and 2×2 transposed convolution (TConv) upscaling operations. The number of channels in each layer from the first scale to the fourth scale are 64, 128, 256 and 512, respectively. Four successive residual blocks are adopted in the downscaling and upscaling of each scale. Inspired by the network architecture design for superresolution in Lim et al. (2017), no activation function is followed by the first and the last convolutional (Conv) layers, as well as the SConv and TConv layers. In addition, each residual block only contains one ReLU activation function.

It is worth noting that the proposed DRUNet is bias-free, which means no bias is used in all the Conv, SConv and TConv layers. The reason is two-fold. First, a bias-free network with ReLU activation and identity skip connection naturally enforces the scaling invariance property of many image restoration tasks, i.e., $f(ax) = af(x)$ holds true for any scalar $a \geq 0$ (please refer to Mohan et al. (2019) for more details). Second, we have empirically observed that, for a network with bias, the bias magnitude would be much larger than that of the filters, which in turn may harm the generalizability.

14.3.2 Training details

It is well known that CNN benefits from the availability of large-scale training data. To enrich the denoiser prior for plug-and-play IR, instead of training on a small dataset that includes the 400 Berkeley segmentation dataset (BSD) images of size 180×180 (Chen and Pock, 2017), we construct a large dataset consisting of 400 BSD images, 4,744 images of the Waterloo Exploration Database (Ma et al., 2017), 900 images from the DIV2K dataset (Agustsson and Timofte, 2017), and 2,750 images from the Flick2K dataset (Lim et al., 2017). Because such a dataset covers a larger image space, the learned model can slightly improve the PSNR results on the BSD68 dataset (Roth and Black, 2009) while having an obvious PSNR gain on testing datasets from a different domain.

As a common setting for Gaussian denoising, the noisy counterpart **y** of clean image **x** is obtained by adding AWGN with noise level σ. Correspondingly, the noise level map is a uniform map filled with σ and has the same spatial size as a noisy image. To handle a wide range of noise levels, the noise level σ is randomly chosen from [0, 50] during training. The network parameters are optimized by minimizing the L1 loss rather than L2 loss between the denoised image and its ground-truth with the Adam algorithm (Kingma and Ba, 2015). Although there is no direct evidence on which loss would result in better performance, it is widely acknowledged that L1 loss is more robust than L2 loss in handling outliers (Bishop, 2006). Regarding denoising, outliers may occur during the sampling of AWGN. In this sense, L1 loss tends to be more stable than L2 loss for denoising network training. The learning rate starts from 1e-4 and then decreases by half every 100,000 iterations, and finally ends once it is smaller than 5e-7. In each iteration during training, 16 patches with patch size of 128×128 were randomly sampled from the training data. We separately learn a denoiser model for grayscale images and color images. It takes about four days to train the model with PyTorch and an Nvidia Titan Xp GPU.

14.3.3 Denoising results

14.3.3.1 Grayscale image denoising

For grayscale image denoising, we compared the proposed DRUNet denoiser with several state-of-the-art denoising methods, including two representative model-based methods (i.e., BM3D (Dabov et al., 2007) and WNNM (Gu et al., 2014)), one CNN-based method which

TABLE 14.1 Average PSNR(dB) results of different methods with noise levels 15, 25 and 50 on the widely-used Set12 and BSD68 (Martin et al., 2001; Roth and Black, 2009; Zhang et al., 2017a) datasets. The best results are highlighted in bold.

Datasets	Noise Level	BM3D	WNNM	DnCNN	IRCNN	FFDNet	DRUNet
Set12	15	32.37	32.70	32.86	32.77	32.75	**33.25**
	25	29.97	30.28	30.44	30.38	30.43	**30.94**
	50	26.72	27.05	27.18	27.14	27.32	**27.90**
BSD68	15	31.08	31.37	31.73	31.63	31.63	**31.91**
	25	28.57	28.83	29.23	29.15	29.19	**29.48**
	50	25.60	25.87	26.23	26.19	26.29	**26.59**

separately learns a single model for each noise level (i.e., DnCNN (Zhang et al., 2017a)) and two CNN-based methods which were trained to handle a wide range of noise levels (i.e., IRCNN (Zhang et al., 2017b) and FFDNet (Zhang et al., 2018a)). The PSNR results of different methods on the widely-used Set12 (Zhang et al., 2017a) and BSD68 (Martin et al., 2001; Roth and Black, 2009) datasets for noise levels 15, 25 and 50 are reported in Table 14.1. It can be seen that DRUNet achieves the best PSNR results for all the noise levels on the two datasets. Specifically, DRUNet has an average PSNR gain of about 0.9 dB over BM3D and surpasses DnCNN, IRCNN and FFDNet by an average PSNR of 0.5 dB on the Set12 dataset and 0.25 dB on the BSD68 dataset. Fig. 14.2 shows the grayscale image denoising results of different methods on image *"Monarch"* from the Set12 dataset with noise level 50. It can be seen that DRUNet can recover much sharper edges than DnCNN, FFDNet.

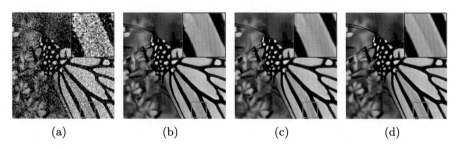

FIGURE 14.2 Grayscale image denoising results of different methods on image *"Monarch"* from Set12 dataset with noise level 50. (a) Noisy (14.78 dB); (b) DnCNN (26.83 dB); (c) FFDNet (26.92 dB); (d) DRUNet (27.31 dB).

14.3.3.2 Color image denoising

Since existing methods mainly focus on grayscale image denoising, we only compare DRUNet with CBM3D, DnCNN, IRCNN and FFDNet for color denoising. Table 14.2 reports the color image denoising results of different methods for noise levels 15, 25 and 50 on CBSD68 (Martin et al., 2001; Roth and Black, 2009; Zhang et al., 2017a), Kodak24 (Franzen, 1999) and McMaster (Zhang et al., 2011) datasets. One can see that DRUNet outperforms the other competing methods by a large margin. It is worth noting that while having a good performance on the CBSD68 dataset, DnCNN does not perform well on the McMaster dataset.

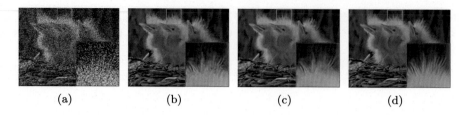

(a)   (b)   (c)   (d)

FIGURE 14.3 Color image denoising results of different methods on image "*163085*" from CBSD68 dataset with noise level 50. (a) Noisy (14.99 dB); (b) DnCNN (28.68 dB); (c) FFDNet (28.75 dB); (d) DRUNet (29.28 dB).

Such a discrepancy highlights the importance of reducing the image domain gap between training and testing for image denoising. The visual results of different methods on image "*163085*" from the CBSD68 dataset with noise level 50 are shown in Fig. 14.3, from which it can be seen that DRUNet can recover more fine details and textures than the competing methods.

TABLE 14.2 Average PSNR(dB) results of different methods for noise levels 15, 25 and 50 on CBSD68 (Martin et al., 2001; Roth and Black, 2009; Zhang et al., 2017a), Kodak24 and McMaster datasets. The best results are highlighted in bold.

Datasets	Noise Level	CBM3D	DnCNN	IRCNN	FFDNet	DRUNet
CBSD68	15	33.52	33.90	33.86	33.87	**34.30**
	25	30.71	31.24	31.16	31.21	**31.69**
	50	27.38	27.95	27.86	27.96	**28.51**
Kodak24	15	34.28	34.60	34.69	34.63	**35.31**
	25	32.15	32.14	32.18	32.13	**32.89**
	50	28.46	28.95	28.93	28.98	**29.86**
McMaster	15	34.06	33.45	34.58	34.66	**35.40**
	25	31.66	31.52	32.18	32.35	**33.14**
	50	28.51	28.62	28.91	29.18	**30.08**

14.3.4 HQS algorithm for plug-and-play IR

As mentioned previously, we adopt the HQS as the variable splitting algorithm for its simplicity and fast convergence. Meanwhile, there is no doubt that parameter setting is always a nontrivial issue (Romano et al., 2017). In other words, careful parameter setting is needed to obtain a good performance. To have a better understanding on the HQS-based plug-and-play IR, we will discuss the general methodology for parameter setting after providing a short review of the HQS algorithm. We then propose a periodical geometric self-ensemble strategy to potentially improve the performance.

14.3.4.1 Half quadratic splitting (HQS) algorithm

HQS adopts the following iterative scheme to solve Eq. (14.2)

$$\begin{cases} \mathbf{x}_k = \arg\min_{\mathbf{x}} \|\mathbf{y} - (\mathbf{x} \otimes \mathbf{k})\downarrow_{\mathbf{s}}\|^2 + \mu\sigma^2\|\mathbf{x} - \mathbf{z}_{k-1}\|^2 & (14.10a) \\[2ex] \mathbf{z}_k = \arg\min_{\mathbf{z}} \frac{1}{2(\sqrt{\lambda/\mu_k})^2}\|\mathbf{z} - \mathbf{x}_k\|^2 + \mathcal{R}(\mathbf{z}). & (14.10b) \end{cases}$$

In particular, the subproblem of (14.10b), from a Bayesian perspective, corresponds to Gaussian denoising on $\mathbf{x}_k$ with noise level $\beta_k \triangleq \sqrt{\lambda/\mu_k}$. Consequently, any Gaussian denoiser can be plugged into the alternating iterations to solve (14.2). To address this, we rewrite (14.10b) as follows

$$\mathbf{z}_k = Denoiser(\mathbf{x}_k, \beta_k). \qquad (14.11)$$

One can make two observations from Eq. (14.11). First, the prior $\mathcal{R}(\cdot)$ can be implicitly specified by a denoiser. For this reason, both the prior and the denoiser for plug-and-play IR are usually termed as denoiser prior. Second, it is interesting to learn a single CNN denoiser to replace Eq. (14.11) so as to exploit the advantages of CNN, such as high flexibility of network design, high efficiency on GPUs, and powerful modeling capacity with deep networks.

14.3.4.2 General methodology for parameter setting

From the alternating iterations between Eq. (14.10a) and Eq. (14.10b), it is easy to see that three adjustable parameters are involved, including penalty parameter μ, regularization parameter λ and the total number of iterations K. Generally, to guarantee that $\mathbf{x}_k$ and $\mathbf{z}_k$ converge to a fixed point, a large μ is needed, which however requires a large K for convergence. Hence, the common way is to adopt the continuation strategy to gradually increase μ, resulting in a sequence of $\mu_1 < \cdots < \mu_k < \cdots < \mu_K$. Nevertheless, a new parameter needs to be introduced to control the step size, making the parameter setting more complicated. According to Eq. (14.11), we can observe that μ controls the noise level $\beta_k (= \sqrt{\lambda/\mu_k})$ in the k-th iteration of the denoiser prior. On the other hand, a noise level range of $[0, 50]$ is supposed to be enough for β_k. Inspired by such domain knowledge, we can instead set β_k and λ to implicitly determine μ_k. Based on the fact that μ_k should be monotonically increasing, we uniformly sample β_k from a large noise level β_1 to a small one β_K in log space. This means that μ_k can be easily determined via $\mu_k = \lambda/\beta_k^2$. Following Zhang et al. (2017b), β_1 is fixed to 49 while β_K is determined by the image noise level β. Since K is user-specified and σ_K has a clear physical meaning, they are practically easy to set. As for the theoretical convergence of plug-and-play IR, please refer to Chan et al. (2017). By far, the remaining parameter for setting is λ. Due to the fact that λ comes from the prior term and thus should be fixed, we can choose the optimal λ by a grid search on a validation dataset. Empirically, λ can yield favorable performance from the range of $[0.19, 0.55]$. In this chapter, we fix it to 0.23 unless otherwise specified. It should be noted that since λ can be absorbed into β and plays the role of controlling the trade-off between data term and prior term, one can implicitly tune λ by multiplying β by a scalar.

14.3.4.3 *Periodical geometric self-ensemble*

Geometric self-ensemble based on flipping and rotation is a commonly-used strategy to boost IR performance (Timofte et al., 2016). It first transforms the input via flipping and rotation to generate 8 images, then gets the corresponding restored images after feeding the model with the 8 images, and finally produces the averaged result after the inverse transformation. While a performance gain can be obtained via geometric self-ensemble, it comes at the cost of increased inference time.

Different from the above method, we instead periodically apply the geometric self-ensemble for every successive 8 iterations. In each iteration, this involves one transformation before denoising and the counterpart inverse transformation after denoising. Note that the averaging step is abandoned because the input of the denoiser prior model varies across iterations. We refer to this method as periodical geometric self-ensemble. Its distinct advantage is that the total inference time will not increase. We empirically found that geometric self-ensemble can generally improve the PSNR by 0.02 dB $\sim$ 0.2 dB.

Based on the above discussion, we summarize the detailed algorithm of deep plug-and-play IR, namely DPIR, in Algorithm 1.

Algorithm 1: Plug-and-play image restoration with deep denoiser prior (DPIR).

Input : Deep denoiser prior model, degraded image $\mathbf{y}$, degradation model
$\mathbf{y} = (\mathbf{x} \otimes \mathbf{k})\downarrow_{\mathbf{s}} + \mathbf{n}$, image noise level σ, β_k of denoiser prior model at k-th
iteration for a total of K iterations, trade-off parameter λ.

Output: Restored image $\mathbf{z}_K$.

1 Initialize $\mathbf{z}_0$ from $\mathbf{y}$, precalculate $\alpha_k = \lambda\sigma^2/\beta_k^2$.
2 **for** $k = 1, 2, \cdots, K$ **do**
3       $\mathbf{x}_k = \arg\min_{\mathbf{x}} \|\mathbf{y} - (\mathbf{x} \otimes \mathbf{k})\downarrow_{\mathbf{s}}\|^2 + \alpha_k\|\mathbf{x} - \mathbf{z}_{k-1}\|^2$; // *Solving data subproblem*
4       $\mathbf{z}_k = Denoiser(\mathbf{x}_k, \beta_k)$; // *Denoising with deep DRUNet denoiser and periodical geometric self-ensemble*
5 **end**

14.4 Deep unfolding image restoration

The early deep unfolding methods can be traced back to Barbu (2009); Samuel and Tappen (2009); Sun and Tappen (2011) where a compact MAP inference based on the gradient descent algorithm was proposed for image denoising. Since then, a flurry of deep unfolding methods based on certain optimization algorithms (e.g., half-quadratic splitting (Afonso et al., 2010), alternating direction method of multipliers (Boyd et al., 2011) and primal-dual (Chambolle and Pock, 2011)) have been proposed to solve different image restoration tasks, such as image denoising (Chen and Pock, 2017; Lefkimmiatis, 2017), image deblurring (Schmidt and Roth, 2014; Kruse et al., 2017), and image compressive sensing (Zhang and Ghanem, 2018). Compared to plain learning-based methods, deep unfolding methods are interpretable and can fuse the degradation constraint into the learning model. However, most of them suffer from one or several of the following drawbacks. (i) The solution of the prior subproblem without using a deep CNN is not powerful enough for good performance. (ii) The data subproblem is not solved by a closed-form solution, which may hinder convergence. (iii) The whole inference is trained in a stage-wise and fine-tuning manner rather than a complete end-to-end manner. It is thus of particular interest to propose a method that overcomes the above mentioned drawbacks.

14.4.1 Deep unfolding network

Once the unfolding optimization, i.e., HQS algorithm, is determined, the next step is to design the deep unfolding image restoration (DUIR) network. Because the unfolding optimization mainly consists of iteratively solving a data subproblem (i.e., Eq. (14.6)) and a prior subproblem (i.e., Eq. (14.7)), DUIR should alternate between a data module $\mathcal{D}$ and a prior module $\mathcal{P}$. In addition, as the solutions of the subproblems also take the respective hyper-parameters α_k and β_k as input, a hyper-parameter module $\mathcal{H}$ is further introduced into DUIR. Fig. 14.4 illustrates the overall architecture of DUIR with K iterations, where K is empirically set to 8 for the speed-accuracy trade-off. Next, more details on $\mathcal{D}$, $\mathcal{P}$ and $\mathcal{H}$ are provided.

14.4.1.1 Data module $\mathcal{D}$

The data module plays the role of Eq. (14.8) which is the closed-form solution of the data subproblem. Intuitively, it aims to find a clearer HR image which minimizes a weighted combination of the data term $\|\mathbf{y} - (\mathbf{z} \otimes \mathbf{k})\downarrow_{\mathbf{s}}\|^2$ and the quadratic regularization term $\|\mathbf{z} - \mathbf{x}_{k-1}\|^2$ with trade-off hyper-parameter α_k. Because the data term corresponds to the degradation model, the data module not only has the advantage of taking the scale factor $\mathbf{s}$ and blur kernel $\mathbf{k}$ as input but also imposes a degradation constraint on the solution. Actually, it is difficult to manually design such a simple but useful multiple-input module. For brevity, Eq. (14.8) is rewritten as

$$\mathbf{z}_k = \mathcal{D}(\mathbf{x}_{k-1}, \mathbf{s}, \mathbf{k}, \mathbf{y}, \alpha_k). \tag{14.12}$$

Note that $\mathbf{x}_0$ is initialized by interpolating $\mathbf{y}$ with scale factor $\mathbf{s}$ via the simplest nearest neighbor interpolation. It should be noted that Eq. (14.12) contains no trainable parameters, which in turn results in better generalizability due to the complete decoupling between data term and prior term. For the implementation, we use PyTorch, where the main FFT and inverse FFT operators can be implemented by `torch.fft.fftn` and `torch.fft.ifftn`, respectively.

14.4.1.2 Prior module $\mathcal{P}$

The prior module aims to obtain a cleaner estimation $\mathbf{x}_k$ by passing $\mathbf{z}_k$ through a denoiser with noise level β_k. Inspired by Zhang et al. (2018a), we propose a deep CNN denoiser that takes the noise level as input

$$\mathbf{x}_k = \mathcal{P}(\mathbf{z}_k, \beta_k). \tag{14.13}$$

The proposed denoiser, namely ResUNet, integrates residual blocks (He et al., 2016) into U-Net (Ronneberger et al., 2015). U-Net is widely used for image-to-image mapping, while ResNet owes its popularity to fast training and its large capacity with many residual blocks.

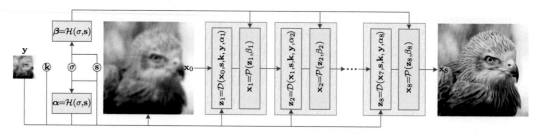

FIGURE 14.4 The overall architecture of the DUIR with $K = 8$ iterations. DUIR can flexibly handle the degradation model $\mathbf{y} = (\mathbf{x} \otimes \mathbf{k}) \downarrow_{\mathbf{s}} + \mathbf{n}$ via a single model as it takes the degraded image $\mathbf{y}$, scale factor $\mathbf{s}$, blur kernel $\mathbf{k}$ and noise level σ as input. Specifically, DUIR consists of three main modules, including the data module $\mathcal{D}$ that makes HR estimation clearer, the prior module $\mathcal{P}$ that makes HR estimation cleaner, and the hyper-parameter module $\mathcal{H}$ that controls the outputs of $\mathcal{D}$ and $\mathcal{P}$.

ResUNet takes the concatenated $\mathbf{z}_k$ and noise level map as input and outputs the denoised image $\mathbf{x}_k$. By doing so, ResUNet can handle various noise levels via a single model, which significantly reduces the total number of parameters. Following the common setting of U-Net, ResUNet involves four scales, each of which has an identity skip connection between downscaling and upscaling operations. Specifically, the numbers of channels in the layers from the first to the fourth scale are set to 64, 128, 256 and 512, respectively. For the downscaling and upscaling operations, 2×2 strided convolution (SConv) and 2×2 transposed convolution (TConv) are adopted, respectively. Note that no activation function is followed by the SConv and TConv layers, as well as the first and the last convolutional layers. For the sake of inheriting the merits of ResNet, a group of 2 residual blocks is adopted in the downscaling and upscaling of each scale. As suggested in Lim et al. (2017), each residual block is composed of two 3×3 convolution layers with ReLU activation in the middle and an identity skip connection summed to its output.

14.4.1.3 Hyper-parameter module $\mathcal{H}$

The hyper-parameter module acts as a 'slide bar' to control the outputs of the data and prior modules. For example, the solution $\mathbf{z}_k$ would gradually approach $\mathbf{x}_{k-1}$ as α_k increases. According to the definition of α_k and β_k, α_k is determined by σ and μ_k, while β_k depends on λ and μ_k. Although it is possible to learn a fixed λ and μ_k, we argue that a performance gain can be obtained if λ and μ_k vary with two key elements, i.e., scale factor $\mathbf{s}$ and noise level σ, that influence the degree of ill-posedness. Let $\boldsymbol{\alpha} = [\alpha_1, \alpha_2, \ldots, \alpha_K]$ and $\boldsymbol{\beta} = [\beta_1, \beta_2, \ldots, \beta_K]$: we use a single module to predict $\boldsymbol{\alpha}$ and $\boldsymbol{\beta}$

$$[\boldsymbol{\alpha}, \boldsymbol{\beta}] = \mathcal{H}(\sigma, \mathbf{s}). \tag{14.14}$$

The hyper-parameter module consists of three fully connected layers with ReLUs as the first two activation functions and Softplus as the last. The number of hidden nodes in each layer is 64. Considering the fact that α_k and β_k should be positive, and Eq. (14.8) should avoid division by extremely small α_k, the output Softplus layer is followed by an extra addition of `1e-6`.

14.4.2 End-to-end training

The end-to-end training aims to learn the trainable parameters of DUIR by minimizing a loss function over a large training data set. Thus, this section mainly describes the training data, loss function and training settings. Following Wang et al. (2018), we use DIV2K (Agustsson and Timofte, 2017) and Flickr2K (Timofte et al., 2017). The degraded images are synthesized via the degradation model $\mathbf{y} = (\mathbf{x} \otimes \mathbf{k}) \downarrow_{\mathbf{s}} + \mathbf{n}$. The scale factors are chosen from $\{1, 2, 3, 4\}$. For the blur kernels, we use anisotropic Gaussian kernels as in Riegler et al. (2015); Shocher et al. (2018); Zhang et al. (2018b) and motion kernels as in Boracchi and Foi (2012). We fix the kernel size to 25×25. For the noise level, we set its range to $[0, 25]$. With regard to the loss function, we adopt the L1 loss for PSNR performance. To optimize the parameters of DUIR, we adopt the Adam solver (Kingma and Ba, 2015) with mini-batch size 128. The learning rate starts from 1×10^{-4} and decays by a factor of 0.5 every 4×10^{4} iterations and finally ends with 3×10^{-6}. It is worth pointing out that due to the infeasibility of parallel computing for different scale factors, each min-batch only involves one random scale factor. The patch size of the clean image is set to 96×96. We train the models with PyTorch on four Nvidia Tesla V100 GPUs in Amazon AWS cloud. It takes about two days to obtain the DUIR model.

14.5 Experiments

To validate the flexibility and effectiveness of DPIR and DUIR, we consider two classical IR tasks, including image deblurring and single image superresolution (SISR). For each task, we will show the quantitative and qualitative results of DPIR and DUIR on three classical testing images, as shown in Fig. 14.5. In addition, we will compare the hand-designed and learned hyper-parameter setting between DPIR and DUIR. Furthermore, we will provide the visual comparison of $\mathbf{x}_k$ and $\mathbf{z}_k$ at intermediate iterations for DPIR and DUIR.

(a)   (b)   (c)

FIGURE 14.5 Three classical testing images. (a) *Butterfly*; (b) *Leaves*; (c) *Starfish*.

14.5.1 Image deblurring

We consider two of the eight real blur kernels from (Levin et al., 2009) for testing which are of size 17×17 and 27×27, respectively. As shown in Table 14.3, we also consider Gaussian noise with different noise levels 2.55(1%) and 7.65(3%). Following the common setting of uniform deblurring, we synthesize the blurry images by first applying a blur kernel and then adding AWGN with noise level σ. For the hyper-parameter setting of DPIR, K and σ_K are set to 8 and σ, respectively, while $\mathbf{z}_0$ is initialized as $\mathbf{y}$.

14.5.1.1 Quantitative and qualitative results

Table 14.3 reports the PSNR(dB) results of DPIR and DUIR on the three testing images. It can be seen that DPIR and DUIR achieve similar results. Note that DPIR has a larger and slower denoiser prior than DUIR, thus DPIR is less efficient than DUIR. The visual results of DPIR and DUIR on the testing images are shown in Fig. 14.6. One can see that both DPIR and DUIR can effectively recover image sharpness and naturalness.

TABLE 14.3 PSNR (dB) results of DPIR and DUIR on the three testing images for image deblurring. The best results are highlighted in bold.

Methods	σ	Butterfly	Leaves	Starfish
The second kernel of size 17×17 from Levin et al. (2009)				
DPIR	2.55	**34.26**	**35.19**	**34.21**
DUIR		33.73	34.35	34.02
DPIR	7.65	29.52	**30.11**	29.83
DUIR		**29.55**	30.02	**29.84**
The fourth kernel of size 27×27 from Levin et al. (2009)				
DPIR	2.55	**34.18**	**35.12**	**33.91**
DUIR		33.58	34.26	33.73
DPIR	7.65	**29.45**	**30.27**	**29.46**
DUIR		29.38	29.94	29.43

14.5.1.2 Hand-designed vs. learned hyper-parameters

Figs. 14.7 and 14.8 illustrate the hyper-parameters, i.e., α and β, of DPIR and DUIR, respectively. One can see that the learned hyper-parameters of DUIR are generally in accordance with the hand-designed hyper-parameters of DPIR. From Figs. 14.7 and 14.8(a), one can see that α is positively correlated with σ. From Figs. 14.7 and 14.8(b), one can see that β has a decreasing tendency with the number of iterations and increases with noise level. This implies that the noise level of intermediate estimation is gradually reduced across iterations and a severe degradation requires a large β_k to tackle with the illposeness.

14.5.1.3 Intermediate results

Figs. 14.9 and 14.10 provide the visual results and PSNR results of $\mathbf{x}_k$ and $\mathbf{z}_k$ at different iterations of DPIR and DUIR on the *Starfish* image from Fig. 14.12. One can observe from Fig. 14.9 that, while the closed-form solution $\mathbf{x}_1$ can handle the distortion of blur, it would also aggravate the strength of noise. The deep denoiser prior plays the role of removing noise,

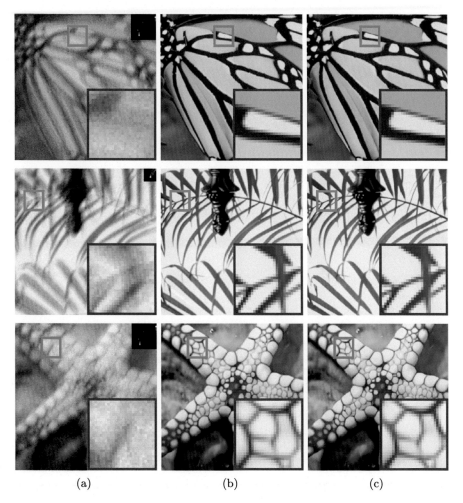

(a)   (b)   (c)

FIGURE 14.6 Visual results of DPIR and DUIR for image deblurring. The blur kernel is visualized in the upper right corner of the blurry image. The noise level is 7.65 (3%). (a) Blurry image; (b) DPIR; (c) DUIR.

leading to a noise-free z_k. As the number of iterations increases, x_7 contains less structured noise than x_1, while z_7 recovers more details and sharper edges than z_1. One can observe from Fig. 14.10 that $\mathcal{D}$ and $\mathcal{P}$ can facilitate each other for iterative and alternating blur removal. Interestingly, z_1 of DPIR significantly differs from z_1 of DUIR, which means that, unlike DPIR, the learned denoiser prior of DUIR is not a Gaussian denoiser prior.

We can have the following observations. First, while Eq. (14.10a) can handle the distortion of blur, it also aggravates the strength of noise compared to its input z_{k-1}. Second, the deep denoiser prior plays the role of removing noise, leading to a noise-free z_k. Third, compared with x_1 and x_2, x_8 contains more fine details, which means Eq. (14.10a) can iteratively recover the details.

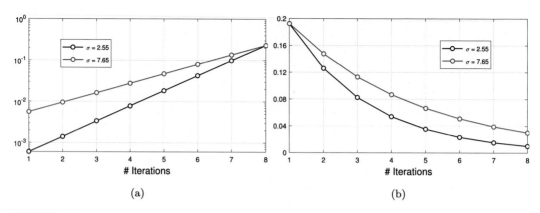

FIGURE 14.7 The hyper-parameters, i.e., (a) α and (b) β, of DPIR for deblurring with respect to different noise levels.

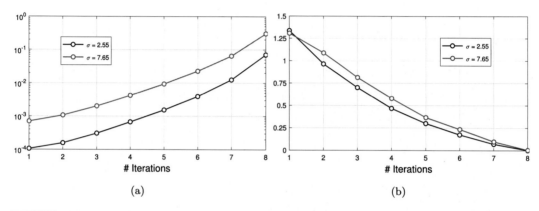

FIGURE 14.8 The hyper-parameters, i.e., (a) α and (b) β, of DUIR for deblurring with respect to different noise levels.

14.5.2 Single image superresolution (SISR)

Although we use the degradation model $\mathbf{y} = (\mathbf{x} \otimes \mathbf{k}) \downarrow_\mathbf{s} + \mathbf{n}$ for SISR, it is worth noting that existing SISR methods are mainly designed for the bicubic degradation model with the formulation $\mathbf{y} = \mathbf{x} \downarrow_s^{bicubic}$, where $\downarrow_s^{bicubic}$ denotes bicubic downsampling with downscaling factor s. However, it has been revealed that these methods would deteriorate seriously if the real degradation model deviates from the assumed one (Efrat et al., 2013; Zhang et al., 2015). Since bicubic degradation is well-studied, it is interesting to investigate its relationship to the classical degradation model. Actually, the bicubic degradation can be approximated by setting a proper blur kernel in the degradation model $\mathbf{y} = (\mathbf{x} \otimes \mathbf{k}) \downarrow_\mathbf{s}$. To achieve this, we adopt the data-driven method to solve the following kernel estimation problem by minimizing the reconstruction error over a large HR/bicubic-LR pair $\{(\mathbf{x}, \mathbf{y})\}$,

$$\mathbf{k}_{bicubic}^{\times \mathbf{s}} = \arg\min_\mathbf{k} \|(\mathbf{x} \otimes \mathbf{k}) \downarrow_\mathbf{s} - \mathbf{y}\|. \tag{14.15}$$

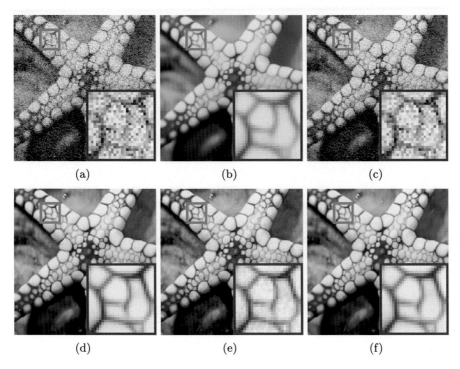

FIGURE 14.9 Estimations in different iterations of DPIR for deblurring on the *Starfish* in Fig. 14.6. (a) x_1; (b) z_1; (c) x_2; (d) z_6; (e) x_7; (f) z_7.

Fig. 14.11 shows the approximated bicubic kernels for scale factors 2, 3 and 4. It should be noted that since the downsamlping operation selects the upper-left pixel for each distinct $s \times s$ patch, the bicubic kernels for scale factors 2, 3 and 4 have a center shift of 0.5, 1 and 1.5 pixels to the upper-left direction, respectively.

For the sake of synthesizing the corresponding testing LR images via the degradation model $y = (x \otimes k) \downarrow_s + n$, blur kernels and noise levels should be provided. Generally, it would be helpful to employ a large variety of blur kernels and noise levels for a thorough evaluation; however, it would also give rise to a burdensome evaluation process. For this reason, we only consider 8 representative and diverse blur kernels, including 4 isotropic Gaussian kernels with different widths (i.e., 0.7, 1.2, 1.6 and 2.0) and 4 anisotropic Gaussian kernels from Zhang et al. (2018b). We do not consider motion blur kernels since it has been pointed out that Gaussian kernels are enough for the SISR task. In order to further analyze the kernel robustness, we will thus separately report the PSNR results for each blur kernel rather than for each type of blur kernel. Although it has been pointed out that the proper blur kernel should vary with scale factor (Zhang et al., 2015), we argue that the 8 blur kernels are diverse enough to cover a large kernel space. For the noise levels, we choose 2.55 (1%) and 7.65 (3%). For the overall parameter setting, K and σ_K are set to 24 and $\max(\sigma, s)$, respectively. For the initialization of z_0, the bicubic interpolation of the LR image is utilized. In particular, since the classical degradation model selects the upper-left pixel for each distinct $s \times s$ patch,

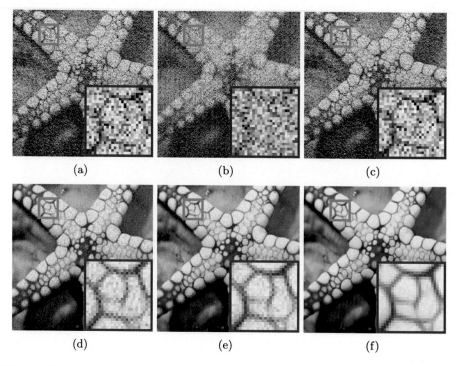

FIGURE 14.10 Estimations in different iterations of DUIR for deblurring on the *Starfish* in Fig. 14.6. (a) x_1; (b) z_1; (c) x_2; (d) z_6; (e) x_7; (f) z_7.

the shift problem should be properly addressed. To tackle this, we adjust z_0 by using grid interpolation.

14.5.2.1 *Quantitative and qualitative comparison*

Table 14.4 reports the average PSNR(dB) results of DPIR and DUIR on the three testing images. From Table 14.4, we can see the PSNR results vary across different blur kernels and the blur kernels that lead to the highest PSNR for each scale factor are different. More specifically, a larger scale factor generally requires a smoother blur kernel. In addition, DUIR outperforms DPIR by a large margin. Such a phenomenon indicates that end-to-end training is more helpful for superresolution than for deblurring. Fig. 14.12 shows the visual results of DPIR and DUIR on the three testing images. It can be observed that both DPIR and DUIR can significantly improve the visual quality. In comparison, DUIR produces sharper edges than DPIR which is possibly due to the end-to-end-training.

14.5.2.2 *Hand-designed vs. learned hyper-parameters*

Figs. 14.13 and 14.14 illustrate the hyper-parameters, i.e., α and β, of DPIR and DUIR for different combinations of scale factor **s** and noise level σ, respectively. One can see that the learned hyper-parameters of DUIR accord with the hand-designed hyper-parameters of

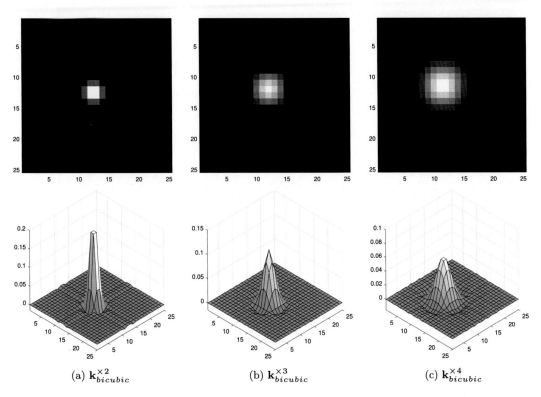

(a) $\mathbf{k}_{bicubic}^{\times 2}$ (b) $\mathbf{k}_{bicubic}^{\times 3}$ (c) $\mathbf{k}_{bicubic}^{\times 4}$

FIGURE 14.11 Approximated bicubic kernels for scale factors 2, 3 and 4 under the degradation model $\mathbf{y} = (\mathbf{x} \otimes \mathbf{k}) \downarrow_{\mathbf{s}}$. Note that these kernels contain negative values.

TABLE 14.4 Average PSNR(dB) results of DPIR and DUIR for different combinations of scale factors, blur kernels and noise levels. The best results are highlighted in bold.

Method	Scale Factor	Noise Level	Blur Kernel							
			▪	●	●	●	●	╱	╲	●
	×2	0	31.79	32.10	31.28	29.63	28.72	28.62	29.45	27.77
	×3	0	26.04	26.82	26.97	26.89	26.73	25.89	26.58	26.49
DPIR	×3	2.55	25.91	26.49	26.23	25.43	24.81	24.44	25.29	24.13
	×3	7.65	25.53	25.67	24.91	23.68	23.14	22.81	23.56	22.33
	×4	0	22.62	23.50	23.74	23.85	23.76	23.18	23.58	23.88
	×2	0	**33.58**	**34.47**	**33.49**	**31.79**	**31.29**	**31.29**	**31.45**	**30.20**
	×3	0	**28.05**	**29.53**	**29.88**	**29.87**	**29.40**	**29.37**	**29.43**	**29.25**
DUIR	×3	2.55	**27.77**	**28.74**	**28.47**	**27.70**	**27.15**	**27.20**	**27.42**	**26.54**
	×3	7.65	**26.87**	**27.13**	**26.43**	**25.47**	**25.15**	**25.03**	**25.24**	**24.47**
	×4	0	**24.71**	**26.38**	**27.02**	**27.30**	**27.16**	**26.76**	**26.69**	**27.28**

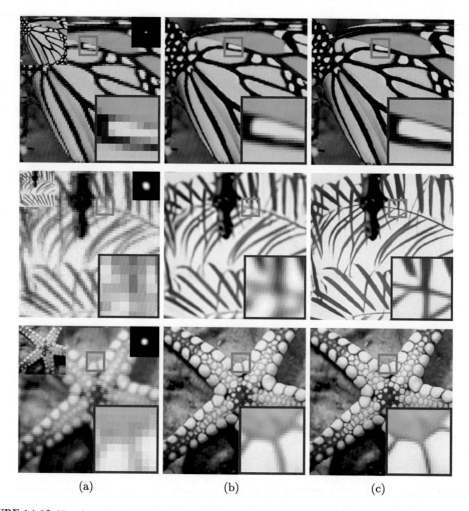

FIGURE 14.12 Visual results of DPIR and DUIR on the three testing images. The blur kernel is visualized in the upper right corner of the LR image. (a) LR image; (b) DPIR; (c) DUIR.

DPIR. From Figs. 14.13 and 14.14(a), one can see that $\boldsymbol{\alpha}$ is positively correlated with σ and varies with $\mathbf{s}$. This actually accords with the definition of α_k. From Figs. 14.13 and 14.14(b), one can see that $\boldsymbol{\beta}$ has a decreasing tendency with the number of iterations and increases with scale factor and noise level. This implies that the noise level of HR estimation is gradually reduced across iterations and complex degradation requires a large β_k to tackle the illposedness.

14.5.2.3 *Intermediate results*

Figs. 14.15 and 14.16 provide the visual results of $\mathbf{x}_k$ and $\mathbf{z}_k$ at different iterations of DPIR and DUIR on the *Starfish* image from Fig. 14.12. One can observe from Fig. 14.15 that, al-

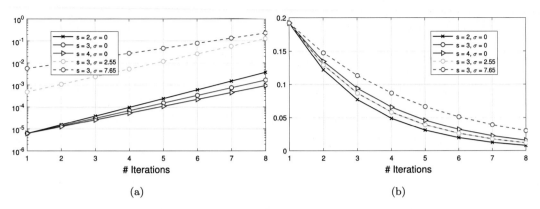

FIGURE 14.13 The hyper-parameters, i.e., (a) α and (b) β, of DPIR for superresolution with respect to different combinations of noise levels and scale factors.

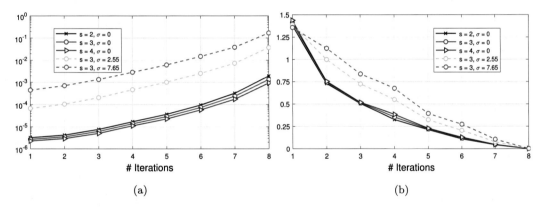

FIGURE 14.14 The hyper-parameters, i.e., (a) α and (b) β, of DUIR for superresolution with respect to different combinations of noise levels and scale factors.

though the LR image contains no noise, the closed-form solution x_1 would introduce severe structured noise. After passing x_1 through the Gaussian denoiser, such structured noise is removed, as can be seen from z_1. Meanwhile, the tiny textures and structures are smoothed out and the edges become blurry. As the number of iterations increases, x_7 contains less structured noise than x_1, while z_7 recovers more details and sharper edges than z_1. One can observe from Fig. 14.16 that, $\mathcal{D}$ and $\mathcal{P}$ can facilitate each other for iterative and alternating blur removal and detail recovery. Interestingly, different from the Gaussian denoiser prior of DPIR, $\mathcal{P}$ can also act as a detail enhancer for high-frequency recovery which is possibly attributed to the task-specific training. In addition, it does not reduce blur kernel induced degradation, which verifies the decoupling between $\mathcal{D}$ and $\mathcal{P}$. As a result, the end-to-end trained DUIR has a task-specific advantage over DPIR.

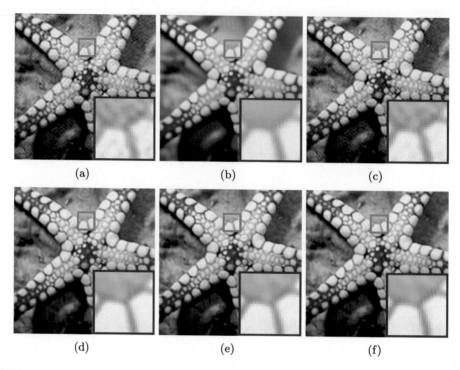

FIGURE 14.15 HR estimations in different iterations of DPIR for superresolution on the *Starfish* in Fig. 14.12. (a) x_1; (b) z_1; (c) x_2; (d) z_6; (e) x_7; (f) z_7.

14.6 Discussion and conclusions

In this chapter, we introduce the deep plug-and-play methods and deep unfolding methods that can integrate model-based methods and learning-based methods for image restoration. Specifically, with the aid of the half-quadratic splitting algorithm which can decouple the data and prior terms, deep plug-and-play methods can replace the prior subproblem with learning-based denoiser prior, while deep unfolding methods learn an end-to-end trainable, iterative network by solving the prior subproblem via neural modules. As a result, both deep plug-and-play methods and deep unfolding methods can inherit the flexibility of model-based methods, while maintaining the advantages of learning-based methods. In order to better understand the working mechanism of both kinds of method, quantitative and qualitative results as well as a comparative analysis of hyper-parameter setting and intermediate results are provided. The results demonstrate that while deep unfolding methods learn similar hyper-parameters to the hand-designed ones of deep plug-and-play methods, the learned prior modules are different from those of the Gaussian denoiser priors. For example, the learned prior module of deep unfolding methods can also enhance the details while the Gaussian denoiser prior of deep plug-and-play methods does not have such a merit.

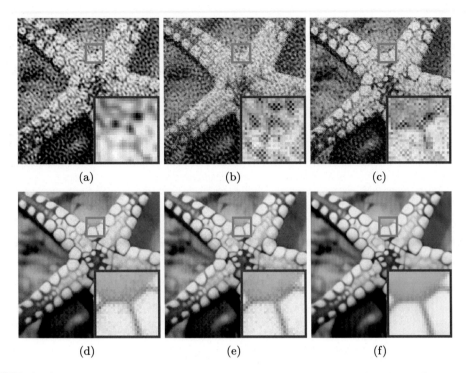

FIGURE 14.16 HR estimations in different iterations of DUIR for superresolution on the *Starfish* in Fig. 14.12. (a) x_1; (b) z_1; (c) x_2; (d) z_6; (e) x_7; (f) z_7.

While the proposed deep plug-and-play methods and deep unfolding methods have shown great promise, they also have several drawbacks for real applications. First, they are nonblind methods which require an accurate estimation of degradation parameters such as the blur kernel. If the estimated blur kernel deviates largely from the ground-truth one, performance would deteriorate seriously. Second, they are designed on the ideal degradation model which rarely matches for real images. If the real-world noise does not follow an additive white Gaussian distribution, this would lead to a drop in performance.

Acknowledgments

This work was partly supported by the ETH Zürich Fund (OK), and a Huawei Technologies Oy (Finland) project.

References

Abdelhamed, A., Brubaker, M.A., Brown, M.S., 2019. Noise flow: noise modeling with conditional normalizing flows. In: IEEE International Conference on Computer Vision, pp. 3165–3173.

Afonso, M.V., Bioucas-Dias, J.M., Figueiredo, M.A., 2010. Fast image recovery using variable splitting and constrained optimization. IEEE Transactions on Image Processing 19 (9), 2345–2356.

Agustsson, E., Timofte, R., 2017. NTIRE 2017 challenge on single image super-resolution: dataset and study. In: IEEE Conference on Computer Vision and Pattern Recognition Workshops, vol. 3, pp. 126–135.

Andrews, H.C., Hunt, B.R., 1977. Digital Image Restoration. Prentice-Hall Signal Processing Series, vol. 1. Prentice-Hall, Englewood Cliffs.

Barbu, A., 2009. Training an active random field for real-time image denoising. IEEE Transactions on Image Processing 18 (11), 2451–2462.

Batson, J., Royer, L., 2019. Noise2self: blind denoising by self-supervision. In: International Conference on Machine Learning, pp. 524–533.

Bishop, C.M., 2006. Pattern Recognition and Machine Learning. Springer.

Boracchi, G., Foi, A., 2012. Modeling the performance of image restoration from motion blur. IEEE TIP 21 (8), 3502–3517.

Boyd, S., Parikh, N., Chu, E., Peleato, B., Eckstein, J., 2011. Distributed optimization and statistical learning via the alternating direction method of multipliers. Foundations and Trends in Machine Learning 3 (1), 1–122.

Brooks, T., Mildenhall, B., Xue, T., Chen, J., Sharlet, D., Barron, J.T., 2019. Unprocessing images for learned raw denoising. In: IEEE Conference on Computer Vision and Pattern Recognition, pp. 11,036–11,045.

Buades, A., Coll, B., Morel, J.M., 2005. A non-local algorithm for image denoising. In: IEEE Conference on Computer Vision and Pattern Recognition, vol. 2, pp. 60–65.

Burger, H.C., Schuler, C.J., Harmeling, S., 2012. Image denoising: can plain neural networks compete with BM3D? In: IEEE Conference on Computer Vision and Pattern Recognition, pp. 2392–2399.

Chambolle, A., Pock, T., 2011. A first-order primal-dual algorithm for convex problems with applications to imaging. Journal of Mathematical Imaging and Vision 40 (1), 120–145.

Chan, S.H., Wang, X., Elgendy, O.A., 2017. Plug-and-play ADMM for image restoration: fixed-point convergence and applications. IEEE Transactions on Computational Imaging 3 (1), 84–98.

Chen, Y., Pock, T., 2017. Trainable nonlinear reaction diffusion: a flexible framework for fast and effective image restoration. IEEE Transactions on Pattern Analysis and Machine Intelligence 39 (6), 1256–1272.

Dabov, K., Foi, A., Katkovnik, V., Egiazarian, K., 2007. Image denoising by sparse 3-d transform-domain collaborative filtering. IEEE Transactions on Image Processing 16 (8), 2080–2095.

Danielyan, A., Katkovnik, V., Egiazarian, K., 2010. Image deblurring by augmented Lagrangian with BM3D frame prior. In: Workshop on Information Theoretic Methods in Science and Engineering, pp. 16–18.

Danielyan, A., Katkovnik, V., Egiazarian, K., 2012. BM3D frames and variational image deblurring. IEEE Transactions on Image Processing 21 (4), 1715–1728.

Dong, C., Loy, C.C., He, K., Tang, X., 2016. Image super-resolution using deep convolutional networks. IEEE Transactions on Pattern Analysis and Machine Intelligence 38 (2), 295–307.

Dong, W., Zhang, L., Shi, G., Li, X., 2013. Nonlocally centralized sparse representation for image restoration. IEEE Transactions on Image Processing 22 (4), 1620–1630.

Efrat, N., Glasner, D., Apartsin, A., Nadler, B., Levin, A., 2013. Accurate blur models vs. image priors in single image super-resolution. In: IEEE International Conference on Computer Vision, pp. 2832–2839.

Egiazarian, K., Katkovnik, V., 2015. Single image super-resolution via BM3D sparse coding. In: European Signal Processing Conference, pp. 2849–2853.

Elad, M., Aharon, M., 2006. Image denoising via sparse and redundant representations over learned dictionaries. IEEE Transactions on Image Processing 15 (12), 3736–3745.

Franzen, R., 1999. Kodak lossless true color image suite. source. http://r0k.us/graphics/kodak. vol. 4.

Gavaskar, R.G., Chaudhury, K.N., 2020. Plug-and-play ista converges with kernel denoisers. IEEE Signal Processing Letters 27, 610–614.

Geman, D., Yang, C., 1995. Nonlinear image recovery with half-quadratic regularization. IEEE Transactions on Image Processing 4 (7), 932–946.

Gu, S., Timofte, R., Van Gool, L., 2018. Integrating local and non-local denoiser priors for image restoration. In: International Conference on Pattern Recognition.

Gu, S., Zhang, L., Zuo, W., Feng, X., 2014. Weighted nuclear norm minimization with application to image denoising. In: IEEE Conference on Computer Vision and Pattern Recognition, pp. 2862–2869.

Guo, S., Yan, Z., Zhang, K., Zuo, W., Zhang, L., 2019. Toward convolutional blind denoising of real photographs. In: IEEE Conference on Computer Vision and Pattern Recognition, pp. 1712–1722.

He, K., Zhang, X., Ren, S., Sun, J., 2016. Deep residual learning for image recognition. In: IEEE Conference on Computer Vision and Pattern Recognition, pp. 770–778.

Heide, F., Steinberger, M., Tsai, Y.T., Rouf, M., Pajak, D., Reddy, D., Gallo, O., Liu, J., Heidrich, W., Egiazarian, K., et al., 2014. Flexisp: a flexible camera image processing framework. ACM Transactions on Graphics 33 (6), 231.

Jain, V., Seung, S., 2009. Natural image denoising with convolutional networks. In: Advances in Neural Information Processing Systems, pp. 769–776.

Kamilov, U.S., Mansour, H., Wohlberg, B., 2017. A plug-and-play priors approach for solving nonlinear imaging inverse problems. IEEE Signal Processing Letters 24 (12), 1872–1876.

Kingma, D., Ba, J., 2015. Adam: a method for stochastic optimization. In: International Conference for Learning Representations.

Krull, A., Buchholz, T.O., Jug, F., 2019. Noise2void-learning denoising from single noisy images. In: IEEE Conference on Computer Vision and Pattern Recognition, pp. 2129–2137.

Kruse, J., Rother, C., Schmidt, U., 2017. Learning to push the limits of efficient fft-based image deconvolution. In: IEEE International Conference on Computer Vision, pp. 4586–4594.

Lefkimmiatis, S., 2017. Non-local color image denoising with convolutional neural networks. In: IEEE Conference on Computer Vision and Pattern Recognition, pp. 3587–3596.

Lehtinen, J., Munkberg, J., Hasselgren, J., Laine, S., Karras, T., Aittala, M., Aila, T., 2018. Noise2noise: learning image restoration without clean data. In: International Conference on Machine Learning, pp. 2965–2974.

Levin, A., Weiss, Y., Durand, F., Freeman, W.T., 2009. Understanding and evaluating blind deconvolution algorithms. In: IEEE Conference on Computer Vision and Pattern Recognition, pp. 1964–1971.

Li, Z., Wu, J., 2019. Learning deep cnn denoiser priors for depth image inpainting. Applied Sciences 9 (6), 1103.

Lim, B., Son, S., Kim, H., Nah, S., Lee, K.M., 2017. Enhanced deep residual networks for single image super-resolution. In: IEEE Conference on Computer Vision and Pattern Recognition Workshops, pp. 136–144.

Liu, D., Wen, B., Fan, Y., Loy, C.C., Huang, T.S., 2018. Non-local recurrent network for image restoration. In: Advances in Neural Information Processing Systems, pp. 1673–1682.

Ma, K., Duanmu, Z., Wu, Q., Wang, Z., Yong, H., Li, H., Zhang, L., 2017. Waterloo exploration database: new challenges for image quality assessment models. IEEE Transactions on Image Processing 26 (2), 1004–1016.

Martin, D., Fowlkes, C., Tal, D., Malik, J., 2001. A database of human segmented natural images and its application to evaluating segmentation algorithms and measuring ecological statistics. In: IEEE International Conference on Computer Vision, vol. 2, pp. 416–423.

Metzler, C.A., Maleki, A., Baraniuk, R.G., 2016. From denoising to compressed sensing. IEEE Transactions on Information Theory 62 (9), 5117–5144.

Mohan, S., Kadkhodaie, Z., Simoncelli, E.P., Fernandez-Granda, C., 2019. Robust and interpretable blind image denoising via bias-free convolutional neural networks. In: International Conference on Learning Representations.

Parikh, N., Boyd, S.P., et al., 2014. Proximal algorithms. Foundations and Trends in Optimization 1 (3), 127–239.

Plötz, T., Roth, S., 2018. Neural nearest neighbors networks. In: Advances in Neural Information Processing Systems, pp. 1087–1098.

Portilla, J., Strela, V., Wainwright, M.J., Simoncelli, E.P., 2003. Image denoising using scale mixtures of Gaussians in the wavelet domain. IEEE Transactions on Image Processing 12 (11), 1338–1351.

Richardson, W.H., 1972. Bayesian-based iterative method of image restoration. JOSA 62 (1), 55–59.

Riegler, G., Schulter, S., Ruther, M., Bischof, H., 2015. Conditioned regression models for non-blind single image super-resolution. In: IEEE International Conference on Computer Vision, pp. 522–530.

Romano, Y., Elad, M., Milanfar, P., 2017. The little engine that could: regularization by denoising (RED). SIAM Journal on Imaging Sciences 10 (4), 1804–1844.

Ronneberger, O., Fischer, P., Brox, T., 2015. U-net: convolutional networks for biomedical image segmentation. In: International Conference on Medical Image Computing and Computer-Assisted Intervention, pp. 234–241.

Roth, S., Black, M.J., 2009. Fields of experts. International Journal of Computer Vision 82 (2), 205–229.

Ryu, E., Liu, J., Wang, S., Chen, X., Wang, Z., Yin, W., 2019. Plug-and-play methods provably converge with properly trained denoisers. In: International Conference on Machine Learning, pp. 5546–5557.

Samuel, K.G., Tappen, M.F., 2009. Learning optimized MAP estimates in continuously-valued MRF models. In: IEEE Conference on Computer Vision and Pattern Recognition, pp. 477–484.

Schmidt, U., Roth, S., 2014. Shrinkage fields for effective image restoration. In: IEEE Conference on Computer Vision and Pattern Recognition, pp. 2774–2781.

Shocher, A., Cohen, N., Irani, M., 2018. "zero-shot" super-resolution using deep internal learning. In: IEEE International Conference on Computer Vision, pp. 3118–3126.

Sun, J., Tappen, M.F., 2011. Learning non-local range Markov random field for image restoration. In: IEEE Conference on Computer Vision and Pattern Recognition, pp. 2745–2752.

Sun, J., Tappen, M.F., 2013. Separable Markov random field model and its applications in low level vision. IEEE Transactions on Image Processing 22 (1), 402–407.

Sun, Y., Liu, J., Kamilov, U., 2019a. Block coordinate regularization by denoising. In: Advances in Neural Information Processing Systems, pp. 380–390.

Sun, Y., Xu, S., Li, Y., Tian, L., Wohlberg, B., Kamilov, U.S., 2019b. Regularized Fourier ptychography using an online plug-and-play algorithm. In: IEEE International Conference on Acoustics, Speech and Signal Processing, pp. 7665–7669.

Tappen, M.F., 2007. Utilizing variational optimization to learn Markov random fields. In: IEEE Conference on Computer Vision and Pattern Recognition, pp. 1–8.

Teodoro, A.M., Bioucas-Dias, J.M., Figueiredo, M.A., 2016. Image restoration and reconstruction using variable splitting and class-adapted image priors. In: IEEE International Conference on Image Processing, pp. 3518–3522.

Timofte, R., Agustsson, E., Van Gool, L., Yang, M.H., Zhang, L., 2017. Ntire 2017 challenge on single image super-resolution: methods and results. In: CVPRW, pp. 114–125.

Timofte, R., Rothe, R., Van Gool, L., 2016. Seven ways to improve example-based single image super resolution. In: IEEE Conference on Computer Vision and Pattern Recognition, pp. 1865–1873.

Tirer, T., Giryes, R., 2018. Image restoration by iterative denoising and backward projections. IEEE Transactions on Image Processing 28 (3), 1220–1234.

Tirer, T., Giryes, R., 2019. Super-resolution via image-adapted denoising cnns: incorporating external and internal learning. IEEE Signal Processing Letters 26 (7), 1080–1084.

Venkatakrishnan, S.V., Bouman, C.A., Wohlberg, B., 2013. Plug-and-play priors for model based reconstruction. In: IEEE Global Conference on Signal and Information Processing, pp. 945–948.

Venkatesh, G., Naresh, Y., Little, S., Oonnor, N.E., 2018. A deep residual architecture for skin lesion segmentation. In: Context-Aware Operating Theaters, Computer Assisted Robotic Endoscopy, Clinical Image-Based Procedures, and Skin Image Analysis. Springer, pp. 277–284.

Wang, X., Yu, K., Wu, S., Gu, J., Liu, Y., Dong, C., Qiao, Y., Loy, C.C., 2018. ESRGAN: enhanced super-resolution generative adversarial networks. In: The European Conference on Computer Vision Workshops.

Xu, L., Ren, J.S., Liu, C., Jia, J., 2014. Deep convolutional neural network for image deconvolution. In: Advances in Neural Information Processing Systems, pp. 1790–1798.

Yair, N., Michaeli, T., 2018. Multi-scale weighted nuclear norm image restoration. In: IEEE Conference on Computer Vision and Pattern Recognition, pp. 3165–3174.

Zamir, S.W., Arora, A., Khan, S., Hayat, M., Khan, F.S., Yang, M.H., Shao, L., 2020. Cycleisp: real image restoration via improved data synthesis. IEEE Conference on Computer Vision and Pattern Recognition.

Zhang, J., Ghanem, B., 2018. ISTA-net: interpretable optimization-inspired deep network for image compressive sensing. In: IEEE Conference on Computer Vision and Pattern Recognition, pp. 1828–1837.

Zhang, K., Zhou, X., Zhang, H., Zuo, W., 2015. Revisiting single image super-resolution under Internet environment: blur kernels and reconstruction algorithms. In: Pacific Rim Conference on Multimedia, pp. 677–687.

Zhang, K., Zuo, W., Chen, Y., Meng, D., Zhang, L., 2017a. Beyond a Gaussian denoiser: residual learning of deep CNN for image denoising. IEEE Transactions on Image Processing, 3142–3155.

Zhang, K., Zuo, W., Gu, S., Zhang, L., 2017b. Learning deep CNN denoiser prior for image restoration. In: IEEE Conference on Computer Vision and Pattern Recognition, pp. 3929–3938.

Zhang, K., Zuo, W., Zhang, L., 2018a. FFDNet: toward a fast and flexible solution for CNN-based image denoising. IEEE TIP 27 (9), 4608–4622.

Zhang, K., Zuo, W., Zhang, L., 2018b. Learning a single convolutional super-resolution network for multiple degradations. In: IEEE Conference on Computer Vision and Pattern Recognition, pp. 3262–3271.

Zhang, L., Wu, X., Buades, A., Li, X., 2011. Color demosaicking by local directional interpolation and nonlocal adaptive thresholding. Journal of Electronic Imaging 20 (2), 1–15.

Zhang, Y., Li, K., Li, K., Zhong, B., Fu, Y., 2019. Residual non-local attention networks for image restoration. In: International Conference on Learning Representations.

Zhang, Z., Liu, Q., Wang, Y., 2018. Road extraction by deep residual u-net. IEEE Geoscience and Remote Sensing Letters 15 (5), 749–753.

Zhao, N., Wei, Q., Basarab, A., Dobigeon, N., Kouamé, D., Tourneret, J.Y., 2016. Fast single image super-resolution using a new analytical solution for ℓ_2-ℓ_2 problems. IEEE Transactions on Image Processing 25 (8), 3683–3697.

Zoran, D., Weiss, Y., 2011. From learning models of natural image patches to whole image restoration. In: IEEE International Conference on Computer Vision, pp. 479–486.

Biographies

Kai Zhang received his Ph.D. degree from School of Computer Science and Technology, Harbin Institute of Technology, China, in 2019. He was a research assistant from July, 2015 to July, 2017 and from July, 2018 to April, 2019 in Department of Computing of The Hong Kong Polytechnic University. He is currently a postdoctoral researcher at Computer Vision Lab, ETH Zurich, Switzerland, working with Prof. Luc Van Gool and Prof. Radu Timofte. His research interests include machine learning and image processing.

Radu Timofte received his PhD degree in Electrical Engineering from the KU Leuven, Belgium in 2013. From 2013 to 2016 he was postdoc in the Computer Vision Lab, ETH Zurich, Switzerland. Since 2016 he has been group leader and lecturer in the same lab. He is also professor and chair of computer science at University of Wurzburg, Germany and the recipient of a 2022 Alexander von Humboldt Professorship Award for Artificial Intelligence. He is an editorial board member of top journals such as IEEE Trans. on Pattern Analysis and Machine Intelligence, Elsevier Neurocomputing, Elsevier Computer Vision and Image Understanding, SIAM Journal on Imaging Sciences and served(s) as an area chair for top conferences such as CVPR 2021, IJCAI 2021, ECCV 2020, ACCV 2020, ICCV 2019. His work received several awards. He is a cofounder of Merantix and a coorganizer of NTIRE, CLIC, AIM, and PIRM events. His current research interests include deep learning, implicit models, compression, tracking, restoration and enhancement.

Visual adversarial attacks and defenses

Changjae Oh, Alessio Xompero, and Andrea Cavallaro

Centre for Intelligent Sensing, Queen Mary University of London, London, United Kingdom

CHAPTER POINTS

- The chapter presents the problem definition of adversarial attacks for visual tasks using both images and videos as input.

- The chapter presents the properties of the adversarial attacks and the types of perturbations.

- The chapter presents the target models and datasets used in the attack scenarios.

- The chapter discusses adversarial attacks for images processing, image classification, semantic segmentation, object detection, object tracking, and video classification.

- The chapter also discusses defenses devised against the adversarial attacks.

15.1 Introduction

Deep neural networks (DNNs) have been shown to be successful in several vision tasks, such as image classification (Krizhevsky et al., 2012; He et al., 2016), object detection (Ren et al., 2017; Redmon and Farhadi, 2017), semantic segmentation (Long et al., 2015; Yu et al., 2017), optical flow estimation (Revaud et al., 2015; Ranjan and Black, 2017), and video classification (Carreira and Zisserman, 2017; Jiang et al., 2018; Ng et al., 2015). However, DNNs are sensitive to perturbations of the input data that produce the so-called *adversarial examples* that induce DNNs to erroneous predictions (Szegedy et al., 2014).

The study of data alterations designed to evade a classifier is not new: techniques that mislead classifiers have been discussed for over two decades and include attacks on fraud detection systems (Bolton et al., 2002), spam filters (Meyer and Whateley, 2004) and on specific classifiers, such as Support Vector Machines (Biggio et al., 2013). More recently, there has been a growing interest in adversarial examples for DNNs in visual tasks (Szegedy et al., 2014).

Advanced Methods and Deep Learning in Computer Vision
https://doi.org/10.1016/B978-0-12-822109-9.00024-2

511

Copyright © 2022 Elsevier Inc. All rights reserved.

Adversarial examples in visual tasks are generated by modifying pixel values with carefully crafted additive noise that is imperceptible to the human eyes (Carlini and Wagner, 2017; Jiang et al., 2019); or that replaces (rectangular or circular) image regions (Brown et al., 2018; Ranjan et al., 2019) or the border of an image (Zajac et al., 2019). Adversarial examples help investigate and improve the robustness of DNN models (Tsipras et al., 2019; Allen-Zhu and Li, 2020; Engstrom et al., 2019; Santurkar et al., 2019) as well as protect private information in images (Li et al., 2019a; Sanchez-Matilla et al., 2020).

An adversarial attack can be targeted or untargeted. *Targeted attacks* modify an image or a video for the DNN model to predict a specified class label, such as an object type (Szegedy et al., 2014) or a predefined object trajectory in subsequent frames (Liang et al., 2020). *Untargeted attacks* modify a source image or video to be classified as any incorrect label other than the original one, or the video perturbations can generate incorrect bounding boxes to mislead a tracker (Liang et al., 2020). Finally, carefully modified stop signs can cause a false negative detection or an incorrect detection of another object type (Lu et al., 2017).

After the problem definition of visual adversarial attacks (Section 15.2), we discuss their main properties (Section 15.3) and the types of perturbation they generate (Section 15.4). Then we overview attack scenarios, models and datasets used to craft adversarial attacks (Section 15.5). Specifically, we cover adversarial attacks for image processing tasks (Section 15.6), image classification (Section 15.7), semantic segmentation and object detection (Section 15.8), object tracking (Section 15.9), and video classification (Section 15.10). Finally, we present strategies to defend DNN models from these attacks (Section 15.11) and conclude the chapter (Section 15.12).

15.2 Problem definition

Let $\mathbf{x}$ be an image or a video and y be a true label associated with $\mathbf{x}$. The label can be a single class for image or video classification, or a region for optical flow, object detection or object tracking.

Let $f(\cdot)$ be a DNN model that maps $\mathbf{x}$ to the label y: $y = f(\mathbf{x})$. An adversarial attack modifies $\mathbf{x}$ to produce an adversarial example $\dot{\mathbf{x}}$ that at test time misleads the model, such that $f(\mathbf{x}) \neq f(\dot{\mathbf{x}})$ (see Fig. 15.1). The adversarial example can be obtained by directly modifying the visual data with a perturbation δ such that $\dot{\mathbf{x}} = g(\mathbf{x}, \delta)$, where $g(\cdot)$ represents the process of adding the perturbation to $\mathbf{x}$ or replacing pixel values in $\mathbf{x}$ with those of δ.

The adversarial perturbation itself may be crafted using machine learning. In such a case, the attacker may design a loss function $\mathcal{L}(\mathbf{x}, y)$ that is used to generate the prediction of the original input with respect to the correct prediction (Goodfellow et al., 2015; Moosavi-Dezfooli et al., 2016; Jiang et al., 2019; Shamsabadi et al., 2020c; Liang et al., 2020). For example, the loss of an adversarial attack may maximize the classification error for a set of training images (Szegedy et al., 2014; Moosavi-Dezfooli et al., 2016; Hosseini and Poovendran, 2018) or may optimize the loss function with the additional objective of modifying the image with an imperceptible perturbation (Szegedy et al., 2014; Goodfellow et al., 2015; Moosavi-Dezfooli et al., 2016; Modas et al., 2019; Shamsabadi et al., 2020c).

Based on knowledge of the target model, $f(\cdot)$, and/or its predictions, y, available to an adversary, an attack can be categorized as white-box or black-box. In a *white-box attack*, the

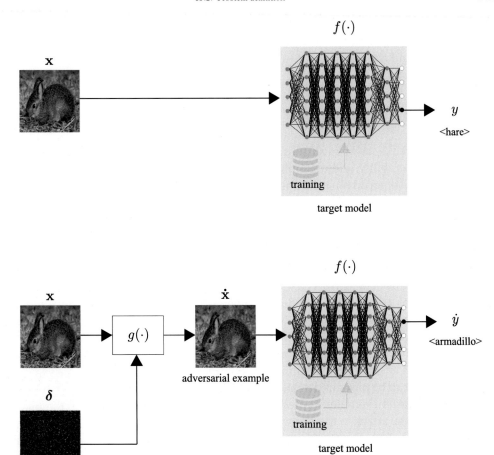

FIGURE 15.1 Illustrative example of adversarial attack for image classification. The original image is classified with the label *hare* by the target model (top), while the perturbed image (adversarial example), obtained with the Basic Iterative Method (Kurakin et al., 2017b) attack, is classified with the label *armadillo* by the same model (bottom). Note that the target model is an illustrative and abstract representation of the Inception V3 classifier (Szegedy et al., 2016) that is trained on ImageNet (Deng et al., 2009). Note also that the magnitude of the perturbation is scaled up to 20 times larger than the real one for visualization purposes.

adversary has full access to the target model (the specific architecture, its parameters and/or the training data), which is used to directly craft the adversarial perturbation. White-box settings help discover the limits of a trained model and evaluate their robustness. In a *black-box attack*, the adversary has no direct access to the architecture or parameters of the target model. Black-box attacks are more realistic than white-box attacks since models (or even their outputs) are not available in real-world applications. With a black box attack the attacker may have no access to the output of $f(\cdot)$ and thus no information about y (no-output attack), or have access only to the label y (classes) generated by adversarial examples submitted to the classifier $f(\cdot)$ (label-output attack); or only to the values of the neurons prior to the last layer, namely the logits/probabilities of adversarial examples submitted to the classifier (distribution-output attack).

15.3 Properties of an adversarial attack

Adversarial attacks can be evaluated based on four main properties, namely effectiveness, robustness, transferability, and noticeability.

The *effectiveness* of an adversarial attack is the degree to which it succeeds in misleading a machine learning model. Effectiveness can be measured as the accuracy of $f(\cdot)$ over a target dataset. The lower the accuracy, the higher the effectiveness of an adversarial attack.

The *robustness* of an adversarial attack is its effectiveness in the presence of a defense that removes the effect of δ prior to the data being processed by $f(\cdot)$. Examples of defense include median filtering (Xu et al., 2018), requantization (Xu et al., 2018), and JPEG compression (Das et al., 2017; Dziugaite et al., 2016; Guo et al., 2018). Robustness can be measured as the difference in accuracy of $f(\cdot)$ over a target dataset when a defense is used with respect to a setting when the defense is not used for $f(\cdot)$. The smaller this difference, the higher the robustness of an adversarial attack.

The *transferability* of an adversarial attack is the extent to which a perturbation δ crafted for a model $f(\cdot)$ is effective in misleading another model $f'(\cdot)$ that was not used to generate δ. Transferability can be measured as the difference in accuracy of $f(\cdot)$ and $f'(\cdot)$ over a target dataset of adversarial examples crafted for $f(\cdot)$. The smaller this difference, the higher the transferability of the attack to the classifier $f'(\cdot)$.

The *noticeability* of an adversarial attack is the extent to which an adversarial perturbation δ can be seen as such by a person looking at an image/video. Noticeability can be measured with a double stimulus test that compares image or video clip pairs, $\{(\mathbf{x}, \dot{\mathbf{x}})\}$; a single stimulus test on the naturalness of image or video clips, $\{\dot{\mathbf{x}}\}$, controlled by the results obtained with the corresponding $\{\mathbf{x}\}$; or with a (reliable) no-reference perceptual quality measure.

In addition to these four main properties, other properties, such as detectability and reversibility, may be considered when analyzing or evaluating adversarial attacks for specific tasks or objectives. The *detectability* of an adversarial attack is the extent to which a defense mechanism is capable of identifying that a perturbation was applied to modify an original image, video, or scene. Detectability, which is related to robustness, can be measured as the proportion of adversarial examples that are detected as such in a given dataset or given scenarios. A (successful) defense can be used to determine the detectability of an attack by comparing the output of $f(\cdot)$ on a given input and on the same input preprocessed with a defense, as different outputs suggest the presence of an attack. Finally, the *reversibility* of an adversarial attack is the extent to which an analysis of the predictions or output labels of $f(\cdot)$ may support the retrieval of the original class of $\dot{\mathbf{x}}$. For instance, the analysis of the frequency of an adversarial-original prediction mapping revealed that untargeted attacks are more reversible than targeted attacks (Li et al., 2021).

15.4 Types of perturbations

An adversarial perturbation can be global or region-based depending on the spatial distribution of δ, and bounded or unbounded based on the amount of change caused to the pixel intensities.

Global perturbations change the intensity of each pixel individually and thus generate high spatial-frequency noise. The nature of these perturbations makes them vulnerable to defenses (e.g., median filtering or JPEG compression) (Dziugaite et al., 2016). Global perturbations can be sparse (Papernot et al., 2016b) or dense (Goodfellow et al., 2015). An extreme case is the *one-pixel* perturbation that modifies a single pixel of the original image (Su et al., 2019).

Region-based perturbations are applied to image areas, such as a frame around the image border, a semantic region, or a rectangular or circular patch. Adversarial patches are noticeable and extremely salient to the target model $f(\cdot)$ (Brown et al., 2018; Ranjan et al., 2019). Selected regions may also be manipulated depending on their expected saliency to the human visual system (Shamsabadi et al., 2020c).

Bounded perturbations restrict the amount of change for the value of each pixel (Goodfellow et al., 2015) or image (Szegedy et al., 2014), generally with an ℓ_p-norm constraint. The bound of the adversarial perturbation can be defined by a parameter ϵ, such that $\|x - \dot{x}\|_p < \epsilon$ (Carlini and Wagner, 2017; Goodfellow et al., 2015; Moosavi-Dezfooli et al., 2016; Modas et al., 2019). Using the ℓ_2-norm, $\|.\|_2$, limits the maximum energy change (Moosavi-Dezfooli et al., 2016), whereas using the ℓ_∞-norm, $\|.\|_\infty$, constrains the maximum change for each pixel (Kurakin et al., 2017a; Li et al., 2019a). Bounded perturbations have generally limited transferability (Xie et al., 2019) and limited robustness against defenses (Xu et al., 2018).

Unbounded perturbations do not limit the intensity changes or attack regions, thus often resulting in noticeably distorted adversarial examples (Hosseini and Poovendran, 2018). To reduce (or avoid) artifacts, *content-based* unbounded perturbations manipulate low-level image features, such as color, texture, or edges, and achieve higher transferability and robustness (Bhattad et al., 2020; Shamsabadi et al., 2020b).

15.5 Attack scenarios

We group models that have been targeted by adversarial attacks in image processing and computer vision tasks, and the corresponding datasets used for crafting and evaluating these attacks for images and videos. Table 15.1 and Table 15.2 summarize the main characteristics of adversarial attacks for images or videos.

15.5.1 Target models

A target DNN model has a specific architecture that consists of layers, whose parameters are learned. The number of layers determines how deep a DNN model is. Increasing the depth of a DNN architecture results in a larger number of parameters and hence a large-scale dataset (over one million images) is needed to learn the parameters values during training. The last layers may predict the logits (values in the range $(-\infty, +\infty)$) or the probabilities

TABLE 15.1 Summary of existing adversarial attacks for tasks that use images as input. KEY – o: white-box, ●: black-box, T: targeted, T̄: untargeted, B: bounded, B̄: unbounded, Opt: optimization-based, Grad: gradient-based, Bound: boundary approximation, GradE: gradient estimation, LocS: local search, RanS: random search, C: classification, S: semantic segmentation, D: object detection, IC: image captioning, E: edge detection.

Reference	Method	box	T	T̄	B	B̄	Approach	Datasets	Task
(Szegedy et al., 2014)	L-BFGS	o	✓		✓		Opt	ImageNet, MNIST, Youtube	C
(Carlini and Wagner, 2017)	CW	o	✓	✓	✓		Opt	MNIST, CIFAR	C
(Goodfellow et al., 2015)	FGSM	o	✓	✓	✓		Grad	MNIST	C
(Kurakin et al., 2017b)	BIM (I-FGSM)	o	✓	✓	✓		Grad	ImageNet	C
(Madry et al., 2018)	PGD	o	✓	✓	✓		Grad	MNIST, CIFAR	C
(Papernot et al., 2016b)	JSMA	o	✓		✓		Grad	MNIST	C
(Moosavi-Dezfooli et al., 2016)	DeepFool	o		✓	✓		Grad	ImageNet, MNIST, CIFAR	C
(Modas et al., 2019)	SparseFool	o		✓	✓		Grad	ImageNet, MNIST, CIFAR	C
(Xie et al., 2019)	DI²-FGSM	o,●	✓		✓		Grad	ImageNet	C
(Tramèr et al., 2018)	E-FGSM	o,●	✓	✓	✓		Grad	ImageNet	C
(Li et al., 2019a)	P-FGSM	o	✓		✓		Grad	ImageNet	C
(Sanchez-Matilla et al., 2020)	RP-FGSM	o	✓		✓		Grad	Places	C
(Moosavi-Dezfooli et al., 2017)	UAP	o		✓	✓		Grad	ImageNet	C
(Mopuri et al., 2017)	Fast Feature Fool	o		✓	✓		Opt	ImageNet, Places-205	C
(Baluja and Fischer, 2018)	ATN	o	✓	✓	✓		Opt	ImageNet, MNIST	C
(Xiao et al., 2018)	AdvGAN	o,●	✓	✓	✓		Opt	ImageNet, MNIST	C,S
(Poursaeed et al., 2018)	GAP	o	✓	✓	✓		Opt	ImageNet, Cityscapes	C
(Mopuri et al., 2018)	NAG	o,●		✓	✓		Opt	ImageNet	C
(Bhattad et al., 2020)	Semantic Manipulation	o	✓			✓	Opt	ImageNet, MSCOCO	C,IC
(Shamsabadi et al., 2020a)	EdgeFool	o	✓			✓	Opt	ImageNet, Places	C
(Papernot et al., 2016a)	SBA	●	✓		✓		Bound	MNIST	C
(Shi et al., 2019)	Curls & Whey	●	✓	✓	✓		Bound	ImageNet	C
(Dong et al., 2019)	TI Attack	●	✓	✓	✓		Bound	ImageNet	C
(Chen et al., 2017)	ZOO	●	✓	✓	✓		GradE	ImageNet, MNIST, CIFAR	C
(Ilyas et al., 2018)	Query-limited Attack	●	✓	✓	✓		GradE	ImageNet	C
(Tu et al., 2019)	AutoZOOM	●	✓	✓	✓		GradE	ImageNet, MNIST, CIFAR	C
(Narodytska and Kasiviswanathan, 2017)	LocSearchAdv	●	✓		✓		GradE	ImageNet, MNIST, CIFAR, SVHN, STL	C
(Brendel et al., 2018)	BA	●	✓	✓	✓		LocS	ImageNet, MNIST, CIFAR	C
(Guo et al., 2019)	SimBA	●	✓	✓	✓		LocS	ImageNet	C
(Hosseini and Poovendran, 2018)	SemanticAdv	●		✓		✓	RanS	CIFAR	C
(Shamsabadi et al., 2020c)	ColorFool	●		✓		✓	RanS	ImageNet, CIFAR, Places	C
(Fischer et al., 2017)	SSA	o	✓	✓	✓		Grad	Cityscapes	S
(Xie et al., 2017)	DAG	o	✓	✓	✓		Grad	VOC	S,D
(Wei et al., 2019)	UEA	o,●	✓			✓	Opt	ImageNet VID	D
(Cosgrove and Yuille, 2020)	Edge Attack	o		✓	✓		Grad	Cityscapes	E

TABLE 15.2 Summary of existing adversarial attacks for tasks that use videos as input. KEY – ○: white-box, ●: black-box, T: targeted, T̄: untargeted, DD: data input dependency, U: universal, B̄: unbounded, B: bounded, R: region-based, Gen: generative, Opt: optimization, C: video classification, ME: motion estimation, OT: object tracking.

Reference	Method	box	T	T̄	DD	U	B	B̄	R	Approach	Data sets	Task
Ranjan et al. (2019)	OFA	○,●		✓		✓	✓		✓	Opt	KITTI	ME
Liang et al. (2020)	FAN	○,●	✓	✓	✓		✓			Gen	OTB, VOT	OT
Jiang et al. (2018)	V-BAD	●	✓	✓	✓		✓			Opt	UCF, HMDB, Kinetics	C
Li et al. (2019b)	C-DUP	○	✓			✓	✓			Gen	UCF, JESTER	C
Lo and Patel (2020)	MultAV	○	✓	✓	✓	✓	✓	✓	✓	Direct, Opt	UCF	C

(values in the range [0,1]) of the predicted labels, or the final labels. As not all layers (e.g., pooling layers) have trainable parameters, in this chapter we will use the term layer only for those with trainable parameters (e.g., convolutional and fully connected layers).

The models that have been attacked, or target models, are LeNet, AlexNet, VGGNets, GoogleNet and variants, ResNets and WideResNets, DenseNets, and MobileNets for image classification; FCN, HED, and Faster R-CNN, for semantic segmentation, edge detection, and object detection, respectively; C3D, CNN + LSTM, I3D, and 3D ResNet-18 for video classification; SiamFC, SiamRPN, SiamRPN + CIR, and SiamRPN ++ for visual object tracking; and FlowNet, FlowNet2, SpyNet, PWC-Net, and Back2Future for motion estimation. We also include two non-DNN models, namely Epic Flow and LDOF, for optical flow estimation.

15.5.1.1 Models for image-based tasks

LeNet is a convolutional neural network (CNN) with 3 convolution layers and 2 fully connected layers for image recognition (LeCun et al., 1998).

AlexNet has 8 convolutional layers and 3 fully connected layers, and the model architecture is similar to LeNet (Krizhevsky et al., 2012).

VGGNet has deeper architectures with 16 and 19 layers, but smaller convolution filter kernels as multiple small convolution filters with fewer parameters outperform a single large filter (Simonyan and Zisserman, 2014).

GoogleNet (Inception-v1) has 22 layers (only layers with parameters are counted) and 9 of these layers consist of multiple convolutional filter sizes (*inception* layer/module) that reduces the computational cost and resources of the architecture (Szegedy et al., 2015). Inception-v2 and Inception-v3 (Szegedy et al., 2016) and Inception-v4 (Szegedy et al., 2017) are variants that further improve accuracy and reduce computational complexity. Improvements include factorizing filter sizes to stacked smaller filter size, increasing the number of filters and their size in each model (wider instead of deeper), label smoothing (regularization term to reduce the confidence of the network on a class), adding reduction blocks, and simplification of different modules (e.g., by choosing different number of filters and their size). Reduction blocks are 1×1 convolutional layers, also known as projection layers, that perform pooling of the feature maps over the channel dimension, thus resulting in a dimensionality reduction.

ResNet may have 18, 50, 101, or 150 layers. Blocks of layers are grouped to learn a function that maps the input with the output as a residual mapping (He et al., 2016).

DenseNet has architectures with 121, 169, 201, and 264 layers, whose (4) dense blocks can contain from 12 to 128 convolutional layers. DenseNet extends the residual mapping of ResNet by providing the feature maps of all previous layers within a block as input to subsequent layers within the same block (dense connection) to reduce the number of parameters, improve the flow of information and gradients throughout the network, and reduce overfitting (Huang et al., 2017).

MobileNet has 28 layers and two hyper-parameters that regulate the width of the network at each layer and the resolution of an input image (Howard et al., 2017). The first 13 layers are depth-wise separable convolution layers that split the standard convolution layer into two layers, a separate layer for filtering and a separate layer for combining the resulting output. This factorization reduces computation and model size. The architecture can be further reduced by removing layers.

WideResNet has higher accuracy than ResNet on image classification tasks with a shallower architecture and a larger width of the residual blocks (Zagoruyko and Komodakis, 2016).

Fully Convolutional Networks (FCN) use CNN models for image classification, such as AlexNet, VGGNet, and GoogleNet, and replace fully connected layers with convolutional layers, followed by an upsampling layer (Long et al., 2015), thus generating an output feature map of the same size as the input. Because of this FCN can be trained end-to-end for (semantic) segmentation.

Holistically-nested Edge Detector (HED) is an FCN-based architecture that uses average pooling to detect edges (Xie and Tu, 2015).

Faster R-CNN is an FCN-based architecture that learns to generate object candidates, followed by localization and classification of the objects among the candidates (Ren et al., 2017). Faster R-CNN consists of pretrained convolution layers, such as 13 convolution layers of VGGNet, where the output is fed into the small network, with two convolution layers, that outputs lower-dimensional features. These features, each of which represents an object candidate, are then fed into two independent fully connected layers to find the location of bounding boxes and their classes for object candidates.

15.5.1.2 Models for video-based tasks

CNN + LSTM is a video classifier that applies 2D architectures (e.g., AlexNet or GoogleNet) to each frame independently using shared parameters through time, and then learns how to integrate information over time by using a recurrent neural network based on the Long-Short Temporal Memory (LSTM) operating on frame-level CNN activations (Ng et al., 2015). The combination of LSTM with the 2D networks allows the video classifier to maintain a constant number of parameters while capturing a global description of the video temporal evolution.

C3D is a spatio-temporal classifier with 8 3D convolutional layers and 2 fully connected layers (Tran et al., 2015). C3D can process 16 frames for action recognition.

Inflated 3D CNN (I3D) is a spatio-temporal architecture, built on top of 2D DNNs for image classification (e.g., InceptionV1), that combines the output of two 3D CNNs, one processing a group of RGB frames and the other processing a group of optical flow predictions among consecutive RGB frames (Carreira and Zisserman, 2017). I3D extends filters and pooling operations from 2D to 3D (inflating). I3D can also use pretrained weights from 2D models as initialization prior to fine-tuning.

3D ResNet is a spatio-temporal architecture that, similarly to I3D, extends image-based ResNet models to the temporal domain using 3D CNNs (Hara et al., 2018). Additional models, such as WideResNet and DenseNet can be extended with 3D CNNs.

SiamFC uses a fully convolutional Siamese architecture consisting of two convolutional neural networks sharing the same parameters. SiamFC is trained to predict a similarity map between a target template (or reference patch) and a search region in a current image via cross-correlation of the features outputted by the two branches. The reference patch is provided or initialized in the first frame (Bertinetto et al., 2016).

SiamRPN is based on the Siamese architecture to extract the features from the template and the search area. SiamRPN adds a region proposal network consisting of two branches, one for foreground-background classification and the other for regression, based on pair-wise cross-correlation (Li et al., 2018).

SiamRPN ++ improves SiamRPN by using a spatially aware sampling strategy that provides a strict translation invariance, as padding introduced within DNNs breaks the invariance. Moreover, SiamRPN ++ adds a depth-wise cross-correlation layer that predicts multichannel correlation features between a template and search patches, exploiting the structure of a ResNet (Li et al., 2019)

SiamRPN + CIR adds Cropping-Inside Residual (CIR) units to eliminate the underlying position bias given by the zero padding. SiamRPN + CIR applies CIR units to different deep and wide networks, such as ResNet and Inception, when used within SiamFC and SiamRPN (Zhang and Peng, 2019).

FlowNet is an autoencoder (encoder-decoder architecture) with 9 convolutional layers and a correlation layer that performs multiplicative patch comparisons between two feature maps. Pairs of images with arbitrary size are provided as input to two branches of the architecture and combined with the correlation layer. The decoder part has 4 upconvolutional layers that perform extension of the feature maps, and concatenates with feature maps from the encoder to refine the optical flow with image details (Dosovitskiy et al., 2015).

FlowNet2 stacks multiple FlowNet architectures to improve the accuracy of the optical flow estimation for both small and large displacements. The first network is FlowNet and the stacked networks use a single branch whose input is given by the concatenation of the flow estimated by the previous network with the two images, the warped image based on the optical flow, and a brightness error. The last stage of FlowNet2 fuses the output of the stacked networks with the output of an autoencoder for small displacements to obtain the final flow (Ilg et al., 2017).

SpyNet (Spatial Pyramid Network) combines a coarse-to-fine strategy, based on a spatial-pyramid formulation, with convolutional neural networks to estimate large motions within the image pyramid. For each level of the pyramid, a convolutional neural network updates the estimated optical flow, as one image of a pair is warped by the estimated optical flow at the coarser level (Ranjan and Black, 2017).

PWC-Net is a coarse-to-fine architecture that replaces the image pyramid with a feature pyramid. For each level, PWC-Net adds a layer that warps the feature from the second image toward the first image using the upsampled flow, and a correlation layer between features of the first image and warped features of the second image to compute partial matching costs for associating a pixel with its corresponding pixels at the next frame (Sun et al., 2018).

Back2Future is a coarse-to-fine architecture that uses both image and feature pyramids of three consecutive frames (past, present, and future) and is trained in an unsupervised way with a photometric loss. The multiframe formulation addresses occlusions and allows adding a linear motion model as a soft temporal constraint (Janai et al., 2018).

Epic Flow is a *classic* optical flow estimation approach that performs a dense matching by edge-preserving interpolation from a sparse set of matches, followed by a variational energy minimization initialized with the dense matches. The sparse-to-dense interpolation relies on an edge-aware geodesic distance, tailored to handle occlusions and motion boundaries (Revaud et al., 2015).

LDOF is a *classic* optical flow estimation approach that combines the matching of local descriptors (i.e., feature vectors extracted from the local area around localized interests points, such as corner points) with variational techniques based on image warping and energy minimization. This approach can handle fast motions of different body parts (Brox and Malik, 2011).

15.5.2 Datasets and labels

We describe the main features of the datasets (and their annotations) used to generate adversarial images and videos. Image datasets include MNIST, SVHN, CIFAR-10, STL-10, ImageNet, Places, COCO. Video datasets include action recognition datasets used for video classification, visual object tracking, and motion estimation. These datasets are KITTI, UCF, JESTER, HMDB, Kinetics, and two benchmarks for tracking (and their variants): Object Tracking Benchmark (OTB) (Wu et al., 2015) and Visual Object Tracking (VOT).

15.5.2.1 Image datasets

MNIST (Modified National Institute of Standards and Technology) has 60,000 training and 10,000 test images (28×28, gray-scale) of handwritten digits (LeCun, 1998).

SVHN (Street View House Numbers) has 600,000 Google Street View images with small cropped digits (Netzer et al., 2011).

CIFAR-10 (Canadian Institute for Advanced Research-10) has 50,000 training and 10,000 test 32×32 images of 10 classes, such as *airplane, automobile, bird, cat, deer, dog, frog, horse, ship*, and *truck* (Krizhevsky et al., 2009).

STL-10 contains 500 training images, 800 test images per class, and 100,000 unlabeled images within 10 classes total (same classes as CIFAR-10). The image resolution is 96 in color (Coates et al., 2011).

ImageNet includes 14,197,122 RGB images with 1000 classes (Deng et al., 2009). Variants of ImageNet include tiny ImageNet that contains a 120,000 images for 200 classes with 64 size.

Places have two scene-centric versions: Places-205, with 2,448,873 images and 205 scene categories, and Places-365, with 1,800,000 images and 365 scene categories (Zhou et al., 2017).

COCO (Common Objects In Context) has 330,000 images, of which over 200,000 are semantically labeled for object detection, semantic segmentation, image classification, and keypoint estimation (Lin et al., 2014).

F-MNIST (Fashion MNIST) contains 60,000 training and 10,000 test 28 gray-scale images with a label for 10 types of clothing, such as *t-shirt, trousers, pullover, dress, coat, sandal, shirt, sneaker, bag,* and *Ankle boot* (Xiao et al., 2017).

VOC (PascalVOC) was used for several image classification, object detection, and semantic segmentation benchmarks, such as VOC2007 and VOC2012 (Everingham et al., 2015). VOC2007 consists of 9963 images with 24,640 annotated objects from 20 classes. VOC2012 consists of 11,530 train/validation images containing 27,450 annotated objects (bounding box) and 6929 segmentations from 20 classes.

LFW (Labeled Faces in the Wild) consists of more than 13,000 photos of faces labeled with the name of the person pictured (Huang et al., 2008). The dataset has one or more distinct photos for 1680 people.

BSDS500 contains 500 natural images with the ground-truth annotations drawn by humans for edge detection (Martin et al., 2004).

15.5.2.2 Video datasets

Cityscapes consists of stereo video sequences from the streets of 50 cities and includes 5,000 images with high quality pixel-level annotations and 20,000 additional images with coarse annotations (Cordts et al., 2016).

HAR (Human Activity Recognition) includes 30 subjects doing activities of daily living (ADL), while carrying a waist-mounted smartphone with embedded inertial sensors (Anguita et al., 2013). Each participant was instructed to perform twice a designed protocol of activities, including 6 actions, namely walking, walking upstairs, walking downstairs, sitting, standing and lying. Performing the protocol takes 192 seconds in total.

UAV has 123 sequences (and a total of 110,000 frames) captured from unmanned aerial vehicles in low-altitude for object tracking (Mueller et al., 2016).

KITTI consists of 389 scenes (image pairs), split into training and testing sets, of a static environment captured by a camera mounted on a moving car (Geiger et al., 2012). The training set includes sparse annotations of the flow between the two images.

UCF has 13,320 YouTube clips of 101 coarse-grained human actions categories split into 25 groups (4–7 clips for each group), sharing common attributes, such as same background or actor (Soomro et al., 2012). The dataset has diverse camera motions, object scales, lighting conditions, backgrounds, and actors involved.

JESTER has 148,092 video clips (of 3 seconds duration, on average) with fine-grained human gestures, divided in 27 categories and performed by a total of 1,376 actors. Examples of gestures are *zooming in with two fingers, pushing hand out,* and *swiping right* (Materzynska et al., 2019).

HMDB contains 6,766 video clips of different actions, split into 51 categories (about 100 clips for each category) and 5 types, such as general facial actions (e.g., smile), facial actions with object manipulation (e.g., smoking), general body movements (e.g., handstand), body movements with object interaction (e.g., pour), and body movements for human interaction (e.g., fencing) (Kuehne et al., 2011).

Kinetics has 306,245 clips of 400 human actions, including single person actions, person-person actions, and person-object actions, depicting a variety of actors and large variations in backgrounds, illuminations, perspective, and action execution (e.g., speed and poses) (Kay et al., 2017). Each action category contains between 400 to 1,000 video clips, whose duration is around 10 seconds.

OTB (or OTB-2015) has 100 short videos with per-frame annotations of bounding boxes and 11 attributes, such as illumination and scale variations, occlusion, nonrigid object deformation, or out-of-view (Wu et al., 2015). OTB-13, the first release, contained 51 sequences.

VOT[1] contains 60 videos of fewer than 1,500 frames at 30 fps with various animals, people, or moving objects, often acquired with a moving camera. Examples of scenes include a person carrying a book, gymnastics, ants moving, flying birds, moving drones, a moving car on the road, fish in a water tank, people playing basketball, soccer, and handball. Per-frame annotations are available of rotated bounding boxes, segmentation masks, and attributes.

15.6 Image processing

In this section, we discuss adversarial attacks that aim at misleading image processing operations, such as *edge detection* (Cosgrove and Yuille, 2020) and *motion estimation* (Ranjan et al., 2019).

Edge Attack is devised to craft adversarial perturbations for a CNN-based edge detection model, HED (Xie and Tu, 2015). HED is trained on BSDS500 and combines the feature maps extracted from each convolution layer to classify whether each pixel in the image belongs to the object boundary or not. Edge Attack estimates the loss function for the edge detection and crafts the perturbations that maximize the gradient of the loss function (Goodfellow et al., 2015).

Optical Flow Attack (OFA) (Ranjan et al., 2019) aims to adversarially perturb two consecutive pairs of images $(\mathbf{x}_k, \mathbf{x}_{k+1})$ in a video such that the labels associated with the optical flow, i.e., the displacement vectors (u, v), are incorrectly predicted (untargeted attack). Because of the limited availability of datasets with dense optical flow annotations, OFA uses the optical flow prediction of the model $f(\cdot)$ as pseudo-annotation to self-supervise the learning of the *region-based* (e.g., circular patch) adversarial perturbation $\boldsymbol{\delta}$. OFA applies (e.g., pastes) the learned adversarial patch onto the image. In addition to perturbing a small number of pixels within a pair of consecutive images, OFA constrains the adversarial perturbation to be quasi-imperceptible, such that

$$\|\mathbf{x}_k - \dot{\mathbf{x}}_k\|_0 + \|\mathbf{x}_{k+1} - \dot{\mathbf{x}}_{k+1}\|_0 < \epsilon, \tag{15.1}$$

where ϵ is a small constant (bounded perturbation), while the predicted output of the optical flow model $f(\cdot, \cdot)$ on the perturbed images is significantly affected, i.e.,

$$\|f(\mathbf{x}_k, \mathbf{x}_{k+1}) - f(\dot{\mathbf{x}}_k, \dot{\mathbf{x}}_{k+1})\| > E, \tag{15.2}$$

where E is a large constant.

---

[1] https://www.votchallenge.net.

The adversarial patch is therefore learned as

$$\boldsymbol{\delta} = \arg\min_{\boldsymbol{\delta}} \mathbb{E}_{(\mathbf{x}_k, \mathbf{x}_{k+1}) \sim \mathcal{X}, l \sim \Omega, t \sim \mathcal{T}} \left[\frac{(u, v) \cdot (\dot{u}, \dot{v})}{\|(u, v)\| \cdot \|(\dot{u}, \dot{v})\|} \right], \tag{15.3}$$

with

$$(u, v) = f(\mathbf{x}_k, \mathbf{x}_{k+1}), \tag{15.4}$$

$$(\dot{u}, \dot{v}) = f(g(\mathbf{x}_k, \boldsymbol{\delta}, l, t(\boldsymbol{\delta})), g(\mathbf{x}_{k+1}, \boldsymbol{\delta}, l, t(\boldsymbol{\delta}))), \tag{15.5}$$

where $\boldsymbol{\delta}$ is the adversarial patch, l is the pixel location sampled over all the locations Ω in the image, $g(\cdot)$ is the operator that replaces the image pixels with the values of $\boldsymbol{\delta}$ at the location l, $t(\boldsymbol{\delta})$ is a 2D transformation sampled from the set of transformations $\mathcal{T}$ (or their combination) and applied to the adversarial patch $\boldsymbol{\delta}$, and $(\dot{u}, \dot{v})$ is the optical flow resulting from the images attacked with the adversarial patch.

OFA works with different types of optical flow models, such as classical (Epic Flow, LDOF), autoencoder-based architectures (FlowNet, FlowNet2), and image-pyramids-based architectures (SpyNet, PWC-Net, Back2Future), in a white-box scenario. In such a scenario, OFA learns a specific adversarial patch for each model independently and of different sizes, while a universal adversarial patch is learned from some of the models and applied to the input videos against other models in a black-box scenario. In both scenarios, the attack on auto-encoder models impacts large image areas even with a small patch size (0.1% of the image resolution). Other types of models are instead more robust to such an attack.

15.7 Image classification

In this section, we discuss and categorize adversarial examples for image classification.

15.7.1 White-box, bounded attacks

White-box bounded attacks commonly generate the perturbations with an ℓ_p-norm constraint so that the distortion of the resulting adversarial example is imperceptible to humans.

L-BFGS (Szegedy et al., 2014) misleads a target classifier with adversarial examples that consist of imperceptible adversarial perturbations added to a normalized input $\mathbf{x} \in [0, 1]$. The L-BFGS attack solves a constrained optimization problem whose goal is to find $\dot{\mathbf{x}} \in [0, 1]$ with an ℓ_2-norm constraint:

$$c\|\dot{\mathbf{x}} - \mathbf{x}\|_2 + \mathcal{L}(\dot{\mathbf{x}}, \dot{y}), \tag{15.6}$$

where c is a hyper-parameter and $\mathcal{L}(\dot{\mathbf{x}}, \dot{y})$ is the cross-entropy loss that measures the difference between the label y predicted by a target classifier from an adversarial input and the target mis-classification label $\dot{y}$. This loss encourages the attack to generate $\dot{\mathbf{x}}$ that can mislead the target classifier. The **Carlini & Wagner (CW)** attack (Carlini and Wagner, 2017) generates an ℓ_p-norm bounded perturbation by solving a constrained optimization problem, similarly to L-BFGS. CW minimizes a loss, within ℓ_p-norm bounded constraints, that measures the difference between the logit value, $\dot{z}_y$, of $\dot{\mathbf{x}}$ belonging to y, the same class of the prediction from $\mathbf{x}$, and the maximum logit value among all the other classes:

$$\min_{\delta} \left(\|\dot{\mathbf{x}} - \mathbf{x}\|_p + c \left(\max_{n=1,\dots,D} \{\dot{z}_n; n \neq y\} - \dot{z}_y \right) \right), \tag{15.7}$$

where D is the total number of labels, $p \in \{0, 2, \infty\}$ and $c > 0$ is a constant determined via linear search to find an optimal value. Minimizing the second term encourages the attack to find $\dot{\mathbf{x}}$ that makes $\dot{z}_n \geq \dot{z}_y$ so that the classification output can avoid y. Unlike L-BFGS that employs cross-entropy loss, CW attack is based on the margin loss that is more efficient in finding the minimally distorted adversarial example (Carlini and Wagner, 2017).

Fast Gradient Sign Method (FGSM) (Goodfellow et al., 2015) is a gradient-based approach that determines a direction of perturbation such that the loss from the target model increases. FGSM estimates the perturbations by computing the gradient of the loss function, $\mathcal{L}(\mathbf{x}, y)$, with respect to a given input $\mathbf{x}$, with a small ϵ to generate imperceptible adversarial perturbations:

$$\dot{\mathbf{x}} = \mathbf{x} + \epsilon \, \mathrm{sgn}\big(\nabla_{\mathbf{x}} \mathcal{L}(\mathbf{x}, y)\big). \tag{15.8}$$

Since the perturbations are generated along the direction of $\nabla_{\mathbf{x}} \mathcal{L}(\mathbf{x}, y)$, the adversarial example can mislead a target model to avoid the original label y, resulting in an untargeted attack. FGSM can also be formulated as a targeted attack by decreasing the loss of the target model with respect to the target label $\dot{y}$ as:

$$\dot{\mathbf{x}} = \mathbf{x} - \epsilon \, \mathrm{sgn}\big(\nabla_{\mathbf{x}} \mathcal{L}(\mathbf{x}, \dot{y})\big), \tag{15.9}$$

which generates a perturbation that misleads the classifier to the target label $\dot{y}$.

Variants of FGSM using gradient-based optimization can also be either untargeted or targeted in a similar manner as shown in Eqs. (15.8) and (15.9). **Basic Iterative Method (BIM)** (or I-FGSM) (Kurakin et al., 2017b) extends FGSM by aggregating the adversarial perturbations for a fixed number of iterations:

$$\dot{\mathbf{x}}_{i+1} = \mathcal{C}_{\dot{\mathbf{x}}_i, \epsilon} \big(\dot{\mathbf{x}}_i + \alpha \, \mathrm{sgn} \left(\nabla_{\mathbf{x}} \mathcal{L} (\dot{\mathbf{x}}_i, y) \right) \big), \tag{15.10}$$

where α controls the magnitude of perturbations at each step and the clipping operation, $\mathcal{C}_{\dot{\mathbf{x}}_i, \epsilon}(\cdot)$, clips the pixel intensities of the adversarial image at time step i to be in the image range. **Projected Gradient Descent (PGD)** (Madry et al., 2018) generalizes BIM without the constraint on the magnitude of perturbation. Instead, in each step, PGD projects the adversarial examples to ℓ_∞ neighbor so that the perturbation can be bounded.

$$\dot{\mathbf{x}}_{i+1} = \mathcal{P}_{\dot{\mathbf{x}}_i, \epsilon}\big(\dot{\mathbf{x}}_i + \alpha \, \mathrm{sgn}\big(\nabla_{\mathbf{x}} \mathcal{L}(\dot{\mathbf{x}}_i, y)\big)\big), \tag{15.11}$$

where $\mathcal{P}_{\dot{\mathbf{x}}_i, \epsilon}$ is the projection operator. **Jacobian-based Saliency Map Attacks (JSMA)** (Papernot et al., 2016b) generates a saliency map, $S(x_{\mathbf{q}}, \dot{y})$, that determines each pixel $x_{\mathbf{q}} \in \mathbf{x}$, at pixel location $\mathbf{q}$ to select the pixels which are likely to be perturbed to obtain the desired changes in classification to a target label $\dot{y}$:

$$S(x_{\mathbf{q}}, \dot{y}) = \begin{cases} 0, & \nabla_{x_{\mathbf{q}}} P_{\dot{y}}(\mathbf{x}) < 0 \text{ or } \sum_{j \neq \dot{y}} \nabla_{x_{\mathbf{q}}} P_j(\mathbf{x}) > 0, \\ \nabla_{x_{\mathbf{q}}} P_{\dot{y}}(\mathbf{x}) \left\| \sum_{j \neq \dot{y}} \nabla_{x_{\mathbf{q}}} P_j(\mathbf{x}) \right\|_1, & \text{otherwise,} \end{cases} \tag{15.12}$$

where $P_{\dot{y}}(\mathbf{x})$ is the predicted softmax probability on $\dot{y}$ prior to the last layer for classification. The algorithm perturbs a pixel $x_{\mathbf{q}}$ with the highest value of $S(x_{\mathbf{q}}, \dot{y})$ to increase (or decrease) the softmax probabilities of the target class. The attack iteratively generates the perturbation until the adversarial image is classified as $\dot{y}$ or a predefined number of pixels has been perturbed.

DeepFool (Moosavi-Dezfooli et al., 2016) is another gradient-based approach that generates l_2-bounded perturbations by estimating the distance of an input at time step i, $\dot{\mathbf{x}}_i$, to the closest decision boundary of a target classifier with the original class y. This process is repeated until $f(\dot{\mathbf{x}}_i) \neq f(\mathbf{x}_i)$. DeepFool produces smaller perturbations compared to the L-BFGS attack and with lower computation cost. **SparseFool** (Modas et al., 2019) employs DeepFool-like adversarial attack with an l_1-norm constraint to the perturbations. SparseFool employs the low mean curvature of the decision boundary to compute adversarial perturbations efficiently with a few pixels.

The aforementioned methods generally focus on the *effectiveness* and *noticeability* of an adversarial attack, while the *transferability* is less considered. Some FGSM variants also investigate further challenges in adversarial attack, such as misleading multiple models or unseen models and protecting the private information, e.g., original class, from the input image. The perturbations crafted from FGSM-based attacks can be overfitted to the target model. To improve the transferability to multiple models, **Diverse Input Iterative FGSM (DI^2-FGSM)** (Xie et al., 2019) applies a random resizing and padding to the input images at each iteration to produce hard and diverse input patterns. **Ensemble FGSM (E-FGSM)** (Tramèr et al., 2018) employs multiple classifiers simultaneously when creating the FGSM-like perturbations (see Eq. (15.9)). E-FGSM can be applicable to multiple models that are used for training, but the perturbations tend to be overfitted to the employed multiple models and not desirable to unseen classifiers. Noting that the true class of images from these FGSM variants can be easily inferred, it is important to protect the privacy of the input label. **Private FGSM (P-FGSM)** Li et al. (2019a) achieves this by discarding the top predicted classes from the class selection to craft perturbation similar to Eq. (15.9), but does not consider the transferability to other models or defenses that can estimate the original label. To satisfy both the transferability and privacy issues, as well as detectability in adversarial attack, **Robust Private FGSM (RP-FGSM)** (Sanchez-Matilla et al., 2020) randomly selects a target model to attack as well as a defense to evade with the strategy of P-FGSM for class selection.

Unlike previous attacks that craft perturbations on a particular image (image-dependent attacks), **Universal Adversarial Perturbation (UAP)** (Moosavi-Dezfooli et al., 2017) is a single perturbation that aims to mislead a model for most images in the training set. UAP accumulates image-dependent perturbations by iteratively applying DeepFool until a certain portion of input images are misclassified. UAP thus requires training data where the target model can be trained to solve a data dependent optimization for producing a single universal perturbation. **Fast Feature Fool** (Mopuri et al., 2017) assumes that there is no access to the original training data. The method thus aims to generate universal perturbations that can fool the features from a target model by optimizing the product of mean activations at multiple layers of the target model when the input is the universal perturbation.

The adversarial attacks discussed so far generate perturbations by solving a constrained optimization problem or updating the feedback from the gradients with respect to the input images. Unlike aforementioned methods, **Adversarial Transformation Networks (ATN)** (Baluja and Fischer, 2018) trains a feed-forward neural network to generate an adversarial example that is similar to the input image. Once trained, the network can generate adversarial examples faster than optimization-based algorithms and gradient-based ones with iterative updates. Similarly, **Adversarial generative adversarial networks (AdvGAN)** (Xiao et al., 2018) learn to craft adversarial perturbations using neural networks, motivated by generative adversarial networks (Goodfellow et al., 2014) that consist of a generator and a discriminator. The generator in AdvGAN generates perturbations, whereas the discriminator, trained to classify the real input image from adversarial images, is added to train the generator to craft an adversarial example similar to the input image. The method can be considered as a semiwhite (or gray) box attack as the trained feed-forward network (generator) can produce adversarial perturbations for any input instances without requiring access to the model itself anymore. With this framework, the black-box attack is also available by replacing a target model by a distilled (or substitute) model. **Generative Adversarial Perturbations (GAP)** (Poursaeed et al., 2018) is a generative model that crafts image-dependent or universal perturbations for untargeted or targeted attacks. Image-dependent perturbations are crafted similar to AdvGAN (Xiao et al., 2018) where the generator outputs the perturbation from an input image. For universal perturbation, the generator takes a fixed pattern, sampled from a uniform distribution, to create the perturbation that is added to clean images to fool the classifier. **Network for Adversary Generation (NAG)** (Mopuri et al., 2018) crafts universal perturbations by employing a GAN-like generative model that models the unknown distribution of adversarial perturbations for a given DNN classifier. Based on the estimated distribution, NAG can generate a wide variety of universal perturbations.

15.7.2 White-box, content-based attacks

Content-based perturbations are crafted by considering image properties, such as image structures (Shamsabadi et al., 2020a), textures or colors (Bhattad et al., 2020). Unlike norm-bounded perturbations (Goodfellow et al., 2014; Szegedy et al., 2014), content-based perturbations are unbounded, i.e., do not constrain the magnitude of adversarial perturbations. This allows content-based adversarial attacks to improve transferability and reduce detectability to defenses.

Semantic Manipulation (Bhattad et al., 2020) presents two adversarial attacks that manipulate the visual descriptors, namely colors and textures, of an input image. The color attack employs image colorization as an adversarial attack. Given an input image converted to grayscale, the attacker learns to colorize the grayscale image with the color seeds from the original image in order to mislead a target classifier. The texture attack transfers texture from a different image to the input image. **EdgeFool** (Shamsabadi et al., 2020a) generates adversarial images with detail enhancement. EdgeFool trains a feed-forward neural network with a multitask loss functions that jointly considers image smoothing and untargeted adversarial attack (Carlini and Wagner, 2017). The image details, extracted from the learned smooth image, are manipulated with traditional detail enhancement technique (Farbman et al., 2008). The detail enhancement output is then fed into the CW loss function (Carlini and Wagner, 2017) so that the EdgeFool framework can generate adversarial images with detail enhancement.

15.7.3 Black-box attacks

Black-box attacks assumed limited or no information about the target model, e.g., only the final output label or the prediction scores. Black-box attack can be performed by training a substitute model that can approximate the target model (decision boundary approximation), generating the perturbation based on the estimated loss gradient by queries that return scores or probabilities of labels from the target model (gradient estimation), manipulating the perturbations of the current step to the correct direction in order to mislead a target model (local search), or performing greedy search until the target model is misled or reaches the maximum iteration (random search).

Decision boundary approximation methods generally consider the transferability of adversarial images. **Substitute Blackbox Attack (SBA)** (Papernot et al., 2016a) trains a substitute model that emulates the original model, and then uses white-box attacks on the trained substitute model to craft adversarial perturbations. By leveraging the transferability of adversarial images, SBA obtains predictions on the synthetic dataset from the targeted model, and then trains a substitute model to imitate the prediction of the targeted model. The trained substitute model is used as a pseudo-target model, where the adversaries can now generate perturbations with a manner of a white-box attack. Since the substitute model is trained based on the assumption of transferability of perturbations, it is important to choose white-box attacks that show high transferability. The **Curls & Whey** (Shi et al., 2019) attack aims to improve the transferability of adversarial images for other target models. First, Curls iteration crafts perturbations along the direction of gradient ascent and descent of the substitute model, which allows the adversarial images to have better transferability by diversifying the generated images. The second step, Whey optimization, reduces the magnitude of perturbations from the crafted adversarial images. Noting that the good transferability of adversarial images allows black-box attacks, **Translation-Invariant (TI)** attack (Dong et al., 2019) employs an ensemble of translated adversarial images rather than optimizing the loss function to estimate a substitute model. TI attack performs black-box attack by crafting transferable adversarial images which are generated for a different white-box classifier with translation-invariant property but have high transferability for black-box attacks.

Gradient estimation-based attacks take a query feedback form, where the adversary iteratively generates the perturbations and asks if the target model is mispredicted until the goal is achieved. By estimating the gradient of the loss function from input queries, the adversary can generate the perturbation based on the direction of the estimated gradient. **Zeroth Order Optimization (ZOO)** (Chen et al., 2017) estimates the gradients of the target model based on the assumption that the adversary has an access to obtain the probability scores of all the classes from the target model. This assumption alleviates the black-box attack easier than the previous works (Papernot et al., 2016a) that can only obtain the label information from the classifier. ZOO generates adversarial perturbations by observing the changes in the probabilities which enables to approximate the gradients. The approximated gradients can then be used for stochastic coordinate descent to perform the adversarial attack. One issue in black-box attack is query-efficiency, where the large number of queries are required to predict unknown target models. **Query-limited Attack** (Ilyas et al., 2018) generates adversarial images with a limited number of queries. The method uses natural evolutional strategies (Salimans et al., 2017) as a gradient estimation technique with the black-box setting. The estimated gradient together with PGD (Madry et al., 2018), used for white-box attacks, can then generate the adversarial example. **Autoencoder based Zeroth Order Optimization (AutoZOOM)** (Tu et al., 2019) also addresses query-efficiency problems. AutoZoom reduces the number of queries required to generate successful adversarial images by an adaptive random gradient estimation strategy in which the attack can be accelerated with an autoencoder either trained offline with unlabeled data or a bilinear resizing operation.

Local search-based approach seeks the correct direction in the perturbations of the current step to mislead a target classifier. **LocSearchAdv** (Narodytska and Kasiviswanathan, 2017) presents a greedy local-search based approach. In each step, LocSearchAdv perturbs a few pixels of the current input and observes the change in probability scores of the target classifier that implicitly approximates the gradient of the loss function. The perturbation is then updated in the next step, based on the approximated gradient. **Boundary Attack (BA)** (Brendel et al., 2018) solely relies on the final decision of the model when crafting the adversarial perturbations. BA starts from the perturbation that is already adversarial, and then iteratively perturbs the image along the boundary between adversarial and the nonadversarial region to find smaller perturbations. **Simple black-box adversarial attacks (SimBA)** (Guo et al., 2019) iteratively samples random perturbations from a predefined orthonormal basis where the perturbations are either added or subtracted to the input image. Assuming that continuous prediction scores are available from the target model, the algorithm chooses to add or subtract the random perturbation which decreases the prediction scores of the adversarial example.

Random search methods craft random perturbations until the adversarial example mislead the target model. **SemanticAdv** (Hosseini and Poovendran, 2018) modifies an input image in HSV color space. The hue and saturation of an input image are perturbed with a random perturbation chosen uniformly from the range of valid values. SemanticAdv draws new perturbations until the classifier is misled or a maximum number of attempts is reached (e.g., 1000). Since the hue and saturation values change by the same amount, adversarial images may look unnatural. Unlike SemanticAdv, **ColorFool** (Shamsabadi et al., 2020c) generates perturbations that consider the naturalness of adversarial images to address the noticeability in addition to the performance. The method first identifies nonsensitive and sensitive regions through semantic segmentation. The perturbations then operate on the *a* and *b* channels of the *Lab* color space: the perturbations of nonsensitive regions are drawn randomly from the whole range of possible values, whereas the perturbations of sensitive regions are chosen randomly from predefined natural-color ranges, defined based on human perception. In addition, SemanticAdv and ColorFool craft unbounded perturbations that achieve better transferability to other models than other black-box attacks, which mainly fix the perturbations to be bounded while focusing on designing the methodology of attack.

15.8 Semantic segmentation and object detection

In this section, we discuss adversarial attacks designed for semantic segmentation and object detection models. Semantic segmentation aims to label the object segments, assigning a label to every pixel in the image. Object detection, unlike semantic segmentation, localizes the objects with bounding boxes and classifies each object.

Semantic Segmentation Attack (SSA) (Fischer et al., 2017) employs BIM (Kurakin et al., 2017b) attack to every pixel in the image or the region with a specific label, to mislead the pixel-wise semantic segmentation task by FCN. **Dense Adversary Generation (DAG)** (Xie et al., 2017) crafts adversarial perturbations for object detection and semantic segmentation tasks performed by Faster R-CNN and FCN, respectively. Note that, in a single image, Faster R-CNN performs classification on multiple object candidates to locate the object as a bounding box, and FCN classifies the label of each pixel in the image. Considering the object candidates and image pixels as multiple targets to mislead, DAG crafts gradient-based perturbations that aim to confuse as many pixels as possible. **Unified and Efficiency Adversary (UEA)** (Wei et al., 2019) generates untargeted adversarial images with a conditional Generative Adversarial Network (cGAN) to mislead object detectors. UEA, together with the adversarial loss used to attack Faster R-CNN in DAG, devises a loss function that integrates a multiscale attention feature loss to concentrate the perturbations more on the object regions. The images crafted with the multiscale attention can mislead Faster R-CNN with fewer number of perturbations than DAG, and improve the transferability of the black-box attack to another object detector, e.g., SSD. An object boundary detector can be attacked as presented in **Edge Attack** (Cosgrove and Yuille, 2020) which employs FGSM (Goodfellow et al., 2015) variants to mislead a HED detector (Xie and Tu, 2015) (see Section 15.6). Edge Attack transfers the crafted adversarial images to image classification and semantic segmentation.

15.9 Object tracking

DNN models tackle single object tracking as a template-based similarity metric problem that searches in each frame for the most similar region to a reference patch, which is given as prior or selected from the first frame. Siamese architectures (Bromley et al., 1994) are used to this end to compare the reference patch with a patch from the searching area in a frame (Bertinetto et al., 2016; Li et al., 2018; Zhang and Peng, 2019; Li et al., 2019). An adversarial attack for object tracking can generate perturbations that modify the reference patch, the searching area, or both.

Fast Attack Network (FAN) (Liang et al., 2020) is, to date, the only adversarial attack for object tracking. FAN can operate as an untargeted or targeted attack. The *untargeted* FAN perturbs the searching area, independently for subsequent frames, such that the response map is maximized in random regions out of the true object trajectory. The *targeted* FAN induces the tracker to follow a different, predefined trajectory by perturbing both the reference patch and the search areas. To avoid early failures of the tracker in the targeted attack, FAN modifies the searching areas (reference patch) such that the distance in the feature space between the adversarial searching areas (or adversarial reference patch) and the specified trajectory areas is minimized.

FAN uses a generative approach to obtain perturbations that are imperceptible to the human eye as well as easy to add to the input videos. During training, the FAN parameters are optimized by alternating between generator and discriminator. The loss function for the discriminator, $\mathcal{D}(\cdot)$, is based on PatchGAN (Isola et al., 2017). The generator $\mathcal{G}(\cdot)$ is trained with a multiobjective loss function, $\mathcal{L}_F$, which accounts for the embedded feature distance, $\mathcal{L}_e$, (targeted attack); a dual-objective drift term, $\beta_1 \mathcal{L}_d + \beta_2 \mathcal{L}_s$, (untargeted attack); and a ℓ_2-norm distance between original images and adversarial images, $\mathcal{L}_u$ (unnoticeability); in addition to the standard term for the generator, $\mathcal{L}_G$ (based on cyclic GAN):

$$\mathcal{L}_F = \mathcal{L}_G + \alpha_1 \mathcal{L}_u + \alpha_2 \mathcal{L}_e + \alpha_3 (\beta_1 \mathcal{L}_d + \beta_2 \mathcal{L}_s), \tag{15.13}$$

where $\alpha_1, \alpha_2, \alpha_3, \beta_1$, and β_2 are hyper-parameters to balance the influence of each term of the multiobjective loss. The hyper-parameters for the targeted and untargeted terms are set to zero when optimizing for either of the attacks. The loss term for the generator is defined as

$$\mathcal{L}_G = \mathbb{E}\left[(\mathcal{D}(\delta + R(\mathbf{x}_k, \mathbf{b}_k) - 1)^2 \right], \tag{15.14}$$

where $\mathbf{x}_t$ is the original video frame at time k, $R(\cdot)$ is the region extractor based on the searching area $\mathbf{b}_k$ at frame k. To craft an imperceptible perturbation, the similarity term in the loss is defined as

$$\mathcal{L}_u = \mathbb{E}_{\mathbf{x}_k \sim \mathcal{X}} \left[\| R(\dot{\mathbf{x}}_k, \mathbf{b}_k) - R(\mathbf{x}_k, \mathbf{b}_k) \|_2 \right], \tag{15.15}$$

where $\mathcal{X}$ denotes the set of frames in an input video. For the targeted attack, $\mathcal{L}_e$ aims to minimize the distance in the feature space between the adversarial reference patch and the specified trajectory

$$\mathcal{L}_e = \mathbb{E}_{\mathbf{x}_k \sim \mathcal{X}, \dot{\boldsymbol{\varepsilon}}_k \sim \mathcal{E}} \left[\| \phi(R(\mathbf{x}_k, \mathbf{b}_k) + \delta) - \phi(\dot{\boldsymbol{\varepsilon}}_k) \|_2 \right], \tag{15.16}$$

where $\dot{\boldsymbol{\varepsilon}}_k$ is the adversarial area of the specified adversarial trajectory $\mathcal{E}$ at frame k, $\phi(\cdot)$ is a function that maps the image area to the embedded feature space, and δ is generated by $\mathcal{G}(\cdot)$ and restricted to the image area selected by $R(\cdot)$. Note that $R(\mathbf{x}_k, \mathbf{b}_k)$ is the reference patch when $k = 0$.

For the untargeted attack, the drift term aims at maximizing the response score outside of the region where the tracker usually has the highest response within a searching area to track the object:

$$\mathcal{L}_d = \frac{1}{\gamma + \| \mathbf{q}_{max}^{+1} - \mathbf{q}_{max}^{-1} \|_2} - \xi, \tag{15.17}$$

while also pushing away the adversarial activation center from the real activation one:

$$\mathcal{L}_s = \min_{\mathbf{q} \in \mathcal{S}^{+1}} \left(\log \left(1 + e^{y_\mathbf{q} s_\mathbf{q}} \right) \right) - \max_{\mathbf{q} \in \mathcal{S}^{-1}} \left(\log \left(1 + e^{y_\mathbf{q} s_\mathbf{q}} \right) \right), \tag{15.18}$$

where $\mathcal{S} = \mathcal{S}^{+1} \cup \mathcal{S}^{-1}$ is the response map with positive and negative labels $y_\mathbf{q} = \{-1, +1\}$ for each pixel location $\mathbf{q}$; $\mathbf{q}_{max}^{+1}$ and $\mathbf{q}_{max}^{-1}$ are the pixel locations with maximum activation scores in positive and negative areas of the response map, respectively; s_q is the response score at pixel location $\mathbf{q}$; γ is a small constant for numerical stability; and ξ controls the offset degree of the activation center.

15.10 Video classification

Attacks for image classification could be naturally extended to the temporal domain to mislead a target model for video classification by operating on each frame independently. However, temporal information can be exploited to create more robust adversarial perturbations using a batch of frames or by processing the frames online, as they are acquired by the camera. In this section, we discuss three adversarial attacks for video classification: C-DUP, V-BAD, and MultAV.

Circular Universal Dual Purpose Perturbations (C-DUP) is a white-box attack that misleads online video classifiers under real-time requirements (Li et al., 2019b). C-DUP aims at inducing the mis-classification of only specific classes, while keeping the recognition of other classes unaltered. C-DUP extends the work of Mopuri for images (Mopuri et al., 2018), and uses a modified GAN to learn offline a single set of (universal) perturbations that can be applied online to an unseen input. The GAN fixes the discriminator with the known pretrained classifier, and only trains the 3D generator to obtain the universal perturbations that are added to the input video clips, while aiming to fool the discriminator. To attack only specific classes, C-DUP uses a loss function with the probability vectors of all classes, as outputted by the discriminator, over all the training data. This dual purpose loss function supports the minimization of the cross-entropy for the target class(es), $\dot{y} \in \mathcal{A} \subset \mathcal{Y}$ (where $\mathcal{Y}$ is the set of class labels) and the maximization of the cross-entropy for all other classes, $\mathcal{Y} \setminus \mathcal{A}$:

$$
\min_{\mathcal{G}} \sum_{o=1}^{W} \left(\sum_{\mathbf{x} \in \mathcal{X} \setminus \mathcal{X}_a} -\log\left[P_y(\mathbf{x} + \rho(\mathcal{G}(\mathbf{h}), o))\right] \right.
$$

$$
\left. + \lambda \sum_{\mathbf{x}_a \in \mathcal{X}_a} -\log\left[1 - P_{\dot{y}}(\mathbf{x}_a + \rho(\mathcal{G}(\mathbf{h}), o))\right] \right), \tag{15.19}
$$

where λ is a weighting parameter that balances the losses, $\mathbf{h}$ is a noise vector from a latent space (e.g., uniform distribution in the range $[-1, 1]$); $\mathcal{G}(\cdot)$ is the generator; P_y is the probability of the unattacked label y; $P_{\dot{y}}$ is the probability of the label under attack $\dot{y}$; $\mathcal{X}_a \subset \mathcal{X}$ is the subset of video clips whose action label is under attack; and $\rho(G(\mathbf{h}), o) = \delta_o$ is the permutation function that applies a circular shift across all the frames ($o = 1, \ldots, W$, where W is the number of frames in a clip) on the generated perturbations. The permutation function enforces C-DUP to be agnostic to the temporal sequence of frames in the clip. To produce these circular perturbations, a postprocessing unit is used between the 3D generator and the discriminator. Finally, the 3D generator constrains the adversarial perturbations to be within a unit ball defined by a bound ϵ. The success of C-DUP on the coarse-grained (UCF) and fine-grained (JESTER) action recognition datasets to mislead the spatio-temporal C3D classifier (Tran et al., 2015) exceeds 80% for target classes while keeping the classification accuracy higher than 80% for the other classes (the original classification accuracy for C3D is 96% and 90% on UCF and JESTER, respectively). A variant of C-DUP generates a single, quasi-imperceptible 2D perturbation to be applied to each frame of a clip. However, this attack is less effective than C-DUP, especially for fine-grained action classification where information about temporal changes is more significant.

V-BAD is a black-box attack with a 3-stage iterative scheme to query the model under attack and obtain, for each iteration, the classification label and corresponding probability (Jiang et al., 2019). V-BAD can be targeted or untargeted. The *targeted* V-BAD aims at fooling the model with a target class returned as top-1, and uses a sample video from the target class as input, while adjusting the size of the perturbation bound. The *untargeted* V-BAD uses the original video as input and keeps the size of the perturbation bound constant. The perturbations are imperceptible and bounded in an ϵ-ball centered at the original video. Tentative perturbations are generated independently for each frame of the original input video (or the adversarial example for each iteration) with an ensemble of three pretrained DNNs and then averaging their outputs. V-BAD exploits the uniform partitions in patches of the input perturbations and limits the number of queries to the video recognition model under attack using a patch-wise gradient estimator. The gradient estimator provides a vector of weights based on an adversarial loss with respect to the probability returned by the model under attack. Patches of perturbations are then rectified using the vector of weights and applied to the input video (or adversarial example at previous iteration) via one-step PGD. V-BAD was used to attack the CNN + LSTM and I3D video classifiers.

MultAV is a set of direct or iterative gradient-based attacks (ℓ_p-norm, ℓ_2-norm PGD) that craft for each frame of a video *multiplicative* perturbations (Lo and Patel, 2020). Similarly to their additive-perturbation counterpart, gradient-based MultAV attacks are bounded. Specifically, MultAV introduces the ratio bound ϵ_m (where m stands for multiplicative) that restricts the pixel-wise ratio between the adversarial example and the original image to make the perturbation imperceptible. Considering as example the ℓ_2-norm bounded attack with additive perturbation (see Eq. (15.10)), the multiplicative counterpart created by the iterative MultAV-ℓ_2 is generated as follows. An adversarial example at iteration $i + 1$ is obtained by multiplying the adversarial example of the previous iteration with a multiplicative step size, α_m, whose exponent is given by the gradient of the loss,

$$\mathbf{x}_{i+1} = \mathcal{C}_{\mathbf{x}_i, \epsilon_m} \left\{ \mathbf{x} \odot \alpha_m^{\frac{\nabla_{\mathbf{x}_i} \mathcal{L}(\mathbf{x}_i, y)}{\|\nabla_{\mathbf{x}_i} \mathcal{L}(\mathbf{x}_i, y)\|_2}} \right\}, \tag{15.20}$$

where $\mathcal{C}_{\mathbf{x}_i, \epsilon_m}\{\cdot\}$ is the clipping operation, ϵ_m is the ratio bound, such that $\|\frac{\mathbf{x}_{i+1}}{\mathbf{x}}\|_2 < (\epsilon_m + 1)$, and $\odot$ is the Hadamard product (element-wise). Alternatively, MultAV applies the same multiplicative principle to patch-based methods, such as rectangular occlusion and adversarial framing, or to pixel-based additive noise. However, these attacks result in perceptible perturbations. MultAV was applied on UCF to mislead the 3D ResNet-18 video classifier.

TABLE 15.3 Summary of defenses against adversarial attacks. KEY – Dec: detection-based method, GradM: gradient masking, ModelR: model robustness, Aux: auxiliary model, Stat: statistical approach, CCheck: consistency checking, NonDiff: nondifferentiable gradient, Van/Exp: Vanishing/exploding gradient, Stoch: Stochastic gradient, AdvT Adversarial training, Reg: regularization, Cert: certified defense, Exp: experiment, C: image classification, D: object detection, S, semantic segmentation, F: face recognition, T: object tracking.

Reference	Method	Goal	Approach	Datasets	Task
Hendrycks and Gimpel (2016)	PCA	Dec	Stat	ImageNet, MNIST, CIFAR	C
Grosse et al. (2017)	Statistical test	Dec	Stat	MNIST	C
Gong et al. (2017)	Binary Classifier	Dec	Aux	MNIST, CIFAR, SVHN	C
Metzen et al. (2017)	Adversary Detector Net.	Dec	Aux	ImageNet, CIFAR	C
Feinman et al. (2017)	KDE & BUE	Dec	CCheck	MNIST, CIFAR	C
Xu et al. (2018)	Feature Squeezing	Dec	CCheck	MNIST, CIFAR	C
Buckman et al. (2018)	Thermometer Encoding	GradM	NonDiff	MNIST, CIFAR, SVHN	C
Guo et al. (2018)	Image Transformations	GradM	NonDiff	ImageNet	C
Papernot et al. (2016c)	Defensive Distillation	GradM	Van/Exp	MNIST, CIFAR	C
Song et al. (2018)	PixelDefend	GradM	Van/Exp	F-MNIST, CIFAR	C
Samangouei et al. (2018)	Defense-GAN	GradM	Van/Exp	F-MNIST	C
Zhou et al. (2020)	A-VAE	GradM	Van/Exp	LFW	F
Dhillon et al. (2018)	Stochastic pruning	GradM	Stoch	CIFAR	C
Xie et al. (2018)	Randomization	GradM	Stoch	ImageNet	C
Gu and Rigazio (2014)	Deep Contractive Net.	ModelR	Reg	MNIST	C
Cisse et al. (2017)	Parseval Net.	ModelR	Reg	CIFAR, SVHN	C
Goodfellow et al. (2015)	AdvTrain	ModelR	AdvT	MNIST	C
Madry et al. (2018)	PGD AdvTrain	ModelR	AdvT	MNIST, CIFAR	C
Tramèr et al. (2018)	Ensemble AdvTrain	ModelR	AdvT	ImageNet	C
Zhang and Wang (2019)	AROD	ModelR	AdvT	PascalVOC, COCO	D
Jia et al. (2020)	RT	ModelR	AdvT	OTB, VOT, UAV	T
Raghunathan et al. (2018)	Single Semidefinite Relax.	ModelR	Cert	MNIST	C
Wong and Kolter (2018)	Deep ReLU Net	ModelR	Cert	F-MNIST, HAR, SVHN	C
Arnab et al. (2018)	RSSM	ModelR	Exp	PascalVOC, CityScapes	S

15.11 Defenses against adversarial attacks

Defenses aim to protect machine learning models from adversarial attacks. A defense may identify an adversarial example (defense by detection) or confuse the gradient feedback from the network loss function to disable the attack (defense by gradient masking). Alternatively, the model itself can be trained with adversarial examples to increase its robustness to attacks (model robustness). Defenses and their properties are summarized in Table 15.3 and discussed below.

15.11.1 Detection

Defenses by detection aim to discriminate an adversarial example and reject it. These defenses may use *statistics* to distinguish adversarial examples, rely on the creation of an *auxiliary model* or on analyzing the prediction *consistency*.

Adversarial examples can be detected by finding differences in the statistical properties (statistics) between adversarial and clean examples. **Principal Component Analysis (PCA)** (Hendrycks and Gimpel, 2016) can be employed to detect adversarial examples relying on the fact that the later principal components of adversarial images have a larger variance than those of clean images. **Statistical Test** (Grosse et al., 2017) employs the fact that the data distributions of adversarial examples and clean ones are different. By using a statistical test, such as maximum mean discrepancy (Gretton et al., 2012), the method examines whether a group of data are adversarial or not. The method also presents an augmented model that is a target model with an extra class used for classifying the adversarial inputs.

Auxiliary models can be trained to discriminate adversarial examples from clean ones. **Binary Classifier** (Gong et al., 2017) is a straightforward approach to build a classifier that detects adversarial examples. The classifier takes an image as input to the classifier and generates a binary label that indicates whether the input is adversarial or not. **Adversary Detector Networks** (Metzen et al., 2017) build a binary classifier where the input for the classifier is the output response from an intermediate layer of the target model.

Other methods detect the adversarial images by checking the *consistency* of the prediction from the input. This can be done by manipulating the input or the target model up to an acceptable limit and then checking whether the output prediction is consistent. **KDE & BUE** (Feinman et al., 2017) addresses detection by investigating the model confidence on adversarial samples by using kernel density estimates (KDE) of training data and Bayesian uncertainty estimates (BUE) in dropout layers. KDE are calculated with the training set in the feature space of the last hidden layer and detect points that lie far from the data manifold. BUE can detect adversarial examples if the uncertainty distribution estimated by the random dropout is different from clean data. **Feature Squeezing** (Xu et al., 2018) performs requantizing images (color bit depth reduction), and spatial filtering (median filtering) to input images. The method assumes that the model prediction from clean images is not affected when requantization or median filtering is applied. If the prediction result becomes different after the requantization or median filtering, the method detects the input as the adversarial image.

15.11.2 Gradient masking

Most adversarial attacks use the gradient of the loss function from the target model to craft perturbations. Gradient masking aims to make the gradient information of a target model not applicable to craft adversarial perturbations. Examples are *nondifferentiable gradient*, *vanishing/exploding gradient*, and *stochastic gradient*.

Thermometer Encoding (Buckman et al., 2018) tackles the linear extrapolation behavior of DNN models by preprocessing the input with discretization to make a defense model nonlinear and *nondifferentiable*. Regular quantization may not be effective to defense as the quantized inputs can be also affected approximately linearly by perturbations. Thermometer Encoding thus learns to project the discretized value, which enables the model to make the input effective for resisting the adversarial perturbations. **Image Transformation** (Guo et al., 2018) presents several image processing approaches, such as cropping and rescaling, bit-depth reduction, and JPEG compression. Total variance minimization also can be used to regularize each small set of pixels in the image. Image quilting, synthesizing images by small patches, is another approach that replaces the local patches in the adversarial image with their nearest-neighbor clean patches. These nondifferentiable transforms make it difficult for the attack to infer the gradient of the loss function from the target model.

Vanishing/exploding gradient makes the gradients of the loss function very small or large when training a defensive model. This approach distracts the adversary from attacking the input images. **Defensive Distillation** (Papernot et al., 2016c) adds flexibility to a classification process so the model is less confident about its prediction. In distillation training, a distilled model is trained to predict the output probabilities of a target model. This target model was trained to achieve high accuracy with learning softer probability distribution by the softmax function with large temperature T: $e^{z_n/T} / \sum_m e^{z_m/T}$. The distilled model is then used for testing with a small T, which makes the model more confident of prediction than using a large T. This prevents the adversary from estimating the gradient since the gradient of the loss from other classes becomes close to zero. **PixelDefend** (Song et al., 2018) trains a generative network on clean data to approximate its distribution and hence encouraging the network to follow the original distribution even when the adversarial images are given as input. This task can be considered as removing the adversarial effect. Moreover, because of the large number of parameters in the generative network, the gradients from the loss function can become very small or large, confusing the adversary from inferring the adversarial perturbations. **Defense-GAN** (Samangouei et al., 2018) is a GAN framework that simultaneously trains a generative network to emulate the data distribution and a discriminative model to predict whether an input is from original data or was artificially crafted. As for PixelDefend, the generative network can work to remove the effect of adversarial perturbations. **Adversarial Variational AutoEncoder (A-VAE)** (Zhou et al., 2020) is a defense against face recognition attack and trains a VAE-based generator with original images to learn the clean data distribution. At inference time, the downsampled input image is fed into the trained VAE, while multiple latent codes are sampled from the encoder to generate decoded images with diverse details. The input for the face recognition model is then chosen by finding the nearest neighbor of the input in the decoded images. This enables A-VAE to find an image that is likely to be predicted as a clean label as well as similar to the input image.

Stochastic gradient confuses the adversary on the target model to attack by applying randomized operations to the input or the network. **Stochastic pruning** (Dhillon et al., 2018) drops a random subset of hidden-layer outputs (or activations) and scales up the rest to compensate. The method retains nodes with probabilities proportional to the magnitude of their activation and scales up the surviving nodes to preserve the dynamic range of the activations in each layer. This approach can make the pretrained models more robust against adversarial examples without fine-tuning. **Randomization** (Xie et al., 2018) employs two randomization operations – random resizing and padding – on the input images in order to mitigate adversarial attacks. Although adding random resizing and random padding shows a little accuracy drop on clean images, the prediction with Randomization shows robustness to norm-bounded attacks.

15.11.3 Model robustness

Robustness of a target model to an adversarial attack can be increased by regularization, adversarial training, or certified defense. *Regularization* of the neural network layers makes the training less sensitive to the input distortions. *Adversarial training* retrains a target model with adversarial examples generated by an attack, in addition to the clean examples. However, there is no guarantee that the retrained model can be robust to other attacks not used for retraining. *Certified defenses* aim to provide a certificate that ensures the robustness of the model within certain bounds of perturbations. This differs from adversarial training and regularization that alleviate adversarial effects in the data/feature domains.

Deep Contractive Network (Gu and Rigazio, 2014) regularizes adversarial attack by applying a layer-wise penalty as a loss that constrains the magnitude of the partial derivative of hidden layers in the target model, in addition to the adversarial loss for the attack. This allows the network to be less sensitive to the changes in an input image, such as adversarial perturbations. **Parseval Network** (Cisse et al., 2017) controls the global Lipschitz constant of the network within the layer-wise regularization. Considering a network as a combination of functions, the network can be robust to small input perturbations by maintaining a small Lipschitz constant for these functions. Parseval Network addresses this with Parseval tight frames that control the spectral norm of the network weight matrix.

Vanilla AdvTrain (Goodfellow et al., 2015) retrains the target classifier with adversarial examples generated by FGSM and whose labels are the same as the original images. The retrained classifier is thus robust to FGSM attack. **PGD AdvTrain** (Madry et al., 2018) extends Vanilla AdvTrain by using the PGD attack that crafts the worst-case adversarial example within an ℓ_∞ bound. **Ensemble AdvTrain** (Tramèr et al., 2018) retrains the target model on the adversarial examples crafted from other pretrained classifiers. This strategy avoids the overfitting problem in Vanilla AdvTrain and the adversarial examples from other classifiers can approximate the worst-case adversarial example. Ensemble AdvTrain is more efficient than previous methods as the retraining process and adversarial attack are separated. **Adversarially robust object detector (AROD)** (Zhang and Wang, 2019) applies FGSM-like attack to classification and localization loss functions that are used in object detection, and crafts adversarial images from each loss. The resulting adversarial image is the one that maximizes the overall loss function (both classification and localization). The same loss is then used for adversarial training. **Robust tracker (RT)** (Jia et al., 2020) estimates the unknown adversarial perturbations, crafted by considering temporal motion in the input videos and learns to eliminate their effects during tracking. The method first defines an adversarial loss that includes the original label, e.g., a bounding box and its class, with a target pseudo label. Given an input image with unknown perturbations, the gradient from the adversarial loss, measured by using the estimated label from the previous frame and its corresponding pseudo label, is subtracted from the input image to reduce the effect of the attack.

Another approach is training to optimize certificates that ensure the model robustness (certified defense). Given a model and input, the verifier outputs a certificate if the image is guaranteed not to be the adversarial one. This can be done by checking whether there is any image, within an ℓ_p distance, that has a different label from the original one. One approach is to train a certificate that approximates the outer bounds of adversarial loss to the target model. **Single Semidefinite Relaxation** generates a certificate that provides an upper bound where no attack is available, given a target model and input (Raghunathan et al., 2018). Similarly, **Deep ReLU Network** (Wong and Kolter, 2018) is a deeper network for training provably robust classifiers that are guaranteed to be robust against any norm-bounded adversarial perturbations on the training set. Since the trained certificate approximates the outer bounds of adversarial loss, unseen adversarial examples can be detected with zero false negatives, but this may also include false positives.

Arnab et al. (2018) investigate the robustness of several modules used in DNNs for semantic segmentation to adversarial attacks, i.e., **Robust Semantic Segmentation Models (RSSM)**. Models for semantic segmentation consist of a pretrained classification model used as backbone and additional layers or modules for better pixel-level localization, such as conditional random fields, dilated convolutions, skip-connections, and multiscale networks. The work attacks semantic segmentation models with FGSM variants (Goodfellow et al., 2015; Kurakin et al., 2017b) and shows that the most accurate model on clean images is not necessarily the most robust one. Conditional random fields, which are commonly used in semantic segmentation for enforcing the structural constraints, include mean-field inference which naturally performs gradient masking, resulting in increases of robustness to untargeted adversarial attacks. RSSM based on skip connections, e.g., ResNet (He et al., 2016) and multiscale formulation are more robust to adversarial attacks than VGG-like models (Simonyan and Zisserman, 2014).

15.12 Conclusions

We presented an extensive overview of adversarial attacks against machine learning models based on DNNs for image processing, image classification, object detection, semantic segmentation, object tracking, and video classification, as well as defenses that protect from these attacks. We introduced the key properties of an adversarial attack, namely effectiveness, robustness, transferability, and noticeability, and we discussed additional properties to consider when evaluating an attack, such as detectability and reversibility. We then categorized methods as bounded and unbounded based on the magnitude of the adversarial perturbation, and we distinguished perturbations as global or region-based. We compared strategies considering the specific task they are designed for (e.g., motion estimation, image or video classification, object tracking) and distinguished white-box and black-box settings. The analysis of the attacks allowed us to define categorized strategies that defend a DNN from adversarial examples. Defenses can detect whether the input is adversarial, or prevent an attack by gradient masking, where the gradient comes from the loss function. Furthermore, we discussed how DNN models can be made more robust through adversarial training.

The interplay between attacks and defenses will be an important direction of future research to help understand the limitations of DNN models and the design of more reliable DNNs. Moreover, another important area of investigation is physical attacks, which alter the appearance of real objects or place adversarial objects in the environment, and corresponding defenses (Eykholt et al., 2018; Sharif et al., 2016; Brown et al., 2018; Kurakin et al., 2017b; Athalye et al., 2018; Ranjan and Black, 2017; Chen et al., 2017; Thys et al., 2019; Xu et al., 2020). DNNs tend to be more vulnerable to adversarial attacks than traditional machine learning models since the end-to-end learnable DNNs architectures make the attack easy and still some properties of DNNs have not yet been fully investigated. Overcoming the vulnerabilities of DNNs to the malicious manipulation of images and videos, either directly on the digital domain or in the physical world, is key for their adoption in real-world applications, such as commerce, security, and authentication, and safety critical autonomous systems, such as self-driving vehicles (Modas et al., 2020).

Acknowledgment

Andrea Cavallaro wishes to thank the Alan Turing Institute (EP/N510129/1), which is funded by the EPSRC, for its support through the project PRIMULA.

References

Allen-Zhu, Z., Li, Y., 2020. Feature purification: how adversarial training performs robust deep learning. arXiv:2005.10190.

Anguita, D., Ghio, A., Oneto, L., Parra, X., Reyes-Ortiz, J., 2013. A public domain dataset for human activity recognition using smartphones. In: Proceedings of the 21st European Symposium on Artificial Neural Networks, Computational Intelligence and Machine Learning.

Arnab, A., Miksik, O., Torr, P.H.S., 2018. On the robustness of semantic segmentation models to adversarial attacks. In: Proceedings of the IEEE Conference on Computer Vision and Pattern Recognition.

Athalye, A., Engstrom, L., Ilyas, A., Kwok, K., 2018. Synthesizing robust adversarial examples. In: Proceedings of the International Conference on Machine Learning.

Baluja, S., Fischer, I., 2018. Learning to attack: adversarial transformation networks. In: Proceedings of the AAAI Conference on Artificial Intelligence.

Bertinetto, L., Valmadre, J., Henriques, J.F., Vedaldi, A., Torr, P.H., 2016. Fully-convolutional Siamese networks for object tracking. In: Proceedings of the European Conference on Computer Vision.

Bhattad, A., Chong, M.J., Liang, K., Li, B., Forsyth, D., 2020. Unrestricted adversarial examples via semantic manipulation. In: Proceedings of the International Conference on Learning Representations.

Biggio, B., Nelson, B., Laskov, P., 2013. Poisoning attacks against support vector machines. In: Proceedings of the International Conference on Machine Learning.

Bolton, R.J., Hand, D.J., et al., 2002. Statistical fraud detection: a review. Statistical Science 17, 235–255.

Brendel, W., Rauber, J., Bethge, M., 2018. Decision-based adversarial attacks: reliable attacks against black-box machine learning models. In: Proceedings of the International Conference on Learning Representations.

Bromley, J., Guyon, I., LeCun, Y., Säckinger, E., Shah, R., 1994. Signature verification using a Siamese time delay neural network. In: Proceedings of the Advances in Neural Information Processing Systems.

Brown, T.B., Mané, D., Roy, A., Abadi, M., Gilmer, J., 2018. Adversarial patch. arXiv:1712.09665.

Brox, T., Malik, J., 2011. Large displacement optical flow: descriptor matching in variational motion estimation. IEEE Transactions on Pattern Analysis and Machine Intelligence 33, 500–513.

Buckman, J., Roy, A., Raffel, C., Goodfellow, I., 2018. Thermometer encoding: one hot way to resist adversarial examples. In: International Conference on Learning Representations.

Carlini, N., Wagner, D., 2017. Towards evaluating the robustness of neural networks. In: Proceedings of the IEEE Symposium on Security and Privacy.

Carreira, J., Zisserman, A., 2017. Quo vadis, action recognition? A new model and the kinetics dataset. In: Proceedings of the IEEE Conference on Computer Vision and Pattern Recognition.

Chen, P.Y., Zhang, H., Sharma, Y., Yi, J., Hsieh, C.J., 2017. Zoo: zeroth order optimization based black-box attacks to deep neural networks without training substitute models. In: Proceedings of the 10th ACM Workshop on Artificial Intelligence and Security.

Cisse, M., Bojanowski, P., Grave, E., Dauphin, Y., Usunier, N., 2017. Parseval networks: improving robustness to adversarial examples. In: International Conference on Machine Learning.

Coates, A., Ng, A., Lee, H., 2011. An analysis of single-layer networks in unsupervised feature learning. In: Proceedings of the Fourteenth International Conference on Artificial Intelligence and Statistics, JMLR Workshop and Conference Proceedings.

Cordts, M., Omran, M., Ramos, S., Rehfeld, T., Enzweiler, M., Benenson, R., Franke, U., Roth, S., Schiele, B., 2016. The cityscapes dataset for semantic urban scene understanding. In: Proceedings of the IEEE Conference on Computer Vision and Pattern Recognition.

Cosgrove, C., Yuille, A., 2020. Adversarial examples for edge detection: they exist, and they transfer. In: Proceedings of the IEEE/CVF Winter Conference on Applications of Computer Vision.

Das, N., Shanbhogue, M., Chen, S., Hohman, F., Chen, L., Kounavis, M.E., Chau, D.H., 2017. Keeping the bad guys out: protecting and vaccinating deep learning with JPEG compression. arXiv:1705.02900.

Deng, J., Dong, W., Socher, R., Li, L.J., Li, K., Fei-Fei, L., 2009. Imagenet: a large-scale hierarchical image database. In: Proceedings of the IEEE Conference on Computer Vision and Pattern Recognition.

Dhillon, G.S., Azizzadenesheli, K., Lipton, Z.C., Bernstein, J.D., Kossaifi, J., Khanna, A., Anandkumar, A., 2018. Stochastic activation pruning for robust adversarial defense. In: International Conference on Learning Representations.

Dong, Y., Pang, T., Su, H., Zhu, J., 2019. Evading defenses to transferable adversarial examples by translation-invariant attacks. In: Proceedings of the IEEE Conference on Computer Vision and Pattern Recognition.

Dosovitskiy, A., Fischer, P., Ilg, E., Häusser, P., Hazirbas, C., Golkov, V., Smagt, P.v.d., Cremers, D., Brox, T., 2015. FlowNet: learning optical flow with convolutional networks. In: Proceedings of the IEEE International Conference on Computer Vision.

Dziugaite, G.K., Ghahramani, Z., Roy, D.M., 2016. A study of the effect of JPG compression on adversarial images. arXiv:1608.00853.

Engstrom, L., Ilyas, A., Santurkar, S., Tsipras, D., Tran, B., Madry, A., 2019. Adversarial robustness as a prior for learned representations. arXiv:1906.00945.

Everingham, M., Eslami, S.A., Van Gool, L., Williams, C.K., Winn, J., Zisserman, A., 2015. The Pascal visual object classes challenge: a retrospective. International Journal of Computer Vision 111, 98–136.

Eykholt, K., Evtimov, I., Fernandes, E., Li, B., Rahmati, A., Xiao, C., Prakash, A., Kohno, T., Song, D., 2018. Robust physical-world attacks on deep learning visual classification. In: Proceedings of the IEEE Conference on Computer Vision and Pattern Recognition.

Farbman, Z., Fattal, R., Lischinski, D., Szeliski, R., 2008. Edge-preserving decompositions for multi-scale tone and detail manipulation. ACM Transactions on Graphics 27, 1–10.

Feinman, R., Curtin, R.R., Shintre, S., Gardner, A.B., 2017. Detecting adversarial samples from artifacts. arXiv:1703.00410.

Fischer, V., Kumar, M.C., Metzen, J.H., Brox, T., 2017. Adversarial examples for semantic image segmentation. In: Proceedings of the International Conference on Machine Learning Workshop.

Geiger, A., Lenz, P., Urtasun, R., 2012. Are we ready for autonomous driving? The KITTI vision benchmark suite. In: Proceedings of the Conference on Computer Vision and Pattern Recognition.

Gong, Z., Wang, W., Ku, W.S., 2017. Adversarial and clean data are not twins. arXiv:1704.04960.

Goodfellow, I., Shlens, J., Szegedy, C., 2015. Explaining and harnessing adversarial examples. In: Proceedings of the International Conference on Learning Representations.

Goodfellow, I.J., Pouget-Abadie, J., Mirza, M., Xu, B., Warde-Farley, D., Ozair, S., Courville, A., Bengio, Y., 2014. Generative adversarial nets. In: Proceedings of the International Conference on Neural Information Processing Systems.

Gretton, A., Borgwardt, K.M., Rasch, M.J., Schölkopf, B., Smola, A., 2012. A kernel two-sample test. The Journal of Machine Learning Research 13, 723–773.

Grosse, K., Manoharan, P., Papernot, N., Backes, M., McDaniel, P., 2017. On the (statistical) detection of adversarial examples. arXiv:1702.06280.

Gu, S., Rigazio, L., 2014. Towards deep neural network architectures robust to adversarial examples. arXiv preprint. arXiv:1412.5068.

Guo, C., Gardner, J., You, Y., Wilson, A.G., Weinberger, K., 2019. Simple black-box adversarial attacks. In: Proceedings of the International Conference on Machine Learning.

Guo, C., Rana, M., Cisse, M., van der Maaten, L., 2018. Countering adversarial images using input transformations. In: Proceedings of the International Conference on Learning Representations.

Hara, K., Kataoka, H., Satoh, Y., 2018. Can spatiotemporal 3D CNNs retrace the history of 2D CNNs and ImageNet? In: Proceedings of the IEEE Conference on Computer Vision and Pattern Recognition.

He, K., Zhang, X., Ren, S., Sun, J., 2016. Deep residual learning for image recognition. In: Proceedings of the IEEE/CVF Conference on Computer Vision and Pattern Recognition.

Hendrycks, D., Gimpel, K., 2016. Early methods for detecting adversarial images. arXiv:1608.00530.

Hosseini, H., Poovendran, R., 2018. Semantic adversarial examples. In: Proceedings of the IEEE Conference on Computer Vision and Pattern Recognition Workshops.

Howard, A.G., Zhu, M., Chen, B., Kalenichenko, D., Wang, W., Weyand, T., Andreetto, M., Adam, H., 2017. Mobilenets: efficient convolutional neural networks for mobile vision applications. arXiv:1704.04861.

Huang, G., Liu, Z., Van Der Maaten, L., Weinberger, K.Q., 2017. Densely connected convolutional networks. In: Proceedings of the IEEE Conference on Computer Vision and Pattern Recognition.

Huang, G.B., Mattar, M., Berg, T., Learned-Miller, E., 2008. Labeled faces in the wild: a database for studying face recognition in unconstrained environments. In: Workshop on Faces in 'Real-Life' Images: Detection, Alignment, and Recognition.

Ilg, E., Mayer, N., Saikia, T., Keuper, M., Dosovitskiy, A., Brox, T., 2017. FlowNet 2.0: evolution of optical flow estimation with deep networks. In: Proceedings of the IEEE Conference on Computer Vision and Pattern Recognition.

Ilyas, A., Engstrom, L., Athalye, A., Lin, J., 2018. Black-box adversarial attacks with limited queries and information. In: Proceedings of the International Conference on Machine Learning.

Isola, P., Zhu, J., Zhou, T., Efros, A.A., 2017. Image-to-image translation with conditional adversarial networks. In: Proceedings of the IEEE Conference on Computer Vision and Pattern Recognition.

Janai, J., Guney, F., Ranjan, A., Black, M., Geiger, A., 2018. Unsupervised learning of multi-frame optical flow with occlusions. In: Proceedings of the European Conference on Computer Vision.

Jia, S., Ma, C., Song, Y., Yang, X., 2020. Robust tracking against adversarial attacks. In: Proceedings of the European Conference on Computer Vision.

Jiang, L., Ma, X., Chen, S., Bailey, J., Jiang, Y., 2019. Black-box adversarial attacks on video recognition models. In: Proceedings of the ACM International Conference on Multimedia.

Jiang, Y., Wu, Z., Wang, J., Xue, X., Chang, S., 2018. Exploiting feature and class relationships in video categorization with regularized deep neural networks. IEEE Transactions on Pattern Analysis and Machine Intelligence 40, 352–364.

Kay, W., Carreira, J., Simonyan, K., Zhang, B., Hillier, C., Vijayanarasimhan, S., Viola, F., Green, T., Back, T., Natsev, P., Suleyman, M., Zisserman, A., 2017. The kinetics human action video dataset. arXiv:1705.06950 [cs.CV].

Krizhevsky, A., Hinton, G., et al., 2009. Learning multiple layers of features from tiny images.

Krizhevsky, A., Sutskever, I., Hinton, G.E., 2012. ImageNet classification with deep convolutional neural networks. In: Proceedings of the Advances in Neural Information Processing Systems.

Kuehne, H., Jhuang, H., Garrote, E., Poggio, T., Serre, T., 2011. HMDB: a large video database for human motion recognition. In: Proceedings of the International Conference on Computer Vision. Stockholm, Sweden.

Kurakin, A., Goodfellow, I., Bengio, S., 2017a. Adversarial examples in the physical world. In: Proceedings of the International Conference on Learning Representations – Workshops. Toulon, France.

Kurakin, A., Goodfellow, I., Bengio, S., 2017b. Adversarial machine learning at scale. In: Proceedings of the International Conference on Learning Representations.

LeCun, Y., 1998. The MNIST database of handwritten digits.

LeCun, Y., Bottou, L., Bengio, Y., Haffner, P., 1998. Gradient-based learning applied to document recognition. Proceedings of the IEEE 86, 2278–2324.

Li, B., Wu, W., Wang, Q., Zhang, F., Xing, J., Yan, J., 2019. SiamRPN++: evolution of Siamese visual tracking with very deep networks. In: Proceedings of the IEEE Conference on Computer Vision and Pattern Recognition.

Li, B., Yan, J., Wu, W., Zhu, Z., Hu, X., 2018. High performance visual tracking with Siamese region proposal network. In: Proceedings of the IEEE Conference on Computer Vision and Pattern Recognition.

Li, C.Y., Sanchez-Matilla, R., Shahin Shamsabadi, A., Mazzon, R., Cavallaro, A., 2021. On the reversibility of adversarial attacks.

Li, C.Y., Shahin Shamsabadi, A., Sanchez-Matilla, R., Mazzon, R., Cavallaro, A., 2019a. Scene privacy protection. In: Proceedings of the IEEE International Conference on Acoustics, Speech and Signal Processing.

Li, S., Neupane, A., Paul, S., Song, C., Krishnamurthy, S.V., Chowdhury, A.K.R., Swami, A., 2019b. Stealthy adversarial perturbations against real-time video classification systems. In: Proceedings of the Network and Distributed Systems Security Symposium.

Liang, S., Wei, X., Yao, S., Cao, X., 2020. Efficient adversarial attacks for visual object tracking. In: Proceedings of the European Conference on Computer Vision.

Lin, T.Y., Maire, M., Belongie, S., Hays, J., Perona, P., Ramanan, D., Dollár, P., Zitnick, C.L., 2014. Microsoft COCO: common objects in context. In: Proceedings of the European Conference on Computer Vision.

Lo, S., Patel, V.M., 2020. MultAV: multiplicative adversarial videos. arXiv:2009.08058.

Long, J., Shelhamer, E., Darrell, T., 2015. Fully convolutional networks for semantic segmentation. In: Proceedings of the IEEE Conference on Computer Vision and Pattern Recognition.

Lu, J., Sibai, H., Fabry, E., 2017. Adversarial examples that fool detectors. arXiv:1712.02494 [cs.CV].

Madry, A., Makelov, A., Schmidt, L., Tsipras, D., Vladu, A., 2018. Towards deep learning models resistant to adversarial attacks. In: Proceedings of the International Conference on Learning Representations.

Martin, D.R., Fowlkes, C.C., Malik, J., 2004. Learning to detect natural image boundaries using local brightness, color, and texture cues. IEEE Transactions on Pattern Analysis and Machine Intelligence 26, 530–549.

Materzynska, J., Berger, G., Bax, I., Memisevic, R., 2019. The jester dataset: a large-scale video dataset of human gestures. In: Proceedings of the IEEE/CVF International Conference on Computer Vision Workshops.

Metzen, J.H., Genewein, T., Fischer, V., Bischoff, B., 2017. On detecting adversarial perturbations. arXiv:1702.04267.

Meyer, T.A., Whateley, B., 2004. SpamBayes: effective open-source, Bayesian based, email classification system. In: Proceedings of the Conference on Email and Anti-Spam.

Modas, A., Moosavi-Dezfooli, S.M., Frossard, P., 2019. Sparsefool: a few pixels make a big difference. In: Proceedings of the IEEE Conference on Computer Vision and Pattern Recognition.

Modas, A., Sanchez-Matilla, R., Frossard, P., Cavallaro, A., 2020. Towards robust sensing for autonomous vehicles: an adversarial perspective. IEEE Signal Processing and Magazine 37, 14–23.

Moosavi-Dezfooli, S.M., Fawzi, A., Fawzi, O., Frossard, P., 2017. Universal adversarial perturbations. In: Proceedings of the IEEE Conference on Computer Vision and Pattern Recognition.

Moosavi-Dezfooli, S.M., Fawzi, A., Frossard, P., 2016. Deepfool: a simple and accurate method to fool deep neural networks. In: Proceedings of the IEEE Conference on Computer Vision and Pattern Recognition.

Mopuri, K.R., Garg, U., Babu, R.V., 2017. Fast feature fool: a data independent approach to universal adversarial perturbations. In: Proceedings of the British Machine Vision Conference.

Mopuri, K.R., Ojha, U., Garg, U., Babu, R.V., 2018. NAG: network for adversary generation. In: Proceedings of the IEEE Conference on Computer Vision and Pattern Recognition.

Mueller, M., Smith, N., Ghanem, B., 2016. A benchmark and simulator for UAV tracking. In: Proceedings of the European Conference on Computer Vision.

Narodytska, N., Kasiviswanathan, S.P., 2017. Simple black-box adversarial attacks on deep neural networks. In: Proceedings of the IEEE Conference on Computer Vision and Pattern Recognition Workshops.

Netzer, Y., Wang, T., Coates, A., Bissacco, A., Wu, B., Ng, A.Y., 2011. Reading digits in natural images with unsupervised feature learning. In: Workshop on Deep Learning and Unsupervised Feature Learning (in conjunction with Neural Information Processing Systems).

Ng, Y.H.J., Hausknecht, M., Vijayanarasimhan, S., Vinyals, O., Monga, R., Toderici, G., 2015. Beyond short snippets: deep networks for video classification. In: Proceedings of the IEEE Conference on Computer Vision and Pattern Recognition.

Papernot, N., McDaniel, P., Goodfellow, I., 2016a. Transferability in machine learning: from phenomena to black-box attacks using adversarial samples. arXiv:1605.07277.

Papernot, N., McDaniel, P., Jha, S., Fredrikson, M., Celik, Z.B., Swami, A., 2016b. The limitations of deep learning in adversarial settings. In: Proceedings of the IEEE European Symposium on Security and Privacy.

Papernot, N., McDaniel, P., Wu, X., Jha, S., Swami, A., 2016c. Distillation as a defense to adversarial perturbations against deep neural networks. In: Proceedings of the IEEE Symposium on Security and Privacy.

Poursaeed, O., Katsman, I., Gao, B., Belongie, S., 2018. Generative adversarial perturbations. In: Proceedings of the IEEE Conference on Computer Vision and Pattern Recognition.

Raghunathan, A., Steinhardt, J., Liang, P., 2018. Certified defenses against adversarial examples. arXiv:1801.09344.

Ranjan, A., Black, M.J., 2017. Optical flow estimation using a spatial pyramid network. In: Proceedings of the IEEE Conference on Computer Vision and Pattern Recognition.

Ranjan, A., Janai, J., Geiger, A., Black, M.J., 2019. Attacking optical flow. In: Proceedings of the IEEE/CVF International Conference on Computer Vision.

Redmon, J., Farhadi, A., 2017. YOLO9000: better, faster, stronger. In: Proceedings of the IEEE Conference on Computer Vision and Pattern Recognition.

Ren, S., He, K., Girshick, R., Sun, J., 2017. Faster R-CNN: towards real-time object detection with region proposal networks. IEEE Transactions on Pattern Analysis and Machine Intelligence 39, 1137–1149.

Revaud, J., Weinzaepfel, P., Harchaoui, Z., Schmid, C., 2015. Epicflow: edge-preserving interpolation of correspondences for optical flow. In: Proceedings of the IEEE Conference on Computer Vision and Pattern Recognition.

Salimans, T., Ho, J., Chen, X., Sidor, S., Sutskever, I., 2017. Evolution strategies as a scalable alternative to reinforcement learning. arXiv:1703.03864.

Samangouei, P., Kabkab, M., Chellappa, R., 2018. Defense-GAN: protecting classifiers against adversarial attacks using generative models. In: Proceedings of the International Conference on Learning Representations.

Sanchez-Matilla, R., Li, C.Y., Shamsabadi, A.S., Mazzon, R., Cavallaro, A., 2020. Exploiting vulnerabilities of deep neural networks for privacy protection. IEEE Transactions on Multimedia 22, 1862–1873.

Santurkar, S., Ilyas, A., Tsipras, D., Engstrom, L., Tran, B., Madry, A., 2019. Image synthesis with a single (robust) classifier. In: Proceedings of the Conference on Neural Information Processing Systems.

Shamsabadi, A.S., Oh, C., Cavallaro, A., 2020a. Edgefool: an adversarial image enhancement filter. In: Proceedings of the IEEE International Conference on Acoustics, Speech and Signal Processing.

Shamsabadi, A.S., Oh, C., Cavallaro, A., 2020b. Semantically adversarial learnable filters. arXiv:2008.06069.

Shamsabadi, A.S., Sanchez-Matilla, R., Cavallaro, A., 2020c. ColorFool: semantic adversarial colorization. In: Proceedings of the IEEE/CVF Conference on Computer Vision and Pattern Recognition.

Sharif, M., Bhagavatula, S., Bauer, L., Reiter, M.K., 2016. Accessorize to a crime: real and stealthy attacks on state-of-the-art face recognition. In: Proceedings of the 2016 ACM SIGSAC Conference on Computer and Communications Security.

Shi, Y., Wang, S., Han, Y., 2019. Curls & whey: boosting black-box adversarial attacks. In: Proceedings of the IEEE Conference on Computer Vision and Pattern Recognition.

Simonyan, K., Zisserman, A., 2014. Very deep convolutional networks for large-scale image recognition. arXiv:1409.1556.

Song, Y., Kim, T., Nowozin, S., Ermon, S., Kushman, N., 2018. Pixeldefend: leveraging generative models to understand and defend against adversarial examples. In: Proceedings of the International Conference on Learning Representations.

Soomro, K., Zamir, A.R., Shah, M., 2012. UCF101: a dataset of 101 human actions classes from videos in the wild. arXiv:1212.0402 [cs.CV].

Su, J., Vargas, D.V., Sakurai, K., 2019. One pixel attack for fooling deep neural networks. IEEE Transactions on Evolutionary Computation 23, 828–841.

Sun, D., Yang, X., Liu, M., Kautz, J., 2018. PWC-net: CNNs for optical flow using pyramid, warping, and cost volume. In: Proceedings of the IEEE Conference on Computer Vision and Pattern Recognition.

Szegedy, C., Ioffe, S., Vanhoucke, V., Alemi, A.A., 2017. Inception-v4, inception-resnet and the impact of residual connections on learning. In: Proceedings of the Thirty-First AAAI Conference on Artificial Intelligence.

Szegedy, C., Liu, W., Jia, Y., Sermanet, P., Reed, S., Anguelov, D., Erhan, D., Vanhoucke, V., Rabinovich, A., 2015. Going deeper with convolutions. In: Proceedings of the IEEE Conference on Computer Vision and Pattern Recognition.

Szegedy, C., Vanhoucke, V., Ioffe, S., Shlens, J., Wojna, Z., 2016. Rethinking the inception architecture for computer vision. In: Proceedings of the IEEE Conference on Computer Vision and Pattern Recognition.

Szegedy, C., Zaremba, W., Sutskever, I., Bruna, J., Erhan, D., Goodfellow, I., Fergus, R., 2014. Intriguing properties of neural networks. In: Proceedings of the International Conference on Learning Representations.

Thys, S., Van Ranst, W., Goedeme, T., 2019. Fooling automated surveillance cameras: adversarial patches to attack person detection. In: Proceedings of the IEEE/CVF Conference on Computer Vision and Pattern Recognition Workshops.

Tramèr, F., Kurakin, A., Papernot, N., Goodfellow, I., Boneh, D., McDaniel, P., 2018. Ensemble adversarial training: attacks and defenses. In: Proceedings of the International Conference on Learning Representations.

Tran, D., Bourdev, L., Fergus, R., Torresani, L., Paluri, M., 2015. Learning spatiotemporal features with 3d convolutional networks. In: Proceedings of the IEEE International Conference on Computer Vision.

Tsipras, D., Santurkar, S., Engstrom, L., Turner, A., Madry, A., 2019. Robustness may be at odds with accuracy. In: Proceedings of the International Conference on Representation Learning.

Tu, C.C., Ting, P., Chen, P.Y., Liu, S., Zhang, H., Yi, J., Hsieh, C.J., Cheng, S.M., 2019. Autozoom: autoencoder-based zeroth order optimization method for attacking black-box neural networks. In: Proceedings of the AAAI Conference on Artificial Intelligence.

Wei, X., Liang, S., Chen, N., Cao, X., 2019. Transferable adversarial attacks for image and video object detection. In: Proceedings of the International Joint Conference on Artificial Intelligence.

Wong, E., Kolter, Z., 2018. Provable defenses against adversarial examples via the convex outer adversarial polytope. In: Proceedings of the International Conference on Machine Learning.

Wu, Y., Lim, J., Yang, M., 2015. Object tracking benchmark. IEEE Transactions on Pattern Analysis and Machine Intelligence 37, 1834–1848.

Xiao, C., Li, B., Zhu, J., He, W., Liu, M., Song, D., 2018. Generating adversarial examples with adversarial networks. In: Proceedings of the Twenty-Seventh International Joint Conference on Artificial Intelligence.

Xiao, H., Rasul, K., Vollgraf, R., 2017. Fashion-MNIST: a novel image dataset for benchmarking machine learning algorithms. arXiv:1708.07747.

Xie, C., Wang, J., Zhang, Z., Ren, Z., Yuille, A., 2018. Mitigating adversarial effects through randomization. In: Proceedings of the International Conference on Learning Representations.

Xie, C., Wang, J., Zhang, Z., Zhou, Y., Xie, L., Yuille, A., 2017. Adversarial examples for semantic segmentation and object detection. In: Proceedings of the IEEE International Conference on Computer Vision.

Xie, C., Zhang, Z., Zhou, Y., Bai, S., Wang, J., Ren, Z., Yuille, A.L., 2019. Improving transferability of adversarial examples with input diversity. In: Proceedings of the Computer Vision and Pattern Recognition.

Xie, S., Tu, Z., 2015. Holistically-nested edge detection. In: Proceedings of the IEEE International Conference on Computer Vision.

Xu, K., Zhang, G., Liu, S., Fan, Q., Sun, M., Chen, H., Chen, P., Wang, Y., Lin, X., 2020. Adversarial t-shirt! Evading person detectors in a physical world. In: Proceedings of the European Conference on Computer Vision.

Xu, W., Evans, D., Qi, Y., 2018. Feature squeezing: detecting adversarial examples in deep neural networks. In: Proceedings of the Network and Distributed System Security Symposium.

Yu, F., Koltun, V., Funkhouser, T., 2017. Dilated residual networks. In: Proceedings of the IEEE Conference on Computer Vision and Pattern Recognition.

Zagoruyko, S., Komodakis, N., 2016. Wide residual networks. In: Proceedings of the British Machine Vision Conference.

Zajac, M., Żołna, K., Rostamzadeh, N., Pinheiro, P.O., 2019. Adversarial framing for image and video classification. In: Proceedings of the AAAI Conference on Artificial Intelligence.

Zhang, H., Wang, J., 2019. Towards adversarially robust object detection. In: Proceedings of the IEEE/CVF International Conference on Computer Vision.

Zhang, Z., Peng, H., 2019. Deeper and wider Siamese networks for real-time visual tracking. In: Proceedings of the IEEE Conference on Computer Vision and Pattern Recognition.

Zhou, B., Lapedriza, A., Khosla, A., Oliva, A., Torralba, A., 2017. Places: a 10 million image database for scene recognition. IEEE Transactions on Pattern Analysis and Machine Intelligence 40, 1452–1464.

Zhou, J., Liang, C., Chen, J., 2020. Manifold projection for adversarial defense on face recognition. In: Proceedings of the European Conference on Computer Vision.

Biographies

Changjae Oh is a Lecturer at the School of Electrical Engineering and Computer Science and the Centre for Intelligent Sensing from Queen Mary University of London, UK. He received his B.S., M.S., and Ph.D. in Electrical and Electronic Engineering from Yonsei University, Seoul, South Korea, in 2011, 2013, and 2018, respectively. From 2018 to 2019, he was a Postdoctoral Research Assistant at Queen Mary University of London, UK.

Alessio Xompero is a Postdoctoral Research Assistant in Multimodal Sensing with the School of Electronic Engineering and Computer Science and the Centre for Intelligent Sensing at Queen Mary University of London, UK. He received the MSc degree in Telecommunication Engineering from the University of Trento, Italy, in 2015, and the Ph.D. in Electronic Engineering from Queen Mary University of London, UK, in 2020.

Andrea Cavallaro is a professor of multimedia signal processing at Queen Mary University of London (QMUL) and a Turing fellow with the Alan Turing Institute, the UK National Institute for Data Science and Artificial Intelligence. He is a fellow of the International Association for Pattern Recognition; director of the QMUL Centre for Intelligent Sensing; Editor-in-Chief of Signal Processing: Image Communication; Senior Area Editor for the IEEE Transactions on Image Processing; Chair of the IEEE Image, Video, and Multidimensional Signal Processing Technical Committee; and an IEEE Signal Processing Society Distinguished Lecturer.

Index

Printed in the United States
by Baker & Taylor Publisher Services